# Intermediate Algebra

**Arnold R. Steffensen and L. Murphy Johnson**
Northern Arizona University

**Scott, Foresman and Company**
Glenview, Illinois   Dallas, Tex.   Oakland, N.J.   Palo Alto, Cal.
Tucker, Ga.   London, England

**To Barbara, Barbara, Becky, Cindy, and Pam**

Cover: Observatory, Palomar Observatory, California Institute of Technology

**Library of Congress Cataloging in Publication Data**

Steffensen, Arnold R 1942–

Intermediate algebra.

Includes index.
1. Algebra. I. Johnson, Lee Murphy, 1934– joint author. II. Title.
QA154.2.S727 512.9 80–25408
ISBN 0–673–15369–X (pbk.)


Printed in the United States of America.

2 3 4 5 6 - WEB - 85 84 83 82 81

# Preface

*Intermediate Algebra* is the second in a series of precalculus worktexts, which includes *Introductory Algebra, College Algebra,* and *Algebra and Trigonometry*. These worktexts can be used in sequence or individually to fit your curriculum needs. *Intermediate Algebra* begins with a review of elementary algebra and extends these ideas to develop each topic carefully and thoroughly. Each section of the book presents an easy-to-follow sequence of material deliberately measured so that students can work through it completely by reading the explanation, following the examples step by step, and doing all the exercises.

## Features of This Book

- **Short, simply worded explanations** geared to students' reading ability and level of mathematical sophistication
- **Key terms in bold type** for easy reference
- **Important rules and procedures** highlighted in color screens
- **Step-by-step examples** with explanations of difficult steps in color
- **Notes of caution** warning students about common mistakes, special problems, and exceptions to the rules
- **Exercises carefully chosen** for both number and type of problems
- **Sufficient space** for working all exercises
- **Continual review** of earlier sections in the last few exercises of each set
- **Word problems** carefully developed and categorized, with numerous and detailed examples
- **Answers to all exercises** immediately following the exercise sets
- **Chapter Summaries** with key words and phrases plus brief reminders
- **Review Exercises** at the end of each chapter
- **Final Review Exercises** concluding the book

## Instructional Flexibility

The series has been successfully class-tested in both individualized instruction and traditional lecture programs, and offers proven flexibility for a number of teaching situations. Short, simply worded explanations, step-by-step examples, space for working all the exercises, and the accessibility of the answers make this text ideal for either instructor-paced or student-paced programs. At the same time, the text offers the traditional lecture class a straightforward linear approach within a convenient worktext format.

## Instructional Resources

A versatile but optional supplementary program allows you to accommodate a variety of individual student needs, particularly in supplying extra practice and testing opportunities.

• **Supplementary Exercises** A package of Supplementary Exercises provides a full page of additional problems for each section of the text. Supplementary Exercises are keyed to chapter sections and all answers are supplied in the Instructor's Guide. Because the exercises in the book have all the answers immediately following the exercise set, you may wish to use the Supplementary Exercises in a traditional lecture setting as problems without answers for homework assignments, practice in class, or review. In an individualized instruction setting, the Supplementary Exercises are useful for students who need further practice on specific topics. In both instructional settings, the Supplementary Exercises are an additional source of quizzes and tests.

• **Testing** For testing, we provide two options: (1) a packet of Chapter Tests and Final Examinations and (2) a bank of Test Items with computer capabilities. The chapter test packet gives you eight different but equivalent tests for each chapter, ready for duplication. You'll find this resource valuable for both traditional and individualized classes in a variety of ways: preparing for class tests, class tests, make-up tests, pretesting, and retesting. Two final examinations are also included in this test packet. Answers to these tests are in the Instructor's Guide.

If you have a large number of students and computer facilities available to you, you may obtain from the publisher a bank of Test Items, which consist of hundreds of test questions for each chapter and answer sheets together with a short BASIC computer program that we use at Northern Arizona University. The program randomly generates chapter tests with answers and can be adapted to various systems with the help of a flowchart included in the test bank. The test bank gives you the flexibility of testing each student with different but equivalent problems or of using the items for testing and retesting in a unit-mastery situation. For instructors who wish to make up their own quizzes and tests or for schools without computer facilities, this test bank serves as another source of test questions.

• **Instructor's Guide** The Instructor's Guide contains a Placement Test and Student Information Sheet to help you determine student placement in the series. Course outlines as well as suggestions for using the book in an individualized instruction or traditional lecture setting are included. The guide also has answers for all of the Supplementary Exercises and the Chapter Test and Final Examination packet.

## Acknowledgments

We extend our gratitude to the instructors in the Individualized Instruction Program at Northern Arizona University who taught from this text and offered many suggestions and criticisms that have been most helpful and appreciated. Special thanks must go to three of these individuals, James Kirk, Joseph Mutter, and Michael Ratliff, for the many hours spent reading the manuscript and for the countless and valuable suggestions for its improvement.

We also express our thanks to Richard Rockwell of Pacific Union College, who class-tested the text, and to the reviewers, Donald L. Evans, Polk Community College, Shelba Morman, North Lake College, Albert D. Otto, Illinois State University, and Gerald Wisnieski, Des Moines Area Community College. In addition, we wish to thank Louis Hoelzle, Bucks County Community College, Vivian Dennis, Frank Lopez, and Sam Tinsley, Eastfield College, Maria Betkowski, Reginald Luke, and Robert Urbanski, Middlesex County College, John Burns, Mount San Antonio College, Howard Penn, Mountain View College, Ronald Schryer and David Nasby, Orange Coast College, Cleon Diers, Carole Fritz, and Paul Moreland, Rio Hondo College, and Richard Spangler, Tacoma Community College, for their suggestions.

To all the staff of Scott, Foresman and Company we are greatly indebted. Special thanks go to our editors, Pamela Carlson, Margaret Prullage, and Michael Timins.

To our typist, Barbara Schaack, we extend our sincere thanks and appreciation for an excellent job.

Last but certainly not least, we are deeply indebted to our wives, Barbara and Barbara, who not only gave many hours of their time to the project, but also gave up many hours of our time as we worked to complete this text.

Arnold R. Steffensen
L. Murphy Johnson

# Contents

## 3 Polynomials and Factoring



## 4 Exponents and Radicals



## 5 Quadratic Equations and Complex Numbers



## 6 Fractional Expressions

## 7 Graphs and Functions



## 8 Systems of Equations and Inequalities



## 9 Logarithms and Exponentials



## 10 Sequences and Series

# 1

# Fundamental Concepts

## 1.1 The Numbers of Algebra: Real Numbers

A **number** is an idea or concept formed when questions about "how many" or "in what order" are asked. Names or symbols for numbers are called **numerals.** The most basic numbers are the **natural** or **counting numbers.** In our number system the symbols for these numbers are

$$1, \quad 2, \quad 3, \quad 4, \quad 5, \quad 6, \quad 7, \cdots.$$

The three dot notation is used to indicate that a sequence or pattern continues. If zero is included with this set of numbers, we have the **whole numbers,**

$$0, \quad 1, \quad 2, \quad 3, \quad 4, \quad 5, \quad 6, \quad 7, \cdots.$$

When the negatives of all the natural numbers are added to the list of whole numbers, we obtain the set of **integers,**

$$\cdots, \quad -4, \quad -3, \quad -2, \quad -1, \quad 0, \quad 1, \quad 2, \quad 3, \quad 4, \cdots,$$

which includes the **negative integers,** $-1, -2, -3, -4, \cdots$, and the **positive integers,** $1, 2, 3, 4, \cdots$.

An excellent means of displaying the integers and showing some of their important properties is a **number line.** We draw a line and indicate that it can be extended infinitely far in both directions by using arrows on each tip. We then select a starting point or **origin** and a unit of length (any unit will do). The origin is labeled zero and unit lengths in both directions are marked off. Points to the right of zero are identified with the positive integers, while points to the left of zero correspond to the negative integers. The resulting figure is called a number line.

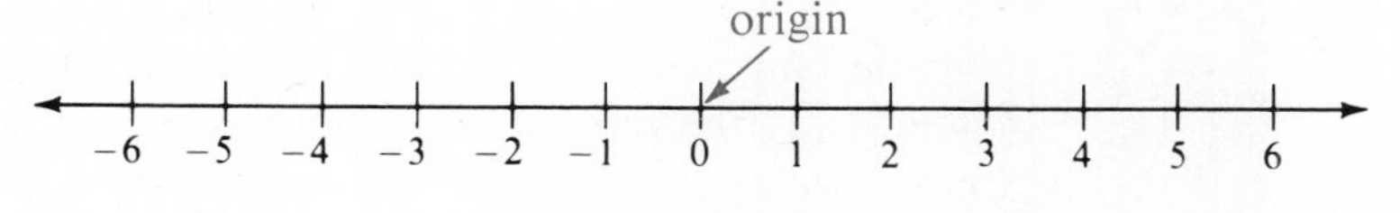


**Figure 1.1**

Numbers that are **equal** correspond to the same point on the line, and numbers that are **unequal** correspond to different points. When two numbers are unequal, the number corresponding to the point on the left of the point corresponding to the other number is **less than** that other number. For example, 3 is to the left of 5 on the number line in Figure 1.1, and we write $3 < 5$. Also, we can see in Figure 1.1 that

$$-4 < -1, \qquad -2 < 4, \qquad -5 < 0, \qquad 0 < 6.$$

We can also say that 5 is **greater than** 3, and write $5 > 3$. Similarly,

$$-1 > -4, \qquad 4 > -2, \qquad 0 > -5, \qquad 6 > 0.$$

We often join the less-than symbol with the equal sign and write $\leq$ which represents the phrase *less than or equal to* (similarly, $\geq$ represents the phrase *greater than or equal to*). Thus we have

$$-3 \leq -3, \qquad -2 \leq 0, \qquad 0 \leq 1, \qquad -1 \leq 1, \qquad 5 \leq 5.$$

The set of all numbers formed by taking quotients of integers is called the set of **rational numbers.** Thus every rational number can be represented as a fraction with numerator and denominator integers (the denominator not zero). Some of the rational numbers are

$$\frac{2}{3}, \quad \frac{-1}{5}, \quad \frac{3}{1}, \quad \frac{12}{5}, \quad \frac{-8}{1}, \quad \frac{28}{-3}.$$

Since every integer can be expressed as the quotient of itself and 1, the integers are also rational numbers:

$$3 = \frac{3}{1}, \qquad -8 = \frac{-8}{1}, \qquad 0 = \frac{0}{1}.$$

Every rational number has many names. For example,

$$\frac{1}{3}, \quad \frac{2}{6}, \quad \frac{3}{9}, \quad \frac{-2}{-6}, \quad \frac{-3}{-9}, \quad \frac{-1}{-3}, \quad \frac{33}{99}, \quad \text{and} \quad \frac{100}{300}$$

are all names for the same rational number. Fractions which are names for the same number or which have the same value are called **equivalent fractions.** The following rule is the most important rule of fractions.

> If the numerator and denominator of a fraction are multiplied or divided by the same nonzero number, the resulting fraction is equivalent to the original fraction.

We build up fractions when we multiply both the numerator and denominator by the same counting number; we reduce fractions when we divide both by the same counting number. When a fraction cannot be reduced further, that is, when 1 is the only natural number which divides both numerator and denominator, the fraction is **reduced to lowest terms.**

This process is easier if we use *prime numbers*. A counting number greater than 1 is called **prime** if its only divisors are 1 and itself.

**EXAMPLE 1**

**(a)** 2 is prime since 1 and 2 are the only divisors of 2.

**(b)** 5 is prime since 1 and 5 are the only divisors of 5.

**(c)** 6 is not prime since 2 (also 3) divides into 6.

The first few primes are

$$2, \quad 3, \quad 5, \quad 7, \quad 11, \quad 13, \quad 17, \quad 19, \quad 23, \cdots.$$

Every counting number greater than 1 is either prime or can be expressed as a product of primes.

**EXAMPLE 2** **(a)** 12 is not prime, but $12 = 2 \cdot 2 \cdot 3$ (2 and 3 are primes).

**(b)** 90 is not prime, but $90 = 2 \cdot 3 \cdot 3 \cdot 5$ (2, 3, and 5 are primes).

**(c)** 165 is not prime, but $165 = 3 \cdot 5 \cdot 11$ (3, 5, and 11 are primes).

Factoring into a product of primes is important when reducing fractions to lowest terms.

**TO REDUCE A FRACTION TO LOWEST TERMS**

1. Factor the numerator and denominator into products of primes.
2. Divide out all common factors.
3. Multiply the remaining factors in the numerator and in the denominator.

**EXAMPLE 3** Reduce each of the following fractions to lowest terms.

**(a)** $\frac{4}{8} = \frac{1 \cdot 2 \cdot 2}{1 \cdot 2 \cdot 2 \cdot 2} = \frac{1}{1 \cdot 2} = \frac{1}{2}$ Divide out the two common factors 2 in numerator and denominator

**(b)** $\frac{-12}{42} = \frac{(-1) \cdot 2 \cdot 2 \cdot 3}{1 \cdot 2 \cdot 3 \cdot 7} = \frac{(-1) \cdot 2}{7} = \frac{-2}{7}$ Divide out common factors 2 and 3 from numerator and denominator

The process of dividing out factors from the numerator and denominator of a fraction is sometimes called **canceling factors,** and is shown by crossing out the common factors. In part (b) of the above example we could have written

$$\frac{-12}{42} = \frac{(-1) \cdot \overset{1}{\cancel{2}} \cdot 2 \cdot \overset{1}{\cancel{3}}}{1 \cdot \underset{1}{\cancel{2}} \cdot \underset{1}{\cancel{3}} \cdot 7} = \frac{(-1) \cdot 1 \cdot 2 \cdot 1}{1 \cdot 1 \cdot 1 \cdot 7} = \frac{-2}{7}.$$

Generally, the 1's are not written and we simplify to

$$\frac{-12}{42} = \frac{(-1) \cdot \cancel{2} \cdot 2 \cdot \cancel{3}}{1 \cdot \cancel{2} \cdot \cancel{3} \cdot 7} = \frac{-2}{7}.$$

When working with fractions, be careful to cancel only factors (numbers that are multiplied) and never cross out numbers that are not factors. For example,

$$\frac{2+4}{2} = \frac{6}{2} = 3,$$

but

$$\frac{\not{2}+4}{\not{2}}$$

is certainly not 3.

Every rational number can be expressed as a fraction (in fact, as many equivalent fractions). In addition, every rational number can be written as a **decimal numeral** (in decimal form). For example, $\frac{3}{8}$ can be written as .375, the result obtained by dividing 3 by 8.

$$\begin{array}{r} .375 \\ 8\overline{)3.000} \\ \underline{2\,4}\phantom{00} \\ 60\phantom{0} \\ \underline{56}\phantom{0} \\ 40 \\ \underline{40} \\ 0 \end{array}$$

Likewise $\frac{3}{11}$ can be expressed as a decimal.

$$\begin{array}{r} .2727\cdots \\ 11\overline{)3.0000} \\ \underline{2\,2}\phantom{000} \\ 80\phantom{00} \\ \underline{77}\phantom{00} \\ 30\phantom{0} \\ \underline{22}\phantom{0} \\ 80 \\ \underline{77} \\ 3 \end{array}$$

The decimal numeral for $\frac{3}{8}$, .375, is called a **terminating decimal** (the sequence of digits comes to an end) while the decimal numeral for $\frac{3}{11}$, $.2727\cdots$, is called a **repeating decimal** (a block of digits, in this case 27, repeats indefinitely). We usually write such numerals by identifying the repeating block and placing a bar over it. For example,

$$\frac{3}{11} = .\overline{27} \qquad \text{and} \qquad \frac{1}{3} = .\overline{3}.$$

In general, every rational number has a decimal representation that either terminates or repeats. This property is often used as the defining characteristic of the rational numbers.

The association of numbers with points on a number line can be expanded to include the rational numbers. The number line in Figure 1.2 has several rational numbers *plotted* on it. (When we identify a point on a number line with a number, we say that we **plot** the number on the line.)

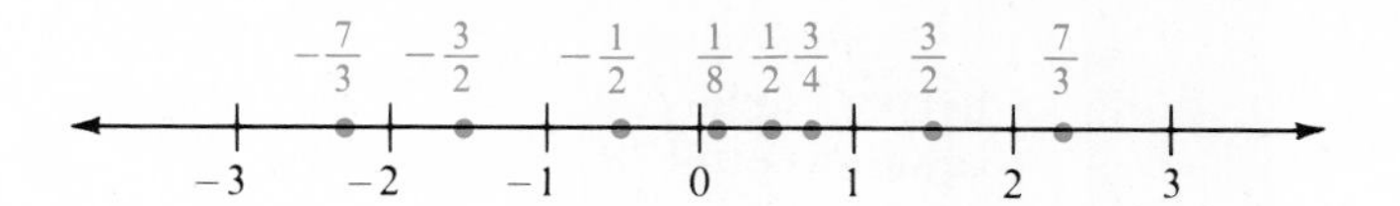


**Figure 1.2**

As with the integers, any rational number to the left (right) of another is **less than (greater than)** the other.

If two numbers are not equal, one of them must be less than the other. Given two integers, we can determine if they are equal, and if not, which is smaller (equivalently, which is larger) than the other. However, this is not quite as easy when we are given two fractions. We will concentrate on positive fractions. (Similar results could be summarized for negative fractions but we have no particular need for them.)

**DEFINITION** Given two positive fractions $\frac{a}{b}$ and $\frac{c}{d}$ ($a$, $b$, $c$, and $d$ integers),

1. $\frac{a}{b} = \frac{c}{d}$ whenever $a \cdot d = b \cdot c$.

2. $\frac{a}{b} > \frac{c}{d}$ whenever $a \cdot d > b \cdot c$.

3. $\frac{a}{b} < \frac{c}{d}$ whenever $a \cdot d < b \cdot c$.

The product $a \cdot d$ is called the **first cross product,** and $b \cdot c$ is called the **second cross product.** Thus, two positive fractions are equal if the cross products are equal, and two unequal fractions are unequal in the same order as the first and second cross products.

EXAMPLE 4 Place the appropriate symbol (=, >, or <) between each pair of fractions.

(a) $\frac{4}{6} \quad \frac{28}{42}$

Since $4 \cdot 42 = 168$ and $6 \cdot 28 = 168$, the fractions are equal.

$$\frac{4}{6} = \frac{28}{42}$$

(b) $\frac{24}{7} \quad \frac{39}{11}$

With $24 \cdot 11 = 264$ and $7 \cdot 39 = 273$, $264 < 273$.

$$\frac{24}{7} < \frac{39}{11}$$

(c) $\frac{15}{9} \quad \frac{31}{20}$

With $15 \cdot 20 = 300$ and $9 \cdot 31 = 279$, $300 > 279$.

$$\frac{15}{9} > \frac{31}{20}$$

Any whole-number product of two identical whole numbers is called a **perfect square.** Either of the identical factors of a perfect square is called a **square root** of the number. Thus, if $x$ and $a$ are whole numbers and

$$x = a \cdot a = a^2$$

then $a$ is called a square root of $x$, and $x$ is a perfect square.

**EXAMPLE 5** **(a)** 16 is a perfect square and the whole-number square root of 16 is 4. By our definition, $-4$ is also a square root of 16 since $(-4)(-4) = 16$.

**(b)** 25 is a perfect square and the whole-number square root of 25 is 5 ($-5$ is also a square root of 25).

The symbol $\sqrt{\ }$ (called a **radical**) is used to denote the **positive** (possibly zero) square root of a number. Thus,

$$\sqrt{16} = 4 \quad \text{and} \quad \sqrt{25} = 5.$$

This symbol by itself does not designate a negative square root. If we want to denote the negative square root of a number, we use $-\sqrt{\ }$. Thus,

$$-\sqrt{9} = -3.$$

The first fifteen perfect squares (except for zero) and their positive and negative square roots are listed in the following table.

| *Perfect square* | *Positive square root* | *Negative square root* |
|---|---|---|
| $N$ | $\sqrt{N}$ | $-\sqrt{N}$ |
| 1 | 1 | −1 |
| 4 | 2 | −2 |
| 9 | 3 | −3 |
| 16 | 4 | −4 |
| 25 | 5 | −5 |
| 36 | 6 | −6 |
| 49 | 7 | −7 |
| 64 | 8 | −8 |
| 81 | 9 | −9 |
| 100 | 10 | −10 |
| 121 | 11 | −11 |
| 144 | 12 | −12 |
| 169 | 13 | −13 |
| 196 | 14 | −14 |
| 225 | 15 | −15 |

In addition to whole-number perfect squares, there are also fractional perfect squares. A fraction which has two identical fractional factors is also called a **perfect square,** and each factor is called a **square root** of the fraction.

**EXAMPLE 6** **(a)** $\frac{4}{9}$ is a perfect square and a square root of $\frac{4}{9}$ is $\frac{2}{3}$. Note also that $-\frac{2}{3}$ is a square root of $\frac{4}{9}$, but $\sqrt{\frac{4}{9}} = \frac{2}{3}$ only. Why?

**(b)** $\frac{25}{144}$ is a perfect square and $\sqrt{\frac{25}{144}} = \frac{5}{12}$.

It is clear that a fraction in lowest terms is a perfect square if the numerator and denominator are whole-number perfect squares. Remember that the radical symbol is *never* used to denote a negative square root.

The attempt to find square roots of whole numbers that are not perfect squares such as

$$2, \quad 3, \quad 5, \quad 6, \quad 7,$$

leads us to another set of numbers. It can be shown that

$$\sqrt{2}, \quad \sqrt{3}, \quad \sqrt{5}, \quad \sqrt{6}, \quad \sqrt{7},$$

are not rational numbers. That is, these numbers (and many others including perhaps the most famous number in mathematics, $\pi$) cannot be represented as fractions. Equivalently, these numbers do not have a terminating or a repeating decimal representation (*approximations* of $\sqrt{2}$ or $\pi$ are sometimes given as terminating decimals or fractions).

Numbers which are not rational numbers are called **irrational numbers,** and the numbers of algebra, known as the **real numbers**, consist of the rational numbers together with the irrational numbers. Every real number can be identified with exactly one point on a number line, and also every point on a line corresponds to exactly one real number.

## EXERCISES 1.1

1. A ________ answers questions such as "how many" or "in what order."
2. A ________ is the name of a number.
3. 1, 2, 3, 4, · · · are called the ________ numbers.
4. 0, 1, 2, 3, 4, · · · are called the ________ numbers.
5. · · · , −3, −2, −1, 0, 1, 2, 3, · · · are called the ________.
6. A line used to display numbers is called a ________.
7. The starting point on a number line, which corresponds to zero, is called the ________.
8. Numbers which correspond to the same point on a number line are called ________.
9. If a number appears to the right of a second number on a number line, it is said to be ________ the second.
10. Numbers formed by taking quotients of integers (division by zero excluded) are called ________ numbers.

11. Two fractions that have the same value or name the same rational number are called ______

12. We build up a fraction when we ______ both numerator and denominator by the same counting number..

13. We reduce a fraction when we ______ both numerator and denominator by the same counting number.

14. A counting number greater than 1 that is divisible only by 1 and itself is called ______.

15. Every counting number greater than 1 is either ______ or can be expressed as a product of primes.

16. The process of canceling factors from a numerator and denominator is really the process of ______ out common factors.

17. Every rational number can be written as a decimal numeral which either (a) ______ or has a (b) ______ block of digits.

18. When we identify the point on a number line associated with a given number, we say we ______ the number.

19. Two positive fractions are equal whenever the ______ are equal.

20. Any whole number which is expressible as a product of two identical whole-number factors is a(n) **(a)** ______ and one of the identical factors is called a(n) **(b)** ______ of the number.

21. The symbol $\sqrt{\phantom{x}}$ is called a(n) **(a)** ______, and by itself it is only used to denote the **(b)** ______ square root of a number.

22. Any number which is not a rational number is called a(n) ______ number.

23. The set of rational numbers together with the set of irrational numbers is called the set of ______ numbers.

24. Write three fractions that are equivalent to $\frac{4}{5}$.

25. Write three fractions that are equivalent to 5.

26. Write three fractions that are equivalent to 1.

27. Write out the first sixteen prime numbers.

**28.** Write out the first sixteen whole-number perfect squares.

**29.** Write an equivalent fraction for $\frac{2}{3}$ with 21 as the denominator. [*Hint:* $3 \cdot ? = 21$]

*Express as a product of primes.*

**30.** $10 =$

**31.** $80 =$

**32.** $45 =$

**33.** $23 =$

**34.** $100 =$

**35.** $56 =$

**36.** $47 =$

**37.** $72 =$

*Reduce to lowest terms.*

**38.** $\frac{10}{15}$

**39.** $\frac{8}{32}$

**40.** $\frac{27}{45}$

**41.** $\frac{72}{72}$

**42.** $\frac{90}{210}$

**43.** $\frac{60}{84}$

**44.** Plot the numbers $\frac{3}{4}, \frac{7}{8}, -\frac{5}{4}, \frac{2}{3}, \frac{5}{2}, -\frac{10}{3}$ on the given number line.

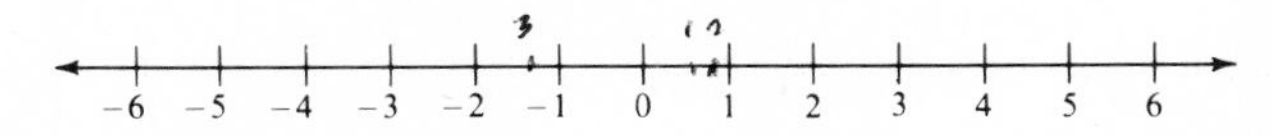


**45.** Insert the proper symbol (=, >, or <) between the two positive fractions.

**(a)** If $ad < bc$ then $\frac{a}{b}$ $\frac{c}{d}$.

**(b)** If $ad = bc$ then $\frac{a}{b}$ $\frac{c}{d}$

**(c)** If $ad > bc$ then $\frac{a}{b}$ $\frac{c}{d}$.

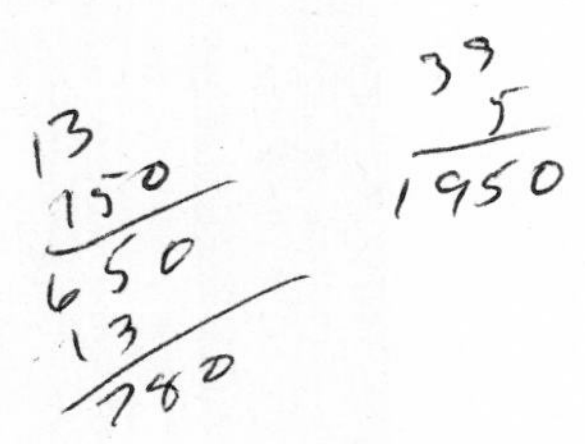

**(d)** The products $ad$ and $bc$ are called ______

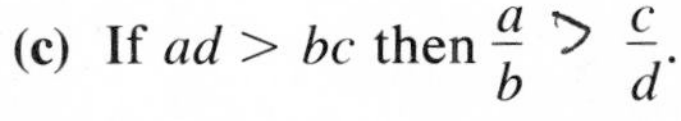

*Place the appropriate symbol (=, >, or <) between the given pair of fractions.*

**46.** $\frac{13}{50}$ $\frac{39}{150}$

**47.** $\frac{5}{7}$ $\frac{60}{63}$

**48.** $\frac{35}{11}$ $\frac{41}{13}$

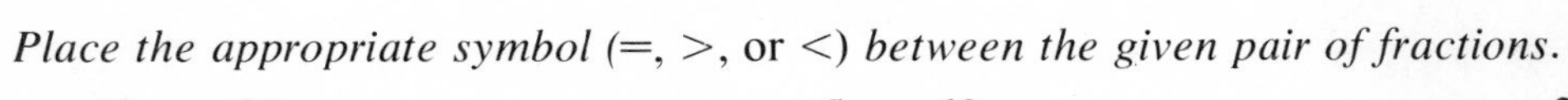

*Evaluate the square roots.*

49. $\sqrt{49} =$ 50. $\sqrt{4} =$ 51. $\sqrt{36} =$ 52. $\sqrt{225} =$

53. $\sqrt{64} =$ 54. $\sqrt{0} =$ 55. $\sqrt{169} =$ 56. $\sqrt{144} =$

57. $\sqrt{1} =$ 58. $\sqrt{196} =$ 59. $\sqrt{121} =$ 60. $\sqrt{81} =$

61. $\sqrt{\dfrac{196}{25}} =$ 62. $\sqrt{\dfrac{1}{121}} =$ 63. $\sqrt{\dfrac{48}{12}} =$
[*Hint:* Reduce the fraction first.]

64. Express $\frac{7}{8}$ as a decimal.

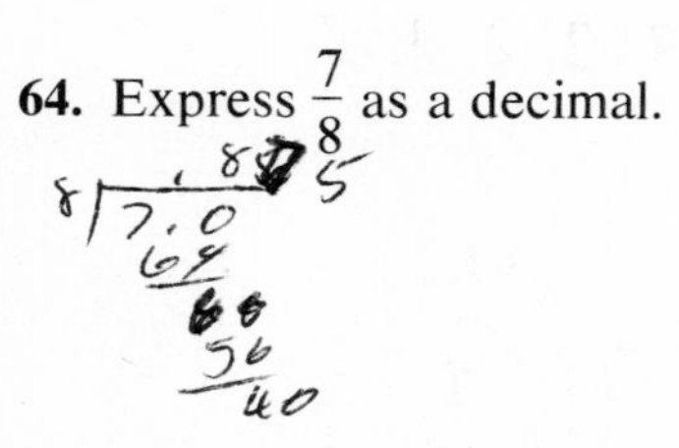

65. Express $\frac{7}{11}$ as a decimal.

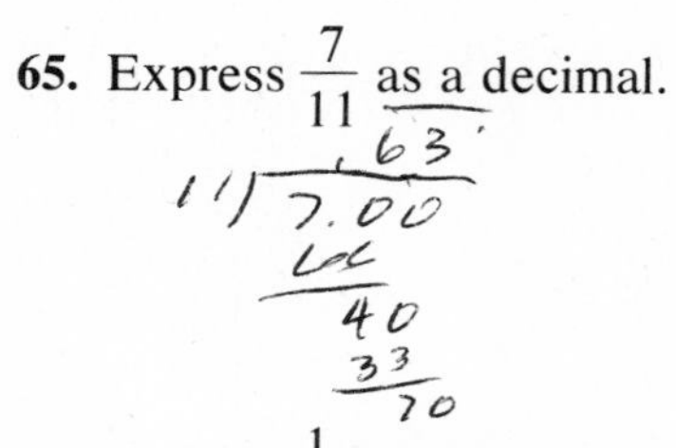

66. Express $\frac{12}{5}$ as a decimal.

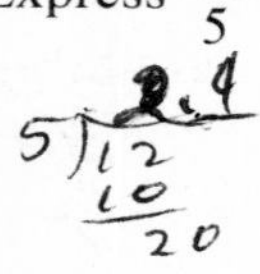

67. Express $\frac{1}{7}$ as a decimal.

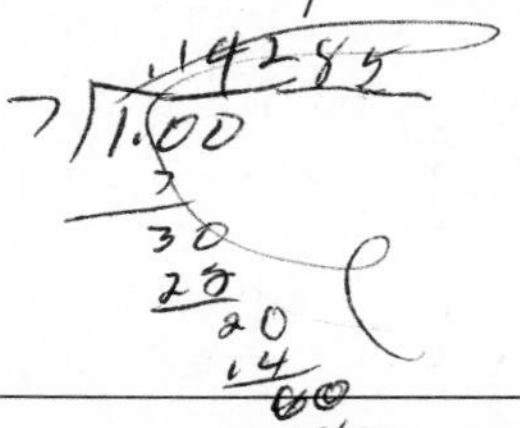

---

ANSWERS: 1. number 2. numeral 3. natural or counting 4. whole 5. integers 6. number line 7. origin 8. equal 9. greater than 10. rational 11. equivalent 12. multiply 13. divide 14. prime 15. prime 16. dividing 17. (a) terminates (b) repeating 18. plot 19. cross products 20. (a) perfect square (b) square root 21. (a) radical (b) positive 22. irrational 23. real 24. answers vary; some are $\frac{8}{10}, \frac{-4}{-5}, \frac{12}{15}$. 25. answers vary; some are $\frac{10}{2}, \frac{15}{3}, \frac{-10}{-2}$. 26. answers vary; some are $\frac{2}{2}, \frac{5}{5}, \frac{-4}{-4}$. 27. 2, 3, 5, 7, 11, 13, 17, 19, 23, 29, 31, 37, 41, 43, 47, 53 28. 0, 1, 4, 9, 16, 25, 36, 49, 64, 81, 100, 121, 144, 169, 196, 225 29. $\frac{14}{21}$ 30. $2 \cdot 5$ 31. $2 \cdot 2 \cdot 2 \cdot 2 \cdot 5$ 32. $3 \cdot 3 \cdot 5$ 33. 23 is prime 34. $2 \cdot 2 \cdot 5 \cdot 5$ 35. $2 \cdot 2 \cdot 2 \cdot 7$ 36. 47 is prime 37. $2 \cdot 2 \cdot 2 \cdot 3 \cdot 3$ 38. $\frac{2}{3}$ 39. $\frac{1}{4}$ 40. $\frac{3}{5}$ 41. 1 42. $\frac{3}{7}$ 43. $\frac{5}{7}$ 44. (number line from −4 to 4 with points $-\frac{10}{3}$, $-\frac{5}{4}$, $\frac{2}{3}$, $\frac{3}{4}$, $\frac{7}{8}$, $\frac{5}{2}$) 45. (a) $<$ (b) $=$ (c) $>$ (d) cross products 46. $=$ 47. $<$ 48. $>$ 49. 7 50. 2 51. 6 52. 15 53. 8 54. 0 55. 13 56. 12 57. 1 58. 14 59. 11 60. 9 61. $\frac{14}{5}$ 62. $\frac{1}{11}$ 63. 2 64. .875 65. $.\overline{63}$ 66. 2.4 67. $.\overline{142857}$

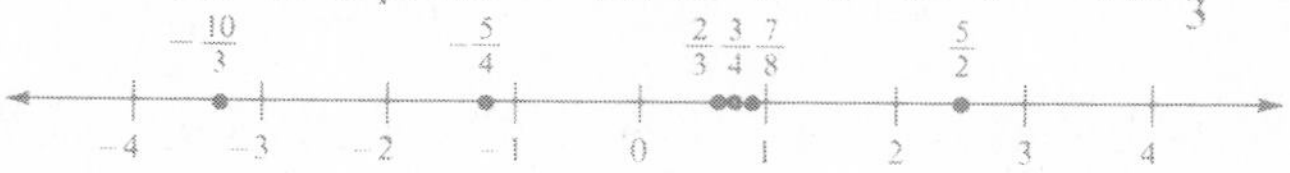

## 1.2 A Review of Operations on Positive Fractions

Although the four basic operations of addition (+), subtraction (—), multiplication (·), and division (÷ or a fraction bar) should be familiar when applied to positive numbers, we will briefly review these operations on fractions. Before beginning, however, let us consider the problem of division by zero. Why is division by zero a problem? In any quotient, such as $8 \div 2$ or $\frac{8}{2}$, $\frac{8}{2}$ is the number that we multiply by 2 to obtain 8. That is,

$$\frac{8}{2} = 4 \quad \text{since} \quad 8 = 2 \cdot 4.$$

Similarly,

$$\frac{20}{4} = 5 \quad \text{since} \quad 20 = 4 \cdot 5.$$

If

$$\frac{20}{0} = \text{a number} \quad \text{then} \quad 20 = 0 \cdot (\text{the number}).$$

But zero times any number is zero and not 20, so there is no number equal to $\frac{20}{0}$. Since we could have used any number besides 20 in our example, no number can be divided by zero except possibly zero itself. What would $\frac{0}{0}$ equal?

$$\frac{0}{0} \quad \text{could be} \quad 5 \quad \text{since} \quad 0 = 0 \cdot 5$$

$$\frac{0}{0} \quad \text{could be} \quad 259 \quad \text{since} \quad 0 = 0 \cdot 259$$

$$\frac{0}{0} \quad \text{could be} \quad \textit{any number} \quad \text{since} \quad 0 = 0 \cdot (\textit{any number})$$

Since it would be very confusing if $\frac{0}{0}$ could be any number we please, we agree never to divide 0 by 0. Hence any division by zero is excluded in mathematics. However, $0 \div 5$ or $\frac{0}{5}$ does make sense and in fact, $\frac{0}{5} = 0$. (Why?)

We now review the four operations on positive fractions.

**TO MULTIPLY TWO (OR MORE) FRACTIONS**

1. Factor all numerators and denominators into products of primes.
2. Place all numerator factors over all denominator factors.
3. Divide out (cancel) all common factors.
4. Multiply the remaining numerator and denominator factors.
5. The resulting fraction is the product (reduced to lowest terms) of the original fractions.

**EXAMPLE 1**

$$\frac{70}{66} \cdot \frac{12}{35} = \frac{2 \cdot 5 \cdot 7}{2 \cdot 3 \cdot 11} \cdot \frac{2 \cdot 2 \cdot 3}{5 \cdot 7} \qquad \text{Factor}$$

$$= \frac{\cancel{2} \cdot \cancel{5} \cdot \cancel{7} \cdot 2 \cdot 2 \cdot \cancel{3}}{\cancel{2} \cdot \cancel{3} \cdot 11 \cdot \cancel{5} \cdot \cancel{7}} \qquad \text{Indicate products and cancel}$$

$$= \frac{4}{11} \qquad \text{The reduced product}$$

**EXAMPLE 2** $\frac{3}{7} \cdot 14 = \frac{3}{7} \cdot \frac{14}{1} = \frac{3}{7} \cdot \frac{2 \cdot 7}{1} = \frac{3 \cdot 2 \cdot \cancel{7}}{\cancel{7} \cdot 1} = \frac{6}{1} = 6$

In order to divide fractions, we need the idea of *reciprocal.* The **reciprocal** of the fraction $\frac{a}{b}$ is the fraction $\frac{b}{a}$. That is, the reciprocal of a fraction is the fraction formed by interchanging the numerator and denominator of the original fraction. The important relationship between a number and its reciprocal is that their product is 1.

$$\frac{a}{b} \cdot \frac{b}{a} = 1$$

Thus, the reciprocal of $\frac{4}{5}$ is $\frac{5}{4}$, the reciprocal of $\frac{1}{3}$ is 3 ($\frac{3}{1} = 3$), and the reciprocal of 7 is $\frac{1}{7}$ ($7 = \frac{7}{1}$). Only one number, 0, does not have a reciprocal. (Why?) When one number is divided by a second, the first is called the **dividend,** the second is called the **divisor,** and the result is called the **quotient.** For example, in $6 \div 3 = 2$, 6 is the dividend, 3 is the divisor, and 2 is the quotient.

**TO DIVIDE TWO FRACTIONS**

1. Replace the divisor by its reciprocal and change the division sign to multiplication.
2. Proceed as in multiplication.

The reason for the above rule is illustrated in the following division problem.

$$\frac{\frac{2}{3}}{\frac{3}{4}} = \frac{\frac{2}{3} \cdot \frac{4}{3}}{\frac{3}{4} \cdot \frac{4}{3}} = \frac{\frac{2}{3} \cdot \frac{4}{3}}{1} = \frac{2}{3} \cdot \frac{4}{3} = \frac{2 \cdot 4}{3 \cdot 3} = \frac{8}{9}$$

**EXAMPLE 3** $\frac{13}{15} \div \frac{39}{5} = \frac{13}{15} \cdot \frac{5}{39}$ — Replace divisor with its reciprocal and multiply

$= \frac{13}{3 \cdot 5} \cdot \frac{5}{3 \cdot 13}$ — Factor

$= \frac{\cancel{13} \cdot \cancel{5}}{3 \cdot \cancel{5} \cdot 3 \cdot \cancel{13}}$ — Indicate products and cancel

$= \frac{1}{9}$ — Canceling is dividing: 1 remains in the numerator, not zero

**EXAMPLE 4**

$$\frac{4}{5} \div 44 = \frac{4}{5} \cdot \frac{1}{44} \qquad \text{Replace divisor with reciprocal and multiply}$$

$$= \frac{2 \cdot 2}{5} \cdot \frac{1}{2 \cdot 2 \cdot 11} \qquad \text{Factor}$$

$$= \frac{\cancel{2} \cdot \cancel{2} \cdot 1}{5 \cdot \cancel{2} \cdot \cancel{2} \cdot 11} \qquad \text{Indicate products and cancel}$$

$$= \frac{1}{55}$$

**EXAMPLE 5**

$$\frac{3}{7} \div \frac{3}{7} = \frac{3}{7} \cdot \frac{7}{3} = \frac{\cancel{3} \cdot \cancel{7}}{\cancel{7} \cdot \cancel{3}} = \frac{1}{1} = 1 \qquad \text{Does this answer seem reasonable?}$$

We can add or subtract fractions only when their denominators are the same. To add or subtract two fractions that have different denominators, we must first convert them to equivalent fractions having the same denominators. For example, to add $\frac{1}{2}$ and $\frac{1}{3}$, we would add $\frac{3}{6}$ and $\frac{2}{6}$, which are equivalent to the given fractions and have the **least common denominator** (LCD) 6. Finding the least common denominator of two or more fractions is an important process.

**TO FIND THE LCD OF TWO (OR MORE) FRACTIONS**

1. Factor each denominator and reduce fractions to lowest terms.
2. If there are no common factors in the denominators, the LCD is the product of *all* denominators.
3. If there are common factors in the denominators, each factor must appear in the LCD as many times as it appears in the denominator where it is found the greatest number of times.

**EXAMPLE 6** Find the LCD of each pair of fractions.

**(a)** $\frac{1}{6}$ and $\frac{4}{15}$.

Factor the denominators.

$$6 = \overset{1}{2} \cdot \overset{1}{3} \quad \text{and} \quad 15 = \overset{1}{3} \cdot \overset{1}{5}.$$

The LCD must consist of one 2, one 3, and one 5. Thus

$$\text{the LCD} = 2 \cdot 3 \cdot 5.$$

For convenience, leave the LCD in factored form.

**(b)** $\frac{13}{90}$ and $\frac{7}{24}$.

Factor the denominators.

$$90 = \overset{1}{2} \cdot \overset{2}{3 \cdot 3} \cdot \overset{1}{5} \quad \text{and} \quad 24 = \overset{3}{2 \cdot 2 \cdot 2} \cdot \overset{1}{3}.$$

The LCD must consist of three 2's, two 3's, and one 5. Thus

$$\text{the LCD} = 2 \cdot 2 \cdot 2 \cdot 3 \cdot 3 \cdot 5.$$

**(c)** $\frac{3}{10}$ and $\frac{5}{21}$.

Factor the denominators.

$$10 = \overset{1}{\boxed{2}} \cdot \overset{1}{\boxed{5}} \quad \text{and} \quad 21 = \overset{1}{\boxed{3}} \cdot \overset{1}{\boxed{7}}.$$

Since there are no common factors, the LCD is the product of all factors. Thus

$$\text{the LCD} = 2 \cdot 3 \cdot 5 \cdot 7.$$

**TO ADD OR SUBTRACT TWO (OR MORE) FRACTIONS**

1. Rewrite the indicated sum or difference with each denominator expressed as a product of prime factors.
2. Reduce fractions if possible and determine the LCD.
3. Multiply the numerator and denominator of each fraction by all those factors present in the LCD but missing in the denominator of the particular fraction. This makes all denominators the same.
4. Place the sum or difference of all numerators over the LCD.
5. Simplify the resulting numerator and reduce the fraction to lowest terms.

**EXAMPLE 7**

$$\frac{5}{6} + \frac{7}{15} = \frac{5}{2 \cdot 3} + \frac{7}{3 \cdot 5} \qquad \text{Factor denominators; the LCD} = 2 \cdot 3 \cdot 5$$

$$= \frac{5 \cdot (5)}{2 \cdot 3 \cdot (5)} + \frac{7 \cdot (2)}{3 \cdot 5 \cdot (2)} \qquad \text{Supply missing factors}$$

$$= \frac{5 \cdot 5 + 7 \cdot 2}{2 \cdot 3 \cdot 5} \qquad \text{Add over LCD}$$

$$= \frac{25 + 14}{2 \cdot 3 \cdot 5} \qquad \text{Simplify}$$

$$= \frac{39}{2 \cdot 3 \cdot 5} \qquad \text{Simplify}$$

$$= \frac{\cancel{3} \cdot 13}{2 \cdot \cancel{3} \cdot 5} \qquad \text{Factor numerator and cancel common factors}$$

$$= \frac{13}{10} \qquad \text{The desired sum}$$

**EXAMPLE 8** Subtract $\frac{7}{12} - \frac{25}{45}$.

We can reduce one of the fractions.

$$\frac{25}{45} = \frac{\cancel{5} \cdot 5}{\cancel{5} \cdot 9} = \frac{5}{9}$$

Then

$$\frac{7}{12} - \frac{5}{9} = \frac{7}{2 \cdot 2 \cdot 3} - \frac{5}{3 \cdot 3} \qquad \text{Factor; the LCD} = 2 \cdot 2 \cdot 3 \cdot 3$$

$$= \frac{7 \cdot (3)}{2 \cdot 2 \cdot 3 \cdot (3)} - \frac{5 \cdot (2 \cdot 2)}{3 \cdot 3 \cdot (2 \cdot 2)} \qquad \text{Supply missing factors}$$

$$= \frac{21 - 20}{2 \cdot 2 \cdot 3 \cdot 3} \qquad \text{Subtract and simplify}$$

$$= \frac{1}{36} \qquad \text{The difference in reduced form}$$

**EXAMPLE 9**

$$2 + \frac{11}{18} = \frac{2}{1} + \frac{11}{2 \cdot 3 \cdot 3} \qquad \text{The LCD} = 2 \cdot 3 \cdot 3$$

$$= \frac{2 \cdot (2 \cdot 3 \cdot 3)}{1 \cdot (2 \cdot 3 \cdot 3)} + \frac{11}{2 \cdot 3 \cdot 3}$$

$$= \frac{36 + 11}{2 \cdot 3 \cdot 3}$$

$$= \frac{47}{18}$$

The answer to Example 9 could be given as the **mixed number** $2\frac{11}{18}$ (read "2 and $\frac{11}{18}$"). The symbol $2\frac{11}{18}$ actually represents $2 + \frac{11}{18}$ or $\frac{47}{18}$. Since $2\frac{11}{18}$ could be confused with $2 \cdot \frac{11}{18}$ (2 *times* $\frac{11}{18}$), it is wise to convert mixed numbers to **improper fractions.**

## EXERCISES 1.2

*Simplify.*

1. $\frac{0}{4} =$
2. $\frac{4}{0} =$
3. $\frac{4}{4} =$
4. $\frac{0}{0} =$

*Multiply.*

5. $\frac{2}{3} \cdot \frac{6}{5} =$
6. $\frac{3}{35} \cdot \frac{20}{9} =$
7. $\frac{2}{7} \cdot \frac{21}{6} =$
8. $\frac{3}{4} \cdot 8 =$

9. $\frac{13}{28} \cdot 0 =$

10. $3\frac{1}{3} \cdot 2\frac{5}{3} =$
[*Hint:* First convert to improper fractions.]

*Divide.*

11. $\frac{1}{3} \div \frac{2}{9} =$

12. $\frac{5}{7} \div \frac{5}{28} =$

13. $\dfrac{\frac{3}{4}}{\frac{3}{2}} =$

14. $\frac{8}{7} \div 4 =$

15. $9 \div \frac{12}{5} =$

16. $3\frac{1}{3} \div 1\frac{7}{8} =$

*Find the LCD of each pair of fractions.*

17. $\frac{7}{15}, \frac{2}{25}$

18. $\frac{1}{7}, \frac{4}{11}$

19. $\frac{5}{12}, \frac{3}{20}$

20. $\frac{2}{3}, 5$
[*Hint:* $5 = \frac{5}{1}$.]

21. $\frac{2}{33}, \frac{4}{77}$

22. $\frac{1}{13}, \frac{7}{20}$

*Add.*

23. $\frac{2}{3} + \frac{1}{6} =$

24. $\frac{5}{12} + \frac{5}{18} =$

25. $6 + \frac{2}{5} =$

26. $\frac{1}{28} + \frac{13}{70} =$

27. $\frac{8}{35} + \frac{2}{21} =$

28. $1\frac{2}{5} + 2\frac{1}{3} =$

*Subtract.*

29. $\frac{3}{4} - \frac{2}{3} =$

30. $\frac{6}{11} - \frac{3}{7} =$

31. $3 - \frac{1}{9} =$

32. $\frac{2}{3} - 0 =$

33. $\frac{11}{15} - \frac{2}{35} =$

34. $4\frac{2}{3} - 1\frac{1}{4} =$

35. What are the counting numbers?

36. What are the integers?

37. What are the rational numbers?

38. How are the rational numbers characterized in decimal notation?

39. What are the irrational numbers?

40. What are the real numbers?

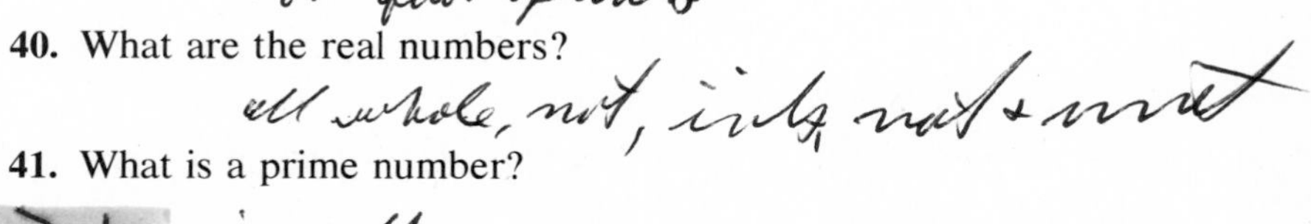

41. What is a prime number?

**42.** Express $\frac{2}{3}$ as a decimal.

**43.** Express $\frac{1}{8}$ as a decimal.

**44.** $\sqrt{\frac{72}{50}} =$

**45.** $\sqrt{\frac{121}{169}} =$

---

ANSWERS: **1.** 0 **2.** undefined **3.** 1 **4.** undefined **5.** $\frac{4}{5}$ **6.** $\frac{4}{21}$ **7.** 1 **8.** 6 **9.** 0 **10.** $\frac{110}{9}$ **11.** $\frac{3}{2}$ **12.** 4 **13.** $\frac{1}{2}$ **14.** $\frac{2}{7}$ **15.** $\frac{15}{4}$ **16.** $\frac{16}{9}$ **17.** $3 \cdot 5 \cdot 5$ **18.** $7 \cdot 11$ **19.** $2 \cdot 2 \cdot 3 \cdot 5$ **20.** 3 **21.** $3 \cdot 7 \cdot 11$ **22.** $2 \cdot 2 \cdot 5 \cdot 13$ **23.** $\frac{5}{6}$ **24.** $\frac{25}{36}$ **25.** $\frac{32}{5}$ **26.** $\frac{31}{140}$ **27.** $\frac{34}{105}$ **28.** $\frac{56}{15}$ **29.** $\frac{1}{12}$ **30.** $\frac{9}{77}$ **31.** $\frac{26}{9}$ **32.** $\frac{2}{3}$ **33.** $\frac{71}{105}$ **34.** $\frac{41}{12}$ **35.** $1, 2, 3, 4, \cdots$ **36.** $\cdots, -4, -3, -2, -1, 0, 1, 2, 3, 4, \cdots$ **37.** quotients of integers **38.** terminating or repeating decimals **39.** numbers which are not rational numbers **40.** the rational numbers together with the irrational **41.** whole numbers greater than one that can be divided only by themselves and one **42.** $.\overline{6}$ **43.** .125 **44.** $\frac{6}{5}$ **45.** $\frac{11}{13}$

## 1.3 Absolute Value and Operations with Signed Numbers

An important concept that is necessary for understanding operations on **signed numbers** (zero together with numbers which have minus or plus signs attached) is **absolute value.**

**DEFINITION** The **absolute value** of a number is:

1. The number itself if the number is positive or zero.
2. The positive number formed by removing the minus sign if the number is negative.

The absolute value of a number is denoted by writing vertical bars on both sides of the number. For example, the absolute value of 4 is denoted by $|4|$.

**EXAMPLE 1** **(a)** $|5| = 5$ or $+5$ **(b)** $|+7| = 7$ or $+7$ **(c)** $|0| = 0$ **(d)** $|-6| = 6$ or $+6$

**(e)** $\left|\frac{1}{3}\right| = \frac{1}{3}$ or $+\frac{1}{3}$ **(f)** $\left|+\frac{3}{4}\right| = \frac{3}{4}$ or $+\frac{3}{4}$ **(g)** $|\sqrt{2}| = \sqrt{2}$ **(h)** $\left|-\frac{7}{8}\right| = \frac{7}{8}$

The absolute value of any number is always greater than or equal to zero and is *never* negative. Often, the absolute value of a number is thought of as the distance from the number to zero on a number line (see Figure 1.3).

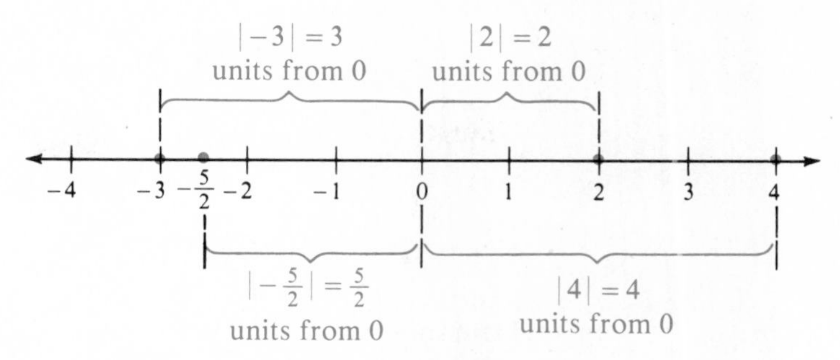


**Figure 1.3**

We now consider the operations of addition, subtraction, multiplication, and division of signed numbers. We assume that these operations on whole numbers and positive fractions (the **numbers of arithmetic**) are well understood. These operations will be extremely important as we develop the rules for operating with two negative numbers or one positive and one negative number. First we consider addition.

**ADDITION OF SIGNED NUMBERS**

1. To add two (or more) positive numbers, add just as we added numbers of arithmetic. The sum is a positive number.
2. To add two (or more) negative numbers, add their absolute values and attach a minus sign to the sum.
3. To add one positive and one negative number:
   **(a)** If the numbers have the same absolute value the sum is zero.
   **(b)** If the numbers do not have the same absolute value, subtract the smaller absolute value from the larger absolute value and give the result the sign of the number with the greater absolute value.

Remember that if a number other than zero has no sign this is the same as having a plus sign attached (+7 and 7 are the same).

**EXAMPLE 2** Add.

**(a)** $(+6) + (+7) = (6) + (7) = 13$

**(b)** $(-6) + (-7) = -(6 + 7) = -13$

**(c)** $(+6) + (-7) = (6) + (-7) = -(7 - 6) = -1$

**(d)** $(-6) + (+7) = (-6) + (7) = +(7 - 6) = +1 = 1$

**(e)** $(+12) + (-12) = (12) + (-12) = 0$

**(f)** $(-3) + (0) = -(3 - 0) = -3$

**(g)** $(0) + (+7) = +(7 + 0) = +7 = 7$

**(h)** $\left(+\frac{1}{2}\right) + \left(-\frac{3}{4}\right) = \left(\frac{1}{2}\right) + \left(-\frac{3}{4}\right) = -\left(\frac{3}{4} - \frac{1}{2}\right) = -\left(\frac{3}{4} - \frac{2}{4}\right) = -\frac{1}{4}$

**(i)** $\left(-\frac{1}{2}\right) + \left(+\frac{3}{4}\right) = \left(-\frac{1}{2}\right) + \left(\frac{3}{4}\right) = +\left(\frac{3}{4} - \frac{1}{2}\right) = +\left(\frac{3}{4} - \frac{2}{4}\right) = +\frac{1}{4} = \frac{1}{4}$

**(j)** $\left(-\frac{1}{2}\right) + \left(-\frac{3}{4}\right) = -\left(\frac{1}{2} + \frac{3}{4}\right) = -\left(\frac{2}{4} + \frac{3}{4}\right) = -\frac{5}{4}$

Subtraction of a signed number is defined as addition of a negative, for example, $5 - 3$ and $5 + (-3)$ both yield 2. We extend this idea to all signed-number subtractions.

**SUBTRACTION OF SIGNED NUMBERS**

To subtract one signed number from another, change the sign of the **subtrahend** (the number being subtracted) and add.

**EXAMPLE 3** Subtract.

**(a)** $(+5) \ -(+\ 3) = (+5)\ +(-\ 3)$ Change sign and add
$= 5 + (-3) = +(5 - 3) = +2 = 2$

**(b)** $(+5) \ -(-\ 3) = (+5)\ +(+\ 3)$ Change sign and add
$= 5 + 3 = 8$

**(c)** $(-5) \ -(+\ 3) = (-5)\ +(-\ 3)$ Change sign and add
$= -(5 + 3) = -8$

**(d)** $(-5) \ -(-\ 3) = (-5)\ +(+\ 3)$ Change sign and add
$= (-5) + 3 = -(5 - 3) = -2$

**(e)** $(+5) - (+5) = 5 + (-5) = 0$

**(f)** $(5) - (-5) = 5 + (+5) = 5 + 5 = 10$

**(g)** $(-5) - (+5) = (-5) + (-5) = -(5 + 5) = -10$

**(h)** $(-5) - (-5) = (-5) + (+5) = (-5) + 5 = 0$

**(i)** $\left(+\frac{2}{3}\right) - \left(+\frac{1}{6}\right) = \left(\frac{2}{3}\right) + \left(-\frac{1}{6}\right) = \frac{4}{6} + \left(-\frac{1}{6}\right) = +\frac{3}{6} = \frac{1}{2}$

**(j)** $\left(-\frac{2}{3}\right) - \left(-\frac{1}{6}\right) = \left(-\frac{2}{3}\right) + \left(\frac{1}{6}\right) = \left(-\frac{4}{6}\right) + \left(\frac{1}{6}\right) = -\frac{3}{6} = -\frac{1}{2}$

Remember that $-(-a) = +a$. Observe how this recurs in the above examples.

**TO MULTIPLY TWO SIGNED NUMBERS**

1. Ignore the signs and multiply just as we multiplied numbers of arithmetic.
2. If both numbers have the same sign (both positive or both negative), the product is positive.
3. If the signs are opposite (one positive and the other negative), the product is negative and we attach a minus sign.
4. If one or both numbers is zero, the product is zero (zero-product rule).

**EXAMPLE 4** Multiply.

(a) $(+4) \cdot (+2) = (4) \cdot (2) = +8 = 8$ — Same signs

(b) $(+4) \cdot (-2) = (4) \cdot (-2) = -(4 \cdot 2) = -8$ — Different signs

(c) $(-4) \cdot (+2) = (-4) \cdot (2) = -(4 \cdot 2) = -8$ — Different signs

(d) $(-4) \cdot (-2) = +(4 \cdot 2) = +8 = 8$ — Same signs

(e) $(0) \cdot (-2) = 0$ — Zero-product rule

(f) $(0) \cdot (0) = 0$

(g) $\left(+\frac{1}{4}\right) \cdot \left(+\frac{3}{5}\right) = \left(\frac{1}{4}\right) \cdot \left(\frac{3}{5}\right) = \frac{3}{20}$

(h) $\left(+\frac{1}{4}\right) \cdot \left(-\frac{3}{5}\right) = \left(\frac{1}{4}\right) \cdot \left(-\frac{3}{5}\right) = -\frac{3}{20}$

(i) $\left(-\frac{1}{4}\right) \cdot \left(+\frac{3}{5}\right) = \left(-\frac{1}{4}\right) \cdot \left(\frac{3}{5}\right) = -\frac{3}{20}$

(j) $\left(-\frac{1}{4}\right) \cdot \left(-\frac{3}{5}\right) = +\left(\frac{1}{4}\right)\left(\frac{3}{5}\right) = \frac{3}{20}$

Simply remember "like signs have a positive product, unlike signs have a negative product."

**TO DIVIDE TWO SIGNED NUMBERS**

1. Ignore the signs and divide just as we divided numbers of arithmetic.
2. If both numbers have the same sign (both positive or both negative), the quotient is positive.
3. If the signs are opposite (one positive and the other negative), the quotient is negative.
4. Zero divided by any number (except zero) is always zero, and division of any number by zero is undefined.

**EXAMPLE 5** Divide.

(a) $(+9) \div (+3) = 9 \div 3 = +3 = 3$ — Same signs

(b) $(+9) \div (-3) = (9) \div (-3) = -(9 \div 3) = -3$ — Different signs

(c) $(-9) \div (+3) = (-9) \div (3) = -(9 \div 3) = -3$ — Different signs

(d) $(-9) \div (-3) = 9 \div 3 = +3 = 3$ — Same signs

(e) $(+9) \div (0)$ — Undefined

(f) $(0) \div (-9) = 0$ — Zero divided by any number is zero

(g) $\left(+\frac{1}{2}\right) \div \left(+\frac{3}{4}\right) = \frac{1}{2} \div \frac{3}{4} = \frac{1}{2} \cdot \frac{4}{3} = \frac{2}{3}$

**(h)** $\left(+\frac{1}{2}\right) \div \left(-\frac{3}{4}\right) = -\left(\frac{1}{2} \div \frac{3}{4}\right) = -\left(\frac{1}{2} \cdot \frac{4}{3}\right) = -\frac{2}{3}$

**(i)** $\left(-\frac{1}{2}\right) \div \left(+\frac{3}{4}\right) = -\left(\frac{1}{2} \div \frac{3}{4}\right) = -\left(\frac{1}{2} \cdot \frac{4}{3}\right) = -\frac{2}{3}$

**(j)** $\left(-\frac{1}{2}\right) \div \left(-\frac{3}{4}\right) = \frac{1}{2} \div \frac{3}{4} = \frac{1}{2} \cdot \frac{4}{3} = \frac{2}{3}$

Simply remember "like signs have a positive quotient, unlike signs have a negative quotient." Also remember that the rules for multiplication and division are essentially the same.

Absolute value expressions often involve operations on signed numbers. It is wise to perform all operations within the absolute value bars before taking the absolute value of the result.

**EXAMPLE 6** **(a)** $|5 - 7| = |-2| = +2 = 2$

**(b)** $|(-2)(-5)| = |10| = +10 = 10$

**(c)** $|(-3)(4) - (-2)| = |-12 - (-2)| = |-12 + 2| = |-10| = +10 = 10$

Two important properties of absolute value are presented next.

> If $a$ and $b$ are any two real numbers,
>
> **1.** $|a \cdot b| = |a| \cdot |b|$. (The absolute value of a product is the product of absolute values.)
>
> **2.** $\left|\frac{a}{b}\right| = \frac{|a|}{|b|}$, $(b \neq 0)$. (The absolute value of a quotient is the quotient of absolute values.)

**EXAMPLE 7** **(a)** $|2 \cdot 3| = |6| = 6$ and $|2| \cdot |3| = 2 \cdot 3 = 6$.

Thus, $|2 \cdot 3| = |2| \cdot |3|$.

**(b)** $|(-3) \cdot 5| = |-15| = 15$ and $|-3| \cdot |5| = 3 \cdot 5 = 15$

Thus, $|(-3) \cdot 5| = |-3| \cdot |5|$.

**(c)** $\left|\frac{-12}{4}\right| = |-3| = 3$ and $\frac{|-12|}{|4|} = \frac{12}{4} = 3$

Thus, $\left|\frac{-12}{4}\right| = \frac{|-12|}{|4|}$.

**CAUTION:** Although the absolute value of a product (or quotient) is the product (or quotient) of the absolute values, similar results do not hold for addition and subtraction. For example, $|(-2) + (3)| = |1| = 1$, but $|-2| + |3| = 2 + 3 = 5$. Thus

$|(-2) + (3)|$ is **not** equal to $|-2| + |3|$.

Similarly,

$|(-2) - 3|$ is **not** equal to $|-2| - |3|$.

## EXERCISES 1.3

*Perform the indicated operations.*

**1.** $(+2) + (+5) =$

**2.** $(+2) + (-5) =$

**3.** $(-2) + (+5) =$

**4.** $(-2) + (-5) =$

**5.** $(+2) - (+5) =$

**6.** $(+2) - (-5) =$

**7.** $(-2) - (+5) =$

**8.** $(-2) - (-5) =$

**9.** $(-2) + (+2) =$

**10.** $(-2) - (-2) =$

**11.** $\left(\frac{1}{9}\right) + \left(-\frac{1}{3}\right) =$

**12.** $\left(-\frac{1}{9}\right) + \left(-\frac{1}{3}\right) =$

**13.** $\left(-\frac{1}{9}\right) - \left(-\frac{1}{3}\right) =$

**14.** $\left(\frac{2}{3}\right) - \left(-\frac{3}{4}\right) =$

**15.** $\left(-\frac{2}{3}\right) - \left(\frac{3}{4}\right) =$

**16.** $\left(-\frac{2}{3}\right) - \left(-\frac{3}{4}\right) =$

**17.** $(+4)(+7) =$

**18.** $(+4)(-7) =$

**19.** $(-4)(+7) =$

**20.** $(-4)(-7) =$

**21.** $\left(\frac{2}{3}\right)\left(-\frac{3}{4}\right) =$

**22.** $\left(-\frac{2}{3}\right)\left(\frac{3}{4}\right) =$

**23.** $\left(-\frac{2}{3}\right)\left(-\frac{3}{4}\right) =$

**24.** $(+15) \div (-3) =$

**25.** $(-15) \div (-3) =$

**26.** $\frac{-15}{+3} =$

**27.** $\frac{-15}{0} =$

**28.** $\frac{0}{-15} =$

**29.** $\frac{-15}{-15} =$

**30.** $\frac{0}{0} =$

**31.** $\left(\frac{2}{3}\right) \div \left(-\frac{3}{4}\right) =$

**32.** $\left(-\frac{2}{3}\right) \div \left(\frac{3}{4}\right) =$

**33.** $\left(-\frac{2}{3}\right) \div \left(-\frac{3}{4}\right) =$

**34.** $|3 - 8| =$

**35.** $|3| - |8| =$

**36.** $|(3)(-8)| =$

**37.** $|3||-8| =$

**38.** $\left|\frac{8}{-2}\right| =$

**39.** $\frac{|8|}{|-2|} =$

**40.** $|(-2) + 1| =$

**41.** $|-2| + |1| =$

**42.** $|(-3)(-2) - (-4)| =$

---

ANSWERS: **1.** +7 or 7 **2.** −3 **3.** +3 or 3 **4.** −7 **5.** −3 **6.** +7 or 7 **7.** −7 **8.** +3 or 3 **9.** 0 **10.** 0 **11.** $-\frac{2}{9}$ **12.** $-\frac{4}{9}$ **13.** $+\frac{2}{9}$ or $\frac{2}{9}$ **14.** $+\frac{17}{12}$ or $\frac{17}{12}$ **15.** $-\frac{17}{12}$ **16.** $\frac{1}{12}$ **17.** 28 **18.** −28 **19.** −28 **20.** 28 **21.** $-\frac{1}{2}$ **22.** $-\frac{1}{2}$ **23.** $\frac{1}{2}$ **24.** −5 **25.** 5 **26.** −5 **27.** undefined **28.** 0 **29.** 1 **30.** undefined **31.** $-\frac{8}{9}$ **32.** $-\frac{8}{9}$ **33.** $\frac{8}{9}$ **34.** 5 **35.** −5 **36.** 24 **37.** 24 **38.** 4 **39.** 4 **40.** 1 **41.** 3 **42.** 10

## 1.4 Properties of the Real Numbers

Two terms used repeatedly in mathematics are *set* and *element*. A **set** is a collection of objects called **elements** of the set. We are generally concerned with sets of numbers. The elements in a set are often listed within braces { }, and capital letters are used to denote sets. For example,

$$N = \{1, 2, 3, 4, \cdots\}$$

is the set of **counting** or **natural numbers,** and

$$W = \{0, 1, 2, 3, 4, \cdots\}$$

is the set of **whole numbers.**

EXAMPLE 1 The set of integers is denoted by

$$I = \{\cdots, -3, -2, -1, 0, 1, 2, 3, \cdots\}$$

and the set of numbers formed by taking quotients of integers (division by zero is excluded) is called the set of rational numbers and denoted by $Q$. The set of real numbers, the rational numbers together with the irrational numbers, is denoted by the letter $R$.

We often use lowercase letters to represent numbers. Such letters are called **variables.** Lengthy verbal statements can often be symbolized by brief algebraic statements using variables. Consider the following examples:

| | | |
|---|---|---|
| Five times a real number | becomes | $5 \cdot x$ |
| Seven more than twice a real number | becomes | $7 + 2 \cdot x$ |

In each case, the variable represents a real number. Other notations for $5 \cdot x$ are $5(x)$, $(5)(x)$ and $5x$ (no multiplication symbol used at all), with the latter preferred. The product of two real numbers $x$ and $y$ would most likely be represented by $xy$ but could also be expressed by $x \cdot y$, $x(y)$, and $(x)(y)$.

We next consider several properties of real numbers relative to the operations of addition and multiplication.

**COMMUTATIVE LAW OF ADDITION**

If $a$ and $b$ are two real numbers,

$$a + b = b + a.$$

That is, changing the order of addition does *not* change the sum.

EXAMPLE 2 **(a)** $15 + 23$ and $23 + 15$ are equal (both equal 38).

**(b)** $\left(\frac{3}{4}\right) + \left(-\frac{1}{3}\right)$ and $\left(-\frac{1}{3}\right) + \left(\frac{3}{4}\right)$ are equal (both equal $\frac{5}{12}$).

**(c)** $2x + 3x$ and $3x + 2x$ are equal.

**COMMUTATIVE LAW OF MULTIPLICATION**

If $a$ and $b$ are two real numbers,

$$a \cdot b = b \cdot a.$$

That is, changing the order of multiplication does *not* change the product.

**EXAMPLE 3** **(a)** $3 \cdot 8$ and $8 \cdot 3$ are equal (both equal 24).

**(b)** $\left(\frac{2}{3}\right)\left(-\frac{3}{5}\right)$ and $\left(-\frac{3}{5}\right)\left(\frac{2}{3}\right)$ are equal (both equal $-\frac{2}{5}$).

**(c)** $(2x)(3x)$ and $(3x)(2x)$ are equal (in fact, as we shall see, both equal $6x^2$).

When the operations of addition and multiplication are combined in numerical expressions, some confusion may result. For example,

$2 \cdot 3 + 4$ could equal $6 + 4$ or 10 if we multiply, then add.
$2 \cdot 3 + 4$ could equal $2 \cdot 7$ or 14 if we add, then multiply.

According to the following rule, the first procedure is correct.

**ORDER OF OPERATIONS**

To evaluate a numerical expression involving the operations of addition, subtraction, multiplication, or division, first do multiplications and divisions in order from left to right, then additions and subtractions in order from left to right.

**EXAMPLE 4** Study the following carefully.

**(a)** $2 + 3 \cdot 4 = 2 + 12 = 14$ First multiply, then add

**(b)** $5 \cdot 6 - 12 \div 3 = 30 - 4 = 26$ First multiply and divide, then subtract

**(c)** $25 \div 5 + 2 \cdot 9 = 5 + 18 = 23$ Divide and multiply first

Suppose that we want to evaluate three times the sum of 2 and 5. If we write $3 \cdot 2 + 5$ and use the above rule, we obtain $6 + 5$ or 11. However, it is clear from the first sentence, that we want 3 times 7 or 21 for the result. The use of **symbols of grouping** (such as parentheses ( ), square brackets [ ], or braces { }) leads us to symbolize our problem correctly as $3 \cdot (2 + 5)$. The grouping symbols contain the expression that must be evaluated first. In this case, we must add before multiplying.

$$3 \cdot (2 + 5) = 3 \cdot (7) = 21$$

Generally, we omit the dot and write $3(2 + 5)$ for $3 \cdot (2 + 5)$. Also, instead of writing $3 \cdot (7)$ we remove the parentheses and write $3 \cdot 7$.

**TO EVALUATE A NUMERICAL EXPRESSION INVOLVING GROUPING SYMBOLS**

1. Evaluate expressions within the grouping symbols first.
2. Begin with the innermost symbols of grouping if more than one set of symbols is present.

**EXAMPLE 5** Study the following carefully.

(a) $3 + (2 \cdot 5) = 3 + 10 = 13$ — Are the parentheses even necessary?

(b) $(4 + 5)2 + 3 = (9)2 + 3$ — Work inside parentheses first
$= 18 + 3$ — Multiply before adding
$= 21$

(c) $[3(8 + 2) + 1]4 = [3(10) + 1]4$ — Innermost parentheses first
$= [30 + 1]4$ — Multiply before adding inside brackets
$= [31]4$ — Combine inside bracket before multiplying
$= 124$

(d) $2\{15 - 2(1 + 3)\} = 2\{15 - 2(4)\}$ — Innermost grouping symbols first
$= 2\{15 - 8\}$ — Multiply before subtracting
$= 2\{7\}$ — Inside first
$= 14$

(e) $5[(-2) - 3(2 + (-4))] = 5[(-2) - 3(-2)]$ — Innermost first
$= 5[(-2) - (-6)]$ — Multiply 3 times (−2)
$= 5[4]$ — Inside brackets
$= 20$

We know how to add two numbers, but how do we add three numbers? For example, what does $2 + 3 + 5$ mean? We can add only two at a time, but since

$$(2 + 3) + 5 = 5 + 5 = 10 \qquad \text{and} \qquad 2 + (3 + 5) = 2 + 8 = 10,$$

the method of grouping is immaterial.

**ASSOCIATIVE LAW OF ADDITION**

If $a$, $b$, and $c$ are three real numbers,

$$(a + b) + c = a + (b + c).$$

That is, changing the grouping symbols in an addition problem does *not* change the sum.

**EXAMPLE 6**

(a) $2 + \left(3 + \frac{1}{2}\right) = (2 + 3) + \frac{1}{2}$ — Both are $\frac{11}{2}$

(b) $2x + (3 + 4x) = 2x + (4x + 3)$ — Commutative law of addition

$= (2x + 4x) + 3$ — Associative law of addition

**ASSOCIATIVE LAW OF MULTIPLICATION**

If $a$, $b$, and $c$ are three real numbers,

$$(a \cdot b) \cdot c = a \cdot (b \cdot c).$$

That is, changing the grouping symbols in a multiplication problem does *not* change the product.

**EXAMPLE 7**

(a) $3\,(4 \cdot (-2)) = (3 \cdot 4) \cdot (-2)$ — Both are $-24$

(b) $5(y \cdot 2) = 5\,(2 \cdot y)$ — Commutative law of multiplication

$= (5 \cdot 2) \cdot y$ — Associative law of multiplication

$= 10y$

The commutative and associative laws allow us to evaluate sums or products by rearranging orders and then adding or multiplying in pairs from left to right.

**TO ADD MORE THAN TWO SIGNED NUMBERS**

1. Add all positive numbers
2. Add all negative numbers
3. Add the resulting pair of numbers (one positive and one negative).

**EXAMPLE 8**

$(-2) + (-3) + (+2) + (+7) + (-5) + (+6) + (+9) + (-1)$

$= [(-2) + (-3) + (-5) + (-1)] + [(+2) + (+7) + (+6) + (+9)]$ — Rearrange and group

$= [-(2 + 3 + 5 + 1)] + [+(2 + 7 + 6 + 9)]$

$= [-11] + [+24] = +[24 - 11] = +13 = 13$

With practice we should be able to skip many of the intermediate steps in the above example by making computations mentally.

**TO MULTIPLY MORE THAN TWO SIGNED NUMBERS**

Multiply in pairs from left to right, keeping track of the appropriate sign.

**EXAMPLE 9** Multiply.

**(a)** $(+3)(-2)(-4) = (-6)(-4) = 24$ — Multiply 3 times $-2$ first

**(b)** $(-1)(-1)(-2)(-3)(-4) = (1)(-2)(-3)(-4)$ — $(-1)(-1) = 1$ first
$= (-2)(-3)(-4)$ — $(1)(-2) = -2$
$= (6)(-4)$ — $(-2)(-3) = 6$
$= -24$

**(c)** $(-2)(-2)(-2) = (4)(-2)$ — $(-2)(-2) = 4$
$= -8$

Notice that products involving an odd number of minus signs are negative while those with an even number are positive.

We now have the basic rules for evaluating numerical expressions. An **algebraic expression** contains variables as well as numbers. It is important to be able to evaluate algebraic expressions when specific values for the variables are given.

**TO EVALUATE AN ALGEBRAIC EXPRESSION**

1. Replace each variable (letter) with its specified value.
2. Proceed as in evaluating numerical expressions.

**EXAMPLE 10** Evaluate $2(a + b) - c$ when $a = 3$, $b = 4$, and $c = 5$.

$2(a + b) - c = 2(3 + 4) - 5$ — Replace each letter with given value
$= 2(7) - 5$ — Evaluate inside parentheses first
$= 14 - 5 = 9$

**EXAMPLE 11** Evaluate $5[12 - 3(a + 1) + b] - c$ when $a = 2$, $b = 7$, and $c = 4$.

$5[12 - 3(2 + 1) + 7] - 4$ — Replace variables with numbers
$= 5[12 - 3(3) + 7] - 4$ — Inside first
$= 5[12 - 9 + 7] - 4$ — Multiply before adding or subtracting
$= 5[10] - 4 = 50 - 4 = 46$

**EXAMPLE 12** Evaluate $2xy + z$ when $x = 3$, $y = 4$, and $z = 5$.

$2(3)(4) + 5$ — Replace with numbers
$= 2 \cdot 12 + 5 = 24 + 5 = 29$

**EXAMPLE 13** Evaluate $\dfrac{ab - 1}{c}$ for the following values.

**(a)** $a = -2$, $b = 3$, $c = 0$

$$\frac{(-2)(3) - 1}{0} = \frac{-6 - 1}{0} = \frac{-7}{0} \quad \text{which is undefined}$$

That is, this expression is meaningless when $c = 0$.

**(b)** $a = 3, b = \frac{1}{3}, c = -5$

$$\frac{(3)\left(\frac{1}{3}\right) - 1}{-5} = \frac{1-1}{-5} = \frac{0}{-5} = 0 \quad \left(\frac{0}{-5} = 0, \text{ but } \frac{-7}{0} \text{ is not defined}\right)$$

Using negative numbers to evaluate certain expressions may produce two signs in front of a number. For example,

when $x = -3$, $1 + x = 1 + (-3)$

when $a = -2$, $3 - a = 3 - (-2)$.

In such cases we use parentheses to avoid writing adjacent signs. The parentheses may be removed and the two signs replaced by one according to the following rule.

If $a$ is any number,

**1.** $+(+a) = +a$.

**2.** $+(-a) = -a$.

**3.** $-(+a) = -a$.

**4.** $-(-a) = +a$.

**EXAMPLE 14** Write without parentheses using only one sign (or no sign).

**(a)** $+(+5) = +5 = 5$

**(b)** $+(-3) = -3$

**(c)** $-(-4) = +4 = 4$

**(d)** $-(+6) = -6$

**(e)** $-[-(-7)] = -[+7] = -7$ Work on innermost parentheses first

A plus sign in front of the parentheses *does not* change the inside sign when the parentheses are removed. A minus sign in front of the parentheses *does* change the inside sign when the parentheses are removed.

**EXAMPLE 15** Evaluate when $a = -2$ and $b = -1$.

**(a)** $a - b = (-2) - (-1) = -2 + 1 = -1$ Use parentheses when substituting

**(b)** $4a + 7 = 4(-2) + 7 = -8 + 7 = -1$.

Using parentheses at the substitution step helps to avoid ambiguous statements. For example, without parentheses above, we would have $4 \cdot -2 + 7$ which is confusing.

**(c)** $b - [-a] = (-1) - [-(-2)] = (-1) - [+2]$
$= -1 - 2 = -3$

**(d)** $a - 2b = (-2) - 2(-1) = -2 - (-2)$ $2(-1) = -2$
$= -2 + 2 = 0$

**(e)** $|-a - b| = |-(-2) - (-1)| = |+2 + 1| = |3| = 3$

The number zero has unique properties when combined with other numbers by the basic operations. The following rule summarizes these properties.

**PROPERTIES OF ZERO**

If $a$ is any real number,

1. $a + 0 = 0 + a = a$.
2. $a \cdot 0 = 0 \cdot a = 0$.
3. $a - 0 = a$.
4. $0 - a = -a$.
5. $0 \div a = \frac{0}{a} = 0$, for $a \neq 0$.
6. $\frac{a}{0}$ is undefined.

## EXERCISES 1.4

1. A collection of objects is called a(n) ______________ of objects.
2. Each object belonging to a set is called a(n) ______________ of that set.
3. Symbolize by brief algebraic statements using variables:
   **(a)** Three more than a number ______________
   **(b)** Two more than eight times a number ______________
4. A letter used to represent a number is called a(n) ______________.
5. The fact that $3 + 2 = 2 + 3$ illustrates the ______________ law of addition.
6. The fact that $3 + (2 + 5) = (3 + 2) + 5$ illustrates the ______________ law of addition.
7. The fact that $3 \cdot 7 = 7 \cdot 3$ illustrates the ______________ law of multiplication.
8. The fact that $3 \cdot (2 \cdot 7) = (3 \cdot 2) \cdot 7$ illustrates the ______________ law of multiplication.
9. When evaluating a numerical expression which does not involve grouping symbols, always perform the operations of **(a)** ______________ and **(b)** ______________ before **(c)** ______________ and **(d)** ______________.
10. $N = \{1, 2, 3, 4, \cdots\}$ is called the set of ______________ numbers.
11. $W = \{0, 1, 2, 3, \cdots\}$ is called the set of ______________ numbers.
12. $P = \{2, 3, 5, 7, 11, 13, 17, 19, \cdots\}$ is called the set of ______________ numbers.
13. True or false: 0 is an element of set $W$ ($W$ as above).

14. True or false: 0 is an element of set $P$ ($P$ as above).

15. Changing the order of addition **(a)** does change the sum **(b)** does not change the sum.

16. Changing the order of multiplication **(a)** does change the product **(b)** does not change the product.

17. Changing the order of subtraction **(a)** does change the difference **(b)** does not change the difference.

18. Changing the order of division **(a)** does change the quotient **(b)** does not change the quotient.

19. When evaluating an expression involving two sets of grouping symbols, evaluate first within the ________________ set of grouping symbols.

20. The sum of two negative numbers is always a ________________ number.

21. The product of two negative numbers is always a ________________ number.

22. The quotient of two negative numbers is always a ________________ number.

23. The product of two numbers with opposite signs is always a ________________ number.

24. The quotient of two numbers with opposite signs is always a ________________ number.

25. The set of integers is given by ________________________.

*Evaluate the following numerical expressions.*

**26.** $2 + 3 \cdot 5 =$

**27.** $2 - 3 + 5 =$

**28.** $2 - (3 + 5) =$

**29.** $4 \cdot 2 - 6 \div 3 =$

**30.** $15 \div 5 \cdot 2 + 3 =$

**31.** $(4 + 2)3 + 7 =$

**32.** $2[3 + 2(1 - 5)] =$

**33.** $-[3 + 2(4 - 6)] =$

**34.** $3[(-1) - 2(3 + (-4))] =$

**35.** $-(-3) =$

**36.** $-(+3) =$

**37.** $+(-3) =$

**38.** $-[-(+2)] =$

**39.** $-[-(-2)] =$

**40.** $-[+(-2)] =$

**41.** $|(3)(-4)| =$

**42.** $|(-4) - (-6)| =$

**43.** $-|4 - 6| =$

**44.** $|(-1) - (-1)| =$

**45.** $[(-2) + (-3)] + (-5) =$

**46.** $(-2) + [(-3) + (-5)] =$

**47.** Why are the answers to Exercises 45 and 46 the same?

**48.** $\left(\frac{1}{2}\right)\left(2 \cdot \frac{7}{8}\right) =$

**49.** $\left(\left(\frac{1}{2}\right) \cdot 2\right) \cdot \frac{7}{8} =$

**50.** Why are the answers to Exercises 48 and 49 the same?

**51.** $(-2) + (-3) + (5) + (3) + (-5) + (2) + (-7) + (4) =$

**52.** $(-2)(-2) + (-3)(+3) + (-1)(+4) + (0)(5) =$

**53.** $(-1.7) + (2.3) - (4.1) - (-3.2) + (-2.5) - (3.2) =$

**54.** $(+3)(-2)(-2)(-1)(-1)(+2)(-1)(+1)(-1) =$

**55.** $(-1)(-1)(-1)(-1)(-1)(-1)(-1)(-1)(-1) =$

**56.** $(-1.2)(-1.2)(-0.8) =$

*Evaluate when* $a = -1$, $b = -2$, *and* $c = 5$.

**57.** $a - b =$

**58.** $2a + b - c =$

**59.** $b - (-a) =$

**60.** $-2a - b =$

**61.** $2(a - b + c) =$

**62.** $\frac{a + b}{c} =$

**63.** $\frac{2c - b}{b - 2a} =$

**64.** $|3a + b| =$

**65.** $|2c + 5b| =$

**66.** $-2a + 0 =$

**67.** $0 - 2a =$

**68.** $\frac{6a - 3b}{2c} =$

**69.** The absolute value of a number $a$ is sometimes defined as follows.

$$|a| = \begin{cases} a, & a \geq 0 \\ -a, & a < 0 \end{cases}$$

Use this to evaluate the following.

**(a)** $|3|$

**(b)** $|-3|$

**(c)** $|0|$

Do you obtain the same results using the definition in Section 1.3?

ANSWERS: **1.** set **2.** element **3.** (a) $3 + x$ (b) $2 + 8 \cdot x$ **4.** variable **5.** commutative **6.** associative **7.** commutative **8.** associative **9.** (a) multiplication (b) division (c) addition (d) subtraction **10.** natural **11.** whole **12.** prime **13.** true **14.** false **15.** does not **16.** does not **17.** does **18.** does **19.** innermost **20.** negative **21.** positive **22.** positive **23.** negative **24.** negative **25.** $\{\cdots, -3, -2, -1, 0, 1, 2, 3, \cdots\}$ **26.** 17 **27.** 4 **28.** −6 **29.** 6 **30.** 9 **31.** 25 **32.** −10 **33.** 1 **34.** 3 **35.** +3 or 3 **36.** −3 **37.** −3 **38.** +2 or 2 **39.** −2 **40.** +2 or 2 **41.** 12 **42.** 2 **43.** −2 **44.** 0 **45.** −10 **46.** −10 **47.** by the associative law of addition **48.** $\frac{7}{8}$ **49.** $\frac{7}{8}$ **50.** by the associative law of multiplication **51.** −3 **52.** −9 **53.** −6.0 **54.** 24 **55.** −1 **56.** −1.152 **57.** 1 **58.** −9 **59.** −3 **60.** 4 **61.** 12 **62.** $-\frac{3}{5}$ **63.** undefined **64.** 5 **65.** 0 **66.** 2 **67.** 2 **68.** 0 **69.** (a) $|3| = 3\ (3 \geq 0)$ (b) $|-3| = -(-3) = 3\ (-3 < 0)$ (c) $|0| = 0\ (0 \geq 0)$

## 1.5 The Distributive Laws: Factoring, Multiplying, and Collecting Like Terms

The operations of addition (or subtraction) and multiplication are related by the distributive laws. Consider the numerical expression $2(3 + 5)$. Since we have learned to work within the parentheses first, we would evaluate it in the following way.

$$2(3 + 5) = 2(8) = 16$$

However, if we "distribute" the multiplier 2 over the sum of $3 + 5$,

$$2(3 + 5) = 2 \cdot 3 + 2 \cdot 5 = 6 + 10 = 16,$$

we obtain the same result. Similarly,

$$4(8 - 3) = 4(5) = 20$$

and

$$4(8 - 3) = 4 \cdot 8 - 4 \cdot 3 = 32 - 12 = 20.$$

Thus, multiplication could also be "distributed over" subtraction. These examples illustrate the following laws.

**DISTRIBUTIVE LAWS**

If $a$, $b$, and $c$ are real numbers,

$$a(b + c) = a \cdot b + a \cdot c \quad \text{and} \quad a(b - c) = a \cdot b - a \cdot c.$$

Since products are not affected by changing the order of multiplication, we also have

$$(b + c)a = b \cdot a + c \cdot a \quad \text{and} \quad (b - c)a = b \cdot a - c \cdot a.$$

Also, multiplication distributes over sums with more than two terms. For example,

$$a(b + c + d) = a \cdot b + a \cdot c + a \cdot d.$$

**EXAMPLE 1**

**(a)** $5(6+2) = 5 \cdot 6 + 5 \cdot 2 = 30 + 10 = 40$. Also, $5(6+2) = 5(8) = 40$.

**(b)** $2(10-4) = 2 \cdot 10 - 2 \cdot 4 = 20 - 8 = 12$. Also, $2(10-4) = 2(6) = 12$.

**(c)** $4(2+3-5) = 4 \cdot 2 + 4 \cdot 3 - 4 \cdot 5 = 8 + 12 - 20 = 0$. Also, $4(2+3-5) = 4(0) = 0$.

**(d)** $-3(5+2) = (-3)(5) + (-3)(2) = (-15) + (-6) = -21$. Also, $-3(5+2) = (-3)(7) = -21$.

**(e)** $-4(-3-5) = (-4)(-3) - (-4)(5) = (12) - (-20) = 12 + 20 = 32$. Also, $-4(-3-5) = (-4)(-8) = 32$.

Although useful for computation, the distributive laws are more useful when we simplify algebraic expressions. An **algebraic expression** involves sums, differences, products, or quotients of numbers and variables. Portions of an expression that are products of numbers and variables and that are separated from the remainder of the expression by plus (or minus) signs are called **terms.** The numbers and letters that are multiplied in a term are called **factors** of the term, and the numerical factor is called the **(numerical) coefficient** of the term. For example,

$$7x, \qquad 2a+5, \qquad x+3y+z, \qquad 2x-5a+3-4x$$

are algebraic expressions with one, two, three, and four terms, respectively. In $2a+5$, the term $2a$ has factors 2 and $a$, and 2 is the coefficient of the term.

Two terms are **similar** or **like terms** if they contain the same variables. In $2x - 5a + 3 - 4x$, the terms $2x$ and $-4x$ are like terms. (Note that the minus sign goes with the term and thus the coefficient of $-4x$ is $-4$.)

If the terms of an expression have a common factor, the distributive law can be used to **remove the common factor** by a process called **factoring.**

**EXAMPLE 2** Use the distributive laws to factor the following.

**(a)** $2x + 2y = 2(x+y)$ — The distributive law in reverse order

Thus the two-term expression $2x + 2y$ can be represented as a single-term expression with factors 2 and $(x+y)$.

**(b)** $5a - 5b = 5(a-b)$ — Distributive law—factor out 5

Thus 5 and $(a-b)$ are the factors of $5a - 5b$.

**(c)** $7a - 7 = 7 \cdot a - 7 \cdot 1$ — Express 7 as $7 \cdot 1$
$= 7(a-1)$ — Factor out 7

**(d)** $6u - 3v + 9 = 3 \cdot 2u - 3 \cdot v + 3 \cdot 3$ — 3 is a common factor
$= 3(2u - v + 3)$ — Factor out 3

**(e)** $-3a + 9b = (-3)(a) - (-3)(3b) = (-3)(a - 3b)$ — Factor out $-3$

We could have factored out $+3$ instead.

$$(3)(-a) + (3)(3b) = 3(-a + 3b).$$

Both of these factorizations are correct but in a given situation one might be preferred.

**(f)** $3x + 8x = (3 + 8)x = 11x$ Factor out $x$

**(g)** $-2y + 5y - 8y = (-2 + 5 - 8)y = (3 - 8)y = -5y$ Factor out $y$

In any factoring problem, we can always check our work by multiplying. For example, since

$$2(x + y) = 2x + 2y, \qquad 5(a - b) = 5a - 5b, \qquad \text{and} \qquad 7(a - 1) = 7a - 7,$$

our factoring in the first three examples above is correct. Multiplying to check factoring is always wise.

When an expression contains like terms, the expression can be simplified by **collecting** the like terms. The last two examples above illustrate this as do the following.

EXAMPLE 3 Use the distributive laws to collect like terms.

**(a)** $2x + 9x = (2 + 9)x$
$= 11x$
Use the distributive law to factor out $x$ and then add 2 and 9

The two like terms $2x$ and $9x$ have been collected to form the single term $11x$.

**(b)** $2y + 5y - y + 4 = (2 + 5 - 1)y + 4$ Distributive law
$= 6y + 4$

The terms $6y$ and 4 *cannot* be collected since they are not like terms. That is, the expressions $6y + 4$ and $10y$ are *not* the same! (This is clear if $y$ is replaced by some number such as 2. Do this!)

**(c)** $3a + 5b - a + 4b = 3a - a + 5b + 4b$ Commutative law
$= 3 \cdot a - 1 \cdot a + 5 \cdot b + 4 \cdot b$ $-1 \cdot a = -a$
$= (3 - 1)a + (5 + 4)b$ Distributive law
$= 2a + 9b$

With practice, the middle two steps can be eliminated.

**(d)** $.07x + x = (.07) \cdot x + 1 \cdot x$
$= (.07 + 1)x$ Distributive law
$= 1.07x$

**(e)** $2a - 5a + a - 3 - a = 2a - 5a + a - a - 3$ Commutative law
$= (2 - 5 + 1 - 1)a - 3$ Factor out $a$
$= -3a - 3$

Why could this be written as $-3(a + 1)$? Why do we *not* collect the terms $-3a$ and $-3$?

**(f)** $3x + 4y - z + 8$ has no like terms to collect

**(g)** $x + z - 4x + 3 - 4z + 2x = x - 4x + 2x + z - 4z + 3$
$= (1 - 4 + 2)x + (1 - 4)z + 3$
$= -x - 3z + 3$

Using the distributive laws and the fact that $-x = (-1) \cdot x$,

$$-(a + b) = (-1)(a + b) = (-1)(a) + (-1)(b) = -a - b,$$
$$-(a - b) = (-1)(a - b) = (-1)(a) - (-1)(b) = -a + b,$$
$$-(-a - b) = (-1)(-a - b) = (-1)(-a) - (-1)(b) = a + b.$$

These observations lead to the next rule.

1. To simplify an expression in which a minus sign precedes a set of parentheses, remove the parentheses and change the sign of every term.
2. When a plus sign precedes a set of parentheses, remove the parentheses and do not change any of the signs of the terms.

**EXAMPLE 4** Simplify by removing parentheses.

(a) $-(x+1)=-x-1$ — Change all signs

(b) $-(x-1)=-x+1$ — Change all signs

(c) $-(-x+1+a)=+x-1-a=x-1-a$

(d) $+(x-1+a)=x-1+a$ — Signs do not change

(e) $x-(2-a)=x-2+a$

The process of removing parentheses is sometimes called **clearing all parentheses.** When the instruction to *simplify* is given, clear all parentheses and collect like terms.

**EXAMPLE 5** Simplify.

(a) $2x - (6x+2) = 2x - 6x - 2$ — Change signs
$= (2-6)x - 2$
$= -4x - 2$ — Collect like terms

(b) $y - (3y-2) + y = y - 3y + 2 + y$ — Change all signs within parentheses
$= -y + 2$ — Collect like terms

(c) $4 + (z-3) - 4z = 4 + z - 3 - 4z$ — Signs remain the same
$= -3z + 1$

(d) $3a - (-a-1) + 7 = 3a + a + 1 + 7$ — Change all signs
$= 4a + 8$ — Collect like terms

When an expression involves more than one set of parentheses (or brackets or braces), clear (remove) the innermost parentheses first. Continue clearing from the inside out until all parentheses have been cleared.

**EXAMPLE 6** Clear all parentheses and collect like terms.

(a) $-3[2-4(1-5)] = -3[2-4(-4)]$ — Combine within inner parentheses
$= -3[2+16]$ — $-4(-4)=+16$
$= -3[18] = -54$

(b) $-2(5-2)-[3(4-2)-2(3-5)] = -2(3)-[3(2)-2(-2)]$
$= -6-[6+4] = -6-10 = -16$

**(c)** $-4[x - 2(x - 3)] = -4[x - 2x + 6]$ Since $-2(x - 3) = -2x + 6$
$= -4[-x + 6]$
$= 4x - 24$ Since $-4[-x + 6] = (-4)(-x) + (-4)(6) = 4x - 24$

**(d)** $a - [2a - (1 - 3a)] = a - [2a - 1 + 3a]$
$= a - [5a - 1]$
$= a - 5a + 1$
$= -4a + 1$

CAUTION: One of the most common mistakes is forgetting to change all signs when a minus sign appears in front of a set of parentheses. For example, $-(x - 2)$ is *not* $-x - 2$.

## EXERCISES 1.5

**1.** State the distributive law of multiplication over addition in symbols.

**2.** State the distributive law of multiplication over subtraction in symbols.

**3.** **(a)** Compute $-2(4 - 7)$. **(b)** Compute $(-2)(4) - (-2)(7)$. **(c)** Why are these two equal?

**4.** **(a)** Compute $-3(5 - 2)$. **(b)** Compute $(-3)(5) - (-3)(2)$. **(c)** Why are these two equal?

*How many terms does the expression have?*

**5.** $2x - y + 3z + 8$ **6.** $ax + by + cz$ **7.** $2w$ **8.** $3(a + 1)$

*Use the distributive laws to factor the following.*

**9.** $2a + 2b$ **10.** $2x - 2y$ **11.** $ca + 3a$

**12.** $2u + 4v - 6w$ **13.** $3b + yb$ **14.** $ax + ay - az$

**15.** $20x - 10y + 40$ **16.** $36x + 6$ **17.** $3x - 6 + 9y$

*Multiply.*

**18.** $2(a + b)$ **19.** $2(x - y)$ **20.** $(c + 3)a$

**21.** $2(u + 2v - 3w)$ **22.** $(3 + y)b$ **23.** $a(x + y - z)$

**24.** $10(2x - y + 4)$ **25.** $6(6x + 1)$ **26.** $3(x - 2 + 3y)$

*Use the distributive laws to collect like terms.*

**27.** $3x + 5x$

**28.** $4z - z$

**29.** $3y - y + 7$

**30.** $4a - a + 3a$

**31.** $4z - 2x + 3z + 1$

**32.** $2y + a - 2y - a$

**33.** $\frac{1}{2}x - \frac{1}{4}x + \frac{3}{4}$

**34.** $\frac{2}{3}a + b + \frac{1}{3}a - b$

**35.** $-\frac{3}{4}x + \frac{1}{4} + \frac{3}{4}x - \frac{1}{4}y$

*Remove parentheses and simplify.*

**36.** $-(x + 3)$

**37.** $-(a - 3)$

**38.** $+(2y - 3)$

**39.** $+(1 - 4z)$

**40.** $-(-a - 1)$

**41.** $-(-x + 3)$

**42.** $-(x - z - 1)$

**43.** $-(x + 3y - 4)$

**44.** $+(-x - y - z)$

**45.** $-(-x - y - z)$

**46.** $-(1 + y) + (y + 1)$

**47.** $-(x + y) - (-x - y)$

**48.** $2x - (-x + 1) + 3$

**49.** $-2[a - 3(a + 2)]$

**50.** $x - [3x - (1 - 2x)]$

*Evaluate when $x = -3$, $y = -1$, and $z = 5$.*

**51.** $3x + 1$

**52.** $-2x - y$

**53.** $x + y + z$

**54.** $|x - y|$

**55.** $-|x + y + z|$

**56.** $|5x + 3z|$

**57.** $-(-x)$

**58.** $-(+x)$

**59.** $+(-x)$

**60.** $x + 0$

**61.** $x \cdot 0$

**62.** $\frac{x}{0}$

**63.** $\frac{0}{x}$

**64.** $x - 0$

**65.** $0 - x$

---

ANSWERS: **1.** $a(b + c) = ab + ac$ **2.** $a(b - c) = ab - ac$ **3.** (a) 6 (b) 6 (c) distributive law **4.** (a) $-9$ (b) $-9$ (c) distributive law **5.** 4 **6.** 3 **7.** 1 **8.** 1 (Note that $3(a + 1)$ has one term but $3a + 3$ has two terms.) **9–17.** answers given in Exercises 18–26 **18–26.** answers given in Exercises 9–17 **27.** $8x$ **28.** $3z$ **29.** $2y + 7$ **30.** $6a$ **31.** $7z - 2x + 1$ **32.** 0 **33.** $\frac{1}{4}x + \frac{3}{4}$ **34.** $a$ **35.** $\frac{1}{4} - \frac{1}{4}y$ **36.** $-x - 3$ **37.** $-a + 3$ **38.** $2y - 3$ **39.** $1 - 4z$ **40.** $a + 1$ **41.** $x - 3$ **42.** $-x + z + 1$ **43.** $-x - 3y + 4$ **44.** $-x - y - z$ **45.** $x + y + z$ **46.** 0 **47.** 0 **48.** $3x + 2$ **49.** $4a + 12$ **50.** $-4x + 1$ **51.** $-8$ **52.** 7 **53.** 1 **54.** 2 **55.** $-1$ **56.** 0 **57.** $-3$ **58.** 3 **59.** 3 **60.** $-3$ **61.** 0 **62.** undefined **63.** 0 **64.** $-3$ **65.** 3

## 1.6 Integer Exponents: Properties of Exponents

Many times, a number or variable is multiplied by itself several times: for example, $2 \cdot 2 \cdot 2 \cdot 2$ or $x \cdot x \cdot x$. In order to avoid lengthy strings of factors such as these, we adopt an **exponential notation.** We write $2 \cdot 2 \cdot 2 \cdot 2$ as $2^4$,

$$\underbrace{2 \cdot 2 \cdot 2 \cdot 2}_{\text{4 factors}} = \underset{\text{base}}{2}^{4} \leftarrow \text{exponent}$$

in which 2 is called the **base,** 4 the **exponent,** and $2^4$ the **exponential expression** (read "2 to the **fourth power**"). Similarly, $x \cdot x \cdot x = x^3$ is called the **third power** or **cube** of $x$. The **square** or **second power** of $a$ is $a^2$. The **first power** of $a$ is $a^1$ which we write simply as $a$.

**DEFINITION** If $a$ is any number and $n$ is a positive integer,

$$a^n = \underbrace{a \cdot a \cdot a \cdots a}_{n \text{ factors}}.$$

**EXAMPLE 1** Write in exponential notation.

**(a)** $\underbrace{7 \cdot 7 \cdot 7}_{\text{3 factors}} = 7^3$

**(b)** $\underbrace{a \cdot a \cdot a \cdot a \cdot a \cdot a}_{\text{6 factors}} = a^6$

**(c)** $\underbrace{3 \cdot 3}_{2} \cdot \underbrace{x \cdot x \cdot x \cdot x \cdot x}_{5} = 3^2 \cdot x^5$

**(d)** $\underbrace{(2y)(2y)(2y)}_{3} = (2y)^3$

**(e)** $\underbrace{(a+b)(a+b)}_{2} = (a+b)^2$

**(f)** $\dfrac{4 \cdot u \cdot u}{z \cdot z \cdot z \cdot z} = \dfrac{4u^2}{z^4}$

**EXAMPLE 2** Write without using exponents.

**(a)** $3^5 = \underbrace{3 \cdot 3 \cdot 3 \cdot 3 \cdot 3}_{\text{5 factors}}$ The product is 243

**(b)** $2x^2 = 2 \cdot \underbrace{x \cdot x}_{2}$ 2 *is not* squared

**(c)** $(2x)^2 = \underbrace{(2x)(2x)}_{2}$ 2 *is* squared

**(d)** $2^3 + 4^3 = 2 \cdot 2 \cdot 2 + 4 \cdot 4 \cdot 4$ This simplifies to 72, not $(2 + 4)^3 = 6^3 = 216$

**(e)** $1^{25} = \underbrace{1 \cdot 1 \cdot 1 \cdots 1}_{\text{25 factors}} = 1$ $1^n = 1$ for any $n$

**(f)** $5^2 - 4^2 = 5 \cdot 5 - 4 \cdot 4$ Why not $(5 - 4)^2$?

**(g)** $\left(-\frac{1}{2}\right)^4 = \left(-\frac{1}{2}\right)\left(-\frac{1}{2}\right)\left(-\frac{1}{2}\right)\left(-\frac{1}{2}\right)$ The product is $\frac{1}{16}$

**CAUTION:** Many of the most common errors made when working with exponents have been indicated in the above examples.

$$2x^2 \neq (2x)^2, \qquad (2+4)^3 \neq 2^3 + 4^3 \qquad (5-4)^2 \neq 5^2 - 4^2$$

In the first case, an exponent is applicable only to the factor immediately next to it; that is, in $2x^2$ only $x$ is squared, not 2. If we want to square $2x$, we must use parentheses: $(2x)^2 = (2x)(2x)$. In the second and third cases, powers of sums or differences are *not* sums or differences of powers!

**EXAMPLE 3** Evaluate when $a = 2$, $b = 1$, $c = -3$.

**(a)** $3a^2 = 3(2)^2 = 3 \cdot 4 = 12$ — Not $(3 \cdot 2)^2 = 6^2 = 36$

**(b)** $a^3 - b^3 = (2)^3 - (1)^3 = 8 - 1 = 7$ — Not $(2-1)^3$

**(c)** $2ab^2c^3 = 2(2)(1)^2(-3)^3 = 2(2)(1)(-27) = (4)(-27) = -108$

**(d)** $3a^2 + 2c^2 = 3(2)^2 + 2(-3)^2 = 3(4) + 2(9) = 12 + 18 = 30$

**(e)** $(2c)^2 - 2c^2 = [(2)(-3)]^2 - 2(-3)^2 = [-6]^2 - 2(9) = 36 - 18 = 18$ — Watch the substitution

**(f)** $a^a + c^b = 2^2 + (-3)^1 = 4 + (-3) = 1$

When we combine terms containing exponential expressions by multiplication, division, or taking powers, our work can be simplified by using the basic properties of exponents. For example,

$$a^4 \cdot a^2 = \underbrace{(a \cdot a \cdot a \cdot a)}_{\text{4 factors}}\underbrace{(a \cdot a)}_{\text{2 factors}} = \underbrace{a \cdot a \cdot a \cdot a \cdot a \cdot a}_{\text{6 factors}} = a^6.$$

When two exponential expressions *with the same base* are multiplied, the product is that base raised to the sum of the exponents on the original pair of expressions.

If $a$ is any number, and $m$ and $n$ are positive integers,

$$a^m \cdot a^n = a^{m+n}.$$

**EXAMPLE 4** **(a)** $a^2 \cdot a^5 = a^{2+5} = a^7$

**(b)** $4^3 \cdot 4^7 = 4^{3+7} = 4^{10}$ — *Not* $16^{10}$

**(c)** $2^2 \cdot 2^3 \cdot 2^5 = 2^{2+3+5} = 2^{10}$ — The rule also applies to more than two factors

**(d)** $2x^2 \cdot x^9 = 2x^{2+9} = 2x^{11}$

Remember this rule by "to multiply powers with the same base, add exponents."

When two powers with the same base are divided, for example

$$\frac{a^7}{a^3} = \frac{\overbrace{a \cdot a \cdot a \cdot a \cdot \not{a} \cdot \not{a} \cdot \not{a}}^{\text{7 factors}}}{\underbrace{\not{a} \cdot \not{a} \cdot \not{a}}_{\text{3 factors}}} = \underbrace{a \cdot a \cdot a \cdot a}_{\text{4 factors}} = a^4,$$

the quotient can be found by raising the base to the difference of the exponents $(7 - 3 = 4)$.

If $a$ is any number except zero and $m$, $n$, and $m - n$ are positive integers, then

$$\frac{a^m}{a^n} = a^{m-n}.$$

**EXAMPLE 5** **(a)** $\frac{a^5}{a^2} = a^{5-2} = a^3$

**(b)** $\frac{4^7}{4^3} = 4^{7-3} = 4^4$

**(c)** $\frac{3^2}{4^3}$ Cannot be simplified using the rule of exponents since the bases are different

**(d)** $\frac{2x^5}{x} = 2x^{5-1}$ $\quad x = x^1$

$= 2x^4$

**(e)** $\frac{3y^2y^5}{y^3} = \frac{3y^{2+5}}{y^3} = \frac{3y^7}{y^3} = 3y^{7-3} = 3y^4$

Remember this rule by "to divide powers with the same base, subtract exponents."

When raising a power to a power, for example,

$$(a^2)^3 = \underbrace{(a^2)\ (a^2)\ (a^2)}_{\text{3 factors}} = a \cdot a \cdot a \cdot a \cdot a \cdot a$$

$$= \underbrace{a \cdot a \cdot a \cdot a \cdot a \cdot a}_{\text{6 factors}} = a^6,$$

the resulting exponential expression can be found by raising the base to the product of the exponents $(2 \cdot 3 = 6)$.

If $a$ is any number, and $m$ and $n$ are positive integers,

$$(a^m)^n = a^{m \cdot n}.$$

**EXAMPLE 6** (a) $(a^5)^2 = a^{5 \cdot 2} = a^{10}$

(b) $(4^7)^3 = 4^{7 \cdot 3} = 4^{21}$

Remember this rule by "to raise a power to a power, multiply exponents."

**CAUTION:** Do not confuse this rule with the rule for multiplying exponential expressions with the same base. In Example 6(a), for example, $(a^5)^2 = a^{10}$, but $a^5 \cdot a^2 = a^7$.

Often, a product or quotient of expressions is raised to a power. For example,

$$(2x^2)^3 = \underbrace{(2x^2)(2x^2)(2x^2)}_{3 \text{ factors}} = \underbrace{2 \cdot 2 \cdot 2}_{3 \text{ factors}} \cdot \underbrace{x^2 \cdot x^2 \cdot x^2}_{3 \text{ factors}} = 2^3 \cdot (x^2)^3$$

$$\left(\frac{3}{y^3}\right)^2 = \underbrace{\frac{3}{y^3} \cdot \frac{3}{y^3}}_{2} = \frac{\overbrace{3 \cdot 3}^{2}}{\underbrace{y^3 \cdot y^3}_{2}} = \frac{3^2}{(y^3)^2}.$$

These illustrate the next rule.

If $a$ and $b$ are any numbers, and $n$ is a positive integer, then

1. $(a \cdot b)^n = a^n \cdot b^n$
2. $\left(\frac{a}{b}\right)^n = \frac{a^n}{b^n}$ ($b$ not zero)

**EXAMPLE 7** (a) $(3y)^4 = 3^4 \cdot y^4 = 3^4y^4$

(b) $(2a^2b^3)^5 = 2^5(a^2)^5(b^3)^5$ Raise each factor to the fifth power

$= 2^5a^{10}b^{15}$ Use power to a power rule

(c) $\left(\frac{3z^2}{u^3}\right)^4 = \frac{(3z^2)^4}{(u^3)^4}$ $\left(\frac{a}{b}\right)^n = \frac{a^n}{b^n}$

$= \frac{3^4(z^2)^4}{(u^3)^4}$ $(a \cdot b)^n = a^nb^n$

$= \frac{3^4z^8}{u^{12}}$ $(a^m)^n = a^{mn}$

(d) $(-2x^2)^3 = (-2)^3(x^2)^3 = -8x^6$

(e) $(2^2 + 3^2)^3$ is *not* $(2^2)^3 + (3^2)^3 = 2^6 + 3^6$ (Why?)

Thus a sum to a power is *not* the sum of powers.

(f) $(3^2 - 2^2)^3$ is *not* $(3^2)^3 - (2^2)^3 = 3^6 - 2^6$ (Why?)

Thus a difference to a power is *not* the difference of powers.

(g) $a^2 \cdot b^3$ is *not* $(ab)^5$ (Substitute $a = 3$, $b = 2$.)

Only exponential expressions with the same base can be combined.

We know that if $a$ is not zero,

$$\frac{a^m}{a^n} = a^{m-n}.$$

If $m = n$, we have

$$\frac{a^m}{a^m} = a^{m-m} = a^0 \qquad \text{and also} \qquad \frac{a^m}{a^m} = 1$$

since any number divided by itself is 1. This suggests the following definition.

**DEFINITION** If $a$ is any number except zero,

$$a^0 = 1.$$

**EXAMPLE 8** **(a)** $5^0 = 1$ **(b)** $17^0 = 1$ **(c)** $0^0$ is not defined

**(d)** $(2x^2y)^0 = 1$ assuming $x \neq 0$ and $y \neq 0$ (Why?)

Considering $a^m/a^n = a^{m-n}$ $(a \neq 0)$ again, what happens when $n > m$? For example, if $n = 5$ and $m = 2$, we have

$$\frac{a^m}{a^n} = \frac{a^2}{a^5} = a^{2-5} = a^{-3}.$$

If we consider this same problem in another way, we have

$$\frac{a^2}{a^5} = \frac{\not{a} \cdot \not{a}}{\not{a} \cdot \not{a} \cdot a \cdot a \cdot a} = \frac{1}{a \cdot a \cdot a} = \frac{1}{a^3}.$$

Thus we conclude that $a^{-3} = \dfrac{1}{a^3}$. This suggests a way to define exponential expressions involving negative integer exponents.

**DEFINITION** If $a \neq 0$ and $n$ is a positive integer ($-n$ is a negative integer), then

$$a^{-n} = \frac{1}{a^n}.$$

**EXAMPLE 9** **(a)** $7^{-3} = \dfrac{1}{7^3} = \dfrac{1}{343}$

**(b)** $3^{-2} = \dfrac{1}{3^2} = \dfrac{1}{9}$

**(c)** $\dfrac{1}{3^{-2}} = \dfrac{1}{\frac{1}{3^2}} = \dfrac{1}{\frac{1}{9}} = 1 \cdot \dfrac{9}{1} = 9 = 3^2$

**(d)** $(-2)^{-3} = \dfrac{1}{(-2)^3} = \dfrac{1}{-8} = -\dfrac{1}{8}$

We often "remove" negative exponents by simply moving an exponential expression involving a negative exponent from denominator to numerator (or numerator to denominator) while simultaneously changing the sign of the exponent. This occurs in (b) and (c) above when $3^{-2}$ becomes $\frac{1}{3^2}$ and $\frac{1}{3^{-2}}$ becomes $\frac{3^2}{1} = 3^2$.

**CAUTION:** Do not make the mistake of concluding that $5^{-2}$ is the same as $-5^2$ or $(-2)(5)$. These are certainly not true since

$$5^{-2} = \frac{1}{25}, \qquad -5^2 = -25, \qquad (-2)(5) = -10.$$

All the rules of exponents developed before considering negative exponents (rules stated in terms of positive integer exponents), are applicable to *all* integer exponents: positive, negative, and zero.

**RULES FOR EXPONENTS**

Let $a$ and $b$ be any two numbers, $m$ and $n$ any two integers.

1. $a^m \cdot a^n = a^{m+n}$
2. $\frac{a^m}{a^n} = a^{m-n} \quad (a \neq 0)$
3. $(a^m)^n = a^{mn}$
4. $(a \cdot b)^n = a^n b^n$
5. $\left(\frac{a}{b}\right)^n = \frac{a^n}{b^n} \quad (b \neq 0)$
6. $a^0 = 1 \quad (a \neq 0)$
7. $a^{-n} = \frac{1}{a^n} \quad (a \neq 0)$
8. $\frac{1}{a^{-n}} = a^n \quad (a \neq 0)$

**EXAMPLE 10** Simplify and express without using negative exponents.

**(a)** $(2y)^{-1} = \frac{1}{(2y)^1} = \frac{1}{2y}$ — $(2y)^{-1}$ is *not* $-2y$

**(b)** $2y^{-1} = 2 \cdot \frac{1}{y^1} = \frac{2}{y}$ — Compare with (a)

**(c)** $x^3 \cdot x^{-2} = x^{3+(-2)}$
$= x^1 = x$ — $a^m a^n = a^{m+n}$

**(d)** $\frac{x^2y^{-3}}{x^{-1}y^4} = x^{2-(-1)}y^{-3-4}$ — $\frac{a^m}{a^n} = a^{m-n}$
$= x^3y^{-7}$
$= x^3 \cdot \frac{1}{y^7} = \frac{x^3}{y^7}$

**(e)** $(-2)^{-4} = \frac{1}{(-2)^4} = \frac{1}{16}$

**(f)** $\frac{1}{a^{-7}} = \frac{1}{\frac{1}{a^7}} = 1 \cdot \frac{a^7}{1} = a^7$

**EXAMPLE 11** Simplify and express without using negative exponents.

**(a)** $\left(\frac{a}{b}\right)^{-1} = \frac{a^{-1}}{b^{-1}}$ $\qquad \left(\frac{a}{b}\right)^n = \frac{a^n}{b^n}$

$$= \frac{\frac{1}{a}}{\frac{1}{b}} = \frac{1}{a} \cdot \frac{b}{1} = \frac{b}{a}$$

**(b)** $(2x^2y^{-3})^4 = 2^4(x^2)^4(y^{-3})^4$ $\qquad (a \cdot b)^n = a^n \cdot b^n$

$$= 16x^8y^{-12} \qquad (a^m)^n = a^{mn}$$

$$= 16x^8 \frac{1}{y^{12}} = \frac{16x^8}{y^{12}}$$

**(c)** $6^0(-3x^2)^{-3} = 1 \cdot (-3x^2)^{-3}$ $\qquad 6^0 = 1$

$$= (-3)^{-3}(x^2)^{-3} \qquad (a \cdot b)^n = a^n \cdot b^n$$

$$= \frac{1}{(-3)^3} x^{2(-3)}$$

$$= \frac{1}{-27} x^{-6}$$

$$= -\frac{1}{27} \cdot \frac{1}{x^6} = -\frac{1}{27x^6}$$

**(d)** $\left(\frac{a^5}{3x^{-3}}\right)^{-2} = \frac{(a^5)^{-2}}{3^{-2}(x^{-3})^{-2}}$

$$= \frac{a^{-10}}{\frac{1}{3^2} \cdot x^6}$$

$$= \frac{\frac{1}{a^{10}}}{\frac{x^6}{9}} = \frac{1}{a^{10}} \cdot \frac{9}{x^6} = \frac{9}{a^{10}x^6}$$

## EXERCISES 1.6

*Write in exponential notation.*

1. $7 \cdot 7 \cdot 7 \cdot 7 \cdot 7$
2. $x \cdot x \cdot x$
3. $4 \cdot 4 \cdot y \cdot y \cdot y \cdot y$
4. $(3a)(3a)$
5. $3 \cdot a \cdot a$
6. $(a + b)(a + b)(a + b)(a + b)$

*Write without using exponents.*

7. $10^4$
8. $x^7$
9. $a^2b^3c^4$
10. $3y^3$
11. $(3y)^3$
12. $a^2 + x^2$

*Square the following.*

13. $5y$
14. $2x^2y$
15. $3a^0 - 3b^0$ ($a$, $b$ not zero)

*Cube the following.*

**16.** $3a^3$

**17.** $\frac{1}{2}y$

**18.** $2x^2y^3z$

*Simplify and write without negative exponents.*

**19.** $x^3 \cdot x^7$

**20.** $a^2 \cdot a^3 \cdot a^5$

**21.** $3y^3 \cdot y^2$

**22.** $\frac{a^8}{a^5}$

**23.** $\frac{3y^3}{y^2}$

**24.** $(a^3)^2$

**25.** $(2x^4)^2$

**26.** $(x^2y^3)^4$

**27.** $\left(\frac{3}{y^2}\right)^3$

**28.** $\frac{x^3}{y^2}$

**29.** $3^0$

**30.** $0^0$

**31.** $(4x^2y)^0 \quad (x \neq 0 \text{ and } y \neq 0)$

**32.** $(a+b)^2(a+b)^3$

**33.** $2x^0 \quad (x \neq 0)$

**34.** $(2x)^0 \quad (x \neq 0)$

**35.** $(3a)^{-1}$

**36.** $3a^{-1}$

**37.** $\frac{2a^2}{a^5}$

**38.** $3x^3x^{-4}$

**39.** $(2x)^{-2}$

**40.** $2x^{-2}$

**41.** $(3a^2b^{-3})^{-1}$

**42.** $(3x^{-2})^3$

**43.** $(3x^{-2})^{-3}$

**44.** $\left(\frac{2x^2}{y^{-3}}\right)^2$

**45.** $\left(\frac{2x^2}{y^{-3}}\right)^{-2}$

**46.** $\frac{a^2b^{-3}}{a^3b^{-2}}$

**47.** $\frac{a^{-3}}{b^{-6}}$

**48.** $\frac{1}{a^{-5}}$

*Evaluate.*

**49.** $2^2 + 3^2$

**50.** $(2+3)^2$

**51.** $2^2 - 3^2$

**52.** $(2-3)^2$

**53.** $(3 \cdot 2)^2$

**54.** $3^2 \cdot 2^2$

**55.** $\frac{3^2}{2^2}$

**56.** $\left(\frac{3}{2}\right)^2$

**57.** $\left(\frac{3}{2}\right)^0$

*Evaluate when $x = -3$, $y = 2$, and $z = -1$.*

**58.** $2x^2$

**59.** $(2x)^2$

**60.** $-2x^2$

**61.** $(-2x)^2$

**62.** $x^2 - y^2 - z^2$

**63.** $-3z^2 - (x + y)$

**64.** $-z^{-3}$

**65.** $-x^2$

**66.** $(-x)^2$

**67.** $x^2 - 4yz$

**68.** $(x - z)^{-3}$

**69.** $x^3 - z^2$

**70.** In the exponential expression $3x^5$, the coefficient of $x^5$ is **(a)** ________, the exponent is **(b)** ________, the base is **(c)** ________, and $3x^5$ represents three times the **(d)** ____________________ power of $x$.

**71.** In the expression $-y + x^2$, the exponent on $x$ is **(a)** ________, the exponent on $y$ is **(b)** ________, the coefficient of the $x^2$ term is **(c)** ______________, the coefficient of the $y$ term is **(d)** ______________, and there are **(e)** ____________________ terms.

*Use the distributive laws to factor the following.*

**72.** $5x - 5y + 5$

**73.** $ax - a$

*Simplify.*

**74.** $3x + 2 - (1 - 3x)$

**75.** $a - [3a - (-1 - a)]$

---

**ANSWERS:** **1.** $7^5$ **2.** $x^3$ **3.** $4^2y^4$ **4.** $(3a)^2$ **5.** $3a^2$ **6.** $(a + b)^4$ **7.** $10 \cdot 10 \cdot 10 \cdot 10$ or 10,000 **8.** $x \cdot x \cdot x \cdot x \cdot x \cdot x \cdot x$ **9.** $a \cdot a \cdot b \cdot b \cdot b \cdot c \cdot c \cdot c \cdot c$ **10.** $3 \cdot y \cdot y \cdot y$ **11.** $3y \cdot 3y \cdot 3y$ or $3 \cdot 3 \cdot 3 \cdot y \cdot y \cdot y$ **12.** $a \cdot a + x \cdot x$ **13.** $25y^2$ **14.** $4x^4y^2$ **15.** 0 **16.** $27a^9$ **17.** $\frac{1}{8}y^3$ **18.** $8x^6y^9z^3$ **19.** $x^{10}$ **20.** $a^{10}$ **21.** $3y^5$ **22.** $a^3$ **23.** $3y$ **24.** $a^6$ **25.** $4x^8$ **26.** $x^8y^{12}$ **27.** $\frac{27}{y^6}$ **28.** $\frac{x^3}{y^2}$ (already simplified) **29.** 1 **30.** undefined **31.** 1 **32.** $(a + b)^5$ **33.** 2 **34.** 1 **35.** $\frac{1}{3a}$ **36.** $\frac{3}{a}$ **37.** $\frac{2}{a^3}$ **38.** $\frac{3}{x}$ **39.** $\frac{1}{4x^2}$ **40.** $\frac{2}{x^2}$ **41.** $\frac{b^3}{3a^2}$ **42.** $\frac{27}{x^6}$ **43.** $\frac{x^6}{27}$ **44.** $4x^4y^6$ **45.** $\frac{1}{4x^4y^6}$ **46.** $\frac{1}{ab}$ **47.** $\frac{b^6}{a^3}$ **48.** $a^5$ **49.** 13 **50.** 25 **51.** $-5$ **52.** 1 **53.** 36 **54.** 36 **55.** $\frac{9}{4}$ **56.** $\frac{9}{4}$ **57.** 1 **58.** 18 **59.** 36 **60.** $-18$ **61.** 36 **62.** 4 **63.** $-2$ **64.** 1 **65.** $-9$ **66.** 9 **67.** 17 **68.** $-\frac{1}{8}$ **69.** $-28$ **70.** (a) 3 (b) 5 (c) $x$ (d) fifth **71.** (a) 2 (b) 1 (c) 1 (d) $-1$ (e) 2 **72.** $5(x - y + 1)$ **73.** $a(x - 1)$ **74.** $6x + 1$ **75.** $-3a - 1$

## 1.7 Percent Notation

Every rational number can be expressed as a fraction that in turn can be written as a decimal numeral that either terminates or has a repeating block of digits. Another common and useful way of expressing a nonnegative rational number is as a percent. Just as *deci* is used to indicate ten (***deci*mal** numerals are based on powers of ten), *cent* indicates one hundred. Thus the term *percent* actually means *per hundred.* The percent symbol % denotes *per hundred,* so that the numeral 48% actually symbolizes

$$48 \text{ per hundred}, \quad 48 \text{ one-hundredths}, \quad 48 \cdot \frac{1}{100}, \quad 48 \cdot (.01), \quad \text{or} \quad .48.$$

The fact that $100\% = 1$ is used to convert a fractional or decimal numeral to a percent numeral.

$$.48 = .48(1) = (.48)(100\%) = (.48)(100)\% = 48\%$$

**TO CONVERT A FRACTION TO A PERCENT**

1. Convert the fraction to a decimal by division.
2. Multiply the decimal by 100 (move the decimal point two places to the right) and attach the percent symbol %.

**EXAMPLE 1** Convert to percent notation.

**(a)** $.67 = (.67)(100)\% = 67\%$

**(b)** $3.28 = (3.28)(100)\% = 328\%$

**(c)** $.\bar{3} = (.\bar{3})(100)\% = 33.\bar{3}\% = 33\frac{1}{3}\%$

(Recall that $.\bar{3} = \frac{1}{3}$. Check this by division.)

**(d)** $\frac{1}{4} = .25$ Divide 1 by 4

$= (.25)(100)\% = 25\%$

**(e)** $\frac{8}{3} = 2.\bar{6}$ Divide 8 by 3

$= (2.\bar{6})(100)\%$

$= 266.\bar{6}\% = 266\frac{2}{3}\%$

(Recall that $.\bar{6} = \frac{2}{3}$. Check this by division.)

The fact that % means 1/100 is used to convert from percent numerals to decimal numerals.

**TO CONVERT A PERCENT TO A DECIMAL**

Multiply the percent by .01 = 1/100 (move the decimal point two places to the left) and remove the percent symbol %.

**EXAMPLE 2** Convert to decimal numerals.

**(a)** 35% = (35)(.01) = .35

**(b)** 225.5% = (225.5)(.01) = 2.255

**(c)** 5% = (5)(.01) = .05

**(d)** .5% = (.5)(.01) = .005

**(e)** .05% = (.05)(.01) = .0005

When making percent conversions, it is helpful to keep one particular simple example in mind in order to remember whether we should multiply by 100 or by 1/100. We know that 50% of some quantity corresponds to .5 or 1/2 of that quantity. To convert 50% to .5, multiply 50 by .01, and to convert .5 to 50%, multiply .5 by 100.

## EXERCISES 1.7

*Convert to decimal numerals.*

**1.** $\frac{5}{8}$ **2.** $\frac{1}{20}$ **3.** $\frac{5}{16}$

**4.** $\frac{2}{5}$ **5.** $\frac{1}{6}$ **6.** $\frac{2}{7}$

*Convert to percent notation.*

**7.** .75 **8.** 2.41 **9.** .08

**10.** .001 **11.** $.5\overline{3}$ **12.** $\frac{1}{6}$

**13.** $\frac{3}{10}$

**14.** 1

**15.** 100

**16.** $\frac{3}{8}$

**17.** 10

**18.** .1

*Convert to decimal notation.*

**19.** 40%

**20.** 23.5%

**21.** $8\frac{1}{2}\%$

**22.** $\frac{1}{2}\%$

**23.** .3%

**24.** .03%

**25.** 100%

**26.** $\frac{1}{3}\%$

**27.** 1%

**28.** If 55% is (55)(.01) = .55 as a decimal, what is $x\%$ as a decimal?

*Simplify and write without negative exponents.*

**29.** $2x^2x^{-5}$

**30.** $\frac{y^3}{y^{-3}}$

**31.** $\left(\frac{x^{-1}}{a^{-2}}\right)^{-2}$

**32.** $5^0 \cdot (x^2a^{-3})^{-4}$

**33.** $\frac{2}{b^{-3}}$

**34.** $3a^0 - (3a)^0 \quad (a \neq 0)$

*Evaluate when $a = -2$ and $b = 3$.*

**35.** $5a^3$

**36.** $(5a)^3$

**37.** $-5a^3$

**38.** $(-5a)^3$ **39.** $(a+b)^3$ **40.** $a^3+b^3$

**41.** $b-a$ **42.** $(2b+3a)^{-2}$ **43.** $a^{-2}$

**44.** $(ab)^{-2}$ **45.** $-2ab$ **46.** $a^0-b^0$

---

ANSWERS: **1.** .625 **2.** .05 **3.** .3125 **4.** .4 **5.** $.1\overline{6}$ **6.** $.\overline{285714}$ **7.** 75% **8.** 241% **9.** 8% **10.** .1% **11.** $53.\overline{3}\%$ or $53\frac{1}{3}\%$ **12.** $16.\overline{6}\%$ or $16\frac{2}{3}\%$ **13.** 30% **14.** 100% **15.** 10,000% **16.** 37.5% or $37\frac{1}{2}\%$ **17.** 1000% **18.** 10% **19.** .4 **20.** .235 **21.** .085 **22.** .005 **23.** .003 **24.** .0003 **25.** 1 **26.** $.00\overline{3}$ **27.** .01 **28.** $(x)(.01)$ **29.** $\frac{2}{x^3}$ **30.** $y^6$ **31.** $\frac{x^2}{a^4}$ **32.** $\frac{a^{12}}{x^8}$ **33.** $2b^3$ **34.** 2 **35.** −40 **36.** −1000 **37.** 40 **38.** 1000 **39.** 1 **40.** 19 **41.** 5 **42.** undefined **43.** $\frac{1}{4}$ **44.** $\frac{1}{36}$ **45.** 12 **46.** 0

## 1.8 Elementary Formulas from Geometry

The important formulas associated with each geometrical figure are presented with the figure. The rectangle, parallelogram, triangle, trapezoid, and circle are plane figures, and the rectangular parallelepiped and sphere are solid figures. All formulas should be memorized.

$l$ = length
$w$ = width

Area = $A = l \cdot w$
Perimeter = $P = 2 \cdot l + 2 \cdot w$
If $l = w$, the rectangle is called a **square.**

**Rectangle**

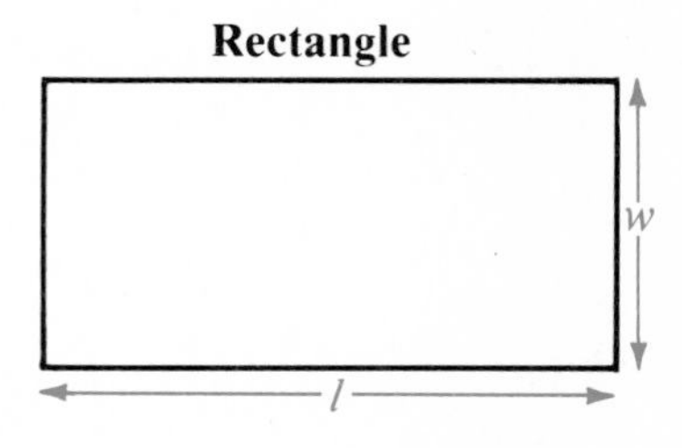


**Figure 1.4**

When calculating areas, the units of measure are square units. For example, a rectangle with width 5 inches and length 9 inches has area $5 \cdot 9 = 45$ square inches. We abbreviate $l = 9$ in, $w = 5$ in, and $A = 45$ in² or $A = 45$ sq in. In addition, we will use ft = feet, yd = yard, mi = mile, m = meter, cm = centimeter, and km = kilometer.

**EXAMPLE 1** Find the perimeter and area of the rectangle with length 3 m and width 2 m shown in Figure 1.5.

$l = 3$ m, $w = 2$ m
$A = l \cdot w = (3 \text{ m}) \cdot (2 \text{ m}) = 6 \text{ m}^2$
$P = 2l + 2w = 2(3 \text{ m}) + 2(2 \text{ m}) = 6 \text{ m} + 4 \text{ m} = 10 \text{ m}$

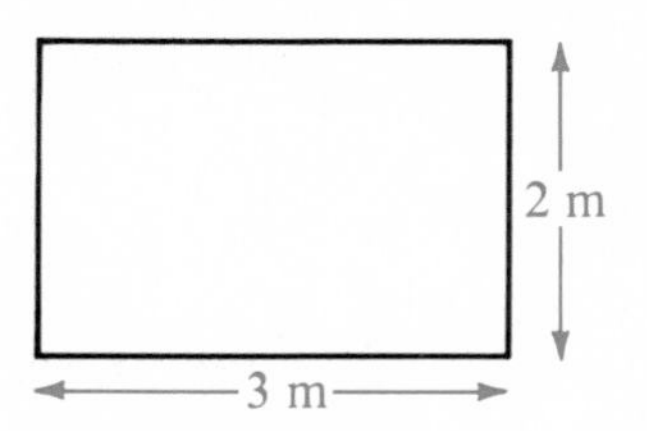


**Figure 1.5**

**EXAMPLE 2** Find the perimeter and area of the square with sides 3.5 ft in Figure 1.6.

$l = 3.5$ ft, $w = 3.5$ ft
$A = l \cdot w = (3.5 \text{ ft}) \cdot (3.5 \text{ ft}) = 12.25 \text{ ft}^2 \approx 12.3 \text{ ft}^2$
$P = 2 \cdot l + 2 \cdot w = 2(3.5 \text{ ft}) + 2(3.5 \text{ ft}) = 7 \text{ ft} + 7 \text{ ft} = 14 \text{ ft}$

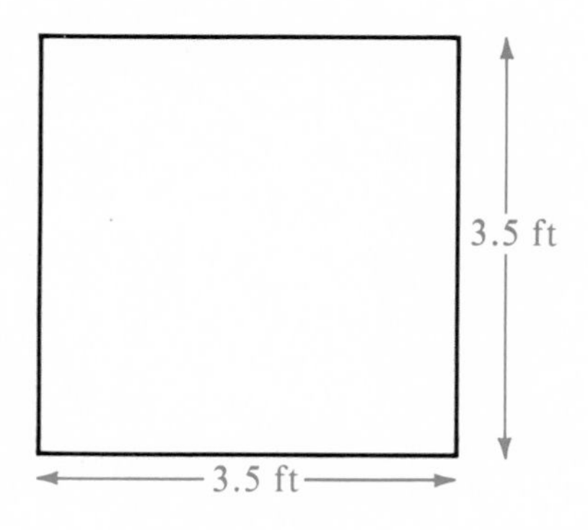


**Figure 1.6**

Generally, we agree to round off answers correct to the same number of decimal places as the least accurate measure given. Thus, in Example 2 we rounded 12.25 ft² to 12.3 ft². We use the symbol $\approx$ to represent the phrase *is approximately equal to* whenever a number is approximated by a rounded decimal form.

$b$ = base
$h$ = altitude
$a$ = side

Area = $\boldsymbol{A = b \cdot h}$
Perimeter = $\boldsymbol{P = 2 \cdot a + 2 \cdot b}$

**Parallelogram**

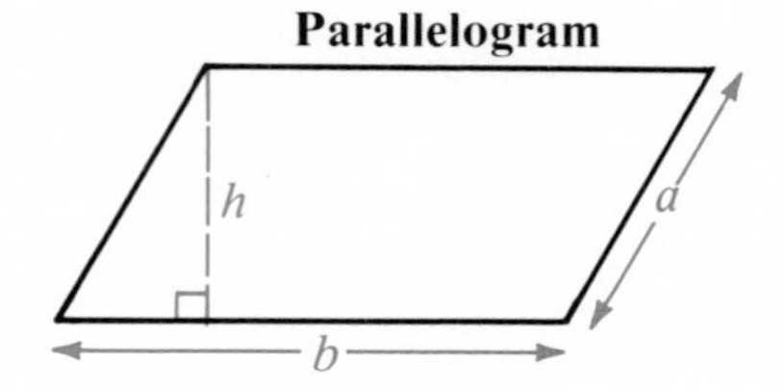


**Figure 1.7**

If all angles formed by the sides of a parallelogram are **right angles** (measure 90°), the parallelogram is actually a rectangle. Notice the similarity between the formulas associated with a parallelogram and a rectangle.

Acute triangle

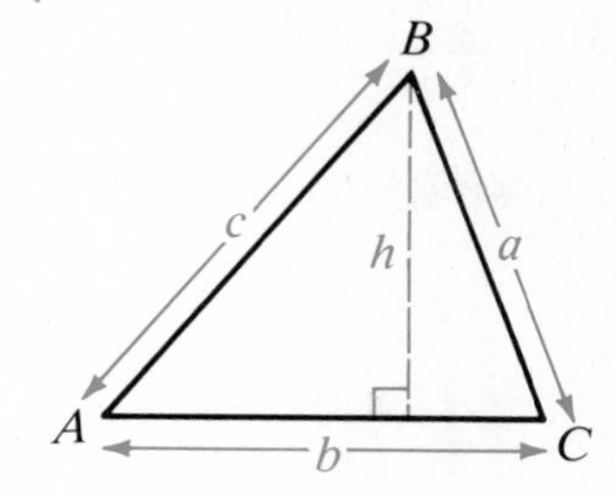


All angles are **acute**—measure less than 90°

Obtuse triangle

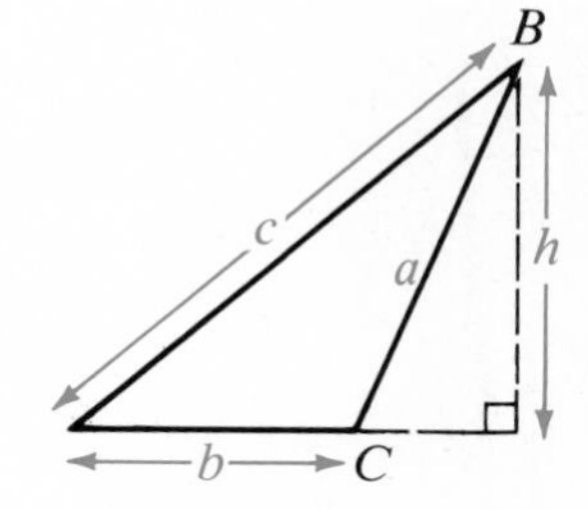


One angle is **obtuse**—measures more than 90°

Right triangle

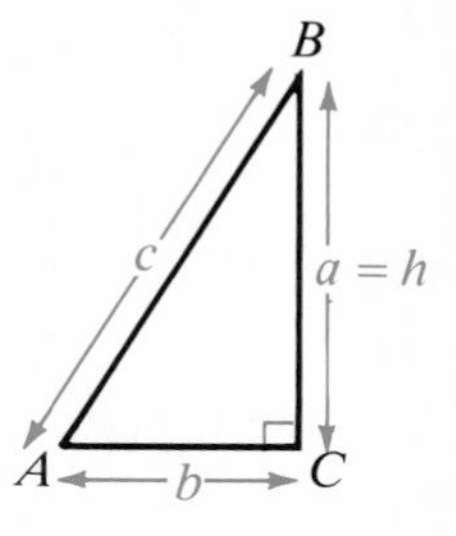


One angle is a **right angle**—measures 90°

Isosceles triangle

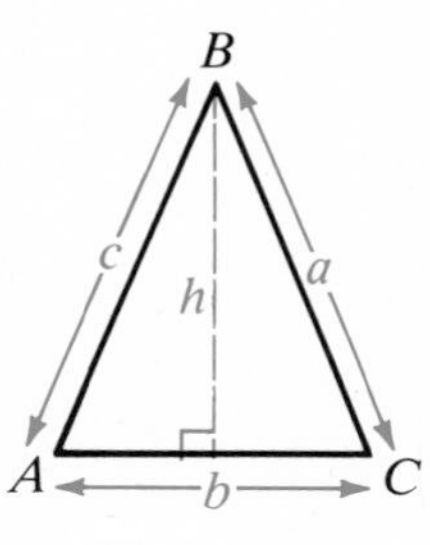


Two sides are equal—$a = c$

Equilateral triangle

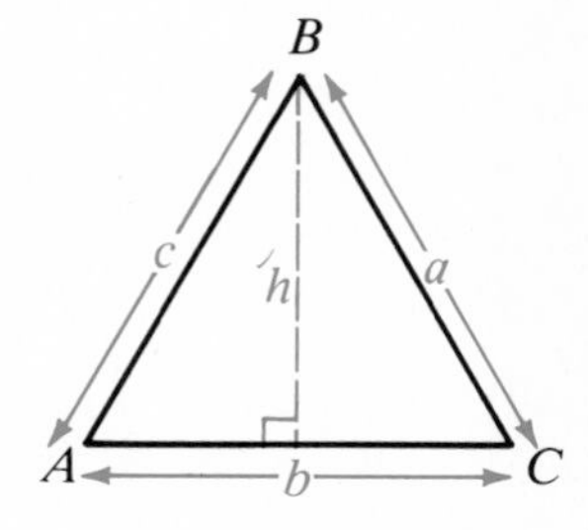


All three sides are equal—$a = b = c$

**Figure 1.8**

$b =$ base
$h =$ altitude or height
$a, b, c =$ sides

Area $= \boldsymbol{A = \frac{1}{2} b \cdot h}$

Perimeter $= \boldsymbol{P = a + b + c}$

The sum of the angles of a triangle is equal to 180°. That is

$$\angle A + \angle B + \angle C = 180°.$$

**EXAMPLE 3** Find the area of the triangle given in Figure 1.9.

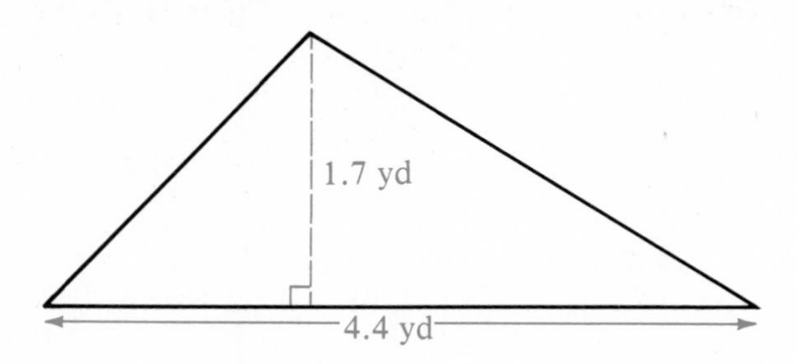


**Figure 1.9**

$$A = \frac{1}{2}b \cdot h = \frac{1}{2}(4.4 \text{ yd})(1.7 \text{ yd})$$
$$= 3.74 \text{ yd}^2$$
$$\approx 3.7 \text{ yd}^2$$

**Trapezoid**

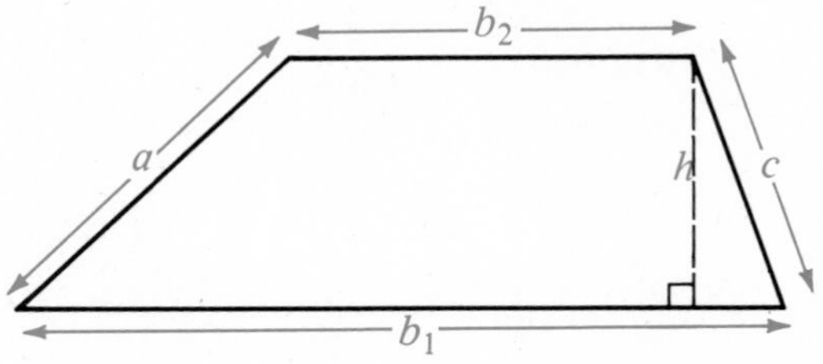

**Figure 1.10**

$b_1, b_2 =$ bases
$h =$ altitude
$a, b_1, c, b_2 =$ sides

$$\text{Area} = A = \frac{1}{2}(b_1 + b_2) \cdot h$$
$$\text{Perimeter} = P = a + b_1 + b_2 + c$$

**Circle**

$r =$ radius
$d =$ diameter $= 2 \cdot r$
Circumference $= c = 2\pi r = \pi \cdot d$
Area $= A = \pi \cdot r^2$

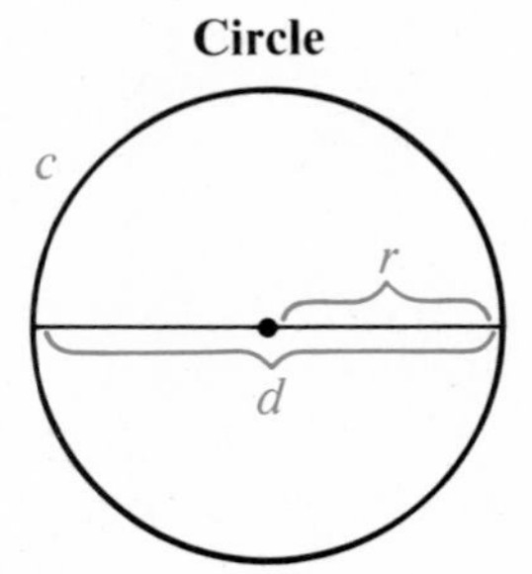


**Figure 1.11**

The ratio of the circumference of any circle to its diameter is the constant irrational number $\pi$, often approximated by 3.14.

**EXAMPLE 4** Find the diameter, circumference, and area of the circle given in Figure 1.12.

$$d = 2 \cdot r = 2(4.2 \text{ cm}) = 8.4 \text{ cm}$$
$$c = \pi \cdot d = \pi(8.4 \text{ cm}) \approx (3.14)(8.4 \text{ cm}) = 26.376 \text{ cm} \approx 26.4 \text{ cm}$$
$$A = \pi \cdot r^2 = \pi(4.2 \text{ cm})^2 \approx (3.14)(17.64 \text{ cm}^2) = 55.3896 \text{ cm}^2 \approx 55.4 \text{ cm}^2$$

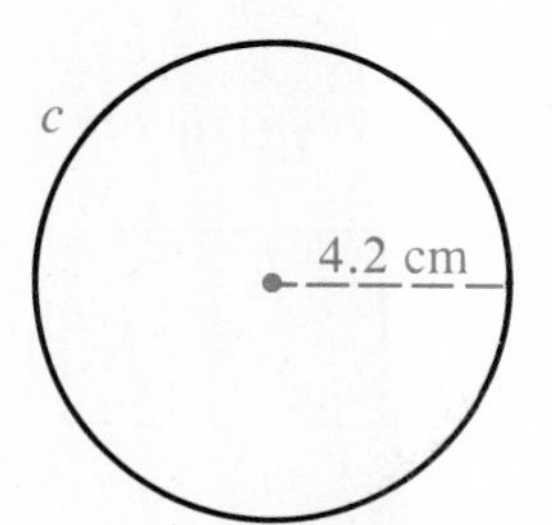


Figure 1.12

Rectangular Parallelepiped

$a, b, c$ = edges
Surface area = $S = 2a \cdot b + 2a \cdot c + 2b \cdot c$
Volume = $V = a \cdot b \cdot c$

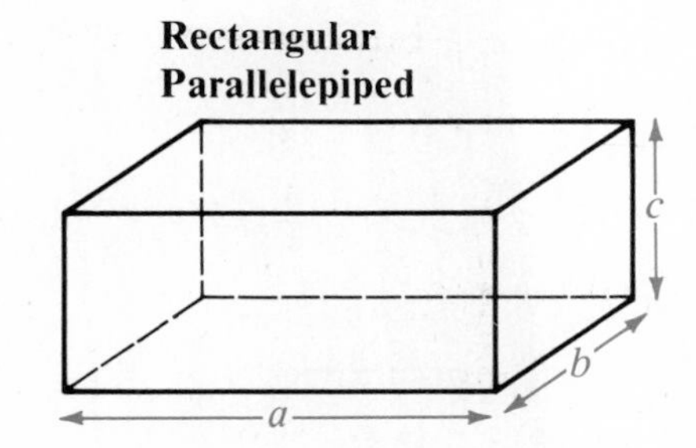


Figure 1.13

Sphere

$r$ = radius
$d$ = diameter
Surface Area = $S = 4\pi \cdot r^2$
Volume = $V = \frac{4}{3}\pi \cdot r^3$

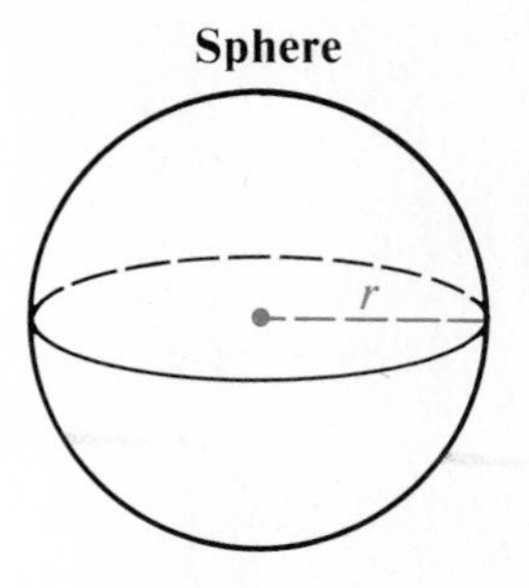


Figure 1.14

Volumes are given in cubic units such as cubic feet, denoted by $\text{ft}^3$ or cu ft, cubic inches, denoted by $\text{in}^3$ or cu in, and cubic meters, denoted by $\text{m}^3$ or cu m.

**EXAMPLE 5** Find the volume and surface area of the sphere with diameter 10 m in Figure 1.15.

Since diameter = $d = 2r = 10$ m, the radius of the sphere is 5 m. Thus,

$$V = \frac{4}{3}\pi \cdot (5 \text{ m})^3 = \frac{4}{3}\pi \cdot 125 \text{ m}^3 = \frac{500\pi}{3} \text{ m}^3 \approx \frac{(500)(3.14)}{3} \text{ m}^3 \approx 523 \text{ m}^3$$

$$S = 4\pi r^2 = 4\pi(5 \text{ m})^2 = 4\pi \cdot 25 \text{ m}^2 = 100\pi \text{ m}^2 \approx 100(3.14) \text{ m}^2 = 314 \text{ m}^2$$

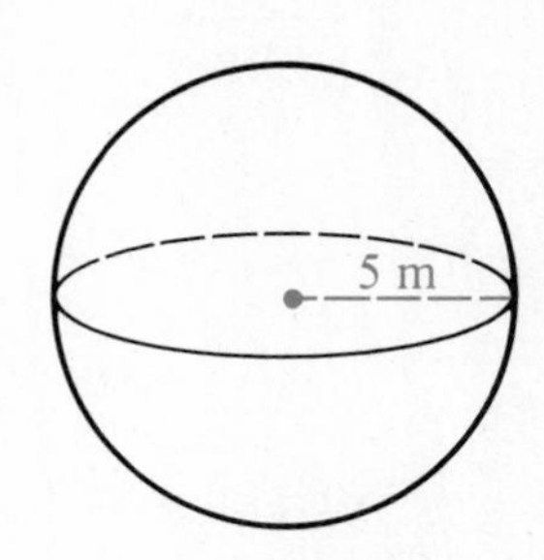


Figure 1.15

## EXERCISES 1.8

*Calculate the requested measures.*

**1.**

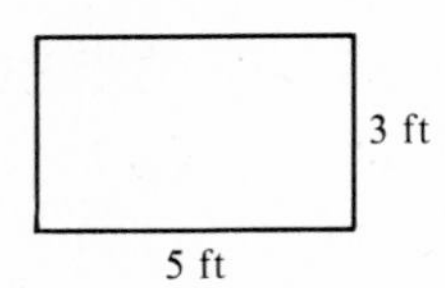


$A =$

$P =$

**2.**

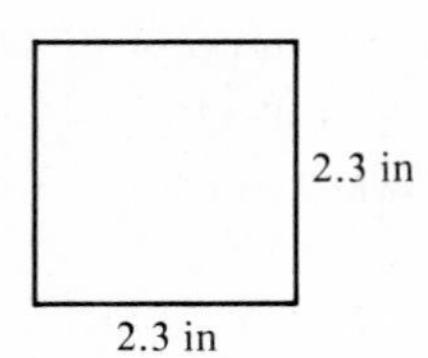


$A =$

$P =$

**3.**

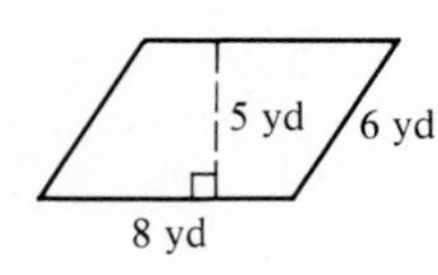


$h =$

$A =$

$P =$

**4.**

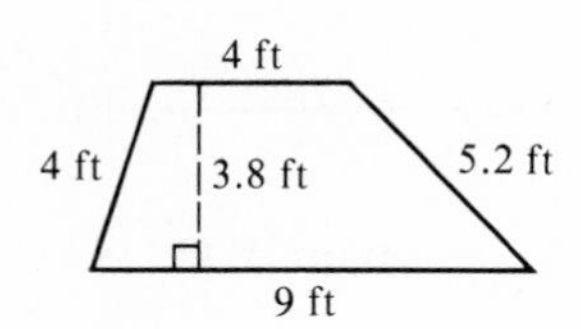


$h =$

$A =$

$P =$

**5.** Use 3.14 for $\pi$.

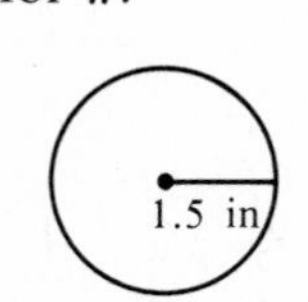


$d =$

$A =$

$c =$

**6.**

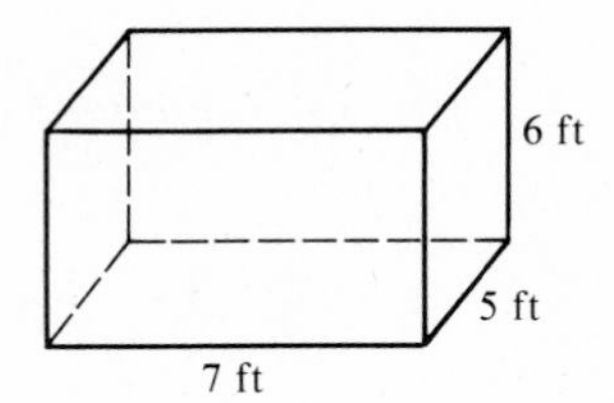


$V =$

$S =$

**7.**

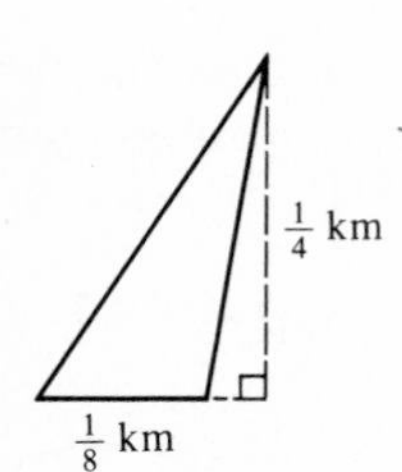


$b =$

$h =$

$A =$

**8.** Use 3.14 for $\pi$.

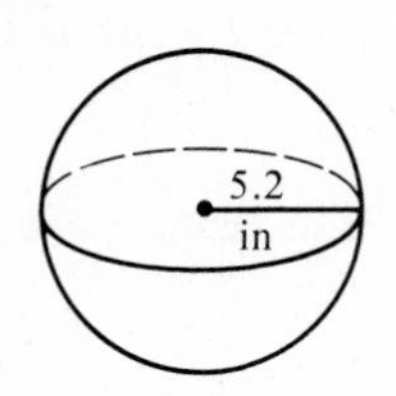


$d =$

$V =$

$S =$

**9.**

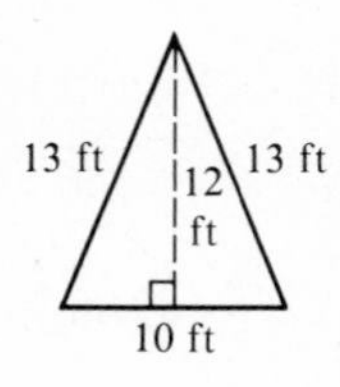


$b =$

$h =$

$A =$

$P =$

**10.** Find the surface area of a sphere of radius 15 cm. (Use $\pi \approx 3.14$)

**11.** Find the volume of a parallelepiped with sides of 6.2 ft, 8.4 ft, and 10.1 ft.

**12.** Find the perimeter of a rectangle of length 4.8 m and width 2.3 m.

**13.** Find the area of a trapezoid with bases 4 yd and 10 yd and altitude 7.2 yd.

**14.** The area of a rectangle is given by the formula ____________________.

**15.** The perimeter of a rectangle is given by the formula ____________________.

**16.** If a rectangle has all sides of the same length, it is called a ____________________.

**17.** The area of a parallelogram is given by the formula ____________________.

**18.** The area of a triangle is given by the formula ____________________.

**19.** The sum of the measures of the angles of a triangle is ____________________.

**20.** If one angle of a triangle measures more than 90°, the triangle is called a(n) ____________________ triangle.

**21.** If one angle of a triangle measures exactly 90°, the triangle is called a(n) ____________________ triangle.

**22.** If all angles of a triangle measure less than 90°, the triangle is called a(n) ____________________ triangle.

**23.** If two sides of a triangle are equal, the triangle is called a(n) ____________________ triangle.

**24.** If all three sides of a triangle are equal, the triangle is called a(n) ____________________ triangle.

**25.** The area of a trapezoid is given by the formula ____________________.

**26.** The area of a circle is given by the formula ____________________.

**27.** The ratio of the circumference of any circle to its diameter is equal to __________.

**28.** The volume of a rectangular parallelepiped is given by the formula __________.

**29.** The surface area of a sphere is given by the formula ____________________.

**30.** The volume of a sphere is given by the formula ____________________.

**31.** Convert $\frac{4}{9}$ to a decimal.

**32.** Convert $\frac{4}{5}$ to a decimal.

**33.** Convert .095 to percent notation.

**34.** Convert 3.22 to percent notation.

**35.** Convert 6% to decimal notation.

**36.** Convert .07% to decimal notation.

*Evaluate when $x = -1$ and $y = -3$.*

**37.** $3x^2$

**38.** $(3x)^2$

**39.** $-3x^2$

**40.** $(-3x)^2$

**41.** $(x + y)^2$

**42.** $x^2 + y^2$

**43.** $y^{-2}$

**44.** $(x^{-3}y^{-1})^{-2}$

**45.** $(x^0 + y^0)^0$

---

ANSWERS: **1.** $A = 15$ ft², $P = 16$ ft **2.** $A \approx 5.3$ in², $P = 9.2$ in **3.** $h = 5$ yd, $A = 40$ yd², $P = 28$ yd **4.** $h = 3.8$ ft, $A \approx 25$ ft², $P \approx 22$ ft **5.** $d = 3$ in, $A \approx 7.1$ in², $c \approx 9.4$ in **6.** $V = 210$ ft³, $S = 214$ ft² **7.** $b = \frac{1}{8}$ km, $h = \frac{1}{4}$ km, $A = \frac{1}{64}$ km² **8.** $d = 10.4$ in, $V \approx 588.7$ in³, $S \approx 339.6$ in² **9.** $b = 10$ ft, $h = 12$ ft, $A = 60$ ft², $P = 36$ ft **10.** 2826 cm² **11.** 526.0 ft³ **12.** 14.2 m **13.** 50 yd² **14.** $A = l \cdot w$ **15.** $P = 2 \cdot l + 2 \cdot w$ **16.** square **17.** $A = b \cdot h$ **18.** $A = \frac{1}{2}b \cdot h$ **19.** 180° **20.** obtuse **21.** right **22.** acute **23.** isosceles **24.** equilateral **25.** $A = \frac{1}{2}(b_1 + b_2) \cdot h$ **26.** $A = \pi r^2$ **27.** $\pi$ **28.** $V = a \cdot b \cdot c$ **29.** $S = 4\pi r^2$ **30.** $V = \frac{4}{3}\pi r^3$ **31.** $.\bar{4}$ **32.** .8 **33.** 9.5% **34.** 322% **35.** .06 **36.** .0007 **37.** 3 **38.** 9 **39.** −3 **40.** 9 **41.** 16 **42.** 10 **43.** $\frac{1}{9}$ **44.** 9 **45.** 1

## 1.9 Conversion from One Unit to Another

The following conversion identities should be memorized.

| *Identities* | *To convert* | *Multiply by* | *To convert* | *Multiply by* |
|---|---|---|---|---|
| 1 ft = 12 in | ft to in | 12 | in to ft | 1/12 |
| 1 yd = 3 ft | yd to ft | 3 | ft to yd | 1/3 |
| 1 mi = 5280 ft | mi to ft | 5280 | ft to mi | 1/5280 |
| 1 ft$^2$ = 144 in$^2$ | sq ft to sq in | 144 | sq in to sq ft | 1/144 |
| 1 yd$^2$ = 9 ft$^2$ | sq yd to sq ft | 9 | sq ft to sq yd | 1/9 |
| 1 ton = 2000 lb | tons to lb | 2000 | lb to tons | 1/2000 |
| 1 lb = 16 oz | lb to oz | 16 | oz to lb | 1/16 |
| 1 hr = 60 min | hr to min | 60 | min to hr | 1/60 |
| 1 min = 60 sec | min to sec | 60 | sec to min | 1/60 |
| 1 yd$^3$ = 27 ft$^3$ | cu yd to cu ft | 27 | cu ft to cu yd | 1/27 |
| 1 ft$^3$ = 1728 in$^3$ | cu ft to cu in | 1728 | cu in to cu ft | 1/1728 |

EXAMPLE 1 **(a)** Convert 8 yd to ft.

$$8 \text{ yd} = 8 \cdot 3 \text{ ft} = 24 \text{ ft}$$

**(b)** Convert 3 mi to in.

$$\begin{aligned} 3 \text{ mi} &= 3 \cdot 5280 \text{ ft} = 15{,}840 \text{ ft} \\ &= 15{,}840 \cdot 12 \text{ in} = 190{,}080 \text{ in} \end{aligned}$$

**(c)** Convert 5 ft$^2$ to in$^2$.

$$5 \text{ ft}^2 = 5 \cdot 144 \text{ in}^2 = 720 \text{ in}^2$$

**(d)** Convert 7000 lb to ton.

$$7000 \text{ lb} = 7000 \cdot \frac{1}{2000} \text{ ton} = \frac{7}{2} \text{ ton} = 3.5 \text{ tons}$$

**(e)** Convert 135 ft$^3$ to yd$^3$.

$$135 \text{ ft}^3 = 135 \cdot \frac{1}{27} \text{ yd}^3 = 5 \text{ yd}^3$$

**(f)** Convert $\frac{5}{2}$ hr to sec.

$$\frac{5}{2} \text{ hr} = \frac{5}{2} \cdot 60 \text{ min} = 150 \text{ min} = 150 \cdot 60 \text{ sec} = 9000 \text{ sec}$$

When converting from one metric unit to another, the conversion factors are always powers of ten or one tenth. A meter (approximately 39 inches) is the standard unit of length, a gram (approximately the weight of a paper clip) is the standard unit of weight, and a liter (slightly larger than a quart) is the standard unit of capacity. The following prefixes may be attached to all three of these units.

| | | | |
|---|---|---|---|
| kilo- = | 1000 | deci- = $\frac{1}{10}$ | = .1 |
| hecto- = | 100 | centi- = $\frac{1}{100}$ | = .01 |
| deka- = | 10 | milli- = $\frac{1}{1000}$ | = .001 |

Thus we have the following metric units.

| | | | | | |
|---|---|---|---|---|---|
| 1 kilometer | (km) = | 1000 meters | 1 kilogram | (kg) = | 1000 grams |
| 1 hectometer | (hm) = | 100 meters | 1 hectogram | (hg) = | 100 grams |
| 1 dekameter | (dam) = | 10 meters | 1 dekagram | (dag) = | 10 grams |
| 1 meter | (m) = | 1 meter | 1 gram | (g) = | 1 gram |
| 1 decimeter | (dm) = | .1 meter | 1 decigram | (dg) = | .1 gram |
| 1 centimeter | (cm) = | .01 meter | 1 centigram | (cg) = | .01 gram |
| 1 millimeter | (mm) = | .001 meter | 1 milligram | (mg) = | .001 gram |

| | | |
|---|---|---|
| 1 kiloliter | (kl) = | 1000 liters |
| 1 hectoliter | (hl) = | 100 liters |
| 1 dekaliter | (dal) = | 10 liters |
| 1 liter | (L) = | 1 liter |
| 1 deciliter | (dl) = | .1 liter |
| 1 centiliter | (cl) = | .01 liter |
| 1 milliliter | (ml) = | .001 liter |

Since one unit is converted to another by multiplying by a power of ten, conversions are made by simply moving the decimal point left or right. The following table is useful.

| 1000 | 100 | 10 | 1 | $\frac{1}{10}$ | $\frac{1}{100}$ | $\frac{1}{1000}$ |
|---|---|---|---|---|---|---|
| kilo-unit | hecto-unit | deka-unit | unit | deci-unit | centi-unit | milli-unit |

To convert hecto-units to deci-units we would multiply by 1000 (moving the decimal point three units to the right). In our table deci-unit is three places to the right of hecto-unit. Similarly, since deka-unit is four positions to the left of milli-unit in the table, to convert 13.3 ml to dal, we move the decimal point four places to the left (multiply by .0001) obtaining .00133 dal.

**EXAMPLE 2** Make the following conversions.

**(a)** 17.2 hm to dm

Move the decimal point three places to the right (multiply by 1000) obtaining 17,200 dm.

**(b)** .03 dg to hg

Move the decimal point three places to the left (multiply by .001) obtaining .00003 hg.

**(c)** 325 cl to kl

Move the decimal point five places to the left (multiply by .00001) obtaining .00325 kl.

**(d)** .045 kg to dag

Move the decimal point two places to the right (multiply by 100) obtaining 4.5 dag.

When working distance-related applied problems, we must often consider various rates of change such as miles per hour (mph or mi/hr), feet per second (ft/sec), feet per minute (ft/min), kilometers per hour (km/hr), etc. Changing from one rate to another involves converting both units of length and units of time. The following examples make the procedure clear.

**EXAMPLE 3** Convert 60 mi/hr to ft/sec.

We first convert 60 mi/hr to ft/hr by multiplying by 5280 (1 mi = 5280 ft).

$$60 \text{ mi/hr} = 60 \cdot 5280 \text{ ft/hr} = 316{,}800 \text{ ft/hr}$$

Next we convert 316,800 ft/hr to ft/min by multiplying by $\frac{1}{60}$.

$$316{,}800 \text{ ft/hr} = 316{,}800 \cdot \frac{1}{60} \text{ ft/min} = 5280 \text{ ft/min}$$

Then we convert 5280 ft/min to ft/sec by multiplying by $\frac{1}{60}$.

$$5280 \text{ ft/min} = 5280 \cdot \frac{1}{60} \text{ ft/sec} = 88 \text{ ft/sec}$$

It is wise to indicate all products and do some canceling in the process. Thus, we would probably write

$$60 \text{ mi/hr} = 60 \cdot 5280 \text{ ft/hr} = \frac{60 \cdot 5280}{60} \text{ ft/min}$$

$$= \frac{\cancel{60} \cdot 5280}{\cancel{60} \cdot 60} \text{ ft/sec} = 88 \text{ ft/sec.}$$

**EXAMPLE 4** Convert 72 km/hr to m/min.

$$72 \text{ km/hr} = 72 \cdot (1000) \text{ m/hr} \qquad \text{Convert km to m}$$

$$= \frac{72 \cdot 1000}{60} \text{ m/min} \qquad \text{Convert hr to min}$$

$$= 1200 \text{ m/min}$$

**EXAMPLE 5** Convert 31,680 ft/min to mi/hr.

$$31{,}680 \text{ ft/min} = 31{,}680 \cdot \frac{1}{5280} \text{ mi/min} \qquad \text{Convert ft to mi}$$

$$= \frac{31{,}680}{5280} \cdot 60 \text{ mi/hr} \qquad \text{Convert min to hr}$$

$$= 360 \text{ mi/hr}$$

## EXERCISES 1.9

*Complete by filling in the blanks.*

**1.** 1 ft = ________ in

**2.** 1 $\text{ft}^2$ = ________ $\text{yd}^2$

**3.** 1 $\text{yd}^3$ = ________ $\text{ft}^3$

**4.** 1 ft = ________ mi

**5.** 1 hr = ________ min

**6.** 1 in = ________ ft

**7.** 1 ton = ________ lb

**8.** 1 $\text{ft}^3$ = ________ $\text{yd}^3$

**9.** 1 ft = ________ yd

**10.** 1 lb = ________ oz

**11.** 1 $\text{in}^3$ = ________ $\text{ft}^3$

**12.** 1 $\text{yd}^2$ = ________ $\text{ft}^2$

**13.** 1 sec = ________ min

**14.** 1 mi = ________ ft

**15.** 1 oz = ________ lb

**16.** 1 $ft^2$ = ________ $in^2$

**17.** 1 min = ________ hr

**18.** 1 yd = ________ ft

**19.** 1 lb = ________ ton

**20.** 1 $ft^3$ = ________ $in^3$

**21.** 1 $in^2$ = ________ $ft^2$

**22.** 1 min = ________ sec

**23.** 8 yd = ________ ft

**24.** 12 yd = ________ in

**25.** 36 in = ________ ft

**26.** 18 ft = ________ yd

**27.** 2 mi = ________ ft

**28.** 7 mi = ________ yd

**29.** 44 $yd^2$ = ________ $ft^2$

**30.** 40 $ft^2$ = ________ $in^2$

**31.** 1440 $in^2$ = ________ $ft^2$

**32.** 36 $ft^2$ = ________ $yd^2$

**33.** 288 $in^2$ = ________ $yd^2$

**34.** 5 hr = ________ min

**35.** 70 min = ________ sec

**36.** 5 sec = ________ min

**37.** 6 hr = ________ sec

**38.** 81 $ft^3$ = ________ $yd^3$

**39.** 6912 $in^3$ = ________ $ft^3$

**40.** 1 km = ________ m

**41.** 1 ml = ________ L

**42.** 1 g = ________ kg

**43.** 1 cm = ________ m

**44.** 1 dag = ________ g

**45.** 1 L = ________ cl

**46.** 1 hl = ________ L

**47.** 1 dg = ________ g

**48.** 1 m = ________ mm

**49.** 1 hm = ________ m

**50.** 1 dal = ________ kl

**51.** 1 cg = ________ g

**52.** 10.5 hm = ________ cm

**53.** .43 cm = ________ km

**54.** .024 hg = ________ cg

**55.** 13.35 mg = ________ dag

**56.** 3.42 cl = ________ hl

**57.** 1.03 kl = ________ cl

**58.** 35 kg = ________ dg

**59.** 41.15 mm = ________ dam

**60.** 425 ml = ________ dl

**61.** 30 mi/hr = **(a)** ________ ft/hr = **(b)** ________ ft/min

**62.** $\frac{1}{6}$ ft/sec = **(a)** ________ ft/min = **(b)** ________ yd/min

**63.** 30 km/hr = **(a)** ________ m/hr = **(b)** ________ m/min

**64.** 6 m/sec = **(a)** ________ m/min = **(b)** ________ cm/min

**65.** Given that 1 km is approximately $\frac{5}{8}$ mi, convert 100 km/hr to mi/hr.

**66.** If the length of a rectangle is 24 ft and the width is 9 inches, what is the area? [*Hint:* In order to calculate the area, the same units of measure for the length and width must be used.]

**67.** What is the radius in cm of a circle with diameter 46 m?

**68.** If a box has sides of length 3 yd, 4 ft, and 6 in, what is the volume in $yd^3$? In $ft^3$? In $in^3$?

**69.** A sphere with radius 9 in has what volume? [Use $\pi \approx 3.14$.]

**70.** Find the perimeter and area of a right triangle with **hypotenuse** (the side opposite the right angle) 13 in and **legs** (the remaining two sides) 5 in and 12 in.

---

ANSWERS: **1.** 12 **2.** $\frac{1}{9}$ **3.** 27 **4.** $\frac{1}{5280}$ **5.** 60 **6.** $\frac{1}{12}$ **7.** 2000 **8.** $\frac{1}{27}$ **9.** $\frac{1}{3}$ **10.** 16 **11.** $\frac{1}{1728}$ **12.** 9 **13.** $\frac{1}{60}$ **14.** 5280 **15.** $\frac{1}{16}$ **16.** 144 **17.** $\frac{1}{60}$ **18.** 3 **19.** $\frac{1}{2000}$ **20.** 1728 **21.** $\frac{1}{144}$ **22.** 60 **23.** 24 **24.** 432 **25.** 3 **26.** 6 **27.** 10,560 **28.** 12,320 **29.** 396 **30.** 5760 **31.** 10 **32.** 4 **33.** $\frac{2}{9}$ **34.** 300 **35.** 4200 **36.** $\frac{1}{12}$ **37.** 21,600 **38.** 3 **39.** 4 **40.** 1000 **41.** .001 **42.** .001 **43.** .01 **44.** 10 **45.** 100 **46.** 100 **47.** .1 **48.** 1000 **49.** 100 **50.** .01 **51.** .01 **52.** 105,000 **53.** .0000043 **54.** 240 **55.** .001335 **56.** .000342 **57.** 103,000 **58.** 350,000 **59.** .004115 **60.** 4.25 **61.** (a) $30 \cdot 5280 = 158{,}400$ (b) $\frac{30 \cdot 5280}{60} = 2640$ **62.** (a) 10 (b) $\frac{10}{3}$ **63.** (a) 30,000 (b) 500 **64.** (a) 360 (b) 36,000 **65.** 62.5 mi/hr **66.** 18 $ft^2$ or 2592 $in^2$ **67.** 2300 cm **68.** $\frac{2}{3}$ $yd^3$ or 18 $ft^3$ or 31,104 $in^3$ **69.** 3052 $in^3$ **70.** $P = 30$ in, $A = 30$ $in^2$

## Chapter 1 Summary

### Words and Phrases for Review

[1.1] natural (counting) number
whole number
integer
number line
equal (=)
unequal ($\neq$)
less than (<)
less than or equal ($\leq$)
greater than (>)
greater than or equal ($\geq$)
rational number
equivalent fractions
prime number
terminating decimal
repeating decimal
cross product
perfect square
square root
radical
irrational number
real number

[1.2] reciprocal
dividend
divisor
quotient
least common denominator (LCD)
mixed number
improper fraction

[1.3] signed number
absolute value
numbers of arithmetic

[1.4] set
element
variable

[1.5] term
factor
coefficient

[1.6] exponential notation
base
exponent
power (first, square, cube, etc.)

[1.7] percent (%)
decimal

[1.8] approximately equal ($\approx$)
acute triangle
obtuse triangle
right triangle
isosceles triangle
equilateral triangle

### Brief Reminders

[1.1] 1. Factors common to the numerator and denominator of a fraction may be divided out or canceled. However, *do not* cancel terms. For example,

$$\frac{\cancel{3} \cdot x}{\cancel{3}} = x \quad \text{but} \quad \frac{3 + x}{3} \text{ is } \textit{not} \text{ equal to } \frac{\cancel{3} + x}{\cancel{3}}.$$

2. The radical symbol by itself designates only the nonnegative square root.

3. When reducing fractions, do not make the numerator 0 when it is actually 1. For example,

$$\frac{7}{7 \cdot x} = \frac{\cancel{7}}{\cancel{7} \cdot x} = \frac{1}{x} \quad \text{and } \textit{not} \quad \frac{0}{x}.$$

[1.2] 1. If $a$ is any number except 0, $\frac{0}{a} = 0$ but $\frac{a}{0}$ is not defined.

2. Fractions to be added or subtracted must first be written with the same LCD.

3. Do *not* find the LCD when multiplying or dividing fractions. However, *do* factor numerators and denominators and cancel common factors.

[1.3] 1. The absolute value of a number is always positive or zero and never negative.

2. If $a$ and $b$ are any numbers, then $|a \cdot b| = |a| \cdot |b|$.

3. If $a$ and $b$ are any numbers and $b \neq 0$, then $\left|\frac{a}{b}\right| = \frac{|a|}{|b|}$.

4. In general, $|a + b|$ is *not* $|a| + |b|$ and $|a - b|$ is *not* $|a| - |b|$.

[1.4] 1. Commutative Laws: $a + b = b + a$ and $a \cdot b = b \cdot a$.

2. Associative Laws: $(a + b) + c = a + (b + c)$ and $(a \cdot b) \cdot c = a \cdot (b \cdot c)$.

3. When simplifying expressions involving grouping symbols, evaluate within the innermost symbols first.

[1.5] 1. Distributive Laws: $a(b + c) = a \cdot b + a \cdot c$ and $a(b - c) = a \cdot b - a \cdot c$.

2. Only like terms can be combined. For example, $5y + 2$ is *not* $7y$ since $5y$ and 2 are not like terms.

3. A minus sign before a set of parentheses changes the sign of *every* term inside when the parentheses are removed.

[1.6] If $a$ and $b$ are any numbers, and $m$ and $n$ are integers,

$$a^m \cdot a^n = a^{m+n} \qquad \left(\frac{a}{b}\right)^n = \frac{a^n}{b^n} \text{ if } b \neq 0$$

$$\frac{a^m}{a^n} = a^{m-n} \text{ if } a \neq 0 \qquad a^0 = 1 \text{ if } a \neq 0$$

$$(a^m)^n = a^{m \cdot n} \qquad a^{-n} = \frac{1}{a^n} \text{ if } a \neq 0$$

$$(a \cdot b)^n = a^n \cdot b^n$$

[1.7] 1. To convert a percent to a decimal, remove the percent symbol and divide by 100. For example,

$$37\% = 37\left(\frac{1}{100}\right) = .37.$$

2. To convert a decimal to a percent, attach the percent symbol and multiply by 100. For example,

$$.045 = .045(100)\% = 4.5\%$$

[1.8] 1. Rectangle: $A = l \cdot w, \quad P = 2 \cdot l + 2 \cdot w$

2. Parallelogram: $A = b \cdot h, \quad P = 2 \cdot a + 2 \cdot b$

3. Triangle: $A = \frac{1}{2} b \cdot h, \quad P = a + b + c$

4. Trapezoid: $A = \frac{1}{2}(b_1 + b_2) \cdot h, \quad P = a + b_1 + b_2 + c$

5. Circle: $d = 2r, \quad c = 2\pi r = \pi d, \quad A = \pi r^2$

6. Rectangular Parallelepiped: $S = 2ab + 2ac + 2bc, \quad V = abc$

7. Sphere: $S = 4\pi r^2, \quad V = \frac{4}{3}\pi r^3$

## CHAPTER 1 REVIEW EXERCISES

[1.1]

1. An invented idea which is used to answer questions such as "how many" and "in what order" is a(n) ________________.
2. A(n) ________________ is the symbol for a number.
3. $\{1, 2, 3, \cdots\}$ is called the set of ________________ numbers.
4. $\{0, 1, 2, 3, \cdots\}$ is called the set of ________________ numbers.
5. $\{\cdots, -2, -1, 0, 1, 2, \cdots\}$ is called the set of ________________.
6. A counting number greater than 1 which is divisible only by itself and 1 is called a(n) ________________ number.
7. A line used to display numbers is called a(n) ________________.
8. The point on a number line corresponding to zero is called the ________.
9. If a number appears to the left of a second number on a number line it is said to be ________________ the second.
10. We build up a fraction when we ________________ both the numerator and denominator by the same counting number.
11. We reduce a fraction when we ________________ both the numerator and denominator by the same counting number.
12. The process of canceling factors from a numerator and denominator is really the process of ________________ out common factors.
13. Every rational number can be written as a decimal numeral that either **(a)** ________________ or has a(n) **(b)** ________________ block of digits.
14. When we identify the point on a number line associated with a given number, we say we ________________ the number.
15. Two positive fractions are equal if the cross products are ________.
16. Any whole number expressible as the product of two identical factors is called a(n) **(a)** ________________ and one of the identical factors is called a(n) **(b)** ________________ of the number.
17. The radical symbol by itself is *only* used to denote the ________ square root of a number.
18. Any number that is not a rational number is called a(n) ________ number.
19. Together the rational and irrational numbers form the set of ________ numbers.

[1.3]

20. The absolute value of a number is always greater than or equal to ________________.

**21.** The sum of two negative numbers is always a ______ number.

**22.** The product of two negative numbers is always a ______ number.

**23.** The quotient of two negative numbers is always a ______ number.

**24.** The product of two numbers with opposite signs is always a ______ number.

**25.** The quotient of two numbers with opposite signs is always a ______ number.

[1.4] **26.** A letter that is used to represent a number is a(n) ______.

**27.** A collection of objects is called a(n) **(a)** ______ of objects, and each object is called a(n) **(b)** ______ of the **(a)** ______.

**28.** The fact that $4 + 5 = 5 + 4$ illustrates the ______ law of addition.

**29.** The fact that $2 + (3 + 5) = (2 + 3) + 5$ illustrates the ______ law of addition.

**30.** When grouping symbols are not used, we always multiply and **(a)** ______ before we add and **(b)** ______.

**31.** When evaluating an expression involving two sets of grouping symbols, always evaluate within the ______ set of grouping symbols first.

[1.5] **32.** The fact that $2(3 + 5) = 2 \cdot 3 + 2 \cdot 5$ illustrates the ______ law.

**33.** The two laws $a(b + c) = ab + ac$ and $a(b - c) = ab - ac$ are called the ______ laws.

**34.** In the expression $x - 3y^2$, the coefficient of $x$ is **(a)** ______, the coefficient of $y^2$ is **(b)** ______, the exponent on $x$ is **(c)** ______, the exponent on $y$ is **(d)** ______, and $y$ is called the **(e)** ______ of the exponential expression $-3y^2$.

[1.6] **35.** When multiplying two powers with the same base, we ______ the exponents.

**36.** When dividing two powers with the same base, we ______ the exponents.

**37.** When raising a power to a power, we ______ the exponents.

**38.** When any number except zero is raised to the zero power, the result is ________.

[1.7] **39.** To convert a decimal numeral to percent notation, multiply by **(a)** ________ (move the decimal point **(b)** ________ places to the **(c)** ________) and attach the percent symbol %.

**40.** To convert a percent numeral to a decimal, multiply by **(a)** ________ (move the decimal point **(b)** ________ places to the **(c)** ________) and remove the percent symbol %.

[1.8] **41.** The sum of the measures of the angles of a triangle is ________.

**42.** If one angle of a triangle measures exactly 90°, the triangle is called a(n) ________ triangle.

**43.** If two sides of a triangle are equal, the triangle is called a(n) ________ triangle.

**44.** If all three sides of a triangle are equal, the triangle is called a(n) ________ triangle.

**45.** The ratio of the circumference of a circle to its diameter is equal to ________.

[1.1] **46.** Reduce $\frac{49}{84}$ to lowest terms.

**47.** Express 108 as a product of primes.

*Insert the proper symbol (=, >, or <) between the two positive fractions.*

**48.** $\frac{5}{4} \quad \frac{7}{6}$

**49.** $\frac{9}{17} \quad \frac{36}{68}$

**50.** $\frac{5}{12} \quad \frac{3}{7}$

*Evaluate.*

**51.** $\sqrt{121}$

**52.** $\sqrt{169}$

**53.** $\sqrt{\frac{196}{49}}$

**54.** $\sqrt{\frac{1}{225}}$

**55.** Express $\frac{3}{8}$ as a decimal.

**56.** Express $\frac{7}{11}$ as a decimal.

[1.2] *Simplify.*

**57.** $\frac{0}{2} =$

**58.** $\frac{2}{0} =$

**59.** $\frac{0}{0} =$

**60.** $\frac{2}{2} =$

[1.3] *Perform the indicated operations.*

**61.** $\frac{4}{9} + \frac{11}{12} =$

**62.** $\frac{2}{3} - \frac{17}{18} =$

**63.** $\left(\frac{3}{4}\right) \cdot \left(-\frac{16}{9}\right) =$

**64.** $\left(-\frac{5}{13}\right) \div \left(-\frac{10}{169}\right) =$

**65.** $(-1)(-2)(-1)(2)(-3) =$

**66.** $|(-2)(-5) + (-3)(3)| =$

**67.** $0 \cdot 7 =$ ________

**68.** $0 - 7 =$ ________

[1.4] *Evaluate when $x = -2$, $y = 3$, and $z = -1$.*

**69.** $3x^2$

**70.** $-3x^2$

**71.** $(3x)^2$

**72.** $(-3x)^2$

**73.** $3x^{-2}$

**74.** $(3x)^{-2}$

**75.** $|3x + 2y|$

**76.** $x - (-z)$

**77.** $(x + y)^2$

**78.** $x^2 + y^2$

**79.** $|x|$

**80.** $|y|$

**81.** $-(-x)$

**82.** $-(+x)$

**83.** $xyz - zyx$

[1.5] *Use the distributive law to factor the following:*

**84.** $3a + 9b$

**85.** $20x - 10y + 50$

**86.** $ax + ay - az$

*Multiply.*

**87.** $4(x - 2y)$

**88.** $-2(4 - 3a)$

**89.** $a(b - 3c)$

*Use the distributive law to collect like terms.*

**90.** $3z - z + 5z$

**91.** $\frac{1}{2}a - \frac{1}{4}a + \frac{3}{4}$

**92.** $3a - x + 2a + 4x$

*Remove parentheses and simplify.*

**93.** $-(a - 3)$

**94.** $-(-x + 2)$

**95.** $-(-y - 5)$

**96.** $-2[a - 3(a + 1)]$

**97.** $2x - [x - 2(x - 2)]$

**98.** $-(y + 2) - (-y - 2)$

[1.6] *Simplify and write without negative exponents.*

**99.** $(2x^2y^{-3})^{-2}$

**100.** $\dfrac{a^2b^{-3}}{a^{-3}b}$

**101.** $\dfrac{1}{x^{-3}}$

**102.** $\left(\dfrac{2x^2}{y^{-1}}\right)^{-3}$

**103.** $2x^2x^5x^{-1}$

**104.** $\dfrac{x^{-7}}{y^{-3}}$

[1.7] **105.** Convert 3% to a decimal.

**106.** Convert .5% to a decimal.

**107.** Convert 100% to a decimal.

**108.** Convert $\frac{7}{4}$ to percent notation.

[1.8] **109.** Find the perimeter of a rectangle of length 4.1 cm and width 2.3 cm.

**110.** Find the area of a trapezoid with bases 3.2 in and 5.8 in and altitude 4.1 in.

**111.** Find the area of a triangle with base 15 ft and altitude 6 ft.

**112.** Find the area of a circle whose diameter is 22 in. (Use $\pi = 3.14$)

*Complete by filling in the blanks.*

[1.9]

**113.** 1 ft = ________ in

**114.** 1 yd = ________ ft

**115.** 1 mi = ________ ft

**116.** 1 $ft^2$ = ________ $in^2$

**117.** 1 ton = ________ lb

**118.** 1 lb = ________ oz

**119.** 1 hr = ________ min

**120.** 1 $yd^3$ = ________ $ft^3$

**121.** 1 $in^2$ = ________ $ft^2$

**122.** 1 ft = ________ yd

**123.** 12 hr = ________ min

**124.** 3 $yd^3$ = ________ $ft^3$

**125.** 33 ft = ________ yd

**126.** 15,840 ft = ________ mi

**127.** 576 $in^2$ = ________ $ft^2$

**128.** 8 ft = ________ in

**129.** 156 in = ________ ft

**130.** 81 $ft^2$ = ________ $yd^2$

**131.** 2 $ft^3$ = ________ $in^3$

**132.** 15 yd = ________ ft

**133.** .23 hl = ________ ml

**134.** 5.13 cl = ________ dal

**135.** 48 dl = ________ kl

**136.** .98 hm = ________ dam

**137.** 10.5 dm = ________ mm

**138.** 18.4 mm = ________ m

**139.** 51 cg = ________ dag

**140.** .06 hg = ________ g

**141.** 4.2 dag = ________ kg

**142.** 32 kg = ________ hg

**143.** 45 mi/hr = **(a)** ________ ft/hr = **(b)** ________ ft/min = **(c)** ________ ft/sec

**144.** 15 m/sec = **(a)** ________ dm/sec = **(b)** ________ dm/min = **(c)** ________ dm/hr

**145.** 80 km/hr = ________ mi/hr

[*Hint:* 1 km $\approx \frac{5}{8}$ mi.]

---

ANSWERS: **1.** number **2.** numeral **3.** natural or counting **4.** whole **5.** integers **6.** prime **7.** number line **8.** origin **9.** less than **10.** multiply **11.** divide **12.** dividing **13.** (a) terminates (b) repeating **14.** plot **15.** equal **16.** (a) perfect square (b) square root **17.** positive **18.** irrational **19.** real **20.** zero **21.** negative **22.** positive **23.** positive **24.** negative **25.** negative **26.** variable **27.** (a) set (b) element **28.** commutative **29.** associative **30.** (a) divide (b) subtract **31.** innermost **32.** distributive **33.** distributive **34.** (a) 1 (b) −3 (c) 1 (d) 2 (e) base **35.** add **36.** subtract **37.** multiply **38.** 1 **39.** (a) 100 (b) 2 (c) right **40.** (a) $\frac{1}{100}$ (b) 2 (c) left **41.** 180° **42.** right **43.** isosceles **44.** equilateral **45.** $\pi$ **46.** $\frac{7}{12}$ **47.** $2 \cdot 2 \cdot 3 \cdot 3 \cdot 3$ **48.** $>$ **49.** $=$ **50.** $<$ **51.** 11 **52.** 13 **53.** 2 **54.** $\frac{1}{15}$ **55.** .375 **56.** $.\overline{63}$ **57.** 0 **58.** undefined **59.** undefined **60.** 1 **61.** $\frac{49}{36}$ **62.** $-\frac{5}{18}$ **63.** $-\frac{4}{3}$ **64.** $\frac{13}{2}$ **65.** 12 **66.** 1 **67.** 0 **68.** −7 **69.** 12 **70.** −12 **71.** 36 **72.** 36 **73.** $\frac{3}{4}$ **74.** $\frac{1}{36}$ **75.** 0 **76.** −3 **77.** 1 **78.** 13 **79.** 2 **80.** 3 **81.** −2 **82.** 2 **83.** 0 **84.** $3(a+3b)$ **85.** $10(2x-y+5)$ **86.** $a(x+y-z)$ **87.** $4x-8y$ **88.** $-8+6a$ **89.** $ab-3ac$ **90.** $7z$ **91.** $\frac{1}{4}a+\frac{3}{4}$ **92.** $5a+3x$ **93.** $-a+3$ **94.** $x-2$ **95.** $y+5$ **96.** $4a+6$ **97.** $3x-4$ **98.** 0 **99.** $\frac{y^6}{4x^4}$ **100.** $\frac{a^5}{b^4}$ **101.** $x^3$ **102.** $\frac{1}{8x^6y^3}$ **103.** $2x^6$ **104.** $\frac{y^3}{x^7}$ **105.** .03 **106.** .005 **107.** 1 **108.** 175% **109.** 12.8 cm **110.** 18.5 $in^2$ **111.** 45 $ft^2$ **112.** 380 $in^2$ **113.** 12 **114.** 3 **115.** 5280 **116.** 144 **117.** 2000 **118.** 16 **119.** 60 **120.** 27 **121.** $\frac{1}{144}$ **122.** $\frac{1}{3}$ **123.** 720 **124.** 81 **125.** 11 **126.** 3 **127.** 4 **128.** 96 **129.** 13 **130.** 9 **131.** 3456 **132.** 45 **133.** 23,000 **134.** .00513 **135.** .0048 **136.** 9.8 **137.** 1050 **138.** .0184 **139.** .051 **140.** 6 **141.** .042 **142.** 320 **143.** (a) 237,600 (b) 3960 (c) 66 **144.** (a) 150 (b) 9000 (c) 540,000 **145.** 50

# 2

# Linear Equations and Inequalities

## 2.1 Linear Equations

A statement that two quantities are equal is called an **equation.** The two quantities, called **members** of the equation, are written with an equal sign (=) between them. The expression to the left of the equal sign is called the **left member** or **left side** of the equation, and the expression to the right is called the **right member** or **right side** of the equation. Some equations are true, some are false, and for some the truth value cannot be determined. For example,

| | |
|---|---|
| $3 + 5 = 8$ | is true |
| $3 + 5 = 3 - 5$ | is false |
| $x + 5 = 8$ | is neither true nor false since the value of $x$ is not known. |

Equations that involve letters called **variables** are important in algebra. If the variable in an equation can be replaced by a number that makes the resulting equation true, that number is called a **solution** or **root** of the equation. The process of determining all solutions or roots of a given equation is called **solving the equation.** Some simple equations may be solved by inspection or by direct observation (for example, it is easy to see that 3 is a solution to the equation $x = 3$), whereas others (for example, $3x + 5 = 7 - x$) require techniques outlined in the material which follows. In this chapter we concentrate on **linear equations** in which the variable is raised to the first power only. For this reason, linear equations are often called **first-degree** equations.

EXAMPLE 1 The linear equation

$$x + 5 = 8$$

has 3 as a solution since, when $x$ is replaced by 3,

$$3 + 5 = 8$$

is a true equation. Note that $-3$ is not a solution since

$$-3 + 5 = 8$$

is false. An equation such as this, which is true for some replacements for the variable (solutions) and false for others, is called a **conditional equation.**

**EXAMPLE 2** The equation

$$x + 5 = 5 + x$$

has many solutions. In fact, every real number is a solution to this equation. (Why?) An equation such as this, which is true for all replacements for the variable (every number is a solution), is called an **identity.**

**EXAMPLE 3** The equation

$$x + 5 = x - 5$$

has no solutions. (Why?) An equation such as this, which is false for all replacements for the variable (no number is a solution), is called a **contradiction.**

## EXERCISES 2.1

1. Give an example of a true equation.

2. Give an example of a false equation.

3. Give an example of an equation that is neither true nor false.

4. Give an example of a conditional equation.

5. Give an example of an identity.

6. Give an example of a contradiction.

7. Give an example of a linear equation.

8. In the equation $x + 1 = 7$, $x + 1$ is called the **(a)** ______________________, 7 is called the **(b)** ______________________, and 6 is called a **(c)** __________ to the equation.

*Solve the following equations.*

9. $x + 2 = 5$

10. $2x = 6$

11. $3 + x = 5 + x$

12. $3 + x = 3 + x$

13. $3 + x = 3 - x$

14. $3 \cdot x = 5 \cdot x$

---

ANSWERS: Answers to Exercises 1–7 will vary. However, one possible answer to each is given.
**1.** $1 + 3 = 3 + 1$ **2.** $1 + 3 = 3 - 1$ **3.** $x - 1 = 3$ **4.** $x + 1 = 5$ **5.** $2 \cdot x = x + x$ **6.** $x + 1 = x$
**7.** $x + 3 = 7$ **8.** **(a)** left member or left side **(b)** right member or right side **(c)** solution or root **9.** 3 **10.** 3
**11.** no solution (contradiction) **12.** any number (identity) **13.** 0 **14.** 0

## 2.2 Solving Equations

Consider the following two equations.

$$x + 3 = 5 \quad \text{and} \quad x = 2$$

Both equations have 2 as a solution, and both can be solved by inspection. However, it is easier to recognize 2 as a solution to the second. Two equations which have exactly the same solutions are called **equivalent equations.** The process of solving an equation is one of transforming the given equation into an equivalent equation which can be solved by direct inspection.

If a given equation is transformed into an equivalent equation with the variable **isolated** on one side, such as $x = 2$, solutions to the original equation are the (obvious) solutions to the new equation. Several rules for solving an equation (transforming it into equivalent equations) follow.

**ADDITION-SUBTRACTION RULE**

If the same number or expression is added to or subtracted from both members of an equation, the resulting equation is equivalent to (has the same solutions as) the original.

**EXAMPLE 1** Solve.

$$x + 3 = 5$$

$$x + 3 - 3 = 5 - 3 \qquad \text{Subtract 3 from both sides to isolate } x$$

$$x + 0 = 2$$

$$x = 2 \qquad \text{Equation equivalent to the original equation } x + 3 = 5$$

Check: $2 + 3 \stackrel{?}{=} 5$ Substitute 2 for $x$ in the original equation

$$5 = 5$$

The solution is 2.

Check all indicated solutions by replacing the variable with the indicated solution throughout the given equation. If the resulting equation is true, the number is a solution.

**EXAMPLE 2** Solve.

$$x - \frac{1}{4} = 3$$

$$x - \frac{1}{4} + \frac{1}{4} = 3 + \frac{1}{4} \qquad \text{Add } \frac{1}{4} \text{ to both sides to isolate } x$$

$$x + 0 = \frac{3 \cdot 4}{4} + \frac{1}{4}$$

$$= \frac{12}{4} + \frac{1}{4} = \frac{13}{4}$$

$$x = \frac{13}{4}$$

Check: $\frac{13}{4} - \frac{1}{4} \stackrel{?}{=} 3$

$$\frac{12}{4} \stackrel{?}{=} 3$$

$$3 = 3$$

The solution is $\frac{13}{4}$.

**MULTIPLICATION-DIVISION RULE**

If each member of an equation is multiplied or divided by the same nonzero number or expression, the resulting equation is equivalent to (has the same solutions as) the original.

**EXAMPLE 3** Solve.

$$3x = 9$$

$$\frac{1}{3} \cdot 3x = \frac{1}{3} \cdot 9 \qquad \text{Multiply both sides by } \frac{1}{3} \text{ to isolate } x$$

$$1 \cdot x = 3$$

$$x = 3$$

Check: $3 \cdot 3 \stackrel{?}{=} 9$

$$9 = 9$$

The solution is 3.

Equivalently, we may divide both sides of the equation by 3.

$$3x = 9$$

$$\frac{3x}{3} = \frac{9}{3} \qquad \text{Divide both sides by 3}$$

$$\frac{3}{3} \cdot x = 3$$

$$1 \cdot x = 3$$

$$x = 3$$

**EXAMPLE 4** Solve.

$$\frac{1}{2}y = 10$$

$$2 \cdot \frac{1}{2}y = 2 \cdot 10 \qquad \text{Multiply both sides by 2}$$

$$1 \cdot y = 20$$

$$y = 20$$

Check: $\frac{1}{2} \cdot 20 \stackrel{?}{=} 10$

$$10 = 10$$

The solution is 20.

**EXAMPLE 5** Solve.

$$\frac{x}{\frac{1}{2}} = 10$$

$$\frac{1}{2} \cdot \frac{x}{\frac{1}{2}} = \frac{1}{2} \cdot 10 \qquad \text{Multiply both sides by } \frac{1}{2}$$

$$\frac{\frac{1}{2}}{\frac{1}{2}} \cdot x = \frac{1}{2} \cdot 10 = 5$$

$$1 \cdot x = 5$$

$$x = 5$$

Check: $\frac{5}{\frac{1}{2}} \stackrel{?}{=} 10$

$$5 \div \frac{1}{2} = \frac{5}{1} \cdot \frac{2}{1} = 10$$

The solution is 5.

In an equation of this type, it may be easier to simplify the left member first.

$$\frac{x}{\frac{1}{2}} = \frac{\frac{x}{1}}{\frac{1}{2}} = \frac{x}{1} \div \frac{1}{2} = \frac{x}{1} \cdot \frac{2}{1} = 2x$$

Thus we are actually solving

$$2x = 10.$$

$$\frac{1}{2} \cdot 2x = \frac{1}{2} \cdot 10 \qquad \text{Multiply both sides by } \frac{1}{2}$$

$$1 \cdot x = 5$$

$$x = 5$$

Study the above examples carefully and observe that in all cases we multiply or divide both sides in such a way that the coefficient of the variable becomes 1.

Many equations are solved by using both the addition-subtraction and the multiplication-division rules.

**EXAMPLE 6** Solve.

$$3x + 7 = 13$$

$$3x + 7 - 7 = 13 - 7 \qquad \text{Subtract 7 from both sides}$$

$$3x = 6$$

$$\frac{1}{3} \cdot 3x = \frac{1}{3} \cdot 6 \qquad \text{Multiply both sides by } \frac{1}{3}$$

$$x = 2$$

Check: $3(2) + 7 \stackrel{?}{=} 13$

$$6 + 7 \stackrel{?}{=} 13$$

$$13 = 13$$

The solution is 2.

Generally, we use the addition-subtraction rule before the multiplication-division rule, as in the above example.

**EXAMPLE 7** Solve.

$$1.2y - 3.6 = 2.4$$
$$1.2y - 3.6 + 3.6 = 2.4 + 3.6 \quad \text{Add 3.6 to both sides}$$
$$1.2y = 6.0$$
$$\frac{1.2y}{1.2} = \frac{6.0}{1.2} \quad \text{Divide both sides by 1.2}$$
$$y = 5$$

Check: $1.2(5) - 3.6 \stackrel{?}{=} 2.4$
$$6.0 - 3.6 \stackrel{?}{=} 2.4$$
$$2.4 = 2.4$$

The solution is 5.

It is sometimes wise to eliminate decimals by multiplying both sides of the equation by the appropriate power of 10. For example, in the above, if we first multiply both sides by 10, we obtain

$$12y - 36 = 24.$$

There might be less chance of making an arithmetic error when solving this equation rather than the original.

The same remarks apply to equations involving fractional coefficients. It might be wise to multiply through first by the least common denominator (LCD) of the fractions.

**EXAMPLE 8** Solve.

$$\frac{1}{3}x + \frac{1}{4} = \frac{1}{2}$$
$$12\left(\frac{1}{3}x + \frac{1}{4}\right) = 12 \cdot \frac{1}{2} \quad \text{Multiply both sides by the LCD 12}$$
$$12 \cdot \frac{1}{3}x + 12 \cdot \frac{1}{4} = 6 \quad \text{Use distributive law}$$
$$4x + 3 = 6$$
$$4x + 3 - 3 = 6 - 3 \quad \text{Subtract 3}$$
$$4x = 3$$
$$\frac{1}{4} \cdot 4x = \frac{1}{4} \cdot 3 \quad \text{Multiply by } \frac{1}{4}$$
$$x = \frac{3}{4}$$

Check: $\frac{1}{3}\left(\frac{3}{4}\right) + \frac{1}{4} \stackrel{?}{=} \frac{1}{2}$
$$\frac{1}{4} + \frac{1}{4} \stackrel{?}{=} \frac{1}{2}$$
$$\frac{2}{4} = \frac{1}{2}$$

The solution is $\frac{3}{4}$.

Compare the above with the following procedure.

$$\frac{1}{3}x + \frac{1}{4} = \frac{1}{2}$$

$$\frac{1}{3}x + \frac{1}{4} - \frac{1}{4} = \frac{1}{2} - \frac{1}{4} \qquad \text{Subtract } \frac{1}{4}$$

$$\frac{1}{3}x = \frac{1}{4}$$

$$3 \cdot \frac{1}{3}x = 3 \cdot \frac{1}{4} \qquad \text{Multiply by 3}$$

$$x = \frac{3}{4}$$

Use the solution procedure you prefer.

## EXERCISES 2.2

*Solve the following equations.*

**1.** $x + 2 = 9$

**2.** $y - 3 = 8$

**3.** $z + 2.1 = 3.8$

**4.** $x - 5.2 = 10$

**5.** $y + \frac{2}{3} = \frac{1}{3}$

**6.** $z - \frac{3}{4} = 4$

**7.** $-2 + x = 4.2$

**8.** $-3.1 - y = 1.5$

**9.** $-\frac{1}{7} + z = \frac{3}{14}$

**10.** $4x = 16$

**11.** $-3y = 27$

**12.** $-48 = -6z$

**13.** $\frac{1}{5}x = 7$

**14.** $\frac{1}{3}y = -2$

**15.** $-\frac{3}{4}z = 9$

**16.** $-\frac{3}{5}x = 12$

**17.** $\frac{y}{\frac{1}{4}} = 20$

**18.** $\frac{z}{\frac{1}{3}} = -9$

**19.** $\frac{x}{\frac{2}{3}} = -18$

**20.** $4y = 12.8$

**21.** $-3y = 9.3$

**22.** $-5x = -2.5$

**23.** $.3y = -3.9$

**24.** $-1.2z = -72$

**25.** $2x + 1 = 9$

**26.** $3y - 2 = 10$

**27.** $-4z + 1 = 17$

**28.** $\frac{1}{4}x + 2 = 3$

**29.** $\frac{2}{5} - y = \frac{3}{5}$

**30.** $12 = -1.2z + 36$

**31.** $5 = \frac{2}{3}x - \frac{1}{3}$

**32.** $-\frac{1}{10} - \frac{3}{5}y = \frac{1}{5}$

**33.** $\frac{1}{2}z + 1.5 = 3$

**34.** Consider the equation $2x - 3 = 19$.
- **(a)** What is the variable?
- **(b)** What is the right member?
- **(c)** What is the left member?
- **(d)** Is this an identity?
- **(e)** Is this a contradiction?
- **(f)** Is this a conditional equation?
- **(g)** Is this equivalent to $x = 11$?
- **(h)** What is the solution?

---

ANSWERS: **1.** 7 **2.** 11 **3.** 1.7 **4.** 15.2 **5.** $-\frac{1}{3}$ **6.** $\frac{19}{4}$ **7.** 6.2 **8.** $-4.6$ **9.** $\frac{5}{14}$ **10.** 4 **11.** $-9$ **12.** 8 **13.** 35 **14.** $-6$ **15.** $-12$ **16.** $-20$ **17.** 5 **18.** $-3$ **19.** $-12$ **20.** 3.2 **21.** $-3.1$ **22.** .5 **23.** $-13$ **24.** 60 **25.** 4 **26.** 4 **27.** $-4$ **28.** 4 **29.** $-\frac{1}{5}$ **30.** 20 **31.** 8 **32.** $-\frac{1}{2}$ **33.** 3 **34.** (a) $x$ (b) 19 (c) $2x - 3$ (d) no (e) no (f) yes (g) yes (h) 11

## 2.3 Solving Equations Involving Like Terms and Parentheses

The members of an equation often contain like terms. These terms should be collected before we apply the addition-subtraction or multiplication-division rules.

**EXAMPLE 1** Solve.

$$3x + 4x = 21$$
$$7x = 21 \qquad \text{Collect like terms}$$
$$\frac{1}{7} \cdot 7x = \frac{1}{7} \cdot 21 \qquad \text{Multiply by } \frac{1}{7}$$
$$x = 3$$

Check: $3(\mathbf{3}) + 4(\mathbf{3}) \stackrel{?}{=} 21$
$9 + 12 \stackrel{?}{=} 21$
$21 = 21$

The solution is 3.

**EXAMPLE 2** Solve.

$$3y - 6 - 8y = 9$$
$$3y - 8y - 6 = 9 \qquad \text{Commutative law}$$
$$-5y - 6 = 9 \qquad \text{Collect like terms}$$
$$-5y - 6 + 6 = 9 + 6 \qquad \text{Add 6}$$
$$-5y = 15$$
$$\left(-\frac{1}{5}\right)(-5y) = \left(-\frac{1}{5}\right)(15) \qquad \text{Multiply by } -\frac{1}{5}$$
$$y = -3$$

Check: $3(-3) - 6 - 8(-3) \stackrel{?}{=} 9$
$-9 - 6 + 24 \stackrel{?}{=} 9$
$9 = 9$

The solution is $-3$.

**EXAMPLE 3** Solve.

$$6x + 1 - 4x = 12 - 2x - 11$$
$$2x + 1 = -2x + 1 \qquad \text{Collect like terms in both members}$$
$$2x + 2x + 1 = -2x + 2x + 1 \qquad \text{Add } 2x \text{ to both sides}$$
$$4x + 1 = 1$$
$$4x + 1 - 1 = 1 - 1 \qquad \text{Subtract 1}$$
$$4x = 0$$
$$\frac{1}{4}(4x) = \frac{1}{4} \cdot 0 \qquad \text{Multiply by } \frac{1}{4}$$
$$x = 0 \qquad \frac{1}{4} \cdot 0 = 0$$

Check: $6(0) + 1 - 4(0) \stackrel{?}{=} 12 - 2(0) - 11$
$1 \stackrel{?}{=} 12 - 11$
$1 = 1$

The solution is 0.

**EXAMPLE 4** Solve.

$$
\begin{aligned}
2x + 3 &= 2x + 8 \\
2x + 3 - 3 &= 2x + 8 - 3 && \text{Subtract 3} \\
2x &= 2x + 5 \\
2x - 2x &= 2x - 2x + 5 && \text{Subtract } 2x \\
0 &= 5
\end{aligned}
$$

Whenever a contradiction is obtained, we know that the original equation is also a contradiction. Thus there is no solution.

**EXAMPLE 5** Solve.

$$
\begin{aligned}
3x + 1 &= 1 + 3x \\
3x + 1 - 1 &= 1 - 1 + 3x && \text{Subtract 1} \\
3x &= 3x
\end{aligned}
$$

Whenever an identity is obtained, we know that the original equation is an identity. Thus, any number is a solution.

Often, an equation contains one or more sets of parentheses.

**TO SOLVE AN EQUATION WHICH INVOLVES PARENTHESES**

**1.** Multiply (using the distributive laws) and clear all parentheses.

**2.** Collect any resulting like terms and solve the equation.

**EXAMPLE 6** Solve.

$$
\begin{aligned}
2(5x + 1) &= 42 \\
2 \cdot 5x + 2 \cdot 1 &= 42 && \text{Clear parentheses using distributive law} \\
10x + 2 &= 42 \\
10x + 2 - 2 &= 42 - 2 && \text{Subtract 2} \\
10x &= 40 \\
\frac{10x}{10} &= \frac{40}{10} && \text{Divide by 10} \\
x &= 4
\end{aligned}
$$

Check:

$$
\begin{aligned}
2(5(4) + 1) &\stackrel{?}{=} 42 \\
2(20 + 1) &\stackrel{?}{=} 42 \\
2(21) &\stackrel{?}{=} 42 \\
42 &= 42
\end{aligned}
$$

The solution is 4.

**EXAMPLE 7** Solve.

$$6(z+1)-4(z-3)=0$$
$$6z+6-4z+12=0 \quad \text{Clear parentheses and watch signs}$$
$$2z+18=0 \quad \text{Collect like terms}$$
$$2z=-18 \quad \text{Subtract 18 from both sides}$$
$$z=-9 \quad \text{Divide both sides by 2}$$

Check: $6(-9+1)-4(-9-3)\stackrel{?}{=}0$
$$6(-8)-4(-12)\stackrel{?}{=}0$$
$$-48+48\stackrel{?}{=}0$$
$$0=0$$

The solution is $-9$.

**EXAMPLE 8** Solve.

$$5-x=4x-5(x+1)$$
$$5-x=4x-5x-5 \quad \text{Clear parentheses}$$
$$5-x=-x-5 \quad \text{Collect like terms}$$
$$5-x+x=-x+x-5 \quad \text{Add } x \text{ to both sides}$$
$$5=-5 \quad \text{Contradiction}$$

There is no solution.

## EXERCISES 2.3

*Solve the following equations.*

**1.** $3x+2x=15$

**2.** $3y-2y=5$

**3.** $\frac{1}{2}z+\frac{1}{4}z=-12$

**4.** $4x+5=x-16$

**5.** $6+5y=5y-2$

**6.** $z+2z=8-2z+7$

**7.** $x - 3 + 4x = 2 + x - 5$

**8.** $2.1y + 45.2 = 3.2 - 8.4y$

**9.** $3z + \frac{3}{2} + \frac{5}{2}z = \frac{1}{2}z + \frac{5}{2}z$

**10.** $3(2x + 3) = 15$

**11.** $20 = 5(y - 1)$

**12.** $-2(3z - 1) = 10$

**13.** $2(x + 3) = 3(x - 7)$

**14.** $4(2y - 1) - 7(y + 3) = 0$

**15.** $3(2z - 1) = 5 - (3z - 2)$

**16.** $4(1 - 2z) - 3 = 8z - 1$

**17.** $\frac{x}{\frac{2}{3}} = -6$

**18.** $\frac{2}{3}x = -6$

**19.** $x - \frac{2}{3} = -6$

---

ANSWERS: **1.** 3 **2.** 5 **3.** −16 **4.** −7 **5.** no solution **6.** 3 **7.** 0 **8.** −4 **9.** $-\frac{3}{5}$ **10.** 1 **11.** 5 **12.** $-\frac{4}{3}$ **13.** 27 **14.** 25 **15.** $\frac{10}{9}$ **16.** $\frac{1}{8}$ **17.** −4 **18.** −9 **19.** $-\frac{16}{3}$

## 2.4 Solving Equations Involving Absolute Value

In Exercise 69 of Section 1.4 we defined absolute value in the following way.

$$|a| = \begin{cases} a & \text{if } a \geq 0 \\ -a & \text{if } a < 0 \end{cases}$$

Some equations involve absolute values. To solve such an equation, for example,

$$|x| = 5,$$

we must solve two related equations, since

$$|x| = \begin{cases} x & \text{if } x \geq 0 \\ -x & \text{if } x < 0. \end{cases}$$

The two equations we must solve are

$$x = 5 \qquad \text{and} \qquad -x = 5.$$

The second equation becomes $x = -5$, giving us the two equations

$$x = 5 \qquad \text{and} \qquad x = -5.$$

Thus, to solve any absolute value equation, we solve two related equations according to the following rule.

> To solve an absolute value equation of the form
>
> $$|Expression| = a \quad (a \geq 0),$$
>
> solve the two related equations
>
> $$Expression = a \qquad \text{and} \qquad Expression = -a.$$

**EXAMPLE 1** Solve $|x| = 12$.

We must solve

$$x = 12 \qquad \text{and} \qquad x = -12.$$

The solutions are 12 and $-12$ (check).

**EXAMPLE 2** Solve $|2x + 1| = 3$.

$$\begin{aligned} 2x + 1 &= 3 \\ 2x + 1 - 1 &= 3 - 1 \\ 2x &= 2 \\ \frac{1}{2}(2x) &= \frac{1}{2}(2) \\ x &= 1 \end{aligned} \qquad \text{and} \qquad \begin{aligned} 2x + 1 &= -3 \\ 2x + 1 - 1 &= -3 - 1 \\ 2x &= -4 \\ \frac{1}{2}(2x) &= \frac{1}{2}(-4) \\ x &= -2 \end{aligned}$$

The solutions are 1 and $-2$ (check).

**EXAMPLE 3** Solve $|1-3x|=0$

We must solve $1-3x=0$ and $1-3x=-0$. But since $-0=0$, we really have only one equation to solve.

$$1-3x=0$$
$$-3x=-1 \quad \text{Subtract 1 from both sides}$$
$$x=\frac{1}{3} \quad \text{Divide both sides by } -3$$

The solution is $\frac{1}{3}$ (check).

**EXAMPLE 4** Solve $|5x-1|=-2$

*Be careful.* Since absolute value is always $\geq 0$, there are no solutions to $|5x-1|=-2$.

## EXERCISES 2.4

*Solve.*

**1.** $|x|=7$

**2.** $|y|=0$

**3.** $|z|=-7$

**4.** $|x+1|=4$

**5.** $|y+1|=-4$

**6.** $|z+1|=0$

**7.** $|2x-5|=9$

**8.** $|5-2x|=9$

**9.** $|2y-5|=0$

**10.** $|5-2y|=0$

**11.** $\left|\frac{1}{2}z-\frac{3}{4}\right|=\frac{1}{4}$

**12.** $|2.2z-1.1|=5.5$

**13.** $3(1 - 2x) = -4(x + 5)$

**14.** $2y - (4y - 1) = 5 - (y + 3)$

---

ANSWERS: **1.** 7, −7 **2.** 0 **3.** no solution **4.** 3, −5 **5.** no solution **6.** −1 **7.** 7, −2 **8.** 7, −2 **9.** $\frac{5}{2}$ **10.** $\frac{5}{2}$ **11.** 2, 1 **12.** 3, −2 **13.** $\frac{23}{2}$ **14.** −1

## 2.5 The Language of Word Problems

Solving a word problem using algebra involves two steps. First, we must translate the words of the problem into an algebraic equation, and second, we must solve the resulting equation. We have learned how to solve several types of equations, and now we concentrate on translating words into equations. Some common terms and their symbolic translations are presented below.

| *Symbol* | *Stands for* |
|---|---|
| $+$ | and, sum, sum of, added to, increased by, more than |
| $-$ | minus, less, subtracted from, less than, diminished by, difference between, difference, decreased by |
| $\cdot$ | times, product, product of, multiplied by, of |
| $\div$ | divided by, quotient of, ratio |
| $=$ | equals, is equal to, is as much as, is, is the same as |

Any letter (we often use $x$) is used to stand for the unknown or desired quantity. Some examples of translations of these phrases follow.

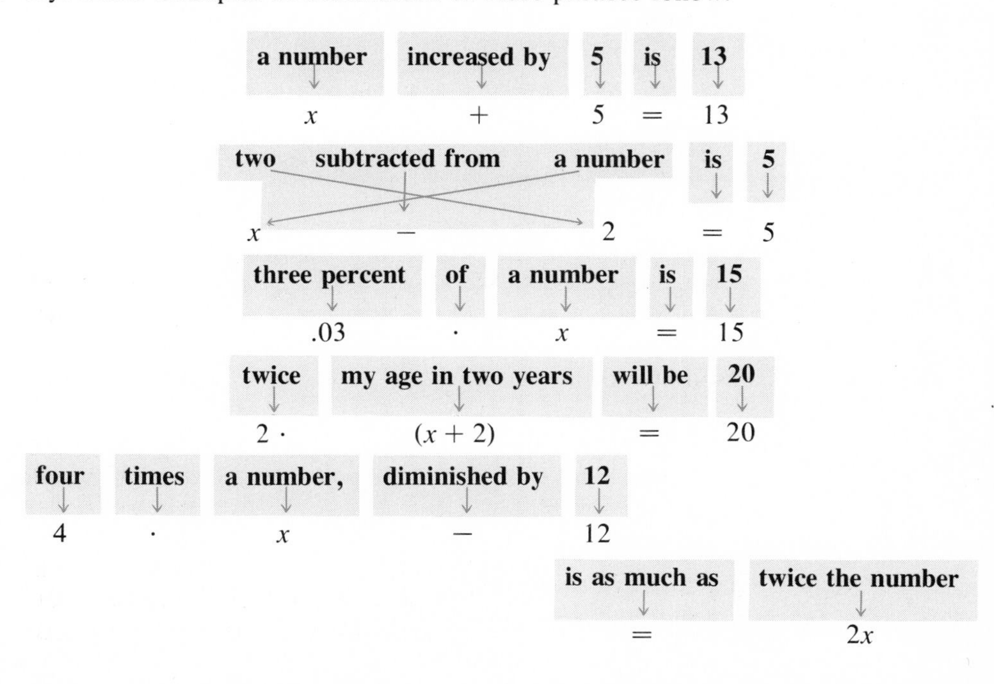

EXAMPLE 1 Translate the word expressions into symbols.

| *Word expression* | *Symbolic translation* |
|---|---|
| The product of a number and 3 is 22. | $x \cdot 3 = 22$ |
| Four times a number is 8. | $4x = 8$ |
| Twice a number, increased by 8, is 11. | $2x + 8 = 11$ |
| Seven is 4 less than three times a number. | $7 = 3x - 4$ |
| Seven is 4 less three times a number. | $7 = 4 - 3x$ |
| One tenth of a number is 6. | $\frac{1}{10} \cdot x = 6$ |
| Twice a number, less five times the number, is the same as 12. | $2x - 5x = 12$ |
| A number is 15% of 28. | $x = (.15)(28)$ |
| A number subtracted from 2 is 5. | $2 - x = 5$ |
| Two subtracted from a number is 8. | $x - 2 = 8$ |
| Twice the sum of a number and 3 is four times the square of the number. | $2(x + 3) = 4x^2$ |
| Eight more than a number is 12. | $8 + x = 12$ |
| My age 3 years ago was 35. (Let $x$ be my present age.) | $x - 3 = 35$ |

## EXERCISES 2.5

*Letting $x$ represent the unknown number, translate the following into symbols.*

**1.** Eight times a number is 72.

**2.** A number subtracted from 3 is 10.

**3.** A number increased by 7 is 21.

**4.** Ten more than a number is 4.

**5.** A number divided by 3 is equal to 4.

**6.** My age in 8 years will be 26.

**7.** A number decreased by 5 is 17.

**8.** Twice a number, increased by 3, is 14.

**9.** Twice the sum of a number and 5 is the same as the number.

**10.** Seven times a number, less 3, is as much as twice the number, plus 5.

**11.** A number less 7 is as much as twice the number.

**12.** Ten is 3 less than four times a number.

**13.** The product of a number and the number plus 3 is equal to 0.

**14.** When 11 is added to three times a number and the result doubled, the resulting quantity is the same as 15 more than four times the number.

**15.** Eight percent of a number is 9.

**16.** A number is 12% of 25.

**17.** A number plus three times itself is 24.

**18.** Solve $|5x + 2| = 12$.

**19.** Solve $7(x - 1) - (3x + 5) = 0$.

---

ANSWERS: **1.** $8x = 72$ **2.** $3 - x = 10$ **3.** $x + 7 = 21$ **4.** $10 + x = 4$ **5.** $\frac{x}{3} = 4$ **6.** $x + 8 = 26$ **7.** $x - 5 = 17$ **8.** $2x + 3 = 14$ **9.** $2(x + 5) = x$ **10.** $7x - 3 = 2x + 5$ **11.** $x - 7 = 2x$ **12.** $10 = 4x - 3$ **13.** $x(x + 3) = 0$ **14.** $2(3x + 11) = 4x + 15$ **15.** $.08 \cdot x = 9$ **16.** $x = (.12)(25)$ **17.** $x + 3x = 24$ **18.** $2, -\frac{14}{5}$ **19.** 3

## 2.6 Word Problems: Number, Age, and Geometry Problems

Solving a word problem is easier if these steps are followed.

1. Read the problem carefully (perhaps several times) and determine what quantity (or quantities) must be found.
2. Represent the unknown quantity (or quantities) by a letter, and write (for example): Let $x =$ *description of unknown.*
3. Make a sketch or diagram whenever possible.
4. Determine which expressions are equal and write an equation.
5. Solve the equation and state the answer to the problem.
6. Check the answer or answers to see if the conditions of the original problem are satisfied. This can often be done mentally.

In this section we consider several types of simple word problems which might be classified as number problems, age problems, and geometry problems.

**EXAMPLE 1** If three times a number is subtracted from 20, the result is twice the number. Find the number.

Let $x =$ the desired number
$20 - 3x =$ three times the number subtracted from 20

$$20 - 3x = 2x \qquad \text{Symbolic translation of the problem}$$
$$20 = 5x$$
$$4 = x$$

The number is 4.

Check: 20 minus $3 \cdot 4$ is indeed twice 4. (Could the solution 4 be checked mentally in the original statement?)

**EXAMPLE 2** The sum of two consecutive integers is 47. Find the integers.

Let $x =$ the first integer
$x + 1 =$ the next consecutive integer (Why?)

$$x + (x + 1) = 47$$
$$2x + 1 = 47$$
$$2x = 46$$
$$x = 23 \ (x + 1 = 24)$$

The integers are 23 and 24. (Check.)

**EXAMPLE 3** Tom is four times as old as Jim, and the difference of their ages is 33 years. How old is each?

Let $x =$ Jim's age
$4x =$ Tom's age (Why?)

$$4x - x = 33 \qquad \text{Why not } x - 4x = 33?$$
$$3x = 33$$
$$x = 11$$
$$4x = 44$$

Jim is 11 years old, and Tom is 44 years old. (Check.)

**EXAMPLE 4** A 20-ft section of rope is cut into two pieces, one 7 ft longer than the other. How long is each piece?

Let $x =$ the length of the shorter piece
$x + 7 =$ the length of the longer piece (Why?)

Make a sketch as in Figure 2.1.

$$x + (x + 7) = 20$$
$$2x + 7 = 20$$
$$2x = 13$$
$$x = \frac{13}{2}$$
$$x + 7 = \frac{27}{2}$$

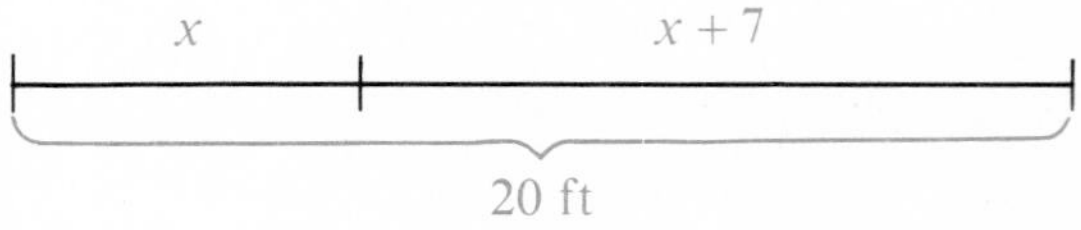

**Figure 2.1**

The pieces are $\frac{13}{2}$ ft and $\frac{27}{2}$ ft. (Check.)

**EXAMPLE 5** The perimeter of a rectangle is 52 m. If the length is 4 m more than the width, find the dimensions.

Recall that the perimeter of a rectangle is given by $P = 2 \cdot l + 2 \cdot w$. It is helpful to make a sketch as in Figure 2.2

Let $\quad x =$ the width of the rectangle
$x + 4 =$ the length of the rectangle

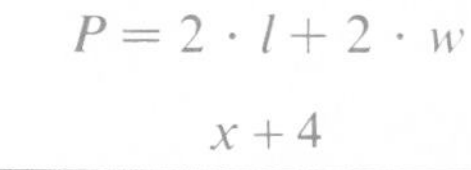

$$52 = 2(x + 4) + 2x$$
$$52 = 2x + 8 + 2x$$
$$44 = 4x$$
$$11 = x$$
$$15 = x + 4$$

**Figure 2.2**

The width is 11 m and the length is 15 m. (Check.)

**EXAMPLE 6** The second angle of a triangle is six times as large as the first angle. If the third angle is 45° more than twice the first, find the measure of each angle.

Recall that the sum of the measures of the angles of any triangle is 180°. Make a sketch as in Figure 2.3.

Let $\quad x =$ the measure of the first angle
$6x =$ the measure of the second angle
$2x + 45 =$ the measure of the third angle

$$x + 6x + (2x + 45) = 180$$
$$9x + 45 = 180$$
$$9x = 135$$
$$x = 15$$
$$6x = 90$$
$$2x + 45 = 75$$

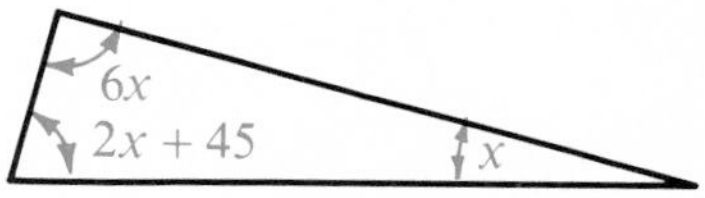

**Figure 2.3**

The measures are 15°, 90°, and 75°. (Check.)

## EXERCISES 2.6

*Solve. (Some problems have been started.)*

**1.** If 3 is added to five times a number, the result is 38. Find the number.

Let $\quad x =$ the number
$5x =$ five times the number

**2.** The sum of two numbers is 24. The larger number is five times the smaller number. Find the two numbers.

Let $x =$ one number
$24 - x =$ other number $[x + (24 - x) = 24]$

**3.** Sam made \$10 more than twice what Pete earned in one month. If together they earned \$760, how much did each earn that month?

Let $x =$ amount Pete earned in one month
$2x + 10 =$ amount Sam earned in one month

**4.** A woman burns up three times as many calories running as she does when walking the same distance. If she runs 2 miles and walks 5 miles and burns up a total of 770 calories, how many calories does she burn up while running 1 mile?

Let $x =$ the number of calories burned up walking one mile
$3x =$ the number of calories burned up running one mile

**5.** The sum of two consecutive odd integers is 76. Find the two integers.
[*Hint:* If one integer is $x$, the other is $x + 2$. Why?]

**6.** Two fifths of a man's income each month goes to taxes. If he pays \$424 in taxes each month, what is his monthly income?

**7.** The sum of three consecutive integers is 126. Find the three integers.

**8.** The sum of Jan's age and Juan's age is 78 years. If Jan is 6 years younger than Juan, how old is each?

**9.** If three times a number is increased by 7, the result is the same as when five times the number is decreased by 35. Find the number.

**10.** If six times a number is increased by six, the result is the same as when seven times the number is decreased by 1. Find the number.

**11.** A board is 11 ft long. It is to be cut into three pieces in such a way that the second piece is twice as long as the first piece and the third piece is 3 ft more than the first piece. Find the length of each piece.

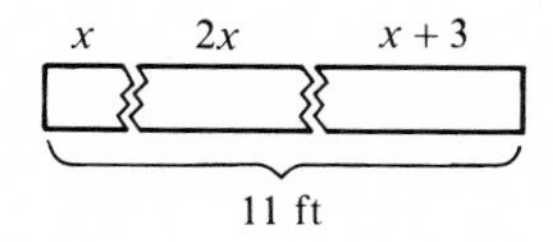


**12.** The perimeter of a rectangle is 86 in. If the length is 19 in more than the width, find the dimensions.

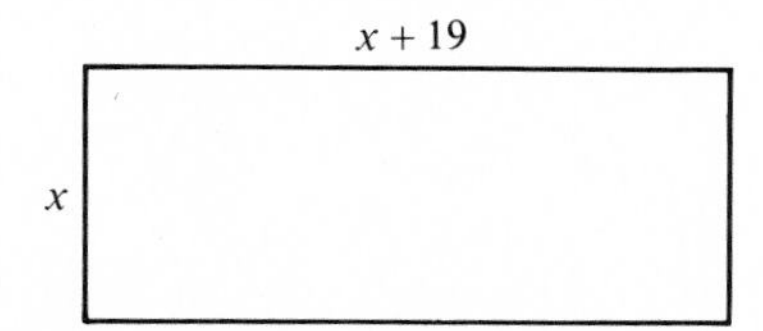

**13.** If the second angle of a triangle is 50° more than the first angle, and the third angle is eleven times the first, determine the measure of each angle.

**14.** If two angles are supplementary (they have measures totaling 180°) and the second is 15° less than twice the first, find the measure of each angle.

**15.** An isosceles triangle has two equal sides called legs and a third side called the base. If each leg of an isosceles triangle is five times the base and the perimeter is 77 m, find the length of the legs and base.

**16.** Find the altitude of a triangle whose base is 10 cm and whose area is 55 cm$^2$.

---

ANSWERS: **1.** 7 **2.** 4, 20 **3.** Pete: \$250, Sam: \$510 **4.** 210 calories **5.** 37, 39 **6.** \$1060 **7.** 41, 42, 43 **8.** Jan: 36, Juan: 42 **9.** 21 **10.** 7 **11.** 2 ft, 4 ft, 5 ft **12.** 12 in by 31 in **13.** 10°, 60°, 110° **14.** 65°, 115° **15.** base: 7 m, legs: 35 m; **16.** 11 cm

## 2.7 Word Problems: Percent Problems

Recall that percent notation can be converted to decimal notation; for example,

$$25\% = .25 \qquad \text{and} \qquad 3\% = .03.$$

When working with percent, the word *of* translates to *multiplied by* or *times*. Two basic statements concerning percent are often made. An example of the first, with its symbolic translation, follows.

| 30% | of | 40 | is | what? | (What is 30% of 40?) |
|---|---|---|---|---|---|
| (.30) | $\cdot$ | 40 | = | $x$ | |

An example of the second, and its translation into symbols, is

| 15 | is | what percent | of | 75? | (What percent of 75 is 15?) |
|---|---|---|---|---|---|
| 15 | = | $x$ | $\cdot$ | 75 | |

Solve this equation.

$$15 = x \cdot 75$$
$$\frac{15}{75} = x$$
$$.2 = x \qquad \text{In percent notation, } x = 20\%.$$

If we study the above examples carefully, the language of percent will be much easier to understand.

**EXAMPLE 1** 3% of 80 is what?

We must solve the following equation.

$$(.03)(80) = x$$
$$2.4 = x$$

**EXAMPLE 2** What percent of 52 is 13?

The equation we must solve is

$$x \cdot 52 = 13$$
$$x = \frac{13}{52} = .25 \qquad x = 25\% \text{ (convert to \% notation).}$$

**EXAMPLE 3** A basketball player made 9 free throws in 15 attempts. What was her free-throw percentage?

In effect, we are asked: 9 is what percent of 15?

$$9 = x \cdot 15$$
$$\frac{9}{15} = x$$
$$.6 = x \qquad \text{In percent notation, } x = 60\%.$$

**EXAMPLE 4** The sales-tax rate in Murphyville is 4%. How much tax would be charged on a purchase of $42?

In effect, we are asked: What is 4 percent of 42?

$$x = (.04)(42)$$
$$x = 1.68$$

The tax is $1.68.

**EXAMPLE 5** A woman received a 6% raise making her new salary $11,130. What was her old salary?

Let $x =$ the woman's old salary

$.06 \cdot x =$ the woman's raise (Why?)

$x + .06 \cdot x =$ the woman's new salary (Why?)

We must solve the following equation.

$$x + .06 \cdot x = 11130$$
$$(1 + .06)x = 11130 \qquad \text{Factor out } x$$
$$(1.06)x = 11130 \qquad \text{Add}$$
$$x = \frac{11130}{1.06} = 10{,}500$$

The woman's old salary was $10,500.

## EXERCISES 2.7

*Solve.*

**1.** 36 is 15% of what?

**2.** 9 is 2% of what?

3. 12% of 240 is what?

**4.** 3% of 180 is what?

**5.** What is $\frac{1}{2}$% of 400?

**6.** 8% of what is 68?

**7.** A baseball player got 42 hits in 140 times at bat. What was his batting average?

**8.** A family spent $180 a month for food. This was 15% of their monthly income. What was their monthly income?

**9.** After a 5% increase in salary, a man's new salary is $8925. What was his old salary?

**10.** A man makes $12,500 a year. If he gets a 7% raise, what will be his new salary?

**11.** If the sales-tax rate is 5%, how much tax would be charged on a purchase of $145?

**12.** What amount of money invested at 4% simple interest will increase to $1248 at the end of one year?
[*Hint:* Let $x$ be the original amount. The simple interest on $x$ in 1 year is $.04 \cdot x$.]

**13.** If five times a number is increased by 6, the result is the same as when seven times the number is decreased by 14. Find the number.

**14.** Mel and Isaac were born in consecutive years. If the sum of their ages is 67 and Mel is the older, how old is each?

**15.** Two angles are *complementary* if their sum is 90°. One of two complementary angles is 30° more than twice the other. Find the measure of each angle.

ANSWERS: 1. 240 2. 450 3. 28.8 4. 5.4 5. 2 6. 850 7. 30% 8. $1200 9. $8500 10. $13,375 11. $7.25 12. $1200 13. 10 14. Mel: 34, Isaac: 33 15. 20°, 70°

## 2.8 Word Problems: Motion Problems

The distance ($d$) that an object travels in a given time ($t$) at a constant or uniform rate ($r$) is given by the following formula.

$$\text{(distance)} = \text{(rate)} \cdot \text{(time)} \qquad \text{or} \qquad d = r \cdot t$$

For example, an automobile traveling at a rate of 55 mph for 3 hours will travel a distance

$$d = r \cdot t = (55)(3) = 165 \text{ miles.}$$

All motion problems depend in some way on the basic formula $d = r \cdot t$.

**EXAMPLE 1** If a runner covers a distance of 96.6 yards in 9.2 seconds, how fast (at what rate) is he running?

Let $r$ = his rate
9.2 = time he runs
96.6 = distance he covers

We must solve the following equation.

$$96.6 = r(9.2)$$

$$\frac{96.6}{9.2} = r$$

$$10.5 = r$$

He is running 10.5 yd/sec.

**EXAMPLE 2** An automobile travels from Phoenix to Flagstaff, a distance of 150 miles, at a rate of 55 miles per hour. How long does it take to make the trip?

Let $t$ = time to make the trip
55 = rate of travel
150 = distance traveled

We must solve the following equation.

$$150 = 55t$$

$$\frac{150}{55} = t$$

$$\frac{30}{11} = t$$

It requires $\frac{30}{11}$ hr or $2\frac{8}{11}$ hr to make the trip.

EXAMPLE 3 A hiker crossed a canyon by walking 4 mph the first 2 hours and 3 mph the second 3 hours. What was the total distance that she hiked?

A sketch like the one in Figure 2.4 is helpful.

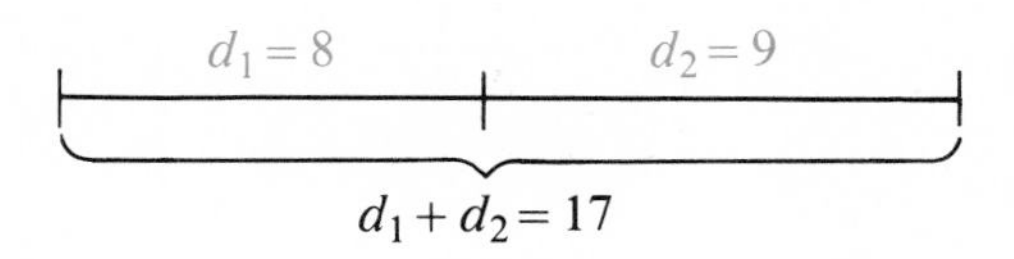


**Figure 2.4**

Let $d_1$ = the distance hiked the first 2 hours
Then $d_1 = (4) \cdot (2) = 8$ miles

Let $d_2$ = the distance hiked the second 3 hours
Then $d_2 = (3) \cdot (3) = 9$ miles

The total distance hiked is $d_1 + d_2 = 8 + 9 = 17$ miles.

EXAMPLE 4 Two cars leave the same point traveling in opposite directions. The second car travels 10 mph faster than the first and after 3 hours they are 300 miles apart. How fast is each car traveling?

Make a sketch like the one in Figure 2.5.

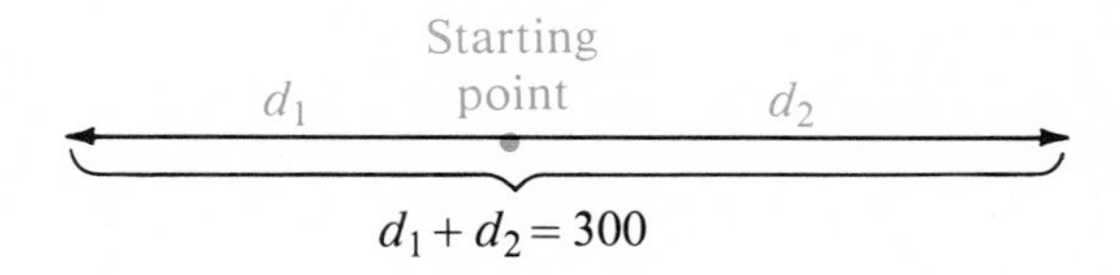


**Figure 2.5**

Let $d_1$ = the distance the first car travels
$r$ = rate of the first car
Then $d_1 = 3 \cdot r$

Let $d_2$ = the distance the second car travels
$r + 10$ = rate of the second car (Why?)
Then $d_2 = 3(r + 10)$

The equation we must solve is

$$d_1 + d_2 = 300 \quad \text{or} \quad \begin{aligned} 3r + 3(r + 10) &= 300 \\ 3r + 3r + 30 &= 300 \\ 6r &= 270 \\ r &= 45 \\ r + 10 &= 55 \end{aligned}$$

The first car is traveling 45 mph and the second 55 mph.

## EXERCISES 2.8

*Solve. (Some problems have been started.)*

**1.** A woman walks for 5 hours at a rate of 4 mph. How far does she walk?

Let $d =$ the distance the woman walks

**2.** A man runs a distance of 12 miles at a rate of 8 miles per hour. How many minutes does he run?
[*Hint:* Notice the units used.]

Let $t =$ the time he runs in minutes

**3.** An automobile traveled from Flagstaff to Las Vegas, a distance of 275 miles. If the time of travel was 5.5 hours, what was the average rate of speed?

Let $r =$ the average rate of speed

**4.** A man taking a trip travels by car for 4 hours at 50 mph and then by boat for 3 hours at 25 mph. What is the total distance that he travels?

**5.** Two trains leave the same city, one traveling north and the other traveling south. If the first train is moving 15 mph faster than the second, and if after 3 hours they are 405 miles apart, how fast is each train traveling?

**6.** Two hikers head towards a mountain 25 miles away. The first travels 4 mph and the second travels 2.5 mph. How far apart will they be after 6 hours?
[*Hint:* See the figure.]

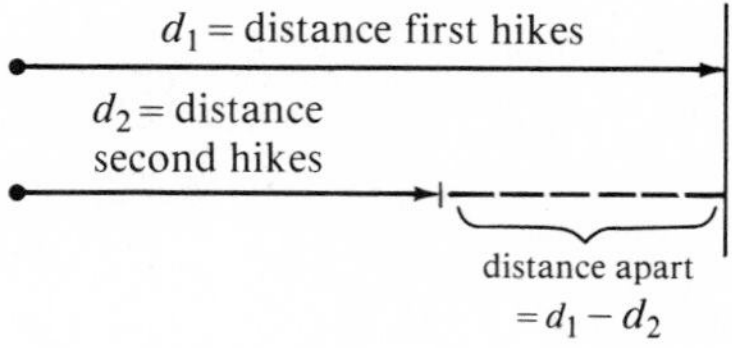

**7.** A woman hikes to the bottom of the Grand Canyon at a rate of 4.5 mph and returns to the top by mule traveling at a rate of 2.5 mph. If the total time of the trip is 7 hours, how long did she walk? What was the total length of the trip?

**8.** Two families leave their homes at 9:30 am, planning to meet for a picnic at a point between them. If one travels at a rate of 55 mph, the other travels at 50 mph and they live 315 miles apart, at what time do they meet?

**9.** The sum of two consecutive even integers is 114. Find the integers.

**10.** Find the base of a triangle with altitude 14 cm and area 77 $cm^2$.

**11.** 5% of what is 16.4?

**12.** 12% of 840 is what?

**13.** After an 8% raise, a man's new salary is $14,472. What was his old salary?

**14.** What amount of money invested at $7\frac{1}{2}\%$ simple interest will increase to \$1612.50 in one year?

---

ANSWERS: **1.** 20 mi **2.** 90 min **3.** 50 mph **4.** 275 mi **5.** 60 mph, 75 mph **6.** 9 mi **7.** 2.5 hr, 22.5 mi **8.** 12:30 pm **9.** 56, 58 **10.** 11 cm **11.** 328 **12.** 100.8 **13.** \$13,400 **14.** \$1500

## 2.9 Solving Literal Equations

Thus far the equations we have solved have contained only one variable. A **literal equation (letter equation)** is an equation which involves two or more variables. Many of the formulas familiar to us, such as

$$A = l \cdot w, \qquad P = 2 \cdot l + 2 \cdot w, \qquad d = r \cdot t, \qquad A = \pi r^2,$$

are literal equations. At times it is desirable to solve a literal equation for a particular variable. For example, the equation

$$d = r \cdot t$$

might be solved for $t$

$$t = \frac{d}{r}$$

or for $r$

$$r = \frac{d}{t}.$$

Remember that the letters in a literal equation, that is, the variables, simply represent numbers. Any of the rules for solving equations involving numerical values are also applicable when solving literal equations, and it is often helpful to solve first a numerical equation which is similar in form to the given literal equation. We illustrate this technique in the examples.

**EXAMPLE 1** Solve $T = C \cdot N$ for $N$.

| | | *Similar numerical example* | |
|---|---|---|---|
| $T = C \cdot N$ | | $5 = 3 \cdot N$ | |
| $\frac{1}{C} \cdot T = \frac{1}{C} \cdot C \cdot N$ | Multiply by $\frac{1}{C}$ | $\frac{1}{3} \cdot 5 = \frac{1}{3} \cdot 3 \cdot N$ | Multiply by $\frac{1}{3}$ |
| $\frac{T}{C} = N$ | | $\frac{5}{3} = N$ | |

Keep in mind that solving a literal equation (just as with any other equation) depends on our ability to isolate the variable on one side of the equation.

**EXAMPLE 2** Solve $F = v + w$ for $v$.

*Similar numerical example*

| | | | |
|---|---|---|---|
| $F = v + w$ | | $5 = v + 7$ | |
| $F - w = v + w - w$ | Subtract $w$ | $5 - 7 = v + 7 - 7$ | Subtract 7 |
| $F - w = v$ | | $-2 = v$ | |

**EXAMPLE 3** Solve $P = 2 \cdot l + 2 \cdot w$ for $w$.

*Similar numerical example*

| | |
|---|---|
| $P = 2 \cdot l + 2 \cdot w$ | $12 = 2 \cdot 3 + 2 \cdot w$ |
| $P - 2 \cdot l = 2 \cdot l - 2 \cdot l + 2 \cdot w$ | $12 - 2 \cdot 3 = 2 \cdot 3 - 2 \cdot 3 + 2 \cdot w$ |
| $P - 2 \cdot l = 2 \cdot w$ | $6 = 2 \cdot w$ |
| $\frac{1}{2}(P - 2 \cdot l) = \frac{1}{2}(2 \cdot w)$ | $\frac{1}{2} \cdot 6 = \frac{1}{2} \cdot 2 \cdot w$ |
| $\frac{P - 2l}{2} = w$ | $3 = w$ |

## EXERCISES 2.9

*Solve.*

1. $a = b + c$ for $c$
2. $a = b + c$ for $b$
3. $a = b \cdot c$ for $b$
4. $a = b \cdot c$ for $c$
5. $a = b \cdot c + d$ for $d$
6. $a = b \cdot c + d$ for $b$
7. $u = 2v + 2w$ for $v$
8. $A = \frac{1}{2}(b_1 + b_2)h$ for $h$
9. $P = 20a + b$ for $a$
10. $C = \frac{5}{9}(F - 32)$ for $F$
11. $V = l \cdot w \cdot h$ for $h$
12. $P = a + b + c$ for $c$

**13.** How far does a train travel from noon until 4:30 P.M. if it is traveling at the uniform rate of 70 mph?

**14.** Two distance runners, Bob and Murph, run the same course with Bob running 1.5 mph slower than Murph. If Murph finishes in 4 hours and Bob takes 5 hours to finish, what is the rate of each? How far do they run?

**15.** The sum of the measures of the angles of a trapezoid is 360°. If the second is 5° more than the first, the third is five times the first, and the fourth is 25° more than four times the first, what is the measure of each angle?

---

ANSWERS: **1.** $c = a - b$ **2.** $b = a - c$ **3.** $b = \frac{a}{c}$ **4.** $c = \frac{a}{b}$ **5.** $d = a - bc$ **6.** $b = \frac{a - d}{c}$
**7.** $v = \frac{u - 2w}{2}$ **8.** $h = \frac{2A}{b_1 + b_2}$ **9.** $a = \frac{P - b}{20}$ **10.** $F = \frac{9}{5}C + 32$ **11.** $h = \frac{V}{l \cdot w}$ **12.** $c = P - a - b$
**13.** 315 mi **14.** Murph: 7.5 mph, Bob: 6 mph, 30 mi **15.** 30°, 35°, 150°, 145°

## 2.10 Solving Inequalities

Statements such as

$$x + 7 < 2, \qquad 3x \geq 9, \qquad 3x - 1 > 11, \qquad 2x + 1 \leq 5 - 7x,$$

are called **inequalities.** A **solution** to an inequality is a number which, when substituted for the variable, makes the inequality true. If the inequality symbols ($<$, $\geq$, $>$, and $\leq$) in the above statements were replaced with the equality symbol ($=$), we would know how to solve the resulting equations. Fortunately, solving an inequality is much like solving an equation since most of the same basic rules apply. There is one exception which will be discussed shortly. If we start with the true inequality

$$5 < 8$$

and add 4 to both sides:

$$5 + 4 < 8 + 4$$
$$9 < 12$$

we obtain another true inequality. Similarly, if we subtract 11 from both sides:

$$5 - 11 < 8 - 11$$
$$-6 < -3$$

we again obtain a true inequality. These observations lead to the following rule.

**ADDITION-SUBTRACTION RULE**

If the same number or expression is added to or subtracted from both sides of an inequality, an **equivalent inequality** (an inequality with exactly the same solutions) is obtained.

This rule is used to solve inequalities in precisely the same way that the corresponding addition-subtraction rule is used to solve equations. As always, the key is to isolate the variable on one side of the inequality.

**EXAMPLE 1** Solve $x + 7 < 2$.

$$x + 7 - 7 < 2 - 7 \qquad \text{Subtract 7 from both sides}$$
$$x < -5$$

The solutions are *all* numbers less than $-5$.

Most of the equations we have solved thus far have had only one or two solutions (identities excepted). Inequalities generally have many solutions.

**EXAMPLE 2** Solve.

$$x - 5 \geq 2$$
$$x - 5 + 5 \geq 2 + 5 \qquad \text{Add 5 to both sides}$$
$$x \geq 7$$

The solutions are *all* numbers greater than or equal to 7. Generally, we will not make this statement but simply indicate the solution by writing $x \geq 7$.

If we start again with the true inequality

$$5 < 8$$

and multiply both sides by 2 we obtain the true inequality

$$10 < 16.$$

However, if we multiply both sides by $-2$ we obtain the false inequality

$$-10 < -16 \qquad -16 \text{ is really less than } -10$$

In order to obtain a true inequality when we multiply by the negative number $-2$, we must reverse the sense of the inequality, that is, change from $<$ to $>$ or $>$ to $<$. These observations form the basis of the next rule.

**MULTIPLICATION-DIVISION RULE**

1. If both sides of an inequality are multiplied or divided by the same *positive* number, the resulting equation is equivalent to the original.
2. If both sides of an inequality are multiplied or divided by the same *negative* number and the sense of the inequality is *reversed,* the resulting inequality is equivalent to the original.

The only substantial difference between solving an equation and solving an inequality concerns multiplying or dividing both sides by a negative number.

**Always reverse the sense of the inequality when multiplying or dividing by a negative number.**

**EXAMPLE 3** Solve.

$$3x \geq 9$$

$$\frac{1}{3}(3x) \geq \frac{1}{3} \cdot 9 \qquad \frac{1}{3} \text{ is positive so inequality remains the same}$$

$$x \geq 3$$

The solution is $x \geq 3$.

**EXAMPLE 4** Solve.

$$-\frac{1}{2} \cdot x < 5$$

$$(-2)\left(-\frac{1}{2}x\right) > (-2)(5) \qquad -2 \text{ is negative so inequality is reversed}$$

$$x > -10$$

As with solving equations, many times we must use a combination of the addition-subtraction and multiplication-division rules.

**EXAMPLE 5** Solve.

$$3x - 1 > 11$$
$$3x - 1 + 1 > 11 + 1 \qquad \text{Add 1}$$
$$3x > 12$$
$$\frac{1}{3} \cdot 3x > \frac{1}{3} \cdot 12 \qquad \text{Multiply by } \textit{positive } \frac{1}{3}$$
$$x > 4$$

**EXAMPLE 6** Solve.

$$5 - 4x > 2 - 3x$$
$$5 - 5 - 4x > 2 - 5 - 3x \qquad \text{Subtract 5}$$
$$-4x > -3 - 3x$$
$$-4x + 3x > -3 - 3x + 3x \qquad \text{Add } 3x$$
$$-x > -3$$
$$(-1)(-x) < (-1)(-3) \qquad \text{Multiply by } -1 \text{ and } \textit{reverse} \text{ inequality}$$
$$x < 3$$

When an inequality involves parentheses, clear all parentheses, collect like terms (if such exist) on each side, and proceed as in previous cases.

**EXAMPLE 7** Solve.

$$x - 3(2 + x) > 2(3x - 1)$$
$$x - 6 - 3x > 6x - 2 \qquad \text{Clear parentheses}$$
$$-2x - 6 > 6x - 2 \qquad \text{Collect like terms}$$
$$-2x - 6 + 6 > 6x - 2 + 6 \qquad \text{Add 6}$$
$$-2x > 6x + 4$$
$$-2x - 6x > 6x - 6x + 4 \qquad \text{Subtract } 6x$$
$$-8x > 4$$
$$\left(-\frac{1}{8}\right)(-8x) < \left(-\frac{1}{8}\right)(4) \qquad \text{Multiply by } -\frac{1}{8} \text{ and reverse the inequality}$$
$$x < -\frac{1}{2}$$

## EXERCISES 2.10

*Are the following statements true or false?*

1. If $x > 9$ then $-3x > -27$.
2. If $x < 4$ then $3x < 12$.
3. If $x \geq 8$ then $x - 1 \geq 7$.
4. If $x < -2$ then $-x > 2$.
5. If $x < 3$ then $x + 5 < 8$.
6. If $x > -1$ then $3 - 2x < 5$.

*Solve.*

7. $x + 3 < 7$
8. $7 < z - 3$
9. $2 - z \leq 9$
10. $x + 2.1 \geq -3.4$
11. $y - \frac{1}{2} < \frac{3}{4}$
12. $2x > 10$
13. $-2x > 10$
14. $\frac{1}{3}y \leq -5$
15. $-\frac{1}{4}z \geq 3$
16. $4 < -3x$
17. $3y - 1 < 8$
18. $1 - 3z < 8$
19. $3 \geq 5x - 12$
20. $5 < 8 - y$
21. $\frac{2}{9} + \frac{1}{9}z < \frac{1}{3}$
22. $2x + 3 > 5x - 3$
23. $-8y - 21 \geq 7 - 15y$

**24.** $5(3 - z) - 10 \le 25$

**25.** $3(2x + 8) \le 4(x - 3)$

**26.** $5(y + 3) + 1 \le y - 4$

**27.** $-4z - 3 + z < 2(z + 12)$

**28.** $U = 5p + d$ for $p$

**29.** $U = 5p + d$ for $d$

**30.** $A = \frac{1}{2}(b_1 + b_2)h$ for $b_1$

**31.** $I = prt$ for $t$

**32.** A woman received a 6% raise, which amounted to $660. What was her old salary? her new salary?

---

ANSWERS: **1.** f **2.** t **3.** t **4.** t **5.** t **6.** t **7.** $x < 4$ **8.** $z > 10$ (or $10 < z$) **9.** $z \ge -7$ **10.** $x \ge -5.5$ **11.** $y < \frac{5}{4}$ **12.** $x > 5$ **13.** $x < -5$ **14.** $y \le -15$ **15.** $z \le -12$ **16.** $x < -\frac{4}{3}$ $\left(\text{or } -\frac{4}{3} > x\right)$ **17.** $y < 3$ **18.** $z > -\frac{7}{3}$ **19.** $x \le 3$ **20.** $y < 3$ **21.** $z < 1$ **22.** $x < 2$ **23.** $y \ge 4$ **24.** $z \ge -4$ **25.** $x \le -18$ **26.** $y \le -5$ **27.** $z > -\frac{27}{5}$ **28.** $p = \frac{U - d}{5}$ **29.** $d = U - 5p$ **30.** $b_1 = \frac{2A - b_2h}{h}$ **31.** $t = \frac{I}{pr}$ **32.** old: $11,000; new: $11,660

## 2.11 Graphing Solutions to Equations and Inequalities

Many times in mathematics when we are working with abstract ideas (such as numbers and equations), it is helpful to picture these ideas. We did this to some degree in the first chapter with a number line. Recall that a **number line** is a line that has been marked off in unit lengths (any unit of length will do) with each point on the line associated with a real number, and conversely. Figure 2.6 is a number line with the **origin** (the point associated with the number zero) identified.

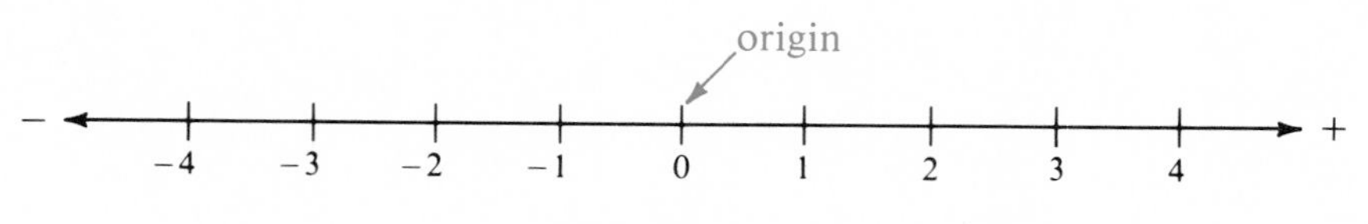


**Figure 2.6**

The positive numbers correspond to points to the right of the origin, while the negative numbers correspond to points to the left of the origin. We think of this line as extending infinitely far in both directions and indicate this by arrows. Thus, no matter what number we consider, we can always associate it with a point on this line and given any point on the line, a real number can be identified with it.

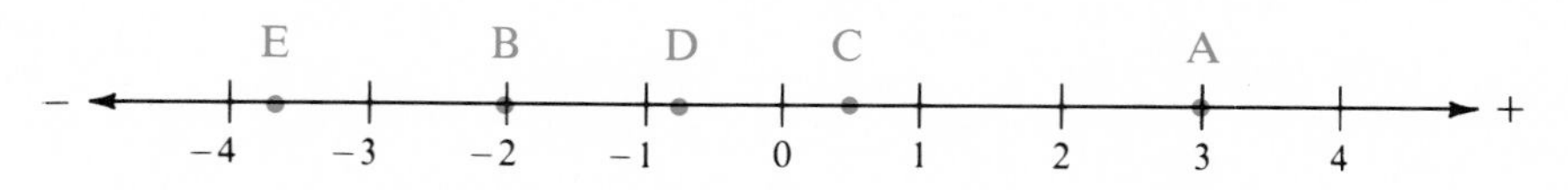


**Figure 2.7**

EXAMPLE 1 On the number line in Figure 2.7, $A$ corresponds to the number 3, $B$ to $-2$, $C$ to $\frac{1}{2}$ ($C$ is half way between 0 and 1), $D$ to $-\frac{3}{4}$, and $E$ to $-\frac{11}{3}$. To find the point associated with the number 12, we extend the line to the right and continue marking unit lengths until the desired position is reached.

Up to this point we have concentrated our efforts on solving equations and inequalities. Now we consider *graphing* equations and inequalities. To graph an equation we *plot* the point or points on a number line which correspond to solutions to the equation. Thus, graphing an equation in one variable requires only a single number line.

EXAMPLE 2 Graph $3x - 1 = 8$.

Solving, we obtain $3x = 9$

$x = 3$ The solution is $x = 3$.

Plotting the point corresponding to 3, we obtain the graph in Figure 2.8.

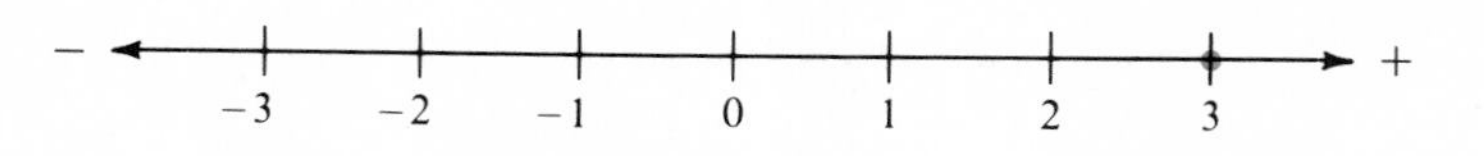


**Figure 2.8**

**EXAMPLE 3** Graph $(x + 1) - (2x - 1) = 0$.

$$
\begin{aligned}
x + 1 - 2x + 1 &= 0 && \text{Change signs of terms in parentheses} \\
-x + 2 &= 0 \\
-x &= -2 \\
x &= 2 && \text{Multiply by } -1
\end{aligned}
$$

The solution is 2, and the graph is shown in Figure 2.9.

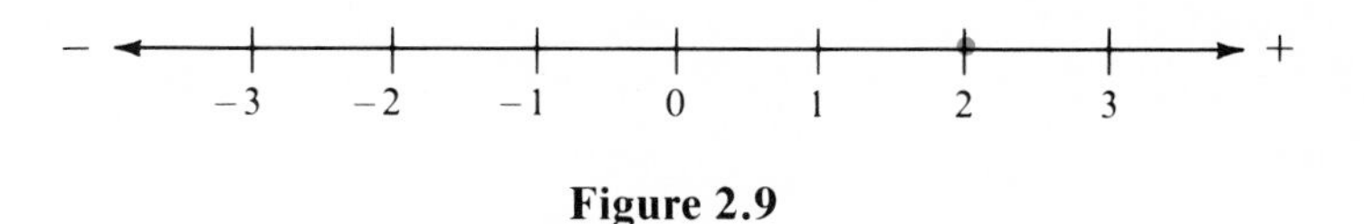


**Figure 2.9**

**EXAMPLE 4** Graph $x + 1 = x - 1$.

If we attempt to solve as in Examples 2 and 3, we obtain the following equation.

$$
\begin{aligned}
x - x + 1 &= x - x - 1 && \text{Subtract } x \\
1 &= -1 && \text{A contradiction}
\end{aligned}
$$

There are no solutions and the graph is the plain number line shown in Figure 2.10.

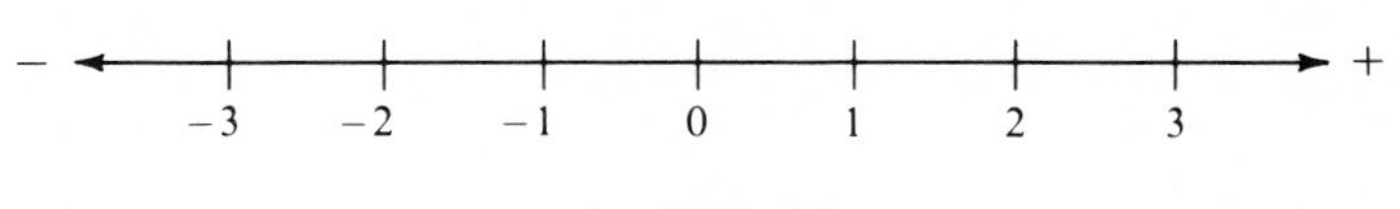


**Figure 2.10**

**EXAMPLE 5** Graph $3x + 1 = 1 + 3x$.

$$
\begin{aligned}
3x - 3x + 1 &= 1 + 3x - 3x && \text{Subtract } 3x \\
1 &= 1 && \text{An identity}
\end{aligned}
$$

Every real number is a solution and the graph is shown in Figure 2.11.

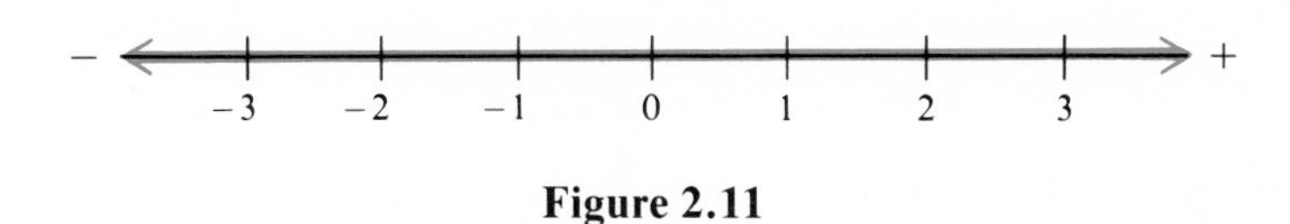


**Figure 2.11**

Obviously, graphing (solutions to) equations on a number line is a relatively simple point-identification procedure. The problem of graphing inequalities and combinations of inequalities is more challenging and interesting. The procedures used are illustrated in the following figures.

The graph of $x > 2$ is shown in Figure 2.12 on p. 114. The open circle at 2 indicates that the number 2 is not included among the solutions while the colored line shows that all points to the right of 2 are included.

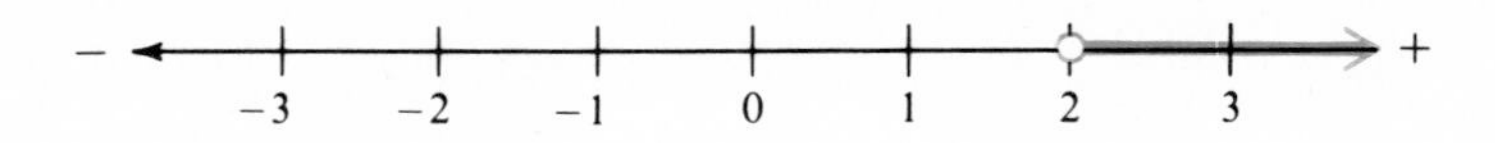


**Figure 2.12**

The graph of $x < 0$ is shown in Figure 2.13.

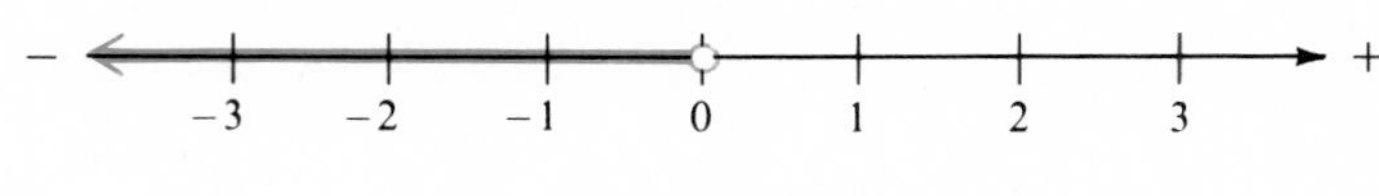


**Figure 2.13**

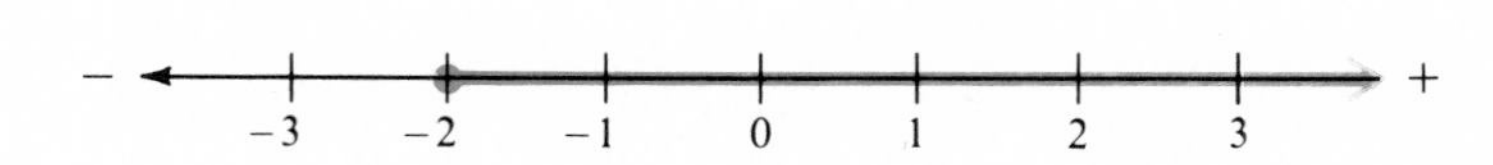


**Figure 2.14**

In the graph of $x \geq -2$ in Figure 2.14, the circle at $-2$ is solid to indicate that the number $-2$ is included in the graph.

Finally, the graph of $x \leq 0$ is given in Figure 2.15.

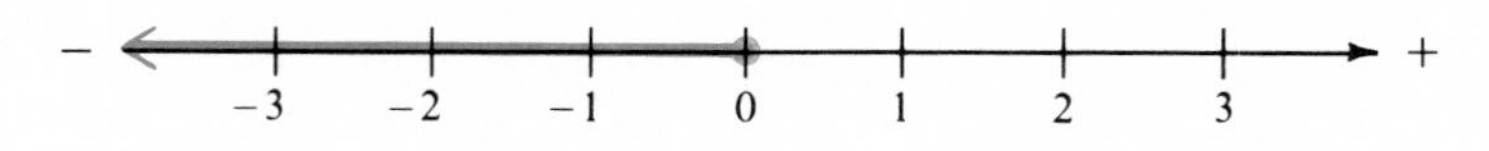


**Figure 2.15**

Most simple inequalities we have solved thus far have a solution, hence a graph, similar to one of the above.

EXAMPLE 6 Graph.

$$
\begin{aligned}
2(x-1) &< 3x-5. & \\
2x-2 &< 3x-5. & \text{Clear parentheses} \\
2x-2+2 &< 3x-5+2 & \text{Add 2} \\
2x &< 3x-3 & \\
2x-3x &< 3x-3x-3 & \text{Subtract } 3x \\
-x &< -3 & \\
x &> 3 & \text{Multiply by} -1 \text{ and reverse inequality}
\end{aligned}
$$

The solution is $x > 3$ and the graph is shown in Figure 2.16.

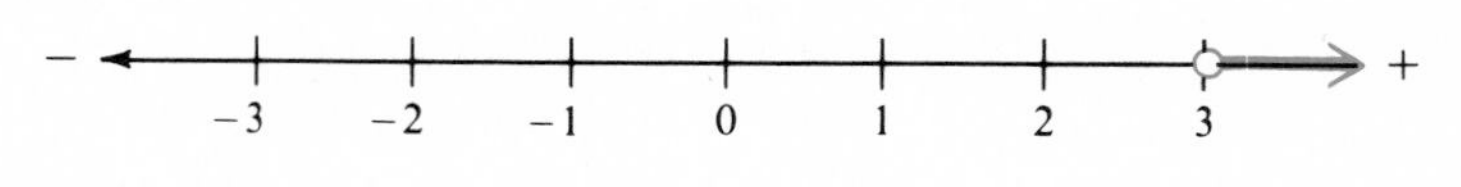


**Figure 2.16**

In mathematics we are often confronted with problems that result in combinations of two simple inequalities. The two most important ways to combine two inequalities is by the words *and* and *or*. For example, the combination

$$x > -2 \quad \textit{and} \quad x < 3$$

could be given by the two inequalities

$$-2 < x \qquad \text{and} \qquad x < 3$$

or by the chain of inequalities

$$-2 < x < 3.$$

The numbers that satisfy this chain of inequalities are solutions to both inequalities: those numbers which are *both* greater than $-2$ *and* less than 3. The graph is shown in Figure 2.17.

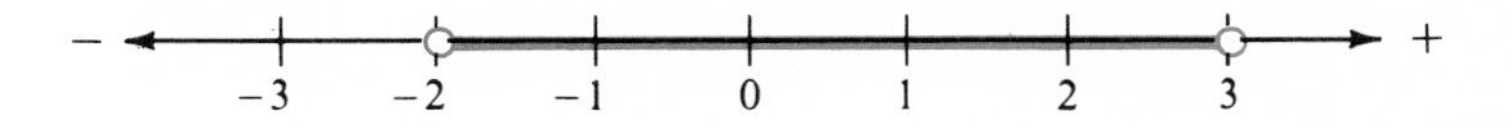


**Figure 2.17**

Similarly, the graph of $-1 \leq x \leq 2$ ($-1 \leq x$ and $x \leq 2$) is shown in Figure 2.18,

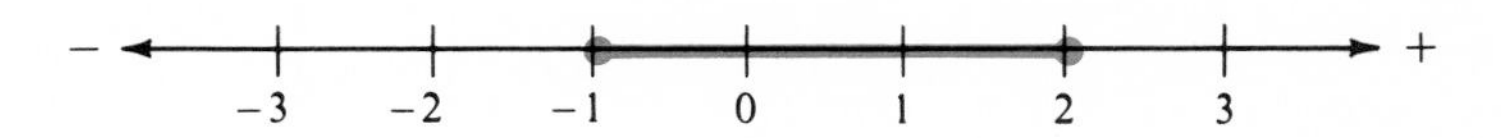


**Figure 2.18**

the graph of $-3 \leq x < 0$ ($-3 \leq x$ and $x < 0$) in Figure 2.19,

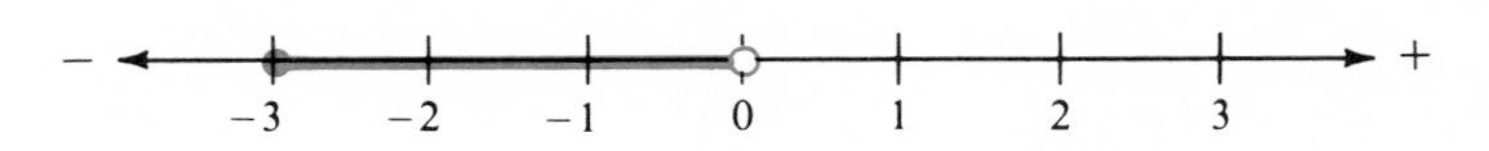


**Figure 2.19**

and the graph of $1 < x \leq 2$ ($1 < x$ and $x \leq 2$) in Figure 2.20.

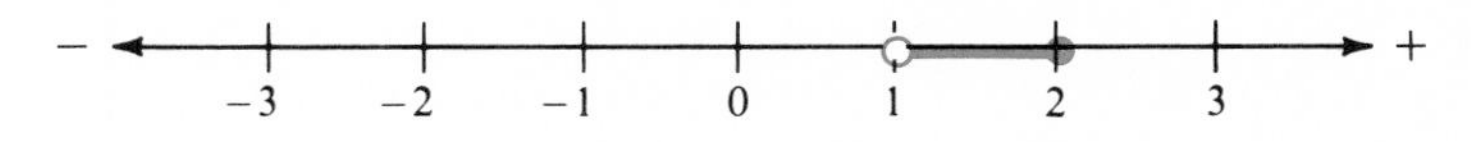


**Figure 2.20**

The combination

$$2 < x \quad \text{and} \quad x < -1$$

has no solution since there are no numbers which are *both* less than $-1$ *and* greater than 2. If we attempt to merge the two inequalities into a single chain as before, we obtain

$$2 < x < -1$$

which indicates that 2 is less than $-1$, a contradiction. Such chains should always be avoided.

A chain of inequalities may be more complex. For example,

$$-1 < 2x + 1 < 3$$

is in reality the combination

$$-1 < 2x + 1 \quad \textit{and} \quad 2x + 1 < 3.$$

Note that $-1$ is indeed less than 3 so that the chain makes sense.

EXAMPLE 7 Solve and graph $-1 < 2x + 1 < 3$.

We must solve two inequalities.

| | | | |
|---|---|---|---|
| $-1 < 2x + 1$ | and | $2x + 1 < 3$ | |
| $-1 - 1 < 2x + 1 - 1$ | and | $2x + 1 - 1 < 3 - 1$ | Subtract 1 |
| $-2 < 2x$ | and | $2x < 2$ | |
| $-1 < x$ | and | $x < 1$ | Divide by 2 |

Rewritten as a single chain, the solution is

$$-1 < x < 1$$

and the graph is shown in Figure 2.21.

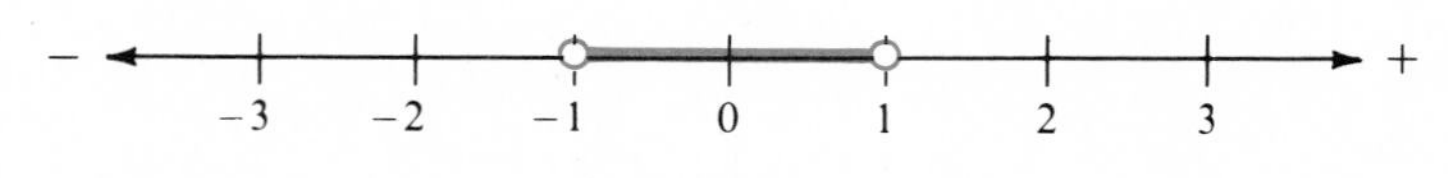


**Figure 2.21**

Alternatively and more compactly, we can solve the original chain by the following method.

$$-1 < 2x + 1 < 3$$

$$-1 - 1 < 2x + 1 - 1 < 3 - 1 \qquad \text{Subtract 1 throughout}$$

$$-2 < 2x < 2$$

$$\frac{1}{2}(-2) < \frac{1}{2}(2x) < \frac{1}{2}(2) \qquad \text{Multiply throughout by } \frac{1}{2}$$

$$-1 < x < 1$$

The second important combination of inequalities involves the connective *or*, for example,

$$x < -1 \quad or \quad x > 2.$$

Do *not* make the mistake of trying to form a single chain when the word *or* is used. Notice that $2 < x < -1$ is a senseless expression (2 is not less than $-1$). The graph of an *or* combination is usually two segments of a number line, such as the graph of $x < -1$ or $x > 2$ shown in Figure 2.22.

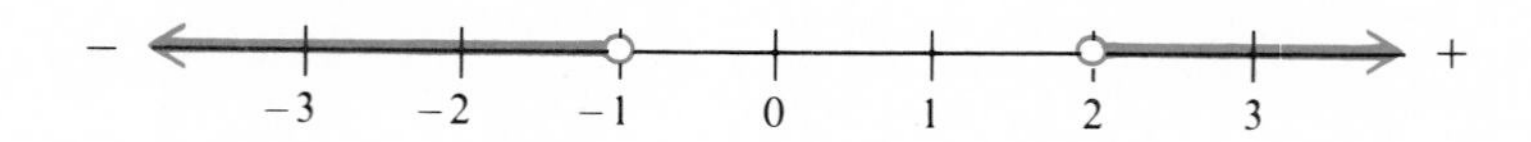


**Figure 2.22**

Similarly, $x < -2$ or $x \geq 1$ is graphed in Figure 2.23.

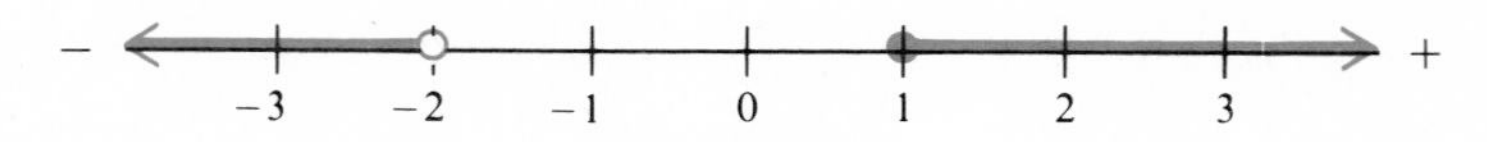


**Figure 2.23**

Also, the graph of $x \leq 0$ or $x > 3$ is shown in Figure 2.24,

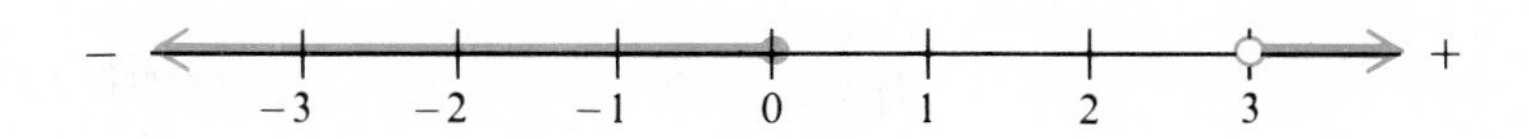


**Figure 2.24**

and the graph of $x \leq -1$ or $x \geq 0$ is in Figure 2.25.

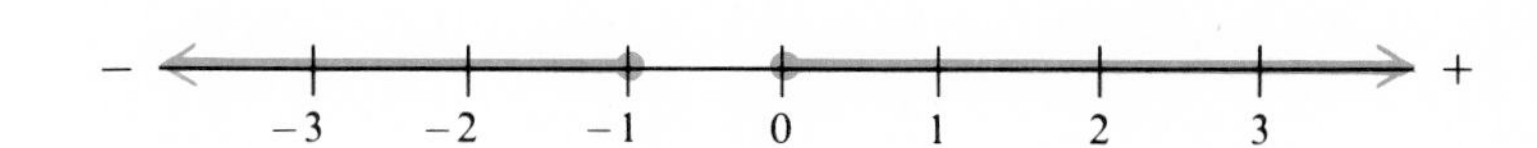


**Figure 2.25**

Finally, the graph of $x \leq 2$ or $x \geq -1$ in Figure 2.26 is the entire number line.

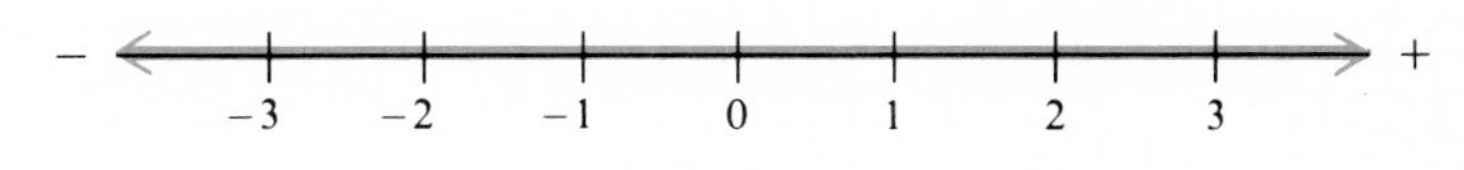


**Figure 2.26**

As with *and* combinations, *or* combinations may be more complex.

EXAMPLE 8 Solve and graph the following combination.

$$\begin{array}{rcrl} 2x + 1 \leq -1 & \text{or} & 2x + 1 > 3 & \\ 2x + 1 - 1 \leq -1 - 1 & \text{or} & 2x + 1 - 1 > 3 - 1 & \text{Subtract 1} \\ 2x \leq -2 & \text{or} & 2x > 2 & \\ x \leq -1 & \text{or} & x > 1 & \text{Divide by 2} \end{array}$$

The solution is

$$x \leq -1 \quad \text{or} \quad x > 1$$

and the graph is shown in Figure 2.27.

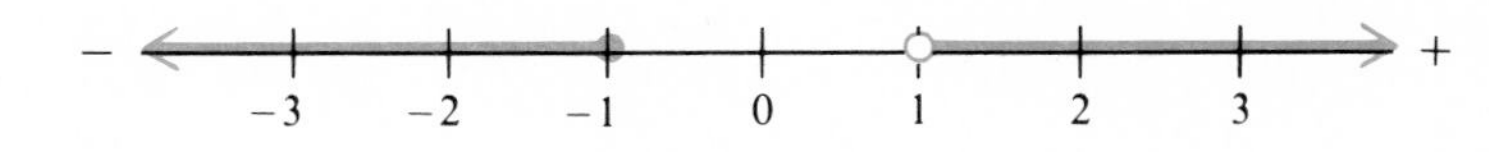


**Figure 2.27**

EXAMPLE 9 Solve and graph.

$$\begin{array}{rcrl} 3 - 4x < -1 & \text{or} & 3 - 4x \geq 9 & \\ -4x < -4 & \text{or} & -4x \geq 6 & \\ x > 1 & \text{or} & x \leq -\dfrac{6}{4} & \text{Reverse} \\ & & x \leq -\dfrac{3}{2} & \end{array}$$

The solution is

$$x > 1 \quad \text{or} \quad x \le -\frac{3}{2}$$

and the graph is given in Figure 2.28.

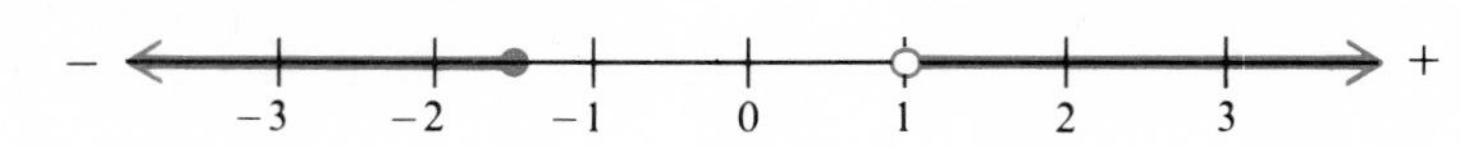


**Figure 2.28**

Remember that an *and* combination (such as $-4 < x$ and $x \le 3$) can be expressed as a single chain of inequalities ($-4 < x \le 3$) and generally has as its graph a single segment of a number line. An *or* combination (such as $x < -1$ or $x > 2$) *cannot* be expressed as a single chain of inequalities (*never write* $2 < x < -1$) and generally has as its graph two separate segments of a number line.

## 2.11 EXERCISES

*Graph the equation on a number line.*

**1.** $x + 3 = 1$

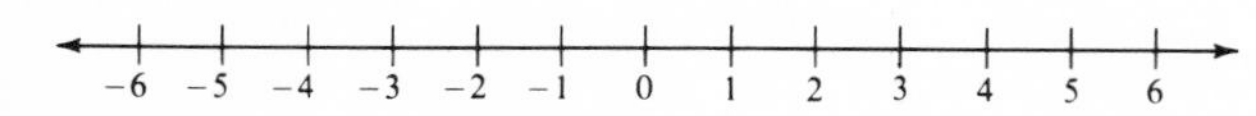

**2.** $3x - 1 = 5$

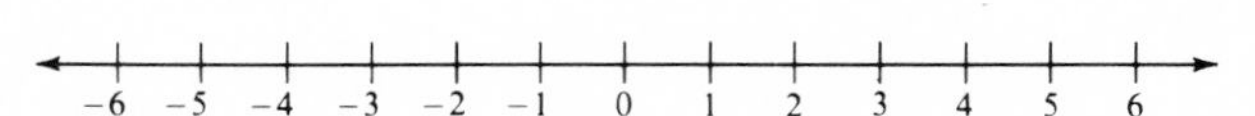

**3.** $2x + 5 = 5 + 2x$

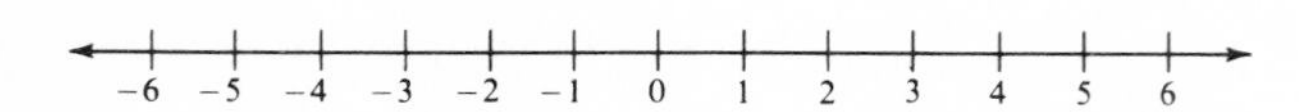

**4.** $1 - x = x - 1$

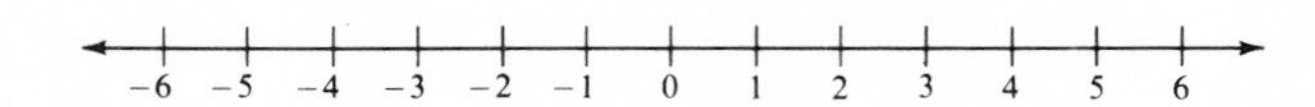

**5.** $3 - x = 2 - x$

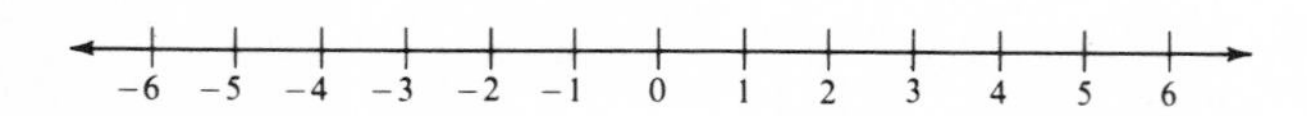

**6.** $2x + 1 = 5x + 1$

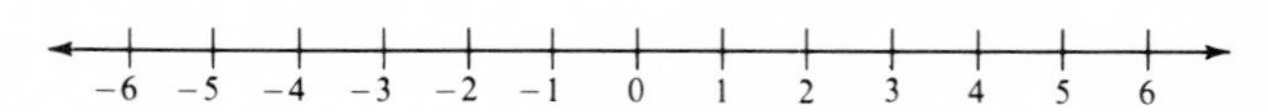

**7.** $(2x - 1) - (3x + 2) = 0$

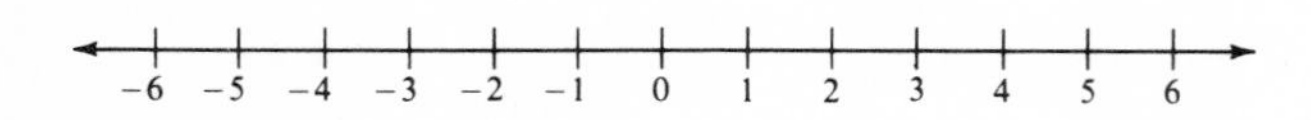

**8.** $|x + 2| = 0$

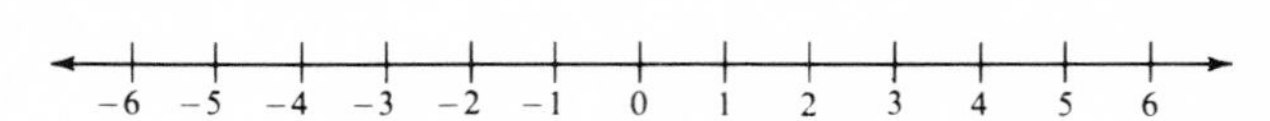

**9.** $|2x - 1| = 5$

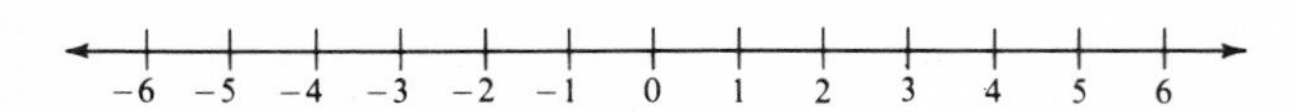

**10.** $|1 - 3x| = -2$

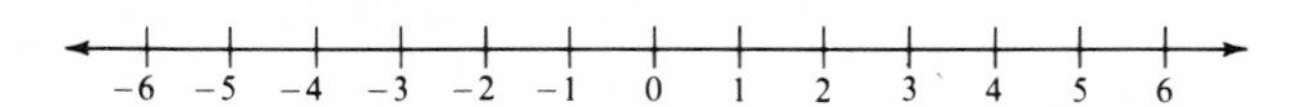

*Graph the inequality on a number line.*

**11.** $x > 4$

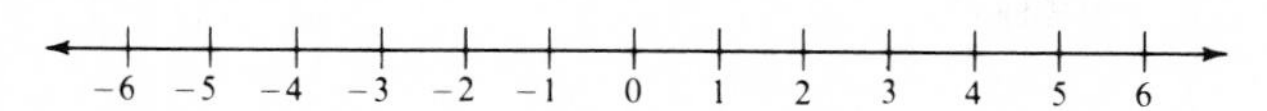


**12.** $x \leq -2$

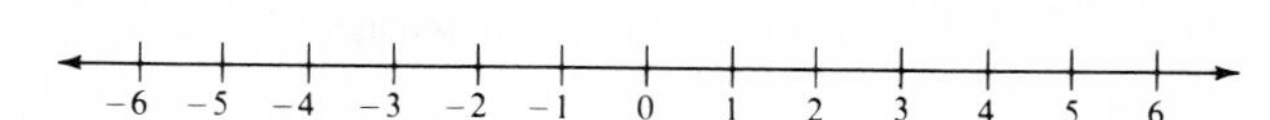


**13.** $2x - 5 < -1$

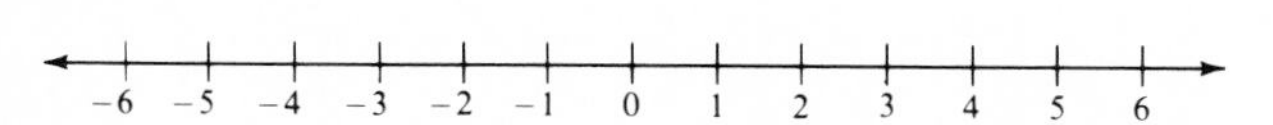


**14.** $3 - 5x \geq -7$

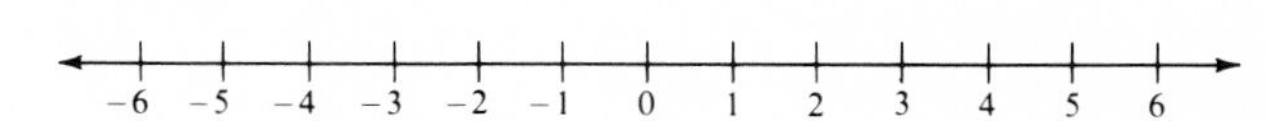


**15.** $3(x + 7) \leq 2x + 18$

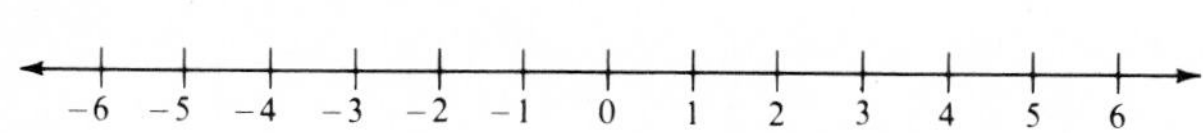


**16.** $x > -3$ and $x < 0$

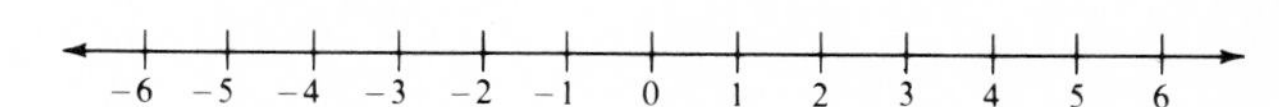


**17.** $-3 < x < 0$

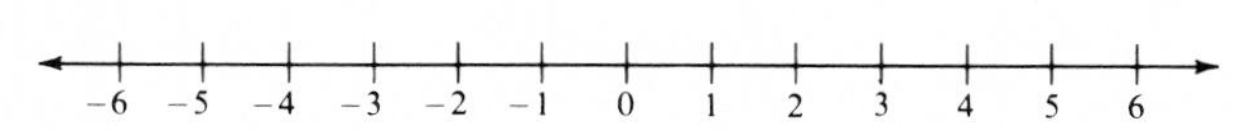


**18.** $x < -3$ or $x > 0$

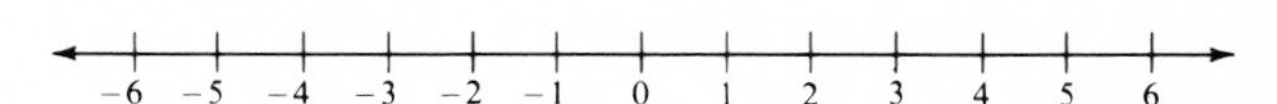


**19.** $0 \leq x \leq 4$

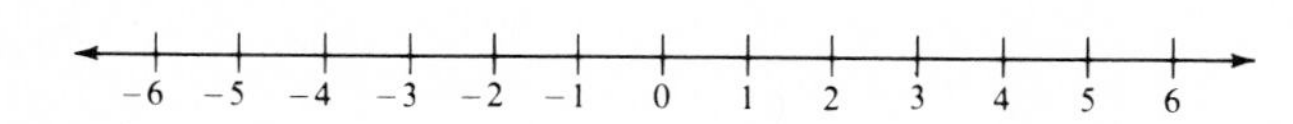


**20.** $x < 2$ or $x \geq 4$

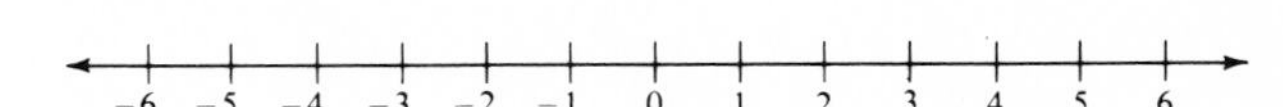


**21.** $-5 \leq 4x + 3 < 5$

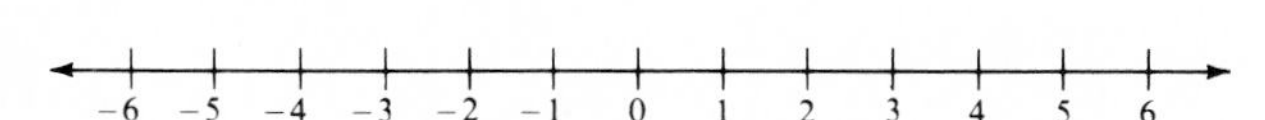


**22.** $4x + 3 < -5$ or $4x + 3 \geq 5$

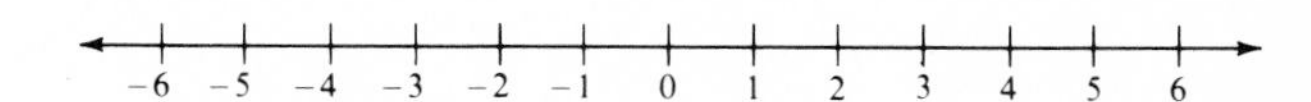


**23.** $0 > 3 - x$ or $3 - x \geq 1$

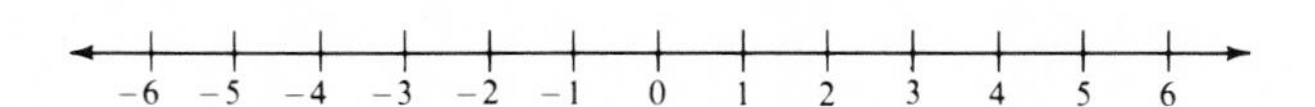


**24.** $0 \leq 3 - x < 1$

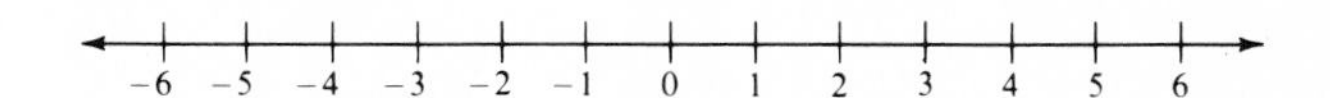


**25.** Two cars leave Phoenix headed in the same direction towards Los Angeles. If one travels 55 mph and the other travels 45 mph, after how many hours will the faster traveler be exactly 40 miles ahead?

ANSWERS:

1. −2 −1 0 1 2
2. −2 −1 0 1 2
3. −2 −1 0 1 2
4. −2 −1 0 1 2
5. −2 −1 0 1 2
6. −2 −1 0 1 2
7. −3 −2 −1 0 1 2
8. −2 −1 0 1 2
9. −3 −2 −1 0 1 2 3
10. −2 −1 0 1 2
11. 0 1 2 3 4
12. −2 −1 0 1 2 3
13. −2 −1 0 1 2
14. −2 −1 0 1 2
15. −3 −2 −1 0 1 2
16. −3 −2 −1 0 1 2
17. −3 −2 −1 0 1 2 3
18. −3 −2 −1 0 1 2 3
19. −2 −1 0 1 2 3 4
20. −2 −1 0 1 2 3 4
21. −2 −1 0 1 2
22. −2 −1 0 1 2
23. 0 1 2 3
24. 0 1 2 3
25. 4 hr

## 2.12 Solving Inequalities Involving Absolute Value

In Section 2.4 we solved equations involving absolute value. Recall that the solutions to an equation such as

$$|x + 1| = 5$$

are the solutions to the two related equations

$$x + 1 = 5 \qquad \text{and} \qquad x + 1 = -5.$$

Solutions to an inequality involving absolute value are also solutions to two related inequalities. For example, to solve

$$|x| > 2$$

we solve

$$x > 2 \qquad \text{and also} \qquad x < -2.$$

A solution to either of these inequalities is a solution to $|x| > 2$. To see that this is the case, find the absolute values of several specific numbers which are greater than 2 (for example 2.3, 14/3, and 10) and find the absolute values of several specific numbers which are less than $-2$ (for example $-2.1, -7$, and $-43/3$). From the graph of $x > 2$ and the graph of $x < -2$ (on the same number line in Figure 2.29), it is clear that any number corresponding to a point in the colored segments has absolute value greater than 2. (Trying several numbers such as $-1$, 0, and 3/2 will show that they are *not* solutions.)

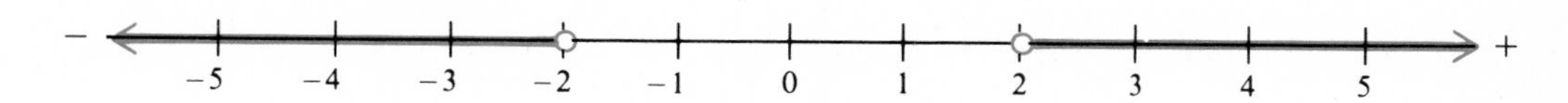


**Figure 2.29**

To solve

$$|x| < 2$$

we solve

$$x > -2 \quad and \quad x < 2$$

which is equivalent to the chain

$$-2 < x < 2.$$

Again, to see that this is reasonable, find the absolute values of several specific numbers between $-2$ and 2 (for example, $-3/2$, 0, and 1.2). From the graph of $-2 < x < 2$ in Figure 2.30 we see that any number corresponding to a point in the colored segment has absolute value less than 2. (Trying several numbers such as $-3, -4$, 3, and 5 will show that they are *not* solutions.)

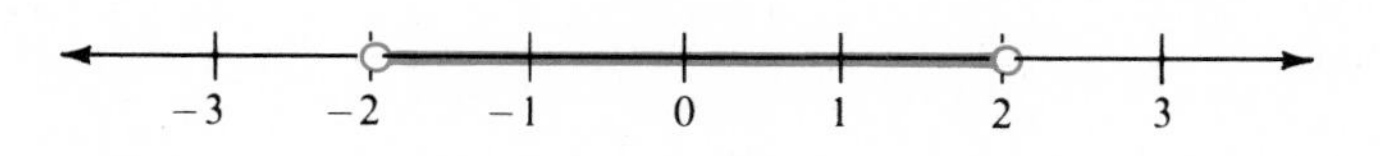


**Figure 2.30**

1. To solve an absolute value inequality of the form $|Expression| > a$ $(a > 0)$, solve the combination of inequalities

$$Expression < -a \quad \textbf{or} \quad Expression > a.$$

2. To solve an absolute value inequality of the form $|Expression| < a$ $(a > 0)$, solve the combination of inequalities

$$Expression < a \quad \textbf{and} \quad Expression > -a.$$

Equivalently, solve

$$-a < Expression < a$$

Similar results hold for $|Expression| \geq a$ and $|Expression| \leq a$.

**EXAMPLE 1** Solve $|x| > 3$.

The solution is

$$x < -3 \ or \ x > 3 \qquad \textit{Expression} \text{ is } x \text{ in this case}$$

**EXAMPLE 2** Solve $|x| \le 5$.

The solution is $x \le 5$ *and* $x \ge -5$ which is equivalent to

$$-5 \le x \le 5 \qquad \textit{Expression} \text{ is } x$$

**EXAMPLE 3** Solve $|x - 2| > 7$.

We must solve

$$\begin{array}{rcrl} x - 2 < -7 & \textit{or} & x - 2 > 7 & \textit{Expression} \text{ is } x - 2 \\ x - 2 + 2 < -7 + 2 & \text{or} & x - 2 + 2 > 7 + 2 & \\ x < -5 & \text{or} & x > 9 & \end{array}$$

The solution is $x < -5$ or $x > 9$. (*Do not write* $9 < x < -5$.)

**EXAMPLE 4** Solve $|x + 3| \le 4$.

We must solve $x + 3 \le 4$ *and* $x + 3 \ge -4$ which is equivalent to

$$\begin{array}{rl} -4 \le x + 3 \le 4 & \textit{Expression} \text{ is } x + 3 \\ -4 - 3 \le x + 3 - 3 \le 4 - 3 & \text{Subtract 3 throughout} \\ -7 \le x \le 1 & \end{array}$$

The solution is $-7 \le x \le 1$. (Equivalently, $x \le 1$ and $x \ge -7$)

**EXAMPLE 5** Solve $|1 - 3x| \ge 7$.

We must solve
$$\begin{array}{rcrl} 1 - 3x \le -7 & \textit{or} & 1 - 3x \ge 7 & \textit{Expression} \text{ is } 1 - 3x \\ -3x \le -8 & \text{or} & -3x \ge 6 & \text{Subtract 1} \\ x \ge \frac{8}{3} & \text{or} & x \le -2 & \text{Reverse inequalities} \end{array}$$

The solution is $x \ge \frac{8}{3}$ or $x \le -2$. (*Do not use* $\frac{8}{3} \le x \le -2$ to describe the solution.)

**CAUTION:** To solve an absolute value inequality, we should *memorize the rule* since the very first step involves translating from the absolute value notation to a combination (either *and* or *or*) of the related inequalities. Do not ignore the absolute value bars and solve $2x + 1 < 5$ when $|2x + 1| < 5$ is given. This is a common mistake.

**EXAMPLE 6** Solve $|1 - x| < -2$.

Notice that $-2$ is negative but that in our rule, the number $a$ is positive. An inequality such as this has no solutions since $|\textit{Expression}|$ is greater than or equal to zero.

The inequality $|1 - x| \le 0$ is actually equivalent to $|1 - x| = 0$, which has 1 as the only solution.

## 2.12 EXERCISES

*Solve.*

**1.** $|x| < 9$

**2.** $|x| > 10$

**3.** $|x| \leq 3$

**4.** $|x| \geq 7$

**5.** $|x| < -1$
[*Hint:* Is this possible?]

**6.** $|x| < 0$

**7.** $|x| \leq 0$

**8.** $|x + 1| < 3$

**9.** $|x - 1| < 3$

**10.** $|x + 4| > 3$

**11.** $|x + 4| \geq 3$

**12.** $|x - 5| \leq 7$

**13.** $|2x + 1| < 9$

**14.** $|2x + 1| \geq 9$

**15.** $|1 - 2x| > 7$

**16.** $|1 - 2x| \leq 7$

**17.** $\left|\frac{x + 1}{2}\right| < 3$

**18.** $\left|\frac{x + 1}{2}\right| > 3$

**19.** $|2x + 5| < -1$

**20.** $|2x + 5| \geq 0$

**21.** $|2x + 5| \leq 0$

**22.** Graph the equation $(2x + 3) - (4x - 1) = 0$ on a number line.

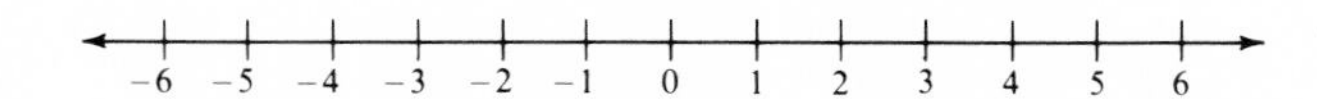

*Graph each inequality on a number line.*

**23.** $4(2x - 3) > 7x - 10$

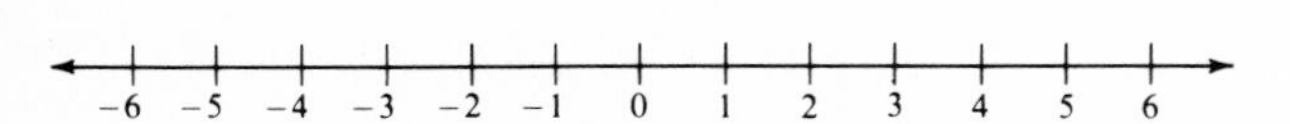


**24.** $0 > 2 - x$ *or* $2 - x \geq 2$

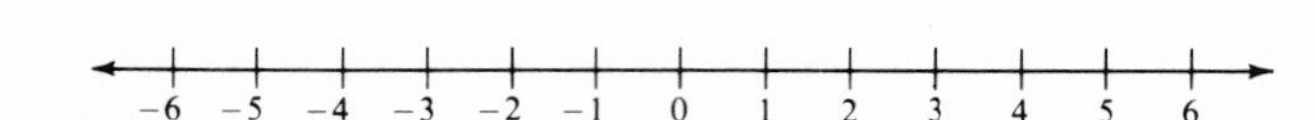


**25.** $-1 \leq x \leq 3$

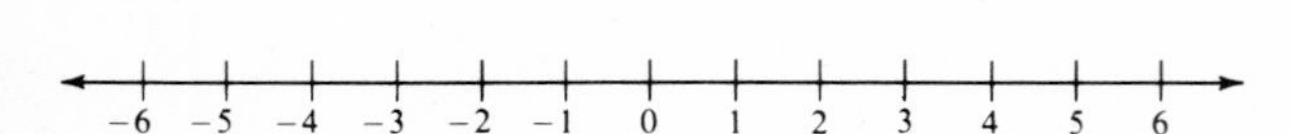


**26.** $x > \frac{1}{2}$ *or* $2x < 1$

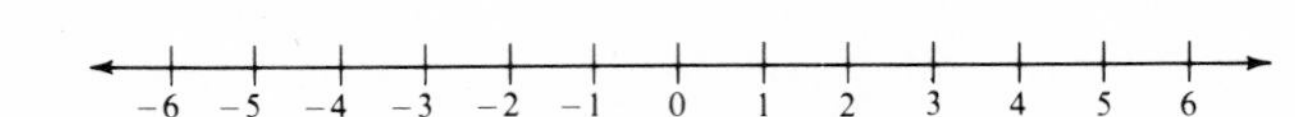


**27.** $5 \leq x \leq 5$

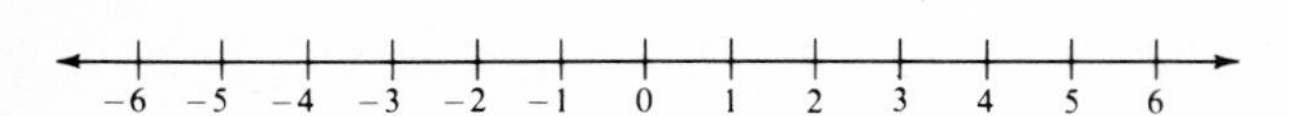


**28.** $x > 2$ *and* $x < -1$

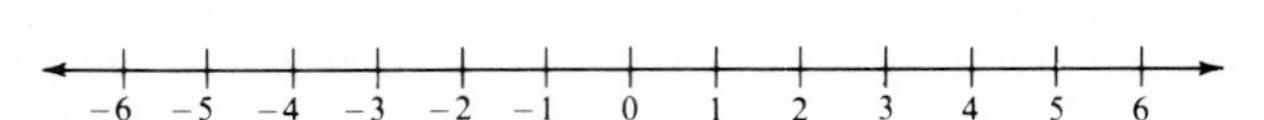


**29.** $|x + 1| < 3$

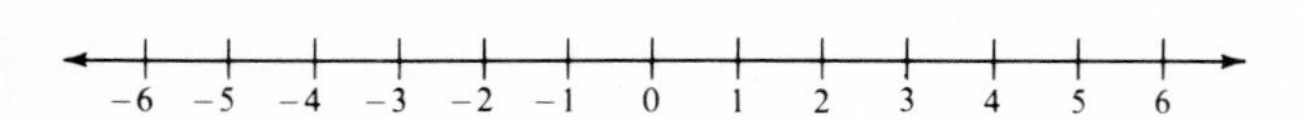


**30.** $|2x - 1| \geq 5$

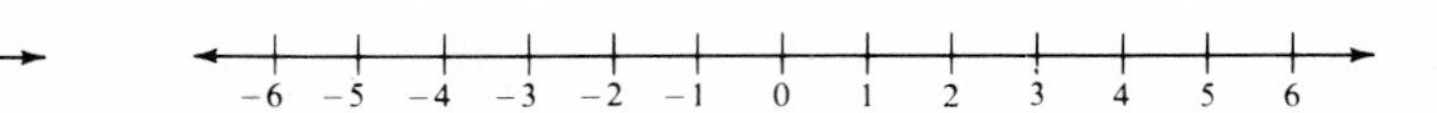


---

**ANSWERS:** **1.** $-9 < x < 9$ ($x > -9$ *and* $x < 9$) **2.** $x < -10$ *or* $x > 10$ (*Do not write* $10 < x < -10$.) **3.** $-3 \leq x \leq 3$ ($x \geq -3$ *and* $x \leq 3$) **4.** $x \leq -7$ *or* $x \geq 7$ **5.** no solution **6.** no solution **7.** 0 **8.** $-4 < x < 2$ **9.** $-2 < x < 4$ **10.** $x < -7$ *or* $x > -1$ **11.** $x \leq -7$ *or* $x \geq -1$ **12.** $-2 \leq x \leq 12$ **13.** $-5 < x < 4$ **14.** $x \leq -5$ *or* $x \geq 4$ **15.** $x > 4$ *or* $x < -3$ **16.** $-3 \leq x \leq 4$ **17.** $-7 < x < 5$ **18.** $x < -7$ *or* $x > 5$ **19.** no solution **20.** Every real number is a solution (identity). **21.** $-\frac{5}{2}$

**22.** 

**23.** 

**24.**

**25.** 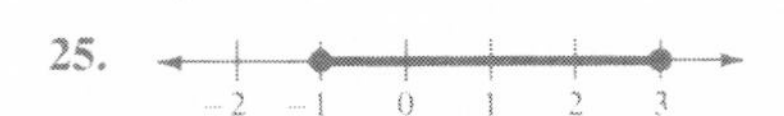


**26.** 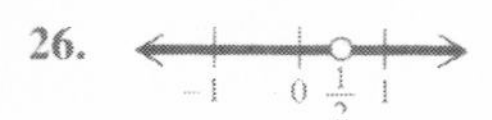


**27.**

**28.**

**29.**

**30.**

## Chapter 2 Summary

### Words and Phrases for Review

[2.1] equation
left member
right member
variable
solution (root)
linear (first-degree) equation
conditional equation
identity
contradiction

[2.2] equivalent equations

[2.4] absolute value

[2.9] literal equation

[2.10] solving inequalities
equivalent inequality

[2.11] number line
origin
plot

[2.12] absolute value inequality

### Brief Reminders

[2.2] **1.** When using the addition-subtraction rule to solve an equation, be sure to add (or subtract) the same expression to both sides.

**2.** When using the multiplication-division rule to solve an equation, be sure to multiply by the correct number. For example, to solve

$$\frac{x}{\frac{1}{2}} = 10$$

multiply by $\frac{1}{2}$ *not* by 2.

**3.** Use the addition-subtraction rule to isolate the variable before using the multiplication-division rule.

[2.3] When solving an equation involving parentheses, first use the distributive law to clear all parentheses. For example,

$$6 - 4(z - 3) = 6 - 4z + 12.$$

[2.4] **1.** $|a| = \begin{cases} a \text{ if } a \geq 0 \\ -a \text{ if } a < 0 \end{cases}$

**2.** To solve an absolute value equation of the form $|Expression| = a$, for $a \geq 0$, solve $Expression = a$ and $Expression = -a$.

[2.6] **1.** Write out complete descriptions of all variables and expressions when solving word problems.

**2.** An accurate sketch is helpful when solving geometry problems.

[2.7] When solving percent-increase problems add the increase to the original amount. For example, an amount of money $x$ plus a 5% increase can be represented by $x + .05x = (1.05)x$.

[2.8] In a motion problem, the basic formula to use is

$$d = r \cdot t \quad [(\text{distance}) = (\text{rate}) \cdot (\text{time})].$$

[2.10] **1.** The same number or quantity can be added (or subtracted) to both sides of an inequality.

**2.** When both sides of an inequality are multiplied (or divided) by a negative number, be sure to *reverse* the inequality.

[2.11] Inequalities connected by the word *and* can generally be written as a single chain of inequalities; those connected by the word *or* cannot be written as a chain.

[2.12] **1.** If $a > 0$ and $|Expression| > a$, then solve

$$Expression < -a \text{ or } Expression > a.$$

**2.** If $a > 0$ and $|Expression| < a$, then solve

$$-a < Expression < a$$

**3.** An inequality such as $|Expression| > -2$ has every number as a solution, while $|Expression| < -2$ has no solutions.

## Chapter 2 Review Exercises

[2.1] **1.** A statement that two quantities are equal is called a(n) ______.

**2.** The equation $x + 1 = x + 1$ is an example of a(n) ______.

**3.** The equation $x + 2 = x + 3$ is an example of a(n) ______.

**4.** Equations that are true for some replacements of the variable and false for others are called ______ equations.

[2.2–2.4] *Solve.*

**5.** $4 - 3x = -5$

**6.** $\frac{1}{3}y = -2$

**7.** $\frac{z}{\frac{1}{4}} = -12$

**8.** $3 - 2(x - 1) = x - 10$

**9.** $4(2y - 4) + 6 = 6y - 2(y - 1)$

**10.** $(2z + 1) - (3z - 7) = 0$

**11.** $|y + 8| = 9$

**12.** $|2z + 1| = 0$

**13.** $|1 - 5x| = -4$

**14.** $(2z + 3) - (z - 1) = 0$

**15.** $4(3x - 5) = 3(4x + 1)$

**16.** $5(z - 1) - (3z - 1) = 0$

[2.6] **17.** If three times a number is increased by 2, the result is the same as 20 less than five times the number. Find the number.

**18.** Find the measures of the angles of a triangle if the second is three times the first and the third is 30° more than twice the first.

**19.** Pete is 9 years older than Rhoda. If the sum of their ages is 87, how old is each?

**20.** After a 7% raise, a woman's new salary is $14,445. What was her old salary?

**21.** 60.5 is 11% of what?

**22.** 7% of 320 is what?

23. A basketball player made 12 shots in 20 attempts. What was her shooting percentage?

[2.8] 24. A man leaves Flagstaff at 8:00 am, traveling by car to Lake Powell, a distance of 135 miles, at a rate of 45 mph. He then drives a boat to Rainbow Bridge, a distance of 50 miles, at a rate of 20 mph. At what time does he reach the bridge?

25. Two trains leave the same town, one traveling east and the other west. If one train is moving 10 mph faster than the other, and if after 2 hours they are 260 miles apart, how fast is each traveling?

[2.9] 26. Solve $B = 40a + w$ for $a$.

27. Solve $a = \frac{1}{2}u \cdot v + s$ for $u$.

[2.10] *Solve.*

28. $x + 9 \leq 2(x - 6)$

29. $y + 2 \leq -8$ or $y + 2 \geq 8$

**30.** $-3 \leq 2z + 1 \leq 7$

[2.11] *Graph each inequality on a number line.*

**31.** $-2 < x \leq 3$

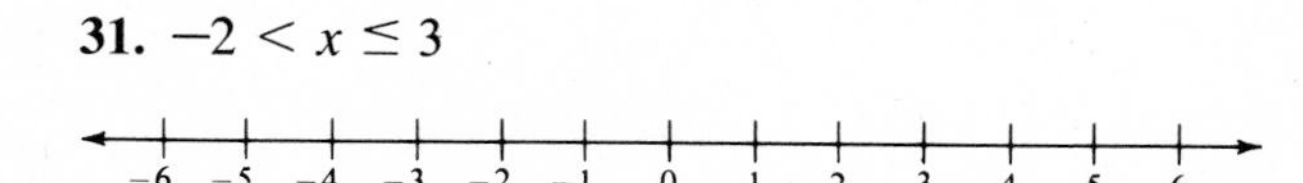

**32.** $x > 3$ or $x \leq -2$

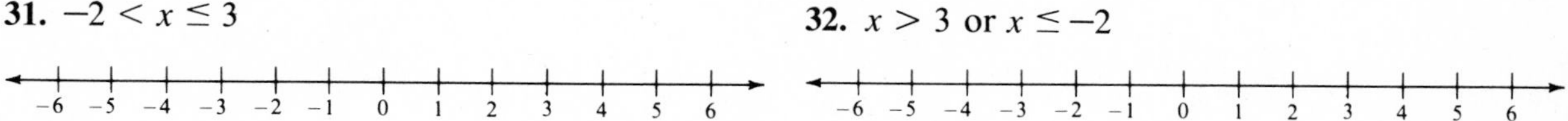

**33.** $|x - 2| \leq 2$

**34.** $|2 - x| \geq 1$

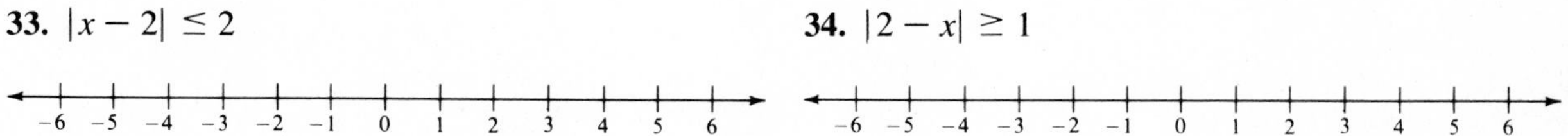

[2.12] *Solve.*

**35.** $|2x - 3| \leq 7$

**36.** $|1 - 3x| < 4$

**37.** $|3y + 4| \geq 28$

---

**ANSWERS:** **1.** equation **2.** identity **3.** contradiction **4.** conditional **5.** 3 **6.** $-6$ **7.** $-3$ **8.** 5 **9.** 3 **10.** 8 **11.** $1, -17$ **12.** $-\frac{1}{2}$ **13.** no solution **14.** $-4$ **15.** no solution **16.** 2 **17.** 11 **18.** 25°, 75°, 80° **19.** Pete: 48, Rhoda: 39 **20.** \$13,500 **21.** 550 **22.** 22.4 **23.** 60% **24.** 1:30 pm **25.** 60 mph, 70 mph **26.** $a = \frac{B - w}{40}$ **27.** $u = \frac{2a - 2s}{v}$ **28.** $x \geq 21$ **29.** $y \leq -10$ or $y \geq 6$ **30.** $-2 \leq z \leq 3$

**31.**

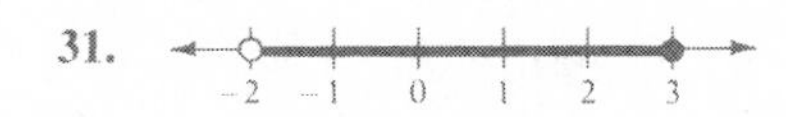

**32.**

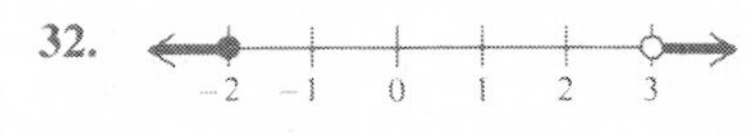

**33.**

**34.**

**35.** $-2 \leq x \leq 5$ **36.** $\frac{5}{3} > x > -1$ **37.** $y \leq -\frac{32}{3}$ or $y \geq 8$

# 3

# Polynomials and Factoring

## 3.1 Polynomials in One Variable

In Chapter 1 we defined a **variable** as a letter which represents a number, and we saw how the use of variables enabled us to express briefly the lengthy statements of word problems. Remember also that a **term** is the product of numbers and variables with only whole number exponents, and that the numerical factor of a term is called the **coefficient.** Any factor of a term which can be divided only by itself and 1 is called a **prime factor.** In this chapter we work with **polynomials,** expressions involving one or more terms which are added. (We use only the word *added* since subtraction is the addition of a negative; for example, $3x - 2 = 3x + (-2)$.)

Some examples of polynomials follow.

$$-11x^2 + x^3 - x + 3, \quad 2a - 3, \quad x + y, \quad 3x^2y, \quad 2, \quad -8a^5b^4 + 6ab - 2, \quad \frac{3}{7}xy - \frac{4}{5}$$

The terms of the first polynomial are

$$-11x^2, \quad x^3, \quad -x, \quad \text{and} \quad 3,$$

while the coefficients of these terms are

$$-11, \quad 1, \quad -1, \quad \text{and} \quad 3,$$

respectively. The terms of $-8a^5b^4 + 6ab - 2$ are

$$-8a^5b^4, \quad 6ab, \quad \text{and} \quad -2$$

and the coefficients are

$$-8, \quad 6, \quad \text{and} \quad -2.$$

A **monomial** is a polynomial with one term, while a **binomial** has two terms and a **trinomial** three terms. A **polynomial in one variable** has the same variable in its terms, while a **polynomial in several variables** involves two or more variables. In this first section we work with polynomials in one variable.

We saw in Chapter 1 that two terms of an expression which have the same variables raised to the same powers are called **like terms.** In the polynomial

$$6y^2 - 2y^3 + 3y - 4y^2 + 1 - 3y - 7,$$

$6y^2$ and $-4y^2$ are like terms, $3y$ and $-3y$ are like terms, and 1 and $-7$ are like terms. A polynomial containing like terms can be simplified by **collecting like terms.** This process involves using the distributive laws.

**EXAMPLE 1** Collect like terms.

$$6y^2 - 2y^3 + 3y - 4y^2 + 1 - 3y - 7$$
$$= 6y^2 - 4y^2 \quad - 2y^3 \quad + 3y - 3y \quad + 1 - 7$$ Commute so that like terms are adjacent
$$= (6 - 4)y^2 - 2y^3 + (3 - 3)y + (1 - 7)$$ Use distributive law
$$= 2y^2 - 2y^3 + 0 \cdot y - 6$$
$$= 2y^2 - 2y^3 - 6$$

**EXAMPLE 2** Collect like terms.

$$-8x^3 + x^3 - 3x + 2 + 5x - 6$$
$$= -8x^3 + x^3 \quad -3x + 5x \quad +2 - 6$$ Commute
$$= (-8 + 1)x^3 + (-3 + 5)x + (2 - 6)$$ Distributive law
$$= -7x^3 + 2x - 4$$

If we write a polynomial so that the term with the least power of the variable is first, followed by the next higher power, and so forth, we say that the polynomial is written in **ascending order.** If, however, we write the term with the highest power first, followed by the next highest power, and so forth, we say the polynomial is in **descending order.** In the following we mostly use descending order.

**EXAMPLE 3**

$$5x^0 - 3x + x^3 - x^6$$ Ascending order
$$-x^6 + x^3 - 3x + 5x^0$$ Descending order

In this example the constant term 5 is written as $5x^0$ since $x^0 = 1$. This is done to show that any constant term comes first in ascending order and last in descending order.

**EXAMPLE 4** Collect like terms and write in descending order.

$$-3a^3 + a^7 - 7a - 4a^7 + a^3 + 1 = a^7 - 4a^7 - 3a^3 + a^3 - 7a + 1$$
$$= (1 - 4)a^7 + (-3 + 1)a^3 - 7a + 1$$
$$= -3a^7 - 2a^3 - 7a + 1$$

## EXERCISES 3.1

*Indicate whether the following are monomials, binomials, or trinomials.*

**1.** $-4x^2 - 3$ ________

**2.** $2xy + 3$ ________

**3.** $7 - abc + 5a^3b$ ________

**4.** $18a^5b^4$ ________

*Indicate the terms and coefficients of the following polynomials.*

**5.** $8x - 7x^3 + 3 - x^2$ **(a)** terms ________ **(b)** coefficients ________

**6.** $5xy^2 - 8x^2y - 17$ **(a)** terms ________ **(b)** coefficients ________

**7.** $9a^2b^4 - \frac{1}{3}$ **(a)** terms ________ **(b)** coefficients ________

**8.** $-14a^5 + b^4 - 7ab$ **(a)** terms ________ **(b)** coefficients ________

*Collect like terms and write in descending order.*

**9.** $5x - 3 - 2x$

**10.** $3y^2 - 6y^3 + 2y^2 - 5$

**11.** $15a^3 - 3 - 6a + 5a^3 - 4 + 4a$

**12.** $-6x^3 + 7x - 13x^5 - 7x + x^3$

**13.** $5y^8 - 6y^4 + 4y^3 - 6 + 7y^8 + 5y^4 + 1$

**14.** $-7 - 3a - 2a^2 + a^4 + 4a^2 + 3$

**15.** $5x^2 - 3x^3 + 6x - 3 + x^3 + x^2 - 3x$

**16.** $5y^3 - 6y^7 - 12y^3 - 8y - 3y^7 + 1$

**17.** $.5a^2 - .2 - 1.3a^2 - .7$

**18.** $\frac{1}{2}x^3 - \frac{1}{3}x - \frac{1}{4}x^3 - \frac{1}{9}x + 1$

---

ANSWERS: **1.** binomial **2.** binomial **3.** trinomial **4.** monomial **5.** (a) $8x, -7x^3, 3, -x^2$ (b) $8, -7, 3, -1$ **6.** (a) $5xy^2, -8x^2y, -17$ (b) $5, -8, -17$ **7.** (a) $9a^2b^4, -\frac{1}{3}$ (b) $9, -\frac{1}{3}$ **8.** (a) $-14a^5, b^4, -7ab$ (b) $-14, 1, -7$ **9.** $3x - 3$ **10.** $-6y^3 + 5y^2 - 5$ **11.** $20a^3 - 2a - 7$ **12.** $-13x^5 - 5x^3$ **13.** $12y^8 - y^4 + 4y^3 - 5$ **14.** $a^4 + 2a^2 - 3a - 4$ **15.** $-2x^3 + 6x^2 + 3x - 3$ **16.** $-9y^7 - 7y^3 - 8y + 1$ **17.** $-.8a^2 - .9$ **18.** $\frac{1}{4}x^3 - \frac{4}{9}x + 1$

## 3.2 Addition and Subtraction of Polynomials in One Variable

**TO ADD POLYNOMIALS**

1. Arrange each in descending order.
2. Indicate the addition using parentheses.
3. Remove parentheses.
4. Collect like terms.

**EXAMPLE 1** Add $-3x + x^4 - 5x^3 + 7$ and $3 + 3x - 5x^4 + 8x^2$.

$$\begin{aligned}
&(x^4 - 5x^3 - 3x + 7) + (-5x^4 + 8x^2 + 3x + 3) && \text{Arrange in descending order}\\
&= x^4 - 5x^3 - 3x + 7 - 5x^4 + 8x^2 + 3x + 3 && \text{Remove parentheses}\\
&= (1 - 5)x^4 - 5x^3 + 8x^2 + (-3 + 3)x + (7 + 3) && \text{Collect like terms}\\
&= -4x^4 - 5x^3 + 8x^2 + 0 \cdot x + 10\\
&= -4x^4 - 5x^3 + 8x^2 + 10
\end{aligned}$$

**TO ADD POLYNOMIALS USING THE COLUMN METHOD**

1. Arrange in descending order with like terms in vertical columns.
2. Add the numerical coefficients down the columns.

**EXAMPLE 2** Add $-3x + x^4 - 5x^3 + 7$ and $3 + 3x - 5x^4 + 8x^2$ using the column method.

$$\begin{array}{rrrrr}
x^4 & - 5x^3 & & - 3x & + 7 \\
\underline{-5x^4} & & \underline{+ 8x^2} & \underline{+ 3x} & \underline{+ 3} \\
-4x^4 & - 5x^3 & + 8x^2 & + 0x & + 10 = -4x^4 - 5x^3 + 8x^2 + 10
\end{array}$$

Leave spaces where terms are missing

**EXAMPLE 3** Add $2 - 3y^2$, $-y^4 + 7y - 3y^3 - 5$, and $-8y^3 + 4y^2 - 7y + 7$ by the column method.

$$\begin{array}{rrrrr}
 & & - 3y^2 & & + 2 \\
-y^4 & - 3y^3 & & + 7y & - 5 \\
 & - 8y^3 & + 4y^2 & - 7y & + 7 \\
\hline
-y^4 & - 11y^3 & + y^2 & + 0 \cdot y & + 4 = -y^4 - 11y^3 + y^2 + 4
\end{array}$$

**TO SUBTRACT POLYNOMIALS**

1. Arrange each in descending order within parentheses.
2. Indicate the subtraction.
3. Remove parentheses by changing all the signs in the polynomial to be subtracted.
4. Proceed as in addition.

To use the column method, change all signs in the polynomial to be subtracted and add.

**EXAMPLE 4** Subtract $7x - 3x^2$ from $4 - 2x - 4x^2$.

$(-4x^2 - 2x + 4) - (-3x^2 + 7x)$

$= -4x^2 - 2x + 4 + 3x^2 - 7x$ Change signs

$= -4x^2 + 3x^2 - 2x - 7x + 4$ Collect like terms

$= (-4 + 3)x^2 + (-2 - 7)x + 4$ Distributive law

$= -x^2 - 9x + 4$

**EXAMPLE 5** Subtract $7x - 3x^2$ from $4 - 2x - 4x^2$ using the column method. Change the signs in $7x - 3x^2$ and add.

$$\begin{array}{rrl} -4x^2 & -2x+4 & \\ +3x^2 & -7x & \text{Change signs} \\ \hline -x^2 & -9x+4 & \text{Add} \end{array}$$

**EXAMPLE 6** Subtract $-3y + 7y^4 + 5y^5 - 2$ from $2y^4 - 8y^5 - 4y + 12y^2 - y^3$.

$(-8y^5 + 2y^4 - y^3 + 12y^2 - 4y) - (5y^5 + 7y^4 - 3y - 2)$

$= -8y^5 + 2y^4 - y^3 + 12y^2 - 4y - 5y^5 - 7y^4 + 3y + 2$ Change signs

$= (-8 - 5)y^5 + (2 - 7)y^4 - y^3 + 12y^2 + (-4 + 3)y + 2$ Commute and distribute

$= -13y^5 - 5y^4 - y^3 + 12y^2 - y + 2$

By the column method we proceed as follows.

$$\begin{array}{rrrrl} -8y^5 & +2y^4 - y^3 + 12y^2 & -4y & & \\ -5y^5 & -7y^4 & +3y & +2 & \text{Change signs} \\ \hline -13y^5 & -5y^4 - y^3 + 12y^2 & -y & +2 & \text{Add} \end{array}$$

**EXAMPLE 7** $(5a^3 + a^2 - 6) + (7a^2 - 3) - (-4a^3 + 7a - 5)$

$= 5a^3 + a^2 - 6 + 7a^2 - 3 + 4a^3 - 7a + 5$

$= 9a^3 + 8a^2 - 7a - 4$ Some steps deleted

By the column method we proceed as follows.

$$\begin{array}{rrrrl} 5a^3 & +a^2 & & -6 & \\ & 7a^2 & & -3 & \\ +4a^3 & & -7a & +5 & \text{Change signs} \\ \hline 9a^3 & +8a^2 & -7a & -4 & \end{array}$$

## EXERCISES 3.2

*Add.*

**1.** $5x - 2$ and $-7x - 8$

**2.** $3y^2 - 5y + 2$ and $-7y^2 + y - 1$

**3.** $5a^3 - 3a + 7$ and $7a^3 - 10a - 8$

**4.** $-3x^4 - 7x^2 + 2$ and $x^3 + 5x - 5$

*Subtract.*

**5.** $3y - 8$ from $2y + 5$

**6.** $-7a^2 + 2a - 5$ from $4a^2 - 5a - 3$

**7.** $9x^3 - 7x^2 - 2x$ from $-7x^2 + 7x - 8$

**8.** $-5y^4 - 6y^2 + 2y - 8$ from $y^4 + y^3 - y^2 + 1$

*Perform the indicated operations.*

**9.** $(3a^3 - 9a^4 + a^5 - 2) + (-6a + 7a^2 - a^3 + 2a^5)$

**10.** $(3a^3 - 9a^4 + a^5 - 2) - (-6a + 7a^2 - a^3 + 2a^5)$

**11.** $(24x - 15x^4 + 10x^2) + (7x^3 - 3x^2 + x^4 - 2)$

**12.** $(17y^3 + y^4 - 2 + 3y) - (8y + 1 - 3y^4 + 8y^3)$

**13.** $(1.3a^2 - 5.2a + 3.7) + (2.3a^2 + 4.8a - 2.8)$

**14.** $\left(\frac{3}{4}x^2 - \frac{1}{3}x - \frac{4}{5}\right) - \left(-\frac{1}{2}x^2 + \frac{4}{9}x - \frac{3}{25}\right)$

**15.** $(4y - 5y^3) + (-7 + 2y - y^2) - (7y^3 + 2 - 3y)$

**16.** $(2 - 2a^4 + a^3) - (a^3 - 3a^4 + 7a + 5) - (14a + a^3 - 3a^2)$

*Add.*

**17.**
$$\begin{array}{r} 5x^4 - 6x^3 - 2x^2 \qquad\qquad + 5 \\ -3x^4 \qquad\qquad + 8x^2 + 3x + 2 \\ \underline{-7x^4 + 8x^3 - 5x^2 \qquad\qquad - 9} \end{array}$$

**18.**
$$\begin{array}{r} 180y^3 - 24y^2 - 85y \qquad\quad \\ -35y^3 + 128y^2 + 99y - 240 \\ \underline{67y^2 - 108y + 560} \end{array}$$

**19.**
$$\begin{array}{r} 3.7a^3 - 8.6a + 9.3 \\ -5.8a^3 + 3.7a + 1.8 \\ \underline{8.5a^3 + 9.2a - 2.2} \end{array}$$

**20.**
$$\begin{array}{r} \frac{2}{7}x^3 - 5x^2 + \frac{3}{5}x + 2 \\ \underline{\frac{3}{14}x^3 - \frac{5}{6}x^2 - \frac{7}{10}x + \frac{3}{2}} \end{array}$$

---

ANSWERS: **1.** $-2x - 10$ **2.** $-4y^2 - 4y + 1$ **3.** $12a^3 - 13a - 1$ **4.** $-3x^4 + x^3 - 7x^2 + 5x - 3$ **5.** $-y + 13$ **6.** $11a^2 - 7a + 2$ **7.** $-9x^3 + 9x - 8$ **8.** $6y^4 + y^3 + 5y^2 - 2y + 9$ **9.** $3a^5 - 9a^4 + 2a^3 + 7a^2 - 6a - 2$ **10.** $-a^5 - 9a^4 + 4a^3 - 7a^2 + 6a - 2$ **11.** $-14x^4 + 7x^3 + 7x^2 + 24x - 2$ **12.** $4y^4 + 9y^3 - 5y - 3$ **13.** $3.6a^2 - .4a + .9$ **14.** $\frac{5}{4}x^2 - \frac{7}{9}x - \frac{17}{25}$ **15.** $-12y^3 - y^2 + 9y - 9$ **16.** $a^4 - a^3 + 3a^2 - 21a - 3$ **17.** $-5x^4 + 2x^3 + x^2 + 3x - 2$ **18.** $145y^3 + 171y^2 - 94y + 320$ **19.** $6.4a^3 + 4.3a + 8.9$ **20.** $\frac{1}{2}x^3 - \frac{35}{6}x^2 - \frac{1}{10}x + \frac{7}{2}$

## 3.3 Addition and Subtraction of Polynomials in Several Variables

When we express a polynomial in two or more variables in ascending or descending order, we must first specify the variable to be considered.

**EXAMPLE 1** Express $7x^2y^3 + 3x^3y + 2 - 5xy^2$ in descending order of powers of $x$.

$$7x^2y^3 + 3x^3y + 2 - 5xy^2 = 3x^3y + 7x^2y^3 - 5xy^2 + 2 \qquad \text{Descending powers of } x$$

We now express the same polynomial in descending order of powers of $y$.

$$7x^2y^3 + 3x^3y + 2 - 5xy^2 = 7x^2y^3 - 5xy^2 + 3x^3y + 2 \qquad \text{Descending powers of } y$$

**EXAMPLE 2** Express $6abc - 5a^2b^3c^2 + 3b^2c^3 - 8a^3c^4$ in descending powers of $a$.

$$6abc - 5a^2b^3c^2 + 3b^2c^3 - 8a^3c^4 = -8a^3c^4 - 5a^2b^3c^2 + 6abc + 3b^2c^3$$

We now express the same polynomial in descending powers of $c$.

$$6abc - 5a^2b^3c^2 + 3b^2c^3 - 8a^3c^4 = -8a^3c^4 + 3b^2c^3 - 5a^2b^3c^2 + 6abc$$

In a polynomial of several variables, two terms are **like terms** if they contain the same variables raised to the same powers.

**EXAMPLE 3** The following pairs of terms are like terms.

| *Term* | *Like term* |
|---|---|
| $7x^3y$ | $-5x^3y$ |
| $-6a^2b^2c$ | $15a^2b^2c$ |
| $8xyz$ | $-5xyz$ |
| $-4a^4b^3c$ | $-8a^4b^3c$ |
| $x^2b^3$ | $15x^2b^3$ |
| $5$ | $-8$ |
| $5a^7x^6y^3$ | $19a^7x^6y^3$ |

**EXAMPLE 4** The following pairs of terms are *not* like terms.

| *Term* | *Unlike term* | |
|---|---|---|
| $7x^3y$ | $-5xy^3$ | Need $x^3y$, not $xy^3$ |
| $-6a^2b^2c$ | $15a^2b^2c^2$ | Need $c$, not $c^2$ |
| $8xyz$ | $-5xy$ | Need $z$ |
| $-4a^4b^3c$ | $-8a^3b^3c$ | Need $a^4$, not $a^3$ |
| $x^2b^3$ | $15x^2y^3$ | Need $b^3$, not $y^3$ |
| $5$ | $-8x$ | Cannot involve $x$ |
| $5a^7x^6y^3$ | $10a^7x^3y^6$ | Need $x^6y^3$, not $x^3y^6$ |

The rules for addition and subtraction of polynomials in several variables are essentially the same as the rules for addition and subtraction of polynomials in a single variable. However, arranging the terms in descending order does not necessarily arrange like terms in the same order so we omit that step.

**TO ADD POLYNOMIALS IN SEVERAL VARIABLES**

1. Place each within parentheses.
2. Indicate the addition.
3. Remove parentheses.
4. Collect like terms.

**EXAMPLE 5** Add $x^4y - 5x^3y^2 - 3xy^3 + 7y$ and $-5x^4y + 8x^2y + 4xy^3 + 3$.

$(x^4y - 5x^3y^2 - 3xy^3 + 7y) + (-5x^4y + 8x^2y + 4xy^3 + 3)$ — Indicate addition

$= x^4y - 5x^3y^2 - 3xy^3 + 7y - 5x^4y + 8x^2y + 4xy^3 + 3$ — Remove parentheses

$= (1 - 5)x^4y - 5x^3y^2 + 8x^2y + (-3 + 4)xy^3 + 7y + 3$ — Collect like terms

$= -4x^4y - 5x^3y^2 + 8x^2y + xy^3 + 7y + 3$

**TO ADD POLYNOMIALS IN SEVERAL VARIABLES USING THE COLUMN METHOD**

1. Arrange like terms in vertical columns.
2. Add the numerical coefficients down the columns.

**EXAMPLE 6** Add $x^4y - 5x^3y^2 - 3xy^3 + 7y$ and $-5x^4y + 8x^2y + 4xy^3 + 3$ using the column method.

$$
\begin{array}{rrrrrr}
x^4y & -\ 5x^3y^2 & & -\ 3xy^3 & +\ 7y & \\
-5x^4y & & +\ 8x^2y & +\ 4xy^3 & & +\ 3 \\
\hline
-4x^4y & -\ 5x^3y^2 & +\ 8x^2y & +\ xy^3 & +\ 7y & +\ 3
\end{array}
$$

**EXAMPLE 7** Add $-3x^3y + 4x^2y^2 - 8x + y$, $5x^2y^2 - 7x + y + 5$, and $9x^3y + 4x - 3y$ using the column method.

$$
\begin{array}{rrrrr}
-3x^3y & +\ 4x^2y^2 & -\ 8x & +\ y & \\
 & 5x^2y^2 & -\ 7x & +\ y & +\ 5 \\
9x^3y & & +\ 4x & -\ 3y & \\
\hline
6x^3y & +\ 9x^2y^2 & -\ 11x & -\ y & +\ 5
\end{array}
$$

**TO SUBTRACT POLYNOMIALS IN SEVERAL VARIABLES**

1. Place each within parentheses.
2. Indicate the subtraction.
3. Remove parentheses by changing all the signs in the polynomial to be subtracted.
4. Proceed as in addition.

To use the column method, change all signs on the polynomial to be subtracted and add.

**EXAMPLE 8** Subtract $-3a^2b^3 + 9ab^2 - 5$ from $4a^2b^3 + 7ab^2 - 3b^2 + 8$.

$$(4a^2b^3 + 7ab^2 - 3b^2 + 8) - (-3a^2b^3 + 9ab^2 - 5)$$

$= 4a^2b^3 + 7ab^2 - 3b^2 + 8 + 3a^2b^3 - 9ab^2 + 5$ Change signs

$= (4 + 3)a^2b^3 + (7 - 9)ab^2 - 3b^2 + (8 + 5)$ Collect like terms

$= 7a^2b^3 - 2ab^2 - 3b^2 + 13$

**EXAMPLE 9** Subtract $-3a^2b^3 + 9ab^2 - 5$ from $4a^2b^3 + 7ab^2 - 3b^2 + 8$ using the column method.

$$\begin{array}{rrrrrl} & 4a^2b^3 & + & 7ab^2 - 3b^2 & + & 8 \\ + & 3a^2b^3 & - & 9ab^2 \phantom{- 3b^2} & + & 5 \\ \hline & 7a^2b^3 & - & 2ab^2 - 3b^2 & + & 13 \end{array}$$

Change signs

Add

**EXAMPLE 10** $(5x^3y^3 - 3x^2y) + (-7x^3y^3 + 4xy) - (9x^2y + 8xy)$

$= 5x^3y^3 - 3x^2y - 7x^3y^3 + 4xy - 9x^2y - 8xy$

$= (5 - 7)x^3y^3 + (-3 - 9)x^2y + (4 - 8)xy$

$= -2x^3y^3 - 12x^2y - 4xy$

## EXERCISES 3.3

*Add.*

1. $3xy + 4$ and $-7xy + 5$

2. $8a^2b - 3ab + 2$ and $9a^2b + 7ab - 5$

3. $-6x^3y + x^2y^2 - 5xy$ and $8x^3y - 10x^2y^2 + 6$

**4.** $10a^3b^3 - 7a^2b - 4ab^2$ and $-9a^3b^3 - 7a^2b + 5ab$

*Subtract.*

**5.** $-2xy + 3y$ from $9xy - 7y$

**6.** $7a^2b + ab - 5$ from $-10a^2b + ab + 8$

**7.** $12x^3y - 7x^2y^2 + 14xy$ from $7x^3y + 6x^2y^2 - 5$

**8.** $-8a^3b^3 + 5a^2b + 10ab^2$ from $6a^3b^3 + 3a^2b - 4ab$

*Perform the indicated operations.*

**9.** $(4x^4y + 6x^3y^2 - 7x^2y^3) + (-12x^4y + 5x^3y^2 + x^2y^3 - 5)$

**10.** $(6a^4b^3 - 7a^3b^2 - 2a^2b) - (-3a^3b^2 + 5a^2b + 10a)$

**11.** $(-10x^4y^3 + 3x^3y^4 - 7x^2z + 5z) + (10x^4y^3 + 9x^3y^4 - 6z)$

**12.** $(5a^5b^5 - 6a^4b^3 + 2b^3c^3 + 6c^2) - (-3a^5b^5 + 7a^3b^4 - 8b^3c^3 - 6c^2)$

**13.** $(7.3x^2y^2 - 8.7xy + 14.2) + (-6.7x^2y^2 + 2.3xy + 5.9)$

**14.** $\left(\frac{2}{9}a^3b^2 + \frac{3}{4}ab - \frac{2}{5}b\right) - \left(\frac{1}{3}a^3b^2 - \frac{1}{8}ab + \frac{3}{10}b\right)$

**15.** $(6x^3y^3 + 5x^2y^2) + (11x^2y^2 - 4xy) - (2x^3y^3 - 2x^2y^2 + 7xy)$

**16.** $(-8a^3b - 5a^2b^2) - (-3a^2b^2 - 10a) - (-7a^3b - 9a^2b^2 + 4a)$

**17.** $(4z - 7z^3 + 12 - 13z^2) + (-5 - 12z^2 + 4z^3 + 11z)$

**18.** $(-5c^3 + 4c^4 - 3c + 10) - (12 + 3c^2 - 2c + 6c^4)$

**19.** $\left(\frac{2}{3}z^2 - \frac{1}{5}z + 2\right) + \left(\frac{1}{2}z^2 + z - \frac{3}{4}\right)$

**20.** $(-5.8c^2 + 6.2c - 4.1) - (2.3c^2 + 4.8c - 1.3)$

---

**ANSWERS:** **1.** $-4xy + 9$ **2.** $17a^2b + 4ab - 3$ **3.** $2x^3y - 9x^2y^2 - 5xy + 6$ **4.** $a^3b^3 - 14a^2b - 4ab^2 + 5ab$ **5.** $11xy - 10y$ **6.** $-17a^2b + 13$ **7.** $-5x^3y + 13x^2y^2 - 14xy - 5$ **8.** $14a^3b^3 - 2a^2b - 10ab^2 - 4ab$ **9.** $-8x^4y + 11x^3y^2 - 6x^2y^3 - 5$ **10.** $6a^4b^3 - 4a^3b^2 - 7a^2b - 10a$ **11.** $12x^3y^4 - 7x^2z - z$ **12.** $8a^5b^5 - 6a^4b^3 - 7a^3b^4 + 10b^3c^3 + 12c^2$ **13.** $.6x^2y^2 - 6.4xy + 20.1$ **14.** $-\frac{1}{9}a^3b^2 + \frac{7}{8}ab - \frac{7}{10}b$ **15.** $4x^3y^3 + 18x^2y^2 - 11xy$ **16.** $-a^3b + 7a^2b^2 + 6a$ **17.** $-3z^3 - 25z^2 + 15z + 7$ **18.** $-2c^4 - 5c^3 - 3c^2 - c - 2$ **19.** $\frac{7}{6}z^2 + \frac{4}{5}z + \frac{5}{4}$ **20.** $-8.1c^2 + 1.4c - 2.8$

## 3.4 Multiplication of Polynomials

**TO MULTIPLY MONOMIALS**

1. Multiply the coefficients.
2. Use the rule

$$a^m \cdot a^n = a^{m+n}$$

to multiply the variables.

**EXAMPLE 1** Multiply $-3x^2y^3$ by $5x^5y^2$.

$$(-3x^2y^3) \cdot (5x^5y^2) = (-3)(5)(x^2 \cdot x^5)(y^3 \cdot y^2)$$

Use the commutative law to change the order of factors

$$= -15x^{2+5}y^{3+2}$$
$$= -15x^7y^5$$

**EXAMPLE 2** Multiply $6ab^2$ by $-a^3$.

$$\begin{aligned}(6ab^2)\cdot(-a^3) &= 6\cdot(-1)(a\cdot a^3)\cdot b^2 && -a^3=(-1)a^3\\ &= -6a^{1+3}b^2 && a=a^1\\ &= -6a^4b^2\end{aligned}$$

**TO MULTIPLY A MONOMIAL BY A BINOMIAL**

1. Use the distributive property of multiplication over addition.
2. Use the rule for multiplying monomials.

**EXAMPLE 3**

$$\begin{aligned}-5x^2y(4x^2y^2-6y) &= (-5x^2y)(4x^2y^2)-(-5x^2y)(6y) && \text{Distributive property}\\ &= (-5)(4)(x^2\cdot x^2)(y\cdot y^2)-(-5)(6)x^2(yy)\\ &= -20x^{2+2}y^{1+2}-(-30)x^2y^{1+1}\\ &= -20x^4y^3+30x^2y^2\end{aligned}$$

If we treat $(x+2y)$ as a single term and multiply $(5x-y)$ by it, we can see how to multiply two binomials.

**EXAMPLE 4**

$$\begin{aligned}(x+2y)(5x-y) &= (x+2y)(5x)+(x+2y)(-y) && \text{Distributive law}\\ &= (x)(5x)+(2y)(5x)+(x)(-y)+(2y)(-y)\\ &= 5x^2+10xy-xy-2y^2\\ &= 5x^2+9xy-2y^2\end{aligned}$$

Notice that each term in the first binomial is multiplied by each term of the second. This motivates the following rule.

**TO MULTIPLY TWO POLYNOMIALS NEITHER OF WHICH IS A MONOMIAL**

1. Multiply each term of one by each term of the other.
2. Collect like terms.

**EXAMPLE 5**

$$\begin{aligned}(5a+2b)(3a-7b) &= \overset{F}{(5a)(3a)}+\overset{O}{(5a)(-7b)}+\overset{I}{(2b)(3a)}+\overset{L}{(2b)(-7b)}\\ &= 15a^2-35ab+6ab-14b^2\\ &= 15a^2-29ab-14b^2\end{aligned}$$

In Example 5, the letters F O I L stand for first terms (F), outside terms (O), inside terms (I), and last terms (L). Remember this word and you will not omit terms when multiplying.

Another method involves writing one polynomial above the other and arranging like terms in vertical columns as we multiply.

**EXAMPLE 6** Multiply by the column method.

$$
\begin{array}{rl}
5a + 2b & \\
3a - 7b & \\
\hline
15a^2 + 6ab \phantom{- 14b^2} & 3a \text{ times each term of the top polynomial} \\
- 35ab - 14b^2 & -7b \text{ times the top polynomial} \\
\hline
15a^2 - 29ab - 14b^2 & \text{Adding}
\end{array}
$$

The column method is especially useful when trinomials or larger polynomials are involved.

**EXAMPLE 7**

$$
\begin{array}{rl}
3x^2 - 5xy + 2y^2 & \\
7x - 8y & \\
\hline
21x^3 - 35x^2y + 14xy^2 \phantom{- 16y^3} & 7x \text{ times each term of the top polynomial} \\
- 24x^2y + 40xy^2 - 16y^3 & -8y \text{ times the top polynomial} \\
\hline
21x^3 - 59x^2y + 54xy^2 - 16y^3 &
\end{array}
$$

**EXAMPLE 8**

$$
\begin{array}{r}
8a^2b^2 + 9ab - 6 \phantom{- 18} \\
10ab + 3 \phantom{- 18} \\
\hline
80a^3b^3 + 90a^2b^2 - 60ab \phantom{- 18} \\
24a^2b^2 + 27ab - 18 \\
\hline
80a^3b^3 + 114a^2b^2 - 33ab - 18
\end{array}
$$

**EXAMPLE 9**

$$
\begin{array}{r}
6x^2 - 7xy + 5y^2 \\
4x^2 + 3xy - 2y^2 \\
\hline
24x^4 - 28x^3y + 20x^2y^2 \\
18x^3y - 21x^2y^2 + 15xy^3 \\
- 12x^2y^2 + 14xy^3 - 10y^4 \\
\hline
24x^4 - 10x^3y - 13x^2y^2 + 29xy^3 - 10y^4
\end{array}
$$

## EXERCISES 3.4

*Multiply the following polynomials.*

**1.** $(3x) \cdot (7y)$

**2.** $(-5x^2) \cdot (-6xy)$

**3.** $(8a^2b)(-9ab^2)$

**4.** $(12a^5b^2)(10a^3b^4)$

**5.** $3xy(7x - 4y)$

**6.** $-4x^3y(2xy - 5)$

**7.** $8a^2b^2(3ab - 2a + 5b)$

**8.** $-7a^3b(3a^2b^2 + 5a^2 - 6b^2)$

**9.** $(x + y)(x + 2y)$

**10.** $(x - y)(x - 2y)$

**11.** $(x - y)(x + 2y)$

**12.** $(x + y)(x - 2y)$

**13.** $(2a + b)(a + 3b)$

**14.** $(2a - b)(a - 3b)$

**15.** $(2a + b)(a - 3b)$

**16.** $(2a - b)(a + 3b)$

**17.** $(2x + 3y)(x + y)$

**18.** $(5x - 2y)(x + 3y)$

**19.** $(3a - 7b)(2a - 5b)$

**20.** $(10a + 3b)(7a - 4b)$

**21.** $\begin{array}{l} 4x + 9y \\ \underline{3x + 5y} \end{array}$

**22.** $\begin{array}{l} 4x - 9y \\ \underline{3x - 5y} \end{array}$

**23.** $\begin{array}{l} 4x - 9y \\ \underline{3x + 5y} \end{array}$

**24.** $\begin{array}{l} 4x + 9y \\ \underline{3x - 5y} \end{array}$

**25.** $\begin{array}{l} 2a^2 + 3b^2 \\ \underline{4a + 5b} \end{array}$

**26.** $\begin{array}{l} 7a^2 - 3ab + 4b^2 \\ \underline{3a - 2b} \end{array}$

**27.** $\begin{array}{l} 2x^2 + 4xy - 5y^2 \\ \underline{9x + 2y} \end{array}$

**28.** $\begin{array}{l} 5x^2y^2 - 3xy - 2 \\ \underline{8xy - 7} \end{array}$

**29.** $\begin{array}{l} 2a - ab + 3b \\ \underline{4a + 2ab - 3b} \end{array}$

**30.** $\begin{array}{l} 7a - 2b + 5 \\ \underline{3a - b + 2} \end{array}$

*Simplify.*

**31.** $(6x^2y^2 - 2xy + 5) + (3x^2y^2 - 10) - (-10x^2y^2 + 5xy - 2)$

**32.** $(3a^3b^3 + 5a^2b^2 - 4) - (-5a^3b^3 + 10ab + 7) - (6a^2b^2 - 3ab + 4)$

---

ANSWERS: **1.** $21xy$ **2.** $30x^3y$ **3.** $-72a^3b^3$ **4.** $120a^8b^6$ **5.** $21x^2y - 12xy^2$ **6.** $-8x^4y^2 + 20x^3y$ **7.** $24a^3b^3 - 16a^3b^2 + 40a^2b^3$ **8.** $-21a^5b^3 - 35a^5b + 42a^3b^3$ **9.** $x^2 + 3xy + 2y^2$ **10.** $x^2 - 3xy + 2y^2$ **11.** $x^2 + xy - 2y^2$ **12.** $x^2 - xy - 2y^2$ **13.** $2a^2 + 7ab + 3b^2$ **14.** $2a^2 - 7ab + 3b^2$ **15.** $2a^2 - 5ab - 3b^2$ **16.** $2a^2 + 5ab - 3b^2$ **17.** $2x^2 + 5xy + 3y^2$ **18.** $5x^2 + 13xy - 6y^2$ **19.** $6a^2 - 29ab + 35b^2$ **20.** $70a^2 - 19ab - 12b^2$ **21.** $12x^2 + 47xy + 45y^2$ **22.** $12x^2 - 47xy + 45y^2$ **23.** $12x^2 - 7xy - 45y^2$ **24.** $12x^2 + 7xy - 45y^2$ **25.** $8a^3 + 10a^2b + 12ab^2 + 15b^3$ **26.** $21a^3 - 23a^2b + 18ab^2 - 8b^3$ **27.** $18x^3 + 40x^2y - 37xy^2 - 10y^3$ **28.** $40x^3y^3 - 59x^2y^2 + 5xy + 14$ **29.** $8a^2 - 2a^2b^2 + 6ab + 9ab^2 - 9b^2$ **30.** $21a^2 - 13ab + 29a + 2b^2 - 9b + 10$ **31.** $19x^2y^2 - 7xy - 3$ **32.** $8a^3b^3 - a^2b^2 - 7ab - 15$

## 3.5 Special Products

Consider the following three products of binomials.

**1.**
$$\begin{array}{r} 2x + 3y \\ 2x - 3y \\ \hline (2x)^2 + 6xy \qquad\quad \\ - 6xy - (3y)^2 \\ \hline (2x)^2 + 0xy - (3y)^2 \\ = 4x^2 - 9y^2 \end{array}$$

**2.**
$$\begin{array}{r} 2x + 3y \\ 2x + 3y \\ \hline (2x)^2 + 6xy \qquad\quad \\ 6xy + (3y)^2 \\ \hline (2x)^2 + 2(6xy) + (3y)^2 \\ = 4x^2 + 12xy + 9y^2 \end{array}$$

**3.**
$$\begin{array}{r} 2x - 3y \\ 2x - 3y \\ \hline (2x)^2 - 6xy \qquad\quad \\ - 6xy + (3y)^2 \\ \hline (2x^2) - 2(6xy) + (3y)^2 \\ = 4x^2 - 12xy + 9y^2 \end{array}$$

In (1), since the middle term is zero, the product is the difference of the two squares. The products (2) and (3) fall into the pattern of the square of the first term, plus or minus twice the product of the first and last terms, plus the square of the last. These are examples of special products given in the next rule.

1. $(a + b)(a - b) = a^2 - b^2$
2. $(a + b)(a + b) = (a + b)^2 = a^2 + 2ab + b^2$
3. $(a - b)(a - b) = (a - b)^2 = a^2 - 2ab + b^2$

CAUTION: Pay special attention to the difference between the first and third formulas: $(a - b)^2$ is not $a^2 - b^2$. Also note that $(a + b)^2 \neq a^2 + b^2$.

We can always multiply directly to obtain these special products, but later on for factoring, it helps to have the formulas in mind.

**EXAMPLE 1** Find the product $(x + 6)(x - 6)$.

*Multiplying*

F O I L

$$(x + 6)(x - 6) = x^2 - 6x + 6x - 36$$
$$= x^2 - 36$$

*Using $(a + b)(a - b) = a^2 - b^2$*

$$(x + 6)(x - 6) = x^2 - 6^2 \qquad a = x,\ b = 6$$
$$= x^2 - 36$$

**EXAMPLE 2** Find the product $(x + 6)(x + 6)$.

*Multiplying*

$$(x + 6)(x + 6) = x^2 + 6x + 6x + 36$$
$$= x^2 + 12x + 36$$

*Using $(a + b)^2 = a^2 + 2ab + b^2$*

$$(x + 6)^2 = x^2 + 2(x)(6) + 6^2 \qquad a = x,\ b = 6$$
$$= x^2 + 12x + 36$$

**EXAMPLE 3** Find the product $(x - 6)(x - 6)$.

*Multiplying*

$$(x - 6)(x - 6) = x^2 - 6x - 6x + 36$$
$$= x^2 - 12x + 36$$

*Using $(a - b)^2 = a^2 - 2ab + b^2$*

$$(x - 6)^2 = x^2 - 2(x)(6) + 6^2 \qquad a = x,\ b = 6$$
$$= x^2 - 12x + 36$$

**EXAMPLE 4** $(3x - 2y)(3x + 2y) = (3x)^2 - (2y)^2$ Use $(a - b)(a + b) = a^2 - b^2$, with $a = 3x$ and $b = 2y$

$= 3^2x^2 - 2^2y^2$ $(ab)^n = a^nb^n$

$= 9x^2 - 4y^2$

**EXAMPLE 5** $(3x + 2y)^2 = (3x)^2 + 2(3x)(2y) + (2y)^2$ Use $(a + b)^2 = a^2 + 2ab + b^2$, with $a = 3x$ and $b = 2y$

$= 9x^2 + 12xy + 4y^2$

**EXAMPLE 6** $(3x - 2y)^2 = (3x)^2 - 2(3x)(2y) + (2y)^2$ Use $(a - b)^2 = a^2 - 2ab + b^2$, with $a = 3x$ and $b = 2y$

$= 9x^2 - 12xy + 4y$

**EXAMPLE 7** $(5a^3 - 4b^2)(5a^3 + 4b^2) = (5a^3)^2 - (4b^2)^2$ Use $(a - b)(a + b) = a^2 - b^2$, with $a = 5a^3$ and $b = 4b^2$

$= 25a^6 - 16b^4$ $(a^m)^n = a^{mn}$

## EXERCISES 3.5

*Use the three rules to multiply the following polynomials.*

**1.** $(x + 3)(x - 3)$

**2.** $(y + 3)^2$

**3.** $(a - 3)^2$

**4.** $(b + 5)^2$

**5.** $(x - 5)^2$

**6.** $(y - 5)(y + 5)$

**7.** $(3a - 5)^2$

**8.** $(3b + 5)^2$

**9.** $(3x + 5)(3x - 5)$

**10.** $(3x + 5y)(3x - 5y)$

**11.** $(3a + 5b)^2$

**12.** $(3a - 5b)^2$

**13.** $(x - 7y)(x + 7y)$

**14.** $(x + 7y)^2$

**15.** $(a - 7b)^2$

**16.** $(8a - 5b)^2$

**17.** $(5x + 9y)^2$

**18.** $(4x - 3y)(4x + 3y)$

**19.** $(3a^2 - 2b^2)(3a^2 + 2b^2)$

**20.** $(a^3 + 5b^2)^2$

*Multiply the following polynomials.*

**21.** $(2x + 3y)(5x - 7y)$

**22.** $(9x - 2y)(4x - y)$

**23.** $(6a + 7b)(a + 2b)$

**24.** $(4a + 11b)(10a - 3b)$

**25.** $(4x - 9y)(4x + 9y)$

**26.** $(10x - 9y)(3x - 5y)$

**27.** $(6a - 5b)(6a - 5b)$

**28.** $(a^2 + 2b^2)(a^2 + 2b^2)$

**29.**
$$\begin{array}{l} 8x^2 - 3y^2 \\ \underline{2x^2 + 7y^2} \end{array}$$

**30.**
$$\begin{array}{l} 6x^2 + 4xy - 9y^2 \\ \underline{3x \quad - 5y \qquad\quad} \end{array}$$

---

ANSWERS: **1.** $x^2 - 9$ **2.** $y^2 + 6y + 9$ **3.** $a^2 - 6a + 9$ **4.** $b^2 + 10b + 25$ **5.** $x^2 - 10x + 25$ **6.** $y^2 - 25$ **7.** $9a^2 - 30a + 25$ **8.** $9b^2 + 30b + 25$ **9.** $9x^2 - 25$ **10.** $9x^2 - 25y^2$ **11.** $9a^2 + 30ab + 25b^2$ **12.** $9a^2 - 30ab + 25b^2$ **13.** $x^2 - 49y^2$ **14.** $x^2 + 14xy + 49y^2$ **15.** $a^2 - 14ab + 49b^2$ **16.** $64a^2 - 80ab + 25b^2$ **17.** $25x^2 + 90xy + 81y^2$ **18.** $16x^2 - 9y^2$ **19.** $9a^4 - 4b^4$ **20.** $a^6 + 10a^3b^2 + 25b^4$ **21.** $10x^2 + xy - 21y^2$ **22.** $36x^2 - 17xy + 2y^2$ **23.** $6a^2 + 19ab + 14b^2$ **24.** $40a^2 + 98ab - 33b^2$ **25.** $16x^2 - 81y^2$ **26.** $30x^2 - 77xy + 45y^2$ **27.** $36a^2 - 60ab + 25b^2$ **28.** $a^4 + 4a^2b^2 + 4b^4$ **29.** $16x^4 + 50x^2y^2 - 21y^4$ **30.** $18x^3 - 18x^2y - 47xy^2 + 45y^3$

## 3.6 Factoring: Common Factor and Grouping

We begin our study of factoring by considering polynomials that have a common factor in each term. A **factor** of a term is a number or power of a variable that is a multiplier of the term. For example,

$$2, \quad x, \quad x^2 \quad \text{are factors of} \quad 2x^2$$

$$2, \quad 3, \quad 6, \quad x, \quad y \quad \text{are factors of} \quad 6xy.$$

Factoring uses the distributive law in the reverse order as used when multiplying, and it is important in applications such as simplifying fractions and solving equations.

**EXAMPLE 1** Since $2x(x - 3y) = 2x^2 - 6xy$, we can factor $2x^2 - 6xy$ by writing

$$2x^2 - 6xy = 2x(x - 3y).$$

That is, 2 and $x$ are common factors of the terms $2x^2$ and $-6xy$.

**EXAMPLE 2** In order to factor $3x^2y + 6xy^2$ we write

$$3x^2y + 6xy^2 = 3 \cdot x \cdot x \cdot y + 2 \cdot 3 \cdot x \cdot y \cdot y = 3xy(x + 2y)$$

since $3xy$ is common to both terms. To check we multiply.

$$3xy(x + 2y) = 3x^2y + 6xy^2$$

CAUTION: $3x^2y + 6xy^2$ could be factored as $3(x^2y + 2xy^2)$ or $3x(xy + 2y^2)$ or, as above, $3xy(x + 2y)$. But only the last expression is **factored completely.** We will always factor completely.

**EXAMPLE 3** $4a^2b^3 - 6a^2b^2 + 8ab^2 = 2ab^2(2ab - 3a + 4)$

We have factored completely since $2ab$, $-3a$, and 4 have no common factors.

Some factoring problems can be solved by grouping terms, factoring the individual polynomials, and then factoring the complete expression, as in the following examples.

**EXAMPLE 4**

$$\begin{aligned} 3x^2 + 6xy + 5x + 10y &= (3x^2 + 6xy) + (5x + 10y) \\ &= 3x(x + 2y) + 5(x + 2y) && \text{Factor } 3x \text{ from first parentheses and 5 from second using the distributive law} \\ &= (3x + 5)(x + 2y) && \text{Since } x + 2y \text{ is a common factor} \end{aligned}$$

**EXAMPLE 5** $6ab^2 - 3ab - 14b + 7 = (6ab^2 - 3ab) - (14b - 7)$ Note the −7, not 7

$= 3ab\ (2b - 1) - 7\ (2b - 1)$ $3ab$ from first parentheses and 7 from second

$= (3ab - 7)\ (2b - 1)$ Since $2b - 1$ is a common factor

**EXAMPLE 6** $4x^2 + 2xy - 2x - y = (4x^2 + 2xy) - (2x + y)$ $2x + y$, not $2x - y$

$= 2x\ (2x + y) - (1)\ (2x + y)$ $2x$ from first parentheses and 1 from second

$= (2x - 1)\ (2x + y)$ Since $2x + y$ is a common factor

Generally, we first try to find any common factors in all the terms of an expression. Once we have done this, if the expression has four terms, we try to **factor by grouping** as illustrated in the preceding three examples.

## EXERCISES 3.6

*Factor using the distributive law.*

**1.** $5x + 10y$

**2.** $6x - 9y$

**3.** $5a^2b - 10ab^2$

**4.** $14a^2b^2 + 21ab$

**5.** $3x^2y^3 + 27xy^2$

**6.** $8x^3y^3 - 12x^2y^2$

**7.** $7a^2b - 14ab^2 + 28ab$

**8.** $-6a^5b^5 - 8a^4b^4 - 4a^3b^2$

**9.** $10xy + 100x + 1000y$

**10.** $12x^3y^2 + 18x^2y^3 - 24x^2y^2$

**11.** $3a(a + b) - 2b(a + b)$

**12.** $5ab(2a - b) + 4a(2a - b)$

**13.** $5x^2 + 2xy + 10x + 4y$

**14.** $7x^2 - 14xy + 4x - 8y$

**15.** $2a^2b + 6ab - 3a - 9$

**16.** $3a^2b^2 - 9ab^2 - 5a + 15$

**17.** $5x^2y - 5xy + 20x - 20$

**18.** $6x^2y^2 + 6xy - 24xy^2 - 24y$

*Multiply.*

**19.** $6a^2b(5a^2 - 4ab + 10b^2)$

**20.** $(5a - 3b)(10a + 7b)$

**21.** $\begin{array}{l} 8x^2 - 4xy + 5y^2 \\ \underline{7x \quad + 3y \qquad\quad} \end{array}$

**22.** $\begin{array}{l} \frac{1}{3}x^2 + \frac{1}{2}xy + \frac{2}{3}y^2 \\ \underline{2x \quad - 3y \qquad\qquad} \end{array}$

**23.** $(3a + 2b)(3a - 2b)$

**24.** $(5a + 2b)^2$

**25.** $(-4x + 3y)^2$

**26.** $(2x^2 - y)(2x^2 + y)$

---

ANSWERS: **1.** $5(x + 2y)$ **2.** $3(2x - 3y)$ **3.** $5ab(a - 2b)$ **4.** $7ab(2ab + 3)$ **5.** $3xy^2(xy + 9)$ **6.** $4x^2y^2(2xy - 3)$ **7.** $7ab(a - 2b + 4)$ **8.** $-2a^3b^2(3a^2b^3 + 4ab^2 + 2)$ **9.** $10(xy + 10x + 100y)$ **10.** $6x^2y^2(2x + 3y - 4)$ **11.** $(3a - 2b)(a + b)$ **12.** $a(5b + 4)(2a - b)$ **13.** $(5x + 2y)(x + 2)$ **14.** $(x - 2y)(7x + 4)$ **15.** $(a + 3)(2ab - 3)$ **16.** $(a - 3)(3ab^2 - 5)$ **17.** $5(x - 1)(xy + 4)$ **18.** $6y(xy + 1)(x - 4)$ **19.** $30a^4b - 24a^3b^2 + 60a^2b^3$ **20.** $50a^2 + 5ab - 21b^2$ **21.** $56x^3 - 4x^2y + 23xy^2 + 15y^3$ **22.** $\frac{2}{3}x^3 - \frac{1}{6}xy^2 - 2y^3$ **23.** $9a^2 - 4b^2$ **24.** $25a^2 + 20ab + 4b^2$ **25.** $16x^2 - 24xy + 9y^2$ **26.** $4x^4 - y^2$

## 3.7 Factoring Trinomials

In order to factor a trinomial into two binomials, we must first recognize the patterns involved in multiplication of binomials. Consider the following product.

$$\begin{aligned}
(5x - 3y)(2x + 7y) &= \underset{\text{F}}{(5x) \cdot (2x)} + \underset{\text{O}}{(5x) \cdot (7y)} + \underset{\text{I}}{(-3y) \cdot (2x)} + \underset{\text{L}}{(-3y) \cdot (7y)} \\
&= (5 \cdot 2)x^2 + (5 \cdot 7)xy + (-3)(2)xy + (-3) \cdot (7)y^2 \\
&= 10x^2 + 35xy - 6xy - 21y^2 \\
&= \underset{\text{F}}{10x^2} + \underset{\text{O}+\text{I}}{29xy} - \underset{\text{L}}{21y^2}
\end{aligned}$$

We see that the first two terms of the binomials multiply

to give $10x^2$ (F)

The last two terms multiply

$$\text{to give } -21y^2 \quad \text{(L)}$$

and the $29xy$ comes from the products $5x \cdot 7y$ and $-3y \cdot 2x$,

$$35xy - 6xy. \quad \text{(O) and (I)}$$

Thus to factor $10x^2 + 29xy - 21y^2$ we must factor $10x^2$, factor $-21y^2$, and determine which products forming the middle terms will add to give $29xy$. There are two ways to factor 10:

$$2 \cdot 5 \qquad \text{and} \qquad 10 \cdot 1$$

four ways to factor $-21$:

$$(-3) \cdot 7, \qquad 3 \cdot (-7), \qquad (-1) \cdot 21, \qquad 1 \cdot (-21)$$

and we must choose the right combination. Unfortunately, such a "right combination" may not even exist since not all trinomials can be factored.

To minimize our work in looking for the factors of a trinomial, we first consider appropriate signs for the factors. If we write the trinomial as

$$ax^2 + bxy + cy^2, \quad a > 0$$

and assume that all common factors have been removed (including $-1$ if necessary to make $a > 0$), then the signs of the binomial factors can be determined. Of course, with $a > 0$ we can make both coefficients of the $x$ terms positive and be concerned only with the coefficients of $y$. The cases are given in the following table. (The $a$, $b$, and $c$ are the coefficients in $ax^2 + bxy + cy^2$.)

| $a$ | $b$ | $c$ | Coefficients of $y$ terms of binomials | Example |
|---|---|---|---|---|
| + | + | + | Both + | $x^2 + 7xy + 12y^2 = (x + 3y)(x + 4y)$ |
| + | − | + | Both − | $x^2 - 7xy + 12y^2 = (x - 3y)(x - 4y)$ |
| + | + | − | One + and one − | $x^2 + xy - 12y^2 = (x + 4y)(x - 3y)$ |
| + | − | − | One + and one − | $x^2 - xy - 12y^2 = (x + 3y)(x - 4y)$ |

We are now prepared to illustrate the factoring technique with several examples. First we consider trinomials in one variable where the constant term replaces the coefficient of $y$.

**EXAMPLE 1** Factor $x^2 + 5x + 6$.

Since all coefficients are positive we put only positive numbers in the table of factors.

| *Factors of a* | *Factors of c* |
|---|---|
| 1, 1 | 6, 1 |
| | 2, 3 |

We now look for the pair of factors which, added, give the $5x$ term.

$x^2 + 5x + 6 = (_x + _)(_x + _)$ — We must fill in the blanks

$= (x + _)(x + _)$ — The coefficients of $x$ are obvious

$\stackrel{?}{=} (x + 6)(x + 1)$ — This does not work because $6x + x = 7x \neq 5x$

$\stackrel{?}{=} (x + 2)(x + 3)$ — This works since $2 \cdot 3 = 6$ and $3x + 2x = 5x$

$x^2 + 5x + 6 = (x + 2)(x + 3)$

To check we multiply.

$$(x + 2)(x + 3) = x^2 + 3x + 2x + 6 = x^2 + 5x + 6$$

**EXAMPLE 2** Factor $x^2 - 5x + 6$.

Since $a > 0$, $b < 0$, and $c > 0$, we have the following table of factors.

| *Factors of a* | *Factors of c* |
|---|---|
| 1, 1 | −6, −1 |
| | −2, −3 |

We now look for the pair of factors which, added, give the $-5x$ term.

$x^2 - 5x + 6 = (x + _)(x + _)$ — The coefficients of $x$ are obvious

$\stackrel{?}{=} (x + (-6))(x + (-1))$ — Does not work

$\stackrel{?}{=} (x + (-2))(x + (-3))$ — This works since $(-2)(-3) = 6$ and $-3x - 2x = -5x$

$x^2 - 5x + 6 = (x - 2)(x - 3)$

To check we multiply.

$$(x - 2)(x - 3) = x^2 - 3x - 2x + 6 = x^2 - 5x + 6$$

**EXAMPLE 3** Factor $x^2 + x - 6$.

Since $a > 0$, $b > 0$, and $c < 0$, we have the following table of factors.

| *Factors of a* | *Factors of c* |
|---|---|
| 1, 1 | −6, 1 |
| | 6, −1 |
| | −3, 2 |
| | 3, −2 |

We now look for the pair of factors which, added, give the $1 \cdot x$ term.

$$x^2 + x - 6 = (x + __)(x + __)$$

$\stackrel{?}{=} (x - 6)(x + 1)$ Does not work but we do need one + and one − to obtain −6

$\stackrel{?}{=} (x - 3)(x + 2)$ $(-3) \cdot 2 = -6$ but $-3x + 2x = -x$

$\stackrel{?}{=} (x + 3)(x - 2)$ $3 \cdot (-2) = -6$ and $3x - 2x = x$

$$x^2 + x - 6 = (x + 3)(x - 2)$$

To check we multiply.

$$(x + 3)(x - 2) = x^2 + 3x - 2x - 6 = x^2 + x - 6$$

**EXAMPLE 4** Factor $x^2 - x - 6$.

Since $a > 0$, $b < 0$, and $c < 0$, we have the same table of factors as in Example 3.

| *Factors of a* | *Factors of c* |
|---|---|
| 1, 1 | −6, 1 |
| | 6, −1 |
| | −3, 2 |
| | 3, −2 |

We now look for the pair of factors which give the $-x$ term.

$$x^2 - x - 6 = (x - 3)(x + 2)$$ After Example 3 this is easy

To check we multiply.

$$(x - 3)(x + 2) = x^2 - 3x + 2x - 6 = x^2 - x - 6$$

Our results would have been similar had the trinomials we were factoring been in two variables. Compare the following with the four preceding examples.

**EXAMPLE 5**

$$x^2 + 5xy + 6y^2 = (x + 2y)(x + 3y)$$
$$x^2 - 5xy + 6y^2 = (x - 2y)(x - 3y)$$
$$x^2 + xy - 6y^2 = (x - 2y)(x + 3y)$$
$$x^2 - xy - 6y^2 = (x + 2y)(x - 3y)$$

Thus, factoring with two or more variables can be as easy as with one variable. We now consider other examples in two variables.

**EXAMPLE 6** Factor $3u^2 + 17uv + 10v^2$.

| *Factors of a* | *Factors of c* |
|---|---|
| 3, 1 | 10, 1 |
| | 1, 10 |
| | 5, 2 |
| | 2, 5 |

Since the factors of $a$ are different, we must try the factors of $c$ in both orders (10, 1 and 1, 10; 5, 2 and 2, 5).

$$3u^2 + 17uv + 10v^2 = (_u + _v)(_u + _v)$$

| | |
|---|---|
| $= (3u + _v)(u + _v)$ | The only factors of 3 are 3 and 1 |
| $\stackrel{?}{=} (3u + 10v)(u + v)$ | Does not work since $3uv + 10uv = 13uv$ |
| $\stackrel{?}{=} (3u + v)(u + 10v)$ | Does not work since $30uv + uv = 31uv$ |
| $\stackrel{?}{=} (3u + 5v)(u + 2v)$ | Does not work since $6uv + 5uv = 11uv$ |
| $\stackrel{?}{=} (3u + 2v)(u + 5v)$ | This works since $15uv + 2uv = 17uv$ |

$$3u^2 + 17uv + 10v^2 = (3u + 2v)(u + 5v)$$

To check we multiply.

$$(3u + 2v)(u + 5v) = 3u^2 + 15uv + 2uv + 10v^2 = 3u^2 + 17uv + 10v^2$$

**EXAMPLE 7** Factor $8x^2 - 13xy + 5y^2$.

| *Factors of a* | *Factors of c* |
|---|---|
| 8, 1 | −1, −5 |
| 1, 8 | |
| 4, 2 | |
| 2, 4 | |

$$8x^2 - 13xy + 5y^2 = (_x + _y)(_x + _y)$$
$$= (_x - y)(_x - 5y)$$

| | |
|---|---|
| $\stackrel{?}{=} (8x - y)(x - 5y)$ | Does not work since $-40xy - xy = -41xy$ |
| $\stackrel{?}{=} (x - y)(8x - 5y)$ | This works because $-5xy - 8xy = -13xy$ |

$$8x^2 - 13xy + 5y^2 = (x - y)(8x - 5y)$$

To check we multiply.

$$(x - y)(8x - 5y) = 8x^2 - 5xy - 8xy + 5y^2$$
$$= 8x^2 - 13xy + 5y^2$$

**EXAMPLE 8** Factor $15z^2w^2 - 23zw - 28$.

| *Factors of a* | *Factors of c* | | |
|---|---|---|---|
| 15, 1 | 28, −1 | and | −28, 1 |
| 5, 3 | 1, −28 | and | −1, 28 |
| | 14, −2 | and | −14, 2 |
| | 2, −14 | and | −2, 14 |
| | 7, −4 | and | −7, 4 |
| | 4, −7 | and | −4, 7 |

Since there are many possibilities in this case, we try to make selections which seem more likely, and keep a record of our trials by crossing out the pairs in the table which fail to produce the factors.

| | |
|---|---|
| $15z^2w^2 - 23zw - 28 = (_zw + _)(_zw - _)$ | One constant is positive and one negative |
| $\stackrel{?}{=} (15zw + 28)(zw - 1)$ | Does not work |
| $\stackrel{?}{=} (5zw + 7)(3zw - 4)$ | Does not work |
| $\stackrel{?}{=} (5zw + 4)(3zw - 7)$ | This works |

$$15z^2w^2 - 23zw - 28 = (5zw + 4)(3zw - 7)$$

To check we multiply.

$$(5zw + 4)(3zw - 7) = 15z^2w^2 - 35zw + 12zw - 28$$
$$= 15z^2w^2 - 23zw - 28$$

**EXAMPLE 9** Factor $-6x^2 - 21xy + 45y^2$.

We first factor out the common factor and make $a > 0$.

$$-6x^2 - 21xy + 45y^2 = -3(2x^2 + 7xy - 15y^2)$$

Next, we list the factors of $a$ and $c$ for the trinomial $2x^2 + 7xy - 15y^2$.

| *Factors of a* | *Factors of c* |
|---|---|
| 2, 1 | 15, −1 and −15, 1 |
| | 1, −15 and −1, 15 |
| | 5, −3 and −5, 3 |
| | 3, −5 and −3, 5 |

$$-6x^2 - 21xy + 45y^2 = -3(2x^2 + 7xy - 15y^2) \quad \text{First factor out the common factor}$$
$$\overset{?}{=} -3(2x + _y)(x + _y)$$
$$\overset{?}{=} -3(2x + 5y)(x - 3y) \quad \text{Does not work}$$
$$\overset{?}{=} -3(2x - 5y)(x + 3y) \quad \text{Does not work}$$
$$\overset{?}{=} -3(2x - 3y)(x + 5y) \quad \text{Works}$$

$$-6x^2 - 21xy + 45y^2 = -3(2x - 3y)(x + 5y)$$

Check this by multiplying.

The methods used in the preceding examples are summarized in the following general rule for factoring a trinomial.

**TO FACTOR A TRINOMIAL OF THE FORM $ax^2 + bxy + cy^2$**

1. Factor out any common factor, including (−1) if the coefficient of $x^2$ is negative.
2. Factor the coefficient, $a$, of $x^2$ and the coefficient, $c$, of $y^2$ in order to determine the combination of coefficients with sum $b$.
3. Use the table below for the signs in the binomial factors.

| $a$ | $b$ | $c$ | Coefficient of $y$ terms of binomials | Example |
|---|---|---|---|---|
| + | + | + | Both + | $x^2 + 5xy + 6y^2 = (x + 2y)(x + 3y)$ |
| + | − | + | Both − | $x^2 - 5xy + 6y^2 = (x - 2y)(x - 3y)$ |
| + | + | − | One + and one − | $x^2 + xy - 6y^2 = (x + 3y)(x - 2y)$ |
| + | − | − | One + and one − | $x^2 - xy - 6y^2 = (x + 2y)(x - 3y)$ |

## EXERCISES 3.7

*Factor and check by multiplying.*

**1.** $x^2 + 6x + 5$

**2.** $x^2 - 6x + 5$

**3.** $x^2 + 4x - 5$

**4.** $x^2 - 4x - 5$

**5.** $y^2 - 4y - 21$

**6.** $y^2 - 10y + 21$

**7.** $y^2 + 4y - 21$

**8.** $y^2 + 10y + 21$

**9.** $u^2 + 2u - 35$

**10.** $u^2 - 12u + 27$

**11.** $w^2 + 4w - 45$

**12.** $w^2 - 2w - 63$

**13.** $x^2 - x - 56$

**14.** $x^2 + 15x + 56$

**15.** $y^2 + 12y + 36$

**16.** $y^2 - 14y + 49$

**17.** $u^2 + 21u + 110$

**18.** $u^2 - u - 110$

**19.** $x^2 + 6xy + 5y^2$

**20.** $x^2 - 6xy + 5y^2$

**21.** $x^2 + 4xy - 5y^2$

**22.** $x^2 - 4xy - 5y^2$

**23.** $u^2 + 15uv + 54v^2$

**24.** $u^2 - 2uv - 35v^2$

**25.** $2w^2 + 19w + 24$

**26.** $2w^2 - 13w - 24$

**27.** $2x^2 + 13xy - 24y^2$

**28.** $2x^2 - 19xy + 24y^2$

29. $2u^2v^2 - 7uv - 30$

30. $5u^2v^2 + 11uv + 2$

31. $-4w^2 - 34w - 70$
[*Hint:* Don't forget the common factor.]

32. $7w^2 - 14w + 7$

33. $5x^2 + 17xy - 12y^2$

34. $6x^2 + 23xy + 20y^2$

35. $8u^2 - 8v^2$

36. $-45u^2 + 150uv - 125v^2$

37. $6x^2 - 19xy - 20y^2$

38. $4x^2 - 32xy + 63y^2$

39. $2u^2 - 3uv - 4uv + 6v^2$

40. $21u^2 + 12uv - 35uv - 20v^2$

---

ANSWERS: If all answers were given, it would be too easy to avoid factoring. Thus, only selected answers are given. *Check all answers by multiplication.* **1.** $(x + 1)(x + 5)$ **2.** $(x - 1)(x - 5)$ **3.** $(x - 1)(x + 5)$ **4.** $(x + 1)(x - 5)$ **5.** $(y + 3)(y - 7)$ **9.** $(u - 5)(u + 7)$ **10.** $(u - 3)(u - 9)$ **12.** $(w + 7)(w - 9)$ **14.** $(x + 7)(x + 8)$ **15.** $(y + 6)(y + 6)$ **17.** $(u + 10)(u + 11)$ **19.** $(x + y)(x + 5y)$ **24.** $(u + 5v)(u - 7v)$ **25.** $(w + 8)(2w + 3)$ **29.** $(uv - 6)(2uv + 5)$ **31.** $(-2)(w + 5)(2w + 7)$ **35.** $8(u - v)(u + v)$ **37.** $(x - 4y)(6x + 5y)$ **38.** $(2x - 7y)(2x - 9y)$ **40.** $(3u - 5v)(7u + 4v)$

## 3.8 Factoring Using Special Formulas

In this section we factor using the following formulas.

| | |
|---|---|
| 1. $a^2 - b^2 = (a + b)(a - b)$ | Difference of squares |
| 2. $a^2 + 2ab + b^2 = (a + b)^2$ | Perfect square |
| 3. $a^2 - 2ab + b^2 = (a - b)^2$ | Perfect square |
| 4. $a^3 + b^3 = (a + b)(a^2 - ab + b^2)$ | Sum of cubes |
| 5. $a^3 - b^3 = (a - b)(a^2 + ab + b^2)$ | Difference of cubes |

We used the first three formulas in Section 3.5 to multiply binomials. The last two formulas are new. All five must be memorized before we can use them effectively in factoring. The following five examples illustrate (in the order given in the rule) the use of the five formulas. Verify each formula by direct multiplication.

**EXAMPLE 1** **(a)** $x^2 - 49 = x^2 - 7^2$

$= (x + 7)(x - 7)$ — $x = a$ and $7 = b$ in $a^2 - b^2 = (a + b)(a - b)$

**(b)** $16u^2 - 25v^2 = 4^2u^2 - 5^2v^2$

$= (4u)^2 - (5v)^2$ — $4^2u^2 = (4u)^2$

$= (4u + 5v)(4u - 5v)$ — $4u = a$ and $5v = b$ in $a^2 - b^2 = (a + b)(a - b)$

**(c)** $18x^6 - 8y^4 = 2(9x^6 - 4y^4)$ — Factor out common factor

$= 2[3^2(x^3)^2 - 2^2(y^2)^2]$

$= 2[(3x^3)^2 - (2y^2)^2]$ — $3x^3 = a$ and $2y^2 = b$

$= 2(3x^3 + 2y^2)(3x^3 - 2y^2)$

**EXAMPLE 2** **(a)** $u^2 + 10uv + 25v^2 = u^2 + 2 \cdot 5uv + 5^2v^2$ — First and last terms are perfect squares

$= u^2 + 2 \cdot u \cdot (5v) + (5v)^2$ — $u = a,\ 5v = b$

$= (u + 5v)^2$

**(b)** $9x^2 + 42xy + 49y^2 = 3^2x^2 + 2 \cdot 3 \cdot 7xy + 7^2y^2$ — $3x = a,\ 7y = b$

$= (3x)^2 + 2(3x)(7y) + (7y)^2$

$= (3x + 7y)^2$

**EXAMPLE 3** **(a)** $u^2 - 10uv + 25v^2 = u^2 - 2 \cdot 5uv + 5^2v^2$

$= u^2 - 2 \cdot u \cdot (5v) + (5v)^2$ — $u = a,\ 5v = b$

$= (u - 5v)^2$

**(b)** $36x^4 - 84x^2y + 49y^2 = 6^2(x^2)^2 - 2 \cdot 6 \cdot 7x^2y + 7^2y^2$

$= (6x^2)^2 - 2 \cdot (6x^2) \cdot (7y) + (7y)^2$ — $6x^2 = a,\ 7y = b$

$= (6x^2 - 7y)^2$

**EXAMPLE 4** (a) $x^3 + 8 = x^3 + 2^3$
$= (x + 2)(x^2 - 2x + 4)$

$x = a$ and $2 = b$ in $a^3 + b^3 = (a + b)(a^2 - ab + b^2)$

(b) $u^3 + 27v^3 = u^3 + 3^3v^3$
$= u^3 + (3v)^3$
$= (u + 3v)[u^2 - u \cdot (3v) + (3v)^2]$
$= (u + 3v)(u^2 - 3uv + 9v^2)$

$u = a,\ 3v = b$

(c) $8x^3 + 125y^6 = 2^3x^3 + 5^3(y^2)^3$
$= (2x)^3 + (5y^2)^3$
$= (2x + 5y^2)[(2x)^2 - (2x)(5y^2) + (5y^2)^2]$
$= (2x + 5y^2)(4x^2 - 10xy^2 + 25y^4)$

$2x = a,\ 5y^2 = b$

**EXAMPLE 5** (a) $y^3 - 64 = y^3 - 4^3$
$= (y - 4)(y^2 + 4y + 16)$

$y = a$ and $4 = b$ in $a^3 - b^3 = (a - b)(a^2 + ab + b^2)$

(b) $u^3 - 27v^3 = u^3 - (3v)^3$
$= (u - 3v)[u^2 + u(3v) + (3v)^2]$
$= (u - 3v)(u^2 + 3uv + 9v^2)$

$u = a,\ 3v = b$

(c) $125x^3y^3 - 64 = 5^3x^3y^3 - 4^3$
$= (5xy)^3 - 4^3$
$= (5xy - 4)[(5xy)^2 + (5xy)(4) + (4)^2]$
$= (5xy - 4)(25x^2y^2 + 20xy + 16)$

$5xy = a,\ 4 = b$

In the next examples, each factorization uses two of the given formulas.

**EXAMPLE 6** (a) $x^6 - 1 = (x^3)^2 - 1^2$

Factor difference of squares first with $x^3 = a$ and $1 = b$

$= (x^3 + 1)(x^3 - 1)$
$= [(x + 1)(x^2 - x + 1)][(x - 1)(x^2 + x + 1)]$

Sum and difference of cubes

$= (x + 1)(x - 1)(x^2 - x + 1)(x^2 + x + 1)$

(b) $x^2 + 2x + 1 - y^2 = (x + 1)^2 - y^2$
$= [(x + 1) + y][(x + 1) - y]$
$= (x + 1 + y)(x + 1 - y)$

$x + 1 = a,\ y = b$

**EXAMPLE 7** $x^4 - 1 = (x^2)^2 - 1^2$ — $x^2 = a,\ 1 = b$
$= (x^2 - 1)(x^2 + 1)$
$= (x - 1)(x + 1)(x^2 + 1)$ — $x^2 + 1$ cannot be factored

## EXERCISES 3.8

*Factor and check by multiplying.*

**1.** $x^2 - 9$

**2.** $x^2 + 6x + 9$

**3.** $y^2 - 6y + 9$

**4.** $y^3 + 8$

**5.** $u^3 - 8$

**6.** $u^2 - 14u + 49$

**7.** $y^2 - 64$

**8.** $y^3 + 64$

**9.** $u^2 - 12uv + 36v^2$

**10.** $u^2 + 12uv + 36v^2$

**11.** $7x^2 - 7y^2$

**12.** $7x^3 + 7y^3$

**13.** $8x^3 + 27y^3$

**14.** $8x^3 - 27y^3$

**15.** $4u^2 + 12uv + 9v^2$

**16.** $4u^2 - 12uv + 9v^2$

**17.** $20x^2 - 45y^2$

**18.** $x^2 + y^2$

**19.** $9u^2 + 6uv + v^2$

**20.** $u^2 + 2uv + 2v^2$

**21.** $40x^3 + 5y^9$

**22.** $49x^4 + 14x^2y^2 + y^4$

**23.** $25u^4 - 9v^6$

**24.** $9u^6 - 30u^3v^2 + 25v^4$

**25.** $125u^3 + 27v^3$

**26.** $5u^3 - 5000v^3$

**27.** $x^2 - 14xy + 48y^2$

**28.** $x^2 - 2xy - 48y^2$

**29.** $6x^2 - 17xy + 10y^2$

**30.** $35x^2 - 2xy - y^2$

Before completing this exercise set, we summarize the factorization techniques that we have learned.

**1.** Factor out any common factor, including (−1) if $a$ is negative, in

$$ax^2 + bxy + cy^2.$$

**2.** To factor a binomial, use

$$a^2 - b^2 = (a + b)(a - b)$$
$$a^3 + b^3 = (a + b)(a^2 - ab + b^2)$$
$$a^3 - b^3 = (a - b)(a^2 + ab + b^2).$$

**3.** To factor a trinomial, use the technique of Section 3.7, but also consider perfect squares using the rules

$$a^2 + 2ab + b^2 = (a + b)^2 \quad \text{and} \quad a^2 - 2ab + b^2 = (a - b)^2.$$

**4.** To factor a polynomial with four terms, try to factor by grouping.

*Use these steps in the rest of this exercise set.*

**31.** $-5x - 5y$

**32.** $27x^2y - 9x^2y^2 + 81xy^2$

**33.** $-10u^2 + 40v^2$

**34.** $7u^3 + 56v^3$

**35.** $2x^2 - 4xy + 2y^2$

**36.** $u^2 + 18uv + 80v^2$

**37.** $u^2 + 22uv + 121v^2$

**38.** $3x^2 - 36xy + 105y^2$

**39.** $4x^2y^2 - 4xy + 1$

**40.** $21u^2 - 4uv - v^2$

**41.** $250x^3 + 54y^3$

**42.** $3u^2 - 2uv - 16v^2$

**43.** $36u^2 + 60uv + 25v^2$

**44.** $x^2 + 5xy + 5y^2$

**45.** $-5x^2 + 70xy - 240y^2$

**46.** $u^6 - v^6$
[*Hint:* Factor difference of squares first.]

**47.** $(x + y)^2 - 9$
[*Hint:* $(x + y) = a$ and $3 = b$ in $a^2 - b^2$.]

**48.** $(x + y)^2 + 2(x + y) + 1$

**49.** $27u^3v^3 - 8$

**50.** $3uv + u - 3v^2 - v$

**51.** $30x^2 + 31xy - 21y^2$

**52.** $x^3y^3 + z^3$

**53.** $9x^6 - 42x^3y + 49y^2$

**54.** $-16x^4 + 64y^2$

**55.** $-18u^2 - 69uv - 60v^2$

**56.** $5x^2 + 15xy + 15y^2$

---

**ANSWERS:** Because of the nature of these problems, only selected answers are given. *Check all answers by multiplication.* **1.** $(x + 3)(x - 3)$ **2.** $(x + 3)^2$ **3.** $(y - 3)^2$ **4.** $(y + 2)(y^2 - 2y + 4)$ **5.** $(u - 2)(u^2 + 2u + 4)$ **6.** $(u - 7)^2$ **7.** $(y + 8)(y - 8)$ **8.** $(y + 4)(y^2 - 4y + 16)$ **9.** $(u - 6v)^2$ **11.** $7(x + y)(x - y)$ **13.** $(2x + 3y)(4x^2 - 6xy + 9y^2)$ **17.** $5(2x + 3y)(2x - 3y)$ **18.** cannot be factored **19.** $(3u + v)^2$ **20.** cannot be factored **21.** $5(2x + y^3)(4x^2 - 2xy^3 + y^6)$ **27.** $(x - 6y)(x - 8y)$ **29.** $(6x - 5y)(x - 2y)$ **30.** $(7x + y)(5x - y)$ **31.** $-5(x + y)$ **32.** $9xy(3x - xy + 9y)$ **33.** $-10(u + 2v)(u - 2v)$ **34.** $7(u + 2v)(u^2 - 2uv + 4v^2)$ **37.** $(u + 11v)^2$ **38.** $3(x - 5y)(x - 7y)$ **39.** $(2xy - 1)^2$ **44.** cannot be factored **46.** $(u + v)(u - v)(u^2 - uv + v^2)(u^2 + uv + v^2)$ **47.** $(x + y + 3)(x + y - 3)$ **48.** $(x + y + 1)^2$ **50.** $(u - v)(3v + 1)$ **51.** $(15x - 7y)(2x + 3y)$ **52.** $(xy + z)(x^2y^2 - xyz + z^2)$ **53.** $(3x^3 - 7y)^2$ **56.** $5(x^2 + 3xy + 3y^2)$—no more factoring possible

## Chapter 3 Summary

### Words and Phrases for Review

[3.1] variable
coefficient
prime factor
polynomial
monomial
binomial
trinomial
ascending order
descending order

[3.6] common factor
factor completely
factor by grouping

### Brief Reminders

[3.1] Only like terms can be collected. For example, $3x^2 - 2x$ cannot be combined since $3x^2$ and $-2x$ are *not* like terms.

[3.2] Change all signs when removing parentheses preceded by a minus sign. For example, $3x^3 - (2x^2 - 5x + 6) = 3x^3 - 2x^2 + 5x - 6$.

[3.4] Write out all details when multiplying polynomials. Use the FOIL or column method.

[3.5] Special products:

**1.** $(a + b)(a - b) = a^2 - b^2$

**2.** $(a + b)(a + b) = (a + b)^2 = a^2 + 2ab + b^2$ (not $a^2 + b^2$)

**3.** $(a - b)(a - b) = (a - b)^2 = a^2 - 2ab + b^2$ (not $a^2 - b^2$)

[3.8] Difference of squares: $a^2 - b^2 = (a + b)(a - b)$
Perfect square: $a^2 + 2ab + b^2 = (a + b)^2$ and $a^2 - 2ab + b^2 = (a - b)^2$
Sum of cubes: $a^3 + b^3 = (a + b)(a^2 - ab + b^2)$
Difference of cubes: $a^3 - b^3 = (a - b)(a^2 + ab + b^2)$

## CHAPTER 3 REVIEW EXERCISES

[3.1] **1.** The polynomial $5x^3 - 3x^2 + 2x + 8$ has (a) ________ terms, and the coefficient of the term $5x^3$ is (b) ________.

**2.** A polynomial with two terms is called a ________________.

[3.3] **3.** The term $-7x^3y$ and the term ____________ in the polynomial $x^5y^3 + 2x^3y^2 - x^3y + 8$ are like terms.

[3.4] **4.** The FOIL method is helpful when multiplying two ____________.

[3.5] **5.** The three special products we should know are:

(a) $(a + b)(a - b) =$ ________________

(b) $(a + b)(a + b) =$ ________________

(c) $(a - b)(a - b) =$ ________________

[3.6] **6.** In any factoring problem, the first step is always to try and factor out any ________________________ factors.

**7.** If an expression has four terms, we should try factoring by________.

[3.7] **8.** When factoring $ax^2 + bxy + cy^2$, the signs of the coefficients of the $y$ terms in the binomial factors are:

(a) ________________ when $a > 0$, $b > 0$, and $c > 0$.

(b) ________________ when $a > 0$, $b > 0$, and $c < 0$.

(c) ________________ when $a > 0$, $b < 0$, and $c > 0$.

(d) ________________ when $a > 0$, $b < 0$, and $c < 0$.

[3.8] **9.** When factoring a binomial, look for the sum of (a) ______, and the difference of (b) __________ and (c) __________ forms.

**10.** Give the factors of each of the following:

(a) $a^2 - b^2 =$ ________________

(b) $a^3 - b^3 =$ ________________

(c) $a^3 + b^3 =$ ________________

(d) $a^2 - 2ab + b^2 =$ ________________

(e) $a^2 + 2ab + b^2 =$ ________________

[3.1] **11.** Collect like terms and write in descending order.
$-3x + 7x^3 - 14 + 7x - 5x^4 + 4x^3 + 10$

[3.2] **12.** Add $4y^2 - 16y + y^3 - 2$ and $-3y + 7y^3 + 2 - y^2$

**13.** Subtract $6y + 5 - 7y^3 + 8y^2$ from $10 - 12y^2 + y^3 + 8y$

[3.3] **14.** Add $5a^2b^2 - 5ab + 14$ and $6a^2b^2 + 7ab - 5$

**15.** Subtract $-10x^3y^2 + 6x^2y^2 - 5xy$ from $5x^3y^2 - 4x^2y^2 + 12$

**16.** $(7a^3b^2 - 3a^2b) + (6a^3b^2 + 5ab^2) - (-2a^3b^2 + 7ab^2)$

[3.4–3.5] *Multiply.*

**17.** $(3x - 5y)(7x + 4y)$

**18.** $(4x - 9y)(2x - 7y)$

**19.** $(4u + 11v)(2u + 3v)$

**20.** $(2u + 9v)(5u - v)$

**21.** $\begin{array}{l} 6x^2y^2 - 7xy + 5 \\ \underline{4xy \quad + 3} \end{array}$

**22.** $\begin{array}{l} 5x^2 + 11xy - 6y^2 \\ \underline{3x \quad - \quad 7y} \end{array}$

**23.** $(2u - 5v)(2u + 5v)$

**24.** $(7u - v)^2$

**25.** $(3x^2 + 2y)^2$

**26.** $(2x^2 - 5y^2)(2x^2 + 5y^2)$

[3.6–3.8] *Factor.*

**27.** $5u^2v^2 - 10uv$

**28.** $-11u^3v^2 - 77u^2v^2 - 88uv^2$

**29.** $x^2 - 4x - 21$

**30.** $2x^2 - x - 10$

**31.** $3y^2 + 23y + 14$

**32.** $6y^2 + y - 40$

**33.** $u^2 + 10uv + 21v^2$

**34.** $6u^2 - 5uv - 6v^2$

**35.** $-5x^2 - 15xy + 50y^2$

**36.** $6x^2 - 54y^2$

**37.** $8u^3 + v^3$

**38.** $8u^3 - v^3$

**39.** $4x^2 - 12xy + 9y^2$

**40.** $9x^2 + 30xy + 25y^2$

**41.** $16u^4 - v^4$

**42.** $u^3v^3w^3 + 1$

**43.** $2x^2 + xy - 8x - 4y$

**44.** $(a - b)^2 - 4$

---

**ANSWERS:** **1.** (a) 4 (b) 5 **2.** binomial **3.** $-x^3y$ **4.** binomials **5.** (a) $a^2 - b^2$ (b) $a^2 + 2ab + b^2$ (c) $a^2 - 2ab + b^2$ **6.** common **7.** grouping **8.** (a) both + (b) one +, one − (c) both − (d) one +, one − **9.** (a) cubes (b) squares (c) cubes **10.** (a) $(a - b)(a + b)$ (b) $(a - b)(a^2 + ab + b^2)$ (c) $(a + b)(a^2 - ab + b^2)$ (d) $(a - b)^2$ (e) $(a + b)^2$ **11.** $-5x^4 + 11x^3 + 4x - 4$ **12.** $8y^3 + 3y^2 - 19y$ **13.** $8y^3 - 20y^2 + 2y + 5$ **14.** $11a^2b^2 + 2ab + 9$ **15.** $15x^3y^2 - 10x^2y^2 + 5xy + 12$ **16.** $15a^3b^2 - 3a^2b - 2ab^2$ **17.** $21x^2 - 23xy - 20y^2$ **18.** $8x^2 - 46xy + 63y^2$ **19.** $8u^2 + 34uv + 33v^2$ **20.** $10u^2 + 43uv - 9v^2$ **21.** $24x^3y^3 - 10x^2y^2 - xy + 15$ **22.** $15x^3 - 2x^2y - 95xy^2 + 42y^3$ **23.** $4u^2 - 25v^2$ **24.** $49u^2 - 14uv + v^2$ **25.** $9x^4 + 12x^2y + 4y^2$ **26.** $4x^4 - 25y^4$ **27.** $5uv(uv - 2)$ **28.** $-11uv^2(u^2 + 7u + 8)$ **29.** $(x - 7)(x + 3)$ **30.** $(2x - 5)(x + 2)$ **31.** $(3y + 2)(y + 7)$ **32.** $(2y - 5)(3y + 8)$ **33.** $(u + 3v)(u + 7v)$ **34.** $(3u + 2v)(2u - 3v)$ **35.** $-5(x + 5y)(x - 2y)$ **36.** $6(x - 3y)(x + 3y)$ **37.** $(2u + v)(4u^2 - 2uv + v^2)$ **38.** $(2u - v)(4u^2 + 2uv + v^2)$ **39.** $(2x - 3y)^2$ **40.** $(3x + 5y)^2$ **41.** $(2u - v)(2u + v)(4u^2 + v^2)$ **42.** $(uvw + 1)(u^2v^2w^2 - uvw + 1)$ **43.** $(2x + y)(x - 4)$ **44.** $(a - b - 2)(a - b + 2)$

# 4

# Exponents and Radicals

## 4.1 Integer Exponents and Scientific Notation

In Chapter 1 the properties of integer exponents were given in detail. We briefly review these properties for use in this chapter. Recall that if we write $3^5$, 3 is called the **base** and 5 the **exponent.**

For real numbers $a$ and $b$ and integers $m$ and $n$,

1. $a^n = \underbrace{a \cdot a \cdot a \cdots a}_{n \text{ factors}} \quad (n \geq 1).$
2. $a^m \cdot a^n = a^{m+n}.$
3. $a^m \div a^n = \dfrac{a^m}{a^n} = a^{m-n} \quad (a \neq 0).$
4. $(a^m)^n = a^{m \cdot n}.$
5. $a^0 = 1 \quad \text{for } a \neq 0.$
6. $a^{-n} = \dfrac{1}{a^n} \quad \text{and} \quad \dfrac{1}{a^{-n}} = a^n \quad \text{if } a \neq 0.$
7. $(a \cdot b)^n = a^n \cdot b^n.$
8. $\left(\dfrac{a}{b}\right)^n = \dfrac{a^n}{b^n} \quad (b \neq 0).$

**EXAMPLE 1** Write in exponential notation.

(a) $\underbrace{5 \cdot 5 \cdot 5 \cdot 5}_{4 \text{ factors}} = 5^4$

(b) $\underbrace{2 \cdot 2 \cdot 2}_{3} \cdot \underbrace{x \cdot x \cdot x \cdot x}_{4} = 2^3 \cdot x^4$

(c) $7 \cdot \underbrace{y \cdot y \cdot y}_{3\ y\text{'s}} = 7y^3$

(d) $\underbrace{(7y) \cdot (7y) \cdot (7y)}_{3 \text{ factors}} = (7y)^3$ Note that $7y^3 \neq (7y)^3$

(e) $\dfrac{(a+b)(a+b)(a+b)}{(a-b)(a-b)} = \dfrac{(a+b)^3}{(a-b)^2}$

**EXAMPLE 2** Write without using exponents.

**(a)** $5x^3 = 5 \cdot \underbrace{x \cdot x \cdot x}_{3\ x\text{'s}}$

**(b)** $(5x)^3 = \underbrace{(5x)(5x)(5x)}_{3\ 5x\text{'s}} = 5 \cdot 5 \cdot 5 \cdot x \cdot x \cdot x = 125 \cdot x \cdot x \cdot x$

**(c)** $2^2 + 3^2 = \underbrace{2 \cdot 2}_{2} + \underbrace{3 \cdot 3}_{2} = 4 + 9 = 13$

**(d)** $(2+3)^2 = \underbrace{(2+3)(2+3)}_{2 \text{ factors } (2+3)} = 5 \cdot 5 = 25$ $\quad 2^2 + 3^2 \neq (2+3)^2$

**(e)** $1^6 = 1 \cdot 1 \cdot 1 \cdot 1 \cdot 1 \cdot 1 = 1$ $\quad 1^n = 1$ for any $n$

**(f)** $\left(\frac{1}{3}\right)^4 = \left(\frac{1}{3}\right)\left(\frac{1}{3}\right)\left(\frac{1}{3}\right)\left(\frac{1}{3}\right) = \frac{1}{81}$

**(g)** $3^{-4} = \frac{1}{3^4} = \frac{1}{3 \cdot 3 \cdot 3 \cdot 3} = \frac{1}{81}$

Rules 2 through 6, reviewed in Examples 3, 4, and 5, involve only one base and cannot be used for expressions such as $a^m \cdot b^n$ and $\frac{a^m}{b^n}$. Rules 2 and 3 concern the product and quotient of exponential expressions.

**EXAMPLE 3** Simplify.

**(a)** $3^2 \cdot 3^3 = 3^{2+3} = 3^5$

**(b)** $5 \cdot x^3 \cdot x^4 = 5x^{3+4} = 5x^7$

**(c)** $2^2 \cdot 2^3 \cdot 2^5 = 2^{2+3+5} = 2^{10}$

**(d)** $\frac{3^3}{3^2} = 3^{3-2} = 3^1 = 3$

**(e)** $\frac{5x^4}{x^2} = 5x^{4-2} = 5x^2$

**(f)** $\frac{7y^2y^4}{y^3} = \frac{7y^{2+4}}{y^3} = \frac{7y^6}{y^3} = 7y^{6-3} = 7y^3$

**(g)** $\frac{5^2}{4^3}$ cannot be simplified using any of the rules, but is equal to $\frac{25}{64}$.

Rule 4 involves powers of exponential expressions, while Rule 5 is for evaluating expressions with zero exponents.

**EXAMPLE 4** **(a)** $(5^2)^3 = 5^{2 \cdot 3} = 5^6 = 15{,}625$

**(b)** $3(x^4)^5 = 3x^{4 \cdot 5} = 3x^{20}$

**(c)** $(35)^0 = 1$

**(d)** $(6x^3y^{10})^0 = 1$ $\quad$ If $x \neq 0$ and $y \neq 0$

Rule 6 shows how to rewrite exponential expressions with negative exponents. In the next example, it is used alone and with the other rules.

**EXAMPLE 5** (a) $5^{-3} = \frac{1}{5^3} = \frac{1}{125}$

(b) $\frac{1}{5^{-3}} = \frac{1}{\frac{1}{5^3}} = 1 \cdot \frac{5^3}{1} = 5^3 = 125$

(c) $\frac{5^2}{5^{-3}} = 5^{2-(-3)} = 5^5 = 3125$

(d) $3^2 \cdot 3^{-4} = 3^{2+(-4)} = 3^{2-4} = 3^{-2} = \frac{1}{3^2} = \frac{1}{9}$

(e) $(4^{-2})^3 = 4^{-6} = \frac{1}{4^6} = \frac{1}{4096}$

Rules 7 and 8 deal with the product and quotient of numbers raised to a single power.

**EXAMPLE 6** Simplify and write answers with positive exponents only.

(a) $(2x)^3 = 2^3 \cdot x^3 = 8x^3$

(b) $\left(\frac{5x^3}{y^2}\right)^2 = \frac{5^2 \cdot x^{3 \cdot 2}}{y^{2 \cdot 2}} = \frac{25x^6}{y^4}$

(c) $(3a)^{-2} = \frac{1}{(3a)^2} = \frac{1}{3^2a^2} = \frac{1}{9a^2}$

(d) $3a^{-2} = 3 \cdot \frac{1}{a^2} = \frac{3}{a^2}$ Note that $(3a)^{-2} \neq 3a^{-2}$

(e) $\left(\frac{6x^3y^{-2}}{x^{-4}y}\right)^2 = (6x^{3-(-4)}y^{-2-1})^2 = (6x^7y^{-3})^2 = \left(\frac{6x^7}{y^3}\right)^2 = \frac{6^2x^{14}}{y^6} = \frac{36x^{14}}{y^6}$

(f) $(7^0a^3b^{-2})^4 = (1 \cdot a^3b^{-2})^4 = (a^3b^{-2})^4 = a^{12}b^{-8} = \frac{a^{12}}{b^8}$

(g) $(7^4a^3b^{-2})^0 = 1$ If $a \neq 0$ and $b \neq 0$

(h) $\left(\frac{x}{y}\right)^{-1} = \frac{x^{-1}}{y^{-1}} = \frac{\frac{1}{x}}{\frac{1}{y}} = \frac{1}{x} \cdot \frac{y}{1} = \frac{y}{x}$

One important application of integer exponents is scientific notation. For example, chemists use the number (Avogadro's number)

602,000,000,000,000,000,000,000

but instead of writing all the zeros, they write

$6.02 \times 10^{23}$.

The second notation is easier to use in computations. Likewise, a scientist might use the number

$$.0000000000000000084$$

but he or she would write

$$8.4 \times 10^{-18}.$$

A number is written in **scientific notation** if it is expressed as the product of a number between 1 and 10 and a power of 10.

**TO WRITE A NUMBER IN SCIENTIFIC NOTATION**

1. Move the decimal point to the position immediately to the right of the first nonzero digit
2. Multiply by a power of ten which is equal in absolute value to the number of decimal places moved. The exponent is positive if the original number is greater than 10 and negative if the number is less than 1.

Thus, the number 78,100 would be written as $7.81 \times 10^4$ and the number .0000027 as $2.7 \times 10^{-6}$ (not $2.7^{-6}$). To check these notations simply multiply.

$$7.81 \times 10^4 = 7.81(10{,}000) = 78{,}100$$

$$2.7 \times 10^{-6} = 2.7\left(\frac{1}{10^6}\right) = \frac{2.7}{1{,}000{,}000} = .0000027$$

EXAMPLE 7 Write in scientific notation.

(a) $5{,}300{,}000 = 5.3 \times 10^6$ (6 places) — Count 6 decimal places

(b) $.0000053 = 5.3 \times 10^{-6}$ (6 places) — Count 6 decimal places

(c) $732{,}100{,}000{,}000 = 7.321 \times 10^{11}$ (11 places)

(d) $.0000000000001 = 1 \times 10^{-13}$ (13 places)

(e) $10 = 1 \times 10^1 = 1 \times 10$ (1 place)

(f) $.1 = 1 \times 10^{-1}$ (1 place)

(g) $6.2 = 6.2 \times 10^0$

Scientific notation not only shortens the notation for certain numbers but also helps in making calculations which involve very large or very small numbers.

**EXAMPLE 8** Perform the indicated operation using scientific notation.

**(a)** $(20{,}000)(3{,}000{,}000) = (2 \times 10^4)(3 \times 10^6)$
$= (2 \cdot 3) \times (10^4 \cdot 10^6)$ Use the commutative rule
$= 6 \times 10^{10}$ $10^4 \cdot 10^6 = 10^{4+6} = 10^{10}$

**(b)** $(8 \times 10^{12})(5 \times 10^8) = (8 \cdot 5) \times (10^{12} \cdot 10^8)$
$= 40 \times 10^{20}$
$= 4 \times 10^1 \times 10^{20}$ Obtain a number between 1 and 10
$= 4 \times 10^{21}$

**(c)** $(2.4 \times 10^{13})(5.0 \times 10^{-6}) = (2.4)(5.0) \times (10^{13} \cdot 10^{-6})$
$= 12 \times 10^7$ $a^m \cdot a^n = a^{m+n}$
$= 1.2 \times 10^1 \times 10^7$ Obtain a number between 1 and 10
$= 1.2 \times 10^8$

**(d)** $(2.4 \times 10^{-13})(5.0 \times 10^6) = 12 \times 10^{-7}$
$= 1.2 \times 10^1 \times 10^{-7}$ Obtain a number between 1 and 10
$= 1.2 \times 10^{-6}$

**(e)** $\dfrac{8.2 \times 10^{-4}}{4.1 \times 10^8} = \left(\dfrac{8.2}{4.1}\right) \times \left(\dfrac{10^{-4}}{10^8}\right)$ $\dfrac{a^m}{a^n} = a^{m-n}$
$= 2.0 \times (10^{-4-8})$
$= 2.0 \times 10^{-12}$

**(f)** $\dfrac{4 \times 10^{15}}{(2 \times 10^{-8})(8 \times 10^6)} = \left(\dfrac{4}{2 \cdot 8}\right) \times \left(\dfrac{10^{15}}{10^{-8}10^6}\right)$
$= \left(\dfrac{4}{16}\right) \times \left(\dfrac{10^{15}}{10^{-2}}\right)$
$= .25 \times (10^{15-(-2)})$
$= .25 \times (10^{17})$
$= 2.5 \times 10^{-1} \times 10^{17}$ Obtain a number between 1 and 10
$= 2.5 \times 10^{16}$

## EXERCISES 4.1

*Write in exponential notation.*

**1.** $5 \cdot 5 \cdot 5 \cdot 5 \cdot 5 \cdot 5$

**2.** $x \cdot x \cdot x \cdot x$

**3.** $3 \cdot 3 \cdot 3 \cdot y \cdot y \cdot y \cdot y$

**4.** $\dfrac{1}{(6a)(6a)}$

**5.** $\dfrac{1}{6 \cdot a \cdot a}$

**6.** $\dfrac{1}{(a+b)(a+b)(a+b)}$

*Write without using exponents.*

**7.** $9^3$

**8.** $x^8$

**9.** $5^3x^4y^2$

**10.** $(4a)^{-2}$

**11.** $4^{-1}a^{-2}$

**12.** $\dfrac{1}{x^2+y^2}$

*Simplify and write without negative exponents.*

**13.** $x^2x^5$

**14.** $5^2 \cdot a^3 \cdot a^4$

**15.** $y^3 \cdot y^6 \cdot y^8$

**16.** $\dfrac{x^{10}}{x^4}$

**17.** $\dfrac{5^3a^5}{5^2a}$

**18.** $(y^2)^5$

**19.** $(2x^3)^4$

**20.** $(3x^2y^3)^2$

**21.** $\left(\dfrac{4x^3}{y^2}\right)^3$

**22.** $(5a^2b^3)^0 \quad (a \neq 0 \text{ and } b \neq 0)$

**23.** $0^0$

**24.** $3x^0 \ (x \neq 0)$

**25.** $(3x)^0 \quad (x \neq 0)$

**26.** $5a^{-1}$

**27.** $(5a)^{-1}$

**28.** $3x^3x^{-2}$

**29.** $\dfrac{5a^3}{a^5}$

**30.** $(6x^2y^{-3})^{-1}$

**31.** $\dfrac{a^2b^{-3}}{a^5b^2}$

**32.** $\dfrac{x^{-3}}{y^{-6}}$

**33.** $\left(\dfrac{a^2b^3}{2a^{-2}b}\right)^{-2}$

**34.** $\left(\dfrac{x^3}{y^{-2}}\right)\left(\dfrac{x^2y^{-1}}{x^{-6}}\right)$

**35.** $\left(\dfrac{3a^{-3}}{b^2}\right)^{-2}$

**36.** $\left(\dfrac{5^0x^{-2}}{3^{-1}y^{-2}}\right)^2$

*Write in scientific notation.*

**37.** 6,800,000

**38.** .000000068

**39.** 127,000,000,000

**40.** .0000000000541

**41.** 10,000,000,000

**42.** .01

*Write without scientific notation.*

**43.** $3.6 \times 10^7$

**44.** $3.6 \times 10^{-7}$

**45.** $6 \times 10^{10}$

**46.** $1 \times 10^5$

**47.** $1 \times 10^{-5}$

**48.** $1 \times 10^1$

*Simplify and give answers in scientific notation.*

**49.** $(3 \times 10^5)(2 \times 10^6)$

**50.** (300,000)(2,000,000)

**51.** $(3 \times 10^{-5})(2 \times 10^{-6})$

**52.** (.00003)(.000002)

**53.** $(2.8 \times 10^{-6})(5.0 \times 10^{10})$

**54.** $(2.8 \times 10^6)(5.0 \times 10^{-10})$

**55.** $(2.8 \times 10^7)(5.0 \times 10^0)$

**56.** (.00000000001)(.000004)

**57.** $\dfrac{4 \times 10^8}{2 \times 10^3}$

**58.** $\dfrac{400{,}000{,}000}{2000}$

**59.** $\dfrac{(4 \times 10^5)(6 \times 10^{-3})}{(8 \times 10^{-7})}$

**60.** $\dfrac{(4 \times 10^7)(6 \times 10^{-5})}{(8 \times 10^{10})}$

---

**ANSWERS:** **1.** $5^6$ **2.** $x^4$ **3.** $3^3y^4$ **4.** $\dfrac{1}{(6a)^2} = (6a)^{-2}$ **5.** $\dfrac{1}{6a^2} = 6^{-1}a^{-2}$ **6.** $\dfrac{1}{(a+b)^3} = (a+b)^{-3}$ **7.** $9 \cdot 9 \cdot 9$ **8.** $x \cdot x \cdot x \cdot x \cdot x \cdot x \cdot x \cdot x$ **9.** $5 \cdot 5 \cdot 5 \cdot x \cdot x \cdot x \cdot x \cdot y \cdot y$ **10.** $\dfrac{1}{(4a)(4a)}$ **11.** $\dfrac{1}{4 \cdot a \cdot a}$ **12.** $\dfrac{1}{x \cdot x + y \cdot y}$ **13.** $x^7$ **14.** $25a^7$ **15.** $y^{17}$ **16.** $x^6$ **17.** $5a^4$ **18.** $y^{10}$ **19.** $16x^{12}$ **20.** $9x^4y^6$ **21.** $\dfrac{64x^9}{y^6}$ **22.** 1 **23.** undefined **24.** 3 **25.** 1 **26.** $\dfrac{5}{a}$ **27.** $\dfrac{1}{5a}$ **28.** $3x$ **29.** $\dfrac{5}{a^2}$ **30.** $\dfrac{y^3}{6x^2}$ **31.** $\dfrac{1}{a^3b^5}$ **32.** $\dfrac{y^6}{x^3}$ **33.** $\dfrac{4}{a^8b^4}$ **34.** $x^{11}y$ **35.** $\dfrac{a^6b^4}{9}$ **36.** $\dfrac{9y^4}{x^4}$ **37.** $6.8 \times 10^6$ **38.** $6.8 \times 10^{-8}$ **39.** $1.27 \times 10^{11}$ **40.** $5.41 \times 10^{-11}$ **41.** $1 \times 10^{10}$ **42.** $1 \times 10^{-2}$ **43.** 36,000,000 **44.** .00000036 **45.** 60,000,000,000 **46.** 100,000 **47.** .00001 **48.** 10 **49.** $6 \times 10^{11}$ **50.** $6 \times 10^{11}$ **51.** $6 \times 10^{-11}$ **52.** $6 \times 10^{-11}$ **53.** $1.4 \times 10^5$ **54.** $1.4 \times 10^{-3}$ **55.** $1.4 \times 10^8$ **56.** $4 \times 10^{-17}$ **57.** $2 \times 10^5$ **58.** $2 \times 10^5$ **59.** $3 \times 10^9$ **60.** $3 \times 10^{-8}$

## 4.2 Radicals

In Chapter 1 a **square root** of a number was defined as one of two equal factors of the number. Thus, 3 is a square root of 9 since $3 \cdot 3 = 3^2 = 9$. Note that $-3$ is also a square root of 9 since $(-3)^2 = 9$. We denote the nonnegative square root of a number by a **radical,** $\sqrt{\ }$. Thus,

$$\sqrt{9} = 3 \qquad \text{and} \qquad -\sqrt{9} = -3.$$

The **positive** square root of a number is called its **principal square root.** To indicate both square roots of 9 in one expression, we write $\pm\sqrt{9}$ or $\pm 3$, where $\pm 3$ means both 3 and $-3$. For example, we indicate the solutions to $x^2 = 9$ as follows.

$$\begin{aligned} x^2 &= 9 \\ x &= \pm\sqrt{9} = \pm 3 \end{aligned}$$

Before considering the basic rules of radicals, recall that the absolute value of a number $a$, denoted $|a|$, is given by

$$|a| = \begin{cases} a, & a \geq 0 \\ -a, & a < 0. \end{cases}$$

Note that $-a$ is positive if $a$ is negative. Thus, if $a = -5$, $-a = -(-5) = 5$ and $|-5| = -(-5) = 5$. In order to simplify radical expressions, we need the following rules for absolute value.

**1.** $|ab| = |a|\,|b|$ **2.** $\left|\frac{a}{b}\right| = \frac{|a|}{|b|}, \quad b \neq 0$

Using absolute value, we can give the rule for finding the square root of a square.

For any real number $a$,

$$\sqrt{a^2} = |a|$$

**EXAMPLE 1** Evaluate the following radicals.

**(a)** $\sqrt{36} = \sqrt{6^2} = |6| = 6$

**(b)** $\sqrt{(-6)^2} = |-6| = 6$

**(c)** $\sqrt{-6^2} = \sqrt{-36}$ is meaningless since no real number squared is negative.

**(d)** $\sqrt{0^2} = |0| = 0$

**(e)** $\sqrt{25x^2} = \sqrt{5^2x^2} = \sqrt{(5x)^2} = |5x| = |5|\,|x| = 5|x|$ $\quad |ab| = |a|\,|b|$

**(f)** $\sqrt{a^4} = \sqrt{(a^2)^2} = |a^2| = a^2$ $\quad a^2 \geq 0$

**(g)** $\sqrt{x^2 + 6x + 9} = \sqrt{(x+3)^2} = |x + 3|$ $\quad$ not $|x| + 3$

**(h)** $\sqrt{x^2 - 6x + 9} = \sqrt{(x-3)^2} = |x - 3|$ $\quad$ not $|x| + 3$

Next, we define the **cube root** of a number to be one of three equal factors of the number. The cube root is denoted by $\sqrt[3]{a}$.

$$(\sqrt[3]{a})(\sqrt[3]{a})(\sqrt[3]{a}) = (\sqrt[3]{a})^3 = a.$$

Unlike square roots, there is only one cube root of a number, and negative numbers have cube roots. For example, $\sqrt[3]{-8} = -2$ since $(-2)^3 = -8$.

**EXAMPLE 2** Evaluate the cube roots.

**(a)** $\sqrt[3]{27} = \sqrt[3]{3^3} = 3$ $\quad 3^3 = 27$

**(b)** $\sqrt[3]{-27} = \sqrt[3]{(-3)^3} = -3$ $\quad (-3)^3 = -27$

**(c)** $\sqrt[3]{125} = \sqrt[3]{5^3} = 5$ $\quad 5^3 = 125$

**(d)** $\sqrt[3]{-64} = \sqrt[3]{(-4)^3} = -4$ $\quad (-4)^3 = -64$

**(e)** $\sqrt[3]{1} = \sqrt[3]{1^3} = 1$ $\quad 1^3 = 1$

**(f)** $\sqrt[3]{0} = \sqrt[3]{0^3} = 0$ $\quad 0^3 = 0$

If $b^2 = a$, then $b$ is a square root of $a$, while if $b^3 = a$, then $b$ is the cube root of $a$. This leads us to the following definition: if

$$b^k = a,$$

where $k$ is a natural number greater than one, then $b$ is a **$k$th root of $a$.** The symbol $\sqrt[k]{a}$ is used to denote the **principal $k$th root of $a$,** which is positive if $a$ has a positive $k$th root and negative if $a$ has no positive $k$th root but does have a negative $k$th root. For example,

$$\sqrt[4]{81} = 3 \quad (\text{not } -3) \qquad \text{but} \qquad \sqrt[3]{-27} = -3.$$

Also note that $\sqrt[4]{-81}$ does not exist since no number raised to the fourth power is negative. In the symbol $\sqrt[k]{a}$, $a$ is called the **radicand** and $k$ is called the **index.** The index must always be specified in radical expressions except in the case of square roots, where the index is not written and is understood to be 2.

The following rule points out the properties of the principal $k$th roots of a number for $k$ both even and odd.

1. $\sqrt[k]{a^k} = |a|$ if $k$ is even.
2. $\sqrt[k]{a^k} = a$ if $k$ is odd.

EXAMPLE 3 Simplify.

**(a)** $\sqrt{(-5)^2} = |-5| = 5$ — $\sqrt{a^2} = |a|$

**(b)** $\sqrt[3]{-27} = \sqrt[3]{(-3)^3} = -3$ — $\sqrt[3]{a^3} = a$

**(c)** $\sqrt[4]{16} = \sqrt[4]{2^4} = |2| = 2$ — $\sqrt[4]{a^4} = |a|$

**(d)** $\sqrt[5]{243} = \sqrt[5]{3^5} = 3$ — $\sqrt[5]{a^5} = a$

**(e)** $\sqrt[5]{-243} = \sqrt[5]{(-3)^5} = -3$

**(f)** $\sqrt[6]{-64}$ is not a real number.

**(g)** $-\sqrt[6]{64} = -\sqrt[6]{2^6} = -|2| = -2$

**(h)** $\sqrt[5]{32x^5} = \sqrt[5]{2^5x^5} = \sqrt[5]{(2x)^5} = 2x$

**(i)** $\sqrt[4]{81y^4} = \sqrt[4]{3^4y^4} = \sqrt[4]{(3y)^4} = |3y| = |3|\,|y| = 3|y|$

**(j)** $\sqrt[3]{27a^{12}} = \sqrt[3]{3^3(a^4)^3} = \sqrt[3]{(3a^4)^3} = 3a^4$

If $k$ is even, then $a$ must be positive (or 0) for $\sqrt[k]{a}$ to be a real number, but when $k$ is odd, $\sqrt[k]{a}$ exists for any real number $a$.

## EXERCISES 4.2

*Simplify.*

**1.** $\sqrt{81}$ **2.** $\sqrt[4]{81}$ **3.** $\sqrt[3]{-64}$

4. $\sqrt[3]{64}$ 5. $\sqrt[6]{64}$ 6. $\sqrt[5]{-32}$

7. $\sqrt[9]{1}$ 8. $\sqrt[9]{0}$ 9. $\sqrt[9]{-1}$

10. $\sqrt[7]{x^7}$ 11. $\sqrt[8]{x^8}$ 12. $\sqrt[5]{(6xy)^5}$

13. $\sqrt[6]{(6xy)^6}$ 14. $\sqrt[4]{-x^4}$ 15. $\sqrt[6]{(-2)^6a^6}$

16. $\sqrt[5]{-32x^5}$ 17. $\sqrt[3]{a^6}$ 18. $\sqrt[4]{16x^{12}}$

19. $\sqrt[3]{64a^3b^6}$ 20. $\sqrt{81x^6y^{10}}$ 21. $\sqrt[5]{-243a^5b^{10}}$

*Simplify and write without negative exponents.*

22. $\left(\frac{3x^5}{y^2}\right)^3$ 23. $\left(\frac{a}{b}\right)^{-2}$ 24. $\left(\frac{2x^{-2}}{x^2y^3}\right)^{-3}$

25. $(4^0a^3b^2)^{-2}$ 26. $5(x^3y^{-3})^0$ ($x \neq 0$ and $y \neq 0$) 27. $\left(\frac{3a^3}{2b^{-2}}\right)^{-2}$

*Simplify and write in scientific notation.*

28. $\frac{1 \times 10^{-7}}{5 \times 10^5}$ 29. $\frac{(4 \times 10^7)(5 \times 10^{-5})}{2 \times 10^{-8}}$ 30. $\frac{.0000025}{500,000,000}$

---

ANSWERS: 1. 9 2. 3 3. −4 4. 4 5. 2 6. −2 7. 1 8. 0 9. −1 10. $x$ 11. $|x|$ 12. $6xy$ 13. $6|xy|$ 14. not a real number unless $x = 0$ 15. $2|a|$ 16. $-2x$ 17. $a^2$ 18. $2|x^3|$ 19. $4ab^2$ 20. $9|x^3y^5|$ 21. $-3ab^2$ 22. $\frac{27x^{15}}{y^6}$ 23. $\frac{b^2}{a^2}$ 24. $\frac{x^{12}y^9}{8}$ 25. $\frac{1}{a^6b^4}$ 26. 5 27. $\frac{4}{9a^6b^4}$ 28. $2 \times 10^{-13}$ 29. $1 \times 10^{11}$ 30. $5 \times 10^{-15}$

## 4.3 Multiplication and Division of Radicals

We now present two rules for multiplying and dividing $k$th roots of expressions.

1. If $\sqrt[k]{a}$ and $\sqrt[k]{b}$ are defined, then
$$\sqrt[k]{a}\sqrt[k]{b} = \sqrt[k]{ab}.$$
2. If $\sqrt[k]{a}$ and $\sqrt[k]{b}$ are defined and $b \neq 0$, then
$$\frac{\sqrt[k]{a}}{\sqrt[k]{b}} = \sqrt[k]{\frac{a}{b}}.$$

The value of the $k$th root of a number depends on whether $k$ is even or odd. The relation between the index, $k$, on the radical and the absolute value bars is summarized as follows.

$\sqrt[\text{even}]{\phantom{x}}$ *Do use* absolute value bars on variables

$\sqrt[\text{odd}]{\phantom{x}}$ *Do not use* absolute value bars on variables

We may remove the absolute value bars from an expression involving a variable when the expression is nonnegative. For example,

$$\sqrt{x^4} = \sqrt{(x^2)^2} = |x^2| = x^2$$

since $x^2 \geq 0$, but

$$\sqrt{x^6} = \sqrt{(x^3)^2} = |x^3|$$

cannot be simplified to $x^3$ since $x^3$ may be negative. However, if we are asked to simplify

$$\sqrt{x}\sqrt{x^5}$$

we assume that $x \geq 0$ because otherwise $\sqrt{x}$ (also $\sqrt{x^5}$) is not defined. Thus, we have sufficient information to remove the absolute value bars.

$$\sqrt{x}\sqrt{x^5} = \sqrt{x^6} = \sqrt{(x^3)^2} = |x^3| = x^3 \qquad x^3 \geq 0 \text{ since } x \geq 0$$

CAUTION: In the examples and exercises which follow, place absolute value bars around variable expressions whenever the index is even. Next, determine whether the enclosed variable expression is nonnegative; if it is, simplify further by removing the bars.

We use the rules both to multiply and to factor and simplify radicals. Care must be taken in each problem to use only values of the variables for which the given radicals are defined.

**EXAMPLE 1** Multiply and simplify.

| | |
|---|---|
| $\sqrt{2x}\sqrt{8x} = \sqrt{2x \cdot 8x}$ | $x \geq 0$ for $\sqrt{2x}$ and $\sqrt{8x}$ to be defined |
| $= \sqrt{16x^2}$ | Look for perfect squares |
| $= \sqrt{16}\sqrt{x^2}$ | $\sqrt{a \cdot b} = \sqrt{a} \cdot \sqrt{b}$ with $a = 16$, $b = x^2$ |
| $= 4\lvert x\rvert$ | Since the index is 2, use absolute value bars |
| $= 4x$ | Since $x \geq 0$ $\lvert x\rvert = x$ |

**EXAMPLE 2** Multiply and simplify.

| | |
|---|---|
| $\sqrt[3]{3a^2} \cdot \sqrt[3]{9a} = \sqrt[3]{(3a^2)(9a)}$ | |
| $= \sqrt[3]{27a^3}$ | Look for perfect cubes |
| $= \sqrt[3]{27}\sqrt[3]{a^3}$ | $\sqrt[3]{a \cdot b} = \sqrt[3]{a} \cdot \sqrt[3]{b}$ |
| $= 3a$ | Since the index is 3, do not use absolute value bars |

**EXAMPLE 3** Factor and simplify.

| | |
|---|---|
| $\sqrt[4]{32x^5} = \sqrt[4]{2 \cdot 2^4 \cdot x \cdot x^4}$ | $x \geq 0$ for $\sqrt[4]{32x^5}$ to be defined |
| $= \sqrt[4]{2^4x^4 \cdot 2x}$ | Look for perfect 4th powers |
| $= \sqrt[4]{(2x)^4}\sqrt[4]{2x}$ | $\sqrt[4]{ab} = \sqrt[4]{a}\sqrt[4]{b}$ with $a = (2x)^4$, $b = 2x$ |
| $= \lvert 2x\rvert\sqrt[4]{2x}$ | Since index is even, use bars |
| $= 2x\sqrt[4]{2x}$ | Since $x \geq 0$, $\lvert 2x\rvert = \lvert 2\rvert\,\lvert x\rvert = 2x$ |

**EXAMPLE 4** Divide and simplify.

| | |
|---|---|
| $\dfrac{\sqrt{27a^3}}{\sqrt{3a}} = \sqrt{\dfrac{27a^3}{3a}}$ | $\dfrac{\sqrt{a}}{\sqrt{b}} = \sqrt{\dfrac{a}{b}}$ with $a = 27x^3$, $b = 3a$ |
| $= \sqrt{9a^2}$ | Simplify first, look for perfect squares |
| $= \sqrt{9}\sqrt{a^2}$ | $\sqrt{ab} = \sqrt{a}\sqrt{b}$ with $a = 9$, $b = a^2$ |
| $= 3\lvert a\rvert$ | Even index, use bars |
| $= 3a$ | $a > 0$ for $\sqrt{3a}$ to be defined |

**EXAMPLE 5** Simplify.

| | |
|---|---|
| $\sqrt[3]{\dfrac{16x^5y}{2x^{-1}y^4}} = \sqrt[3]{\dfrac{8x^6}{y^3}}$ | $\dfrac{a^m}{a^n} = a^{m-n}$ |
| $= \dfrac{\sqrt[3]{8x^6}}{\sqrt[3]{y^3}}$ | |
| $= \dfrac{\sqrt[3]{8}\sqrt[3]{(x^2)^3}}{\sqrt[3]{y^3}}$ | Look for perfect cubes |
| $= \dfrac{2x^2}{y}$ | Do not use absolute value bars (Why ?) |

**EXAMPLE 6** Simplify.

$$\begin{aligned}\sqrt[5]{3a^2b^3}\,\sqrt[5]{81a^3b^3} &= \sqrt[5]{3^5a^5b^6} && 81 = 3^4\\ &= \sqrt[5]{3^5a^5b^5\cdot b} && \text{Look for perfect fifth powers}\\ &= \sqrt[5]{3^5}\,\sqrt[5]{a^5}\,\sqrt[5]{b^5}\,\sqrt[5]{b}\\ &= 3ab\,\sqrt[5]{b}\end{aligned}$$

**EXAMPLE 7** Simplify.

$$\begin{aligned}\frac{\sqrt[6]{256x^3y^{10}}}{\sqrt[6]{2x^{-4}y^3}} &= \sqrt[6]{\frac{256x^3y^{10}}{2x^{-4}y^3}} && x>0 \text{ and } y>0 \text{ and } \frac{\sqrt[6]{a}}{\sqrt[6]{b}} = \sqrt[6]{\frac{a}{b}} \text{ to}\\ &= \sqrt[6]{128x^7y^7} && \text{simplify expression under the radical}\\ &= \sqrt[6]{2^6\cdot 2\cdot x^6\cdot x\cdot y^6\cdot y} && \text{Look for perfect sixth powers}\\ &= 2|xy|\,\sqrt[6]{2xy} && \text{Even index}\\ &= 2xy\,\sqrt[6]{2xy} && \text{Since } x>0 \text{ and } y>0\end{aligned}$$

**EXAMPLE 8** Simplify.

$$\begin{aligned}\sqrt{288x^4(x+y)^2} &= \sqrt{144\cdot 2x^4(x+y)^2}\\ &= \sqrt{12^2\cdot 2\cdot (x^2)^2(x+y)^2} && \text{Look for perfect squares}\\ &= 12\,|x^2|\,|x+y|\,\sqrt{2} && \sqrt{a^2} = |a| \text{ with } a = x^2 \text{ and } a = x+y\\ &= 12x^2\,|x+y|\,\sqrt{2} && |x^2| = x^2 \text{ since } x^2 \geq 0\end{aligned}$$

## EXERCISES 4.3

*Simplify.*

1. $\sqrt{98}$
2. $5\sqrt{3}\,\sqrt{27}$
3. $3\sqrt[3]{16}$
4. $7\sqrt[3]{6}\sqrt[3]{9}$
5. $8\sqrt[4]{81}$
6. $3\sqrt[4]{12}\,\sqrt[4]{32}$
7. $\sqrt{27x^2y^2}$
8. $\sqrt[3]{6ab}\,\sqrt[3]{18a^2b^3}$
9. $\sqrt[5]{243x^5y^{10}}$
10. $\dfrac{\sqrt[4]{25a^3b^2}}{\sqrt[4]{5^{-2}a^{-1}b^{-2}}}$
11. $\sqrt[3]{\dfrac{250(x+y)^5}{2(x+y)}}$
12. $\dfrac{\sqrt{27a^3b^3}}{\sqrt{3a^{-1}b}}$
13. $\sqrt[5]{\dfrac{256x^{10}y^{20}}{2x^4y^6}}$
14. $\dfrac{\sqrt[3]{625a^{-3}b^6}}{\sqrt[3]{5a^3b^{-3}}}$
15. $\sqrt[4]{\dfrac{3(x+y)^5}{3^{-3}(x+y)^{-2}}}$

**16.** $\sqrt{5x^2 + 10x + 5}$ **17.** $\sqrt[4]{-16a^4}$ **18.** $\sqrt[3]{-16}$

---

ANSWERS: **1.** $7\sqrt{2}$ **2.** 45 **3.** $6\sqrt[3]{2}$ **4.** $21\sqrt[3]{2}$ **5.** 24 **6.** $6\sqrt[4]{24}$ **7.** $3|xy|\sqrt{3}$ **8.** $3ab\sqrt[3]{4b}$ **9.** $3xy^2$ **10.** $5a|b|$ **11.** $5(x+y)\sqrt[3]{x+y}$ **12.** $3a^2|b|$ **13.** $2xy^2\sqrt[5]{4xy^4}$ **14.** $\frac{5b^3}{a^2}$ **15.** $3(x+y)\sqrt[4]{(x+y)^3}$ **16.** $|x+1|\sqrt{5}$ **17.** meaningless **18.** $-2\sqrt[3]{2}$

## 4.4 Addition and Subtraction of Radicals

In this section we use the rules of the previous section to simplify radical expressions that involve addition and subtraction. There are no rules for addition and subtraction similar to the rules for multiplication and division,

$$\sqrt{a}\sqrt{b} = \sqrt{ab} \quad \text{and} \quad \frac{\sqrt{a}}{\sqrt{b}} = \sqrt{\frac{a}{b}}.$$

For example,

$$5 = \sqrt{25} = \sqrt{9+16} \neq \sqrt{9} + \sqrt{16} = 3 + 4 = 7 \qquad 5 \neq 7$$
$$4 = \sqrt{16} = \sqrt{25-9} \neq \sqrt{25} - \sqrt{9} = 5 - 3 = 2 \qquad 4 \neq 2$$

Therefore, $\sqrt{a+b} \neq \sqrt{a} + \sqrt{b}$ and $\sqrt{a-b} \neq \sqrt{a} - \sqrt{b}$. But even without rules like these, many radical expressions can be simplified by collecting like terms.

**EXAMPLE 1** Add and simplify.

$$6\sqrt{125} + 3\sqrt{20} = 6\sqrt{25 \cdot 5} + 3\sqrt{4 \cdot 5} \qquad \text{Factor to obtain perfect squares}$$
$$= 6\sqrt{25}\sqrt{5} + 3\sqrt{4}\sqrt{5} \qquad \sqrt{a \cdot b} = \sqrt{a} \cdot \sqrt{b}$$
$$= 6 \cdot 5 \cdot \sqrt{5} + 3 \cdot 2 \cdot \sqrt{5} \qquad 6\sqrt{25 \cdot 5} = 6 \cdot 5\sqrt{5} \text{ not } (6+5)\sqrt{5}$$
$$= 30\sqrt{5} + 6\sqrt{5}$$
$$= (30 + 6)\sqrt{5} \qquad \text{Distributive law}$$
$$= 36\sqrt{5}$$

**EXAMPLE 2** Subtract and simplify.

$$5\sqrt{147} - 9\sqrt{75} = 5\sqrt{49 \cdot 3} - 9\sqrt{25 \cdot 3} \qquad \text{Factor to obtain perfect squares}$$
$$= 5\sqrt{49}\sqrt{3} - 9\sqrt{25}\sqrt{3} \qquad \sqrt{a \cdot b} = \sqrt{a} \cdot \sqrt{b}$$
$$= 5 \cdot 7 \cdot \sqrt{3} - 9 \cdot 5 \cdot \sqrt{3}$$
$$= 35\sqrt{3} - 45\sqrt{3}$$
$$= (35 - 45)\sqrt{3}$$
$$= -10\sqrt{3}$$

**EXAMPLE 3** Simplify.

$$5\sqrt{\frac{128}{9}} + 6\frac{\sqrt{150}}{\sqrt{27}} = 5\frac{\sqrt{64 \cdot 2}}{\sqrt{9}} + 6\sqrt{\frac{150}{27}}$$

9 is a perfect square but 27 is not; we combine and simplify first

$$= 5\frac{\sqrt{64}\sqrt{2}}{3} + 6\sqrt{\frac{50}{9}}$$

Now we have a perfect square in the denominator

$$= \frac{5 \cdot 8\sqrt{2}}{3} + \frac{6\sqrt{25}\sqrt{2}}{\sqrt{9}}$$

$$= \frac{40\sqrt{2}}{3} + \frac{6 \cdot 5\sqrt{2}}{3}$$

$$= \frac{40\sqrt{2}}{3} + \frac{30\sqrt{2}}{3}$$

$$= \left(\frac{40}{3} + \frac{30}{3}\right)\sqrt{2}$$

$$= \frac{70}{3}\sqrt{2}$$

**EXAMPLE 4** Simplify.

$$-3\sqrt[3]{135} + 2\sqrt[3]{625} - 6\sqrt[3]{40} = -3\sqrt[3]{27 \cdot 5} + 2\sqrt[3]{125 \cdot 5} - 6\sqrt[3]{8 \cdot 5}$$

Factor to obtain perfect cubes

$$= -3\sqrt[3]{27}\sqrt[3]{5} + 2\sqrt[3]{125}\sqrt[3]{5} - 6\sqrt[3]{8}\sqrt[3]{5}$$

$$= -3 \cdot 3 \cdot \sqrt[3]{5} + 2 \cdot 5 \cdot \sqrt[3]{5} - 6 \cdot 2 \cdot \sqrt[3]{5}$$

$$= -9\sqrt[3]{5} + 10\sqrt[3]{5} - 12\sqrt[3]{5}$$

$$= -11\sqrt[3]{5}$$

**EXAMPLE 5** Simplify.

$$4\sqrt[3]{8x^4y^5} - 3\sqrt[3]{27x^4y^5} = 4\sqrt[3]{2^3x^3y^3xy^2} - 3\sqrt[3]{3^3x^3y^3xy^2}$$

Factor to obtain perfect cubes

$$= 4\sqrt[3]{(2xy)^3} \cdot \sqrt[3]{xy^2} - 3\sqrt[3]{(3xy)^3}\sqrt[3]{xy^2}$$

$$= 4(2xy)\sqrt[3]{xy^2} - 3(3xy)\sqrt[3]{xy^2}$$

$$= 8xy\sqrt[3]{xy^2} - 9xy\sqrt[3]{xy^2}$$

$$= -xy\sqrt[3]{xy^2}$$

## EXERCISES 4.4

*Simplify.*

**1.** $\sqrt{8} + 3\sqrt{50}$

**2.** $6\sqrt{125} - 8\sqrt{45}$

3. $3\sqrt{16} - 8\sqrt{121}$

4. $-5\sqrt{150} + 10\sqrt{24}$

5. $-6\sqrt{147} - 3\sqrt{75}$

6. $\frac{1}{2}\sqrt{32} - \frac{1}{3}\sqrt{162}$

7. $4\sqrt[3]{81} + 5\sqrt[3]{24}$

8. $9\sqrt[3]{250} - 4\sqrt[3]{128}$

9. $3\sqrt[4]{48} + \sqrt[4]{243}$

10. $\sqrt[4]{625} - \sqrt[3]{125}$

11. $8\sqrt{\frac{16}{9}} + 5\frac{\sqrt{50}}{\sqrt{32}}$

12. $\frac{-3\sqrt{250}}{\sqrt{18}} - 8\frac{\sqrt{270}}{\sqrt{24}}$

13. $2\frac{\sqrt{125}}{\sqrt[3]{27}} - \frac{8\sqrt{20}}{\sqrt[4]{81}}$

14. $-6\sqrt[3]{\frac{81}{8}} + 5\frac{\sqrt[3]{24}}{\sqrt[3]{64}}$

15. $-7\sqrt[3]{\frac{80}{54}} - \frac{\sqrt[3]{10}}{\sqrt[3]{16}}$

16. $\sqrt[4]{\frac{32}{81}} - \frac{\sqrt[4]{162}}{\sqrt[4]{625}}$

17. $5\sqrt{8x^3} - 7\sqrt{18x^3}$

18. $6\sqrt{48y^4} + 4\sqrt{12y^4}$

19. $-7\sqrt[3]{40a^5} + 10a\sqrt[3]{135a^2}$

20. $-8\sqrt{75a^2b^5} + 3b\sqrt{147a^2b^3}$

**21.** $2\sqrt[3]{27x^4y^6} - 4y\sqrt[3]{125x^4y^3}$

**22.** $\sqrt{5x^2 + 10x + 5} + \sqrt{20x^2 + 40x + 20}$

**23.** $\dfrac{5 + \sqrt{125}}{10}$

**24.** $\dfrac{8 - \sqrt{48}}{6}$

**25.** $3\sqrt{45} - 2\sqrt{20} + 4\sqrt{27}$

**26.** $8\sqrt[3]{54} - 4\sqrt[3]{16} - 5\sqrt[3]{250}$

**27.** $5x\sqrt[3]{16xy^3} + 3\sqrt[3]{54x^4y^3} - xy\sqrt[3]{2x}$

**28.** $\sqrt{125xy} \cdot \sqrt{20x^3y^3}$

**29.** $2\sqrt[3]{4a^2b} \cdot \sqrt[3]{10a^2b^2}$

**30.** $\dfrac{\sqrt{125a^6b^{-2}}}{\sqrt{5ab^{-6}}}$

---

**ANSWERS:** **1.** $17\sqrt{2}$ **2.** $6\sqrt{5}$ **3.** $-76$ **4.** $-5\sqrt{6}$ **5.** $-57\sqrt{3}$ **6.** $-\sqrt{2}$ **7.** $22\sqrt[3]{3}$ **8.** $29\sqrt[3]{2}$ **9.** $9\sqrt[4]{3}$ **10.** $0$ **11.** $\dfrac{203}{12}$ **12.** $-17\sqrt{5}$ **13.** $-2\sqrt{5}$ **14.** $\dfrac{-13\sqrt[3]{3}}{2}$ **15.** $\dfrac{-31\sqrt[3]{5}}{6}$ **16.** $\dfrac{\sqrt[4]{2}}{15}$ **17.** $-11x\sqrt{2x}$ $(x \geq 0)$ **18.** $32y^2\sqrt{3}$ **19.** $16a\sqrt[3]{5a^2}$ **20.** $-19|a|b^2\sqrt{3b}$ **21.** $-14xy^2\sqrt[3]{x}$ **22.** $3|x+1|\sqrt{5}$ **23.** $\dfrac{1+\sqrt{5}}{2}$ **24.** $\dfrac{4-2\sqrt{3}}{3}$ **25.** $5\sqrt{5} + 12\sqrt{3}$ **26.** $-9\sqrt[3]{2}$ **27.** $18xy\sqrt[3]{2x}$ **28.** $50x^2y^2$ **29.** $4ab\sqrt[3]{5a}$ **30.** $5a^2b^2\sqrt{a}$

## 4.5 Rationalizing Denominators

When simplifying some fractional radical expressions our work is easier if we rationalize denominators. By **rationalizing the denominator,** we mean transforming the fraction into an equivalent expression with no radicals in the denominator. In this section we first work with fractions that have one term in the denominator, and then consider denominators with two terms.

**EXAMPLE 1** Rationalize the denominator.

$$\frac{\sqrt{7}}{\sqrt{5}} = \frac{\sqrt{7}}{\sqrt{5}} \cdot \frac{\sqrt{5}}{\sqrt{5}} \qquad \text{Multiply by } 1 = \frac{\sqrt{5}}{\sqrt{5}}$$

$$= \frac{\sqrt{7} \cdot \sqrt{5}}{\sqrt{5^2}} \qquad \text{The denominator is now a perfect square}$$

$$= \frac{\sqrt{35}}{5} \qquad \text{The denominator is now a rational number}$$

**EXAMPLE 2** Rationalize the denominator.

$$\frac{\sqrt{3x}}{\sqrt{8y}} = \frac{\sqrt{3x}}{\sqrt{4 \cdot 2y}} \qquad x \geq 0 \text{ and } y > 0$$

$$= \frac{\sqrt{3x}}{2\sqrt{2y}} \cdot \frac{\sqrt{2y}}{\sqrt{2y}} \qquad \text{Simplify and multiply by } 1 = \frac{\sqrt{2y}}{\sqrt{2y}} \text{ to make the denominator a perfect square}$$

$$= \frac{\sqrt{3x} \cdot \sqrt{2y}}{2\sqrt{(2y)^2}}$$

$$= \frac{\sqrt{6xy}}{2(2y)} \qquad \text{Since } y > 0,\ \sqrt{(2y)^2} = |2y| = 2y$$

$$= \frac{\sqrt{6xy}}{4y} \qquad \text{Denominator is now rationalized}$$

**EXAMPLE 3** Rationalize the denominator.

$$\frac{\sqrt[3]{18}}{\sqrt[3]{10}} = \sqrt[3]{\frac{18}{10}} = \sqrt[3]{\frac{9}{5}} \qquad \text{Simplify first}$$

$$= \frac{\sqrt[3]{9}}{\sqrt[3]{5}} \cdot \frac{\sqrt[3]{5^2}}{\sqrt[3]{5^2}} \qquad \text{This will make the denominator a perfect cube}$$

$$= \frac{\sqrt[3]{9}\,\sqrt[3]{25}}{\sqrt[3]{5^3}}$$

$$= \frac{\sqrt[3]{225}}{5}$$

**TO RATIONALIZE THE DENOMINATOR OF A FRACTION WITH ONE TERM IN THE DENOMINATOR**

1. Simplify the fraction.
2. Multiply the numerator and denominator of the fraction by a radical which makes the denominator a perfect $k$th root.

**EXAMPLE 4** Rationalize the denominator.

$$\frac{\sqrt{40x^3}}{\sqrt{15xy}} = \sqrt{\frac{\cancel{5} \cdot 8x^2 \cdot \cancel{x}}{\cancel{5} \cdot 3\cancel{x}y}} \qquad x > 0 \text{ and } y > 0$$

$$= \sqrt{\frac{8x^2}{3y}} \qquad \text{Simplify first}$$

$$= \frac{\sqrt{8x^2}}{\sqrt{3y}} \cdot \frac{\sqrt{3y}}{\sqrt{3y}}$$

$$= \frac{\sqrt{4x^2 \cdot 2}\,\sqrt{3y}}{\sqrt{(3y)^2}} \qquad \text{Now have a perfect square in the denominator}$$

$$= \frac{2x\sqrt{2}\sqrt{3y}}{3y} \qquad \sqrt{(3y)^2} = |3y| = 3y \text{ since } y > 0$$

$$= \frac{2x\sqrt{6y}}{3y}$$

**EXAMPLE 5** Rationalize the denominator.

$$\frac{\sqrt[3]{5a}}{\sqrt[3]{9b}} = \frac{\sqrt[3]{5a}}{\sqrt[3]{9b}} \cdot \frac{\sqrt[3]{3b^2}}{\sqrt[3]{3b^2}} \qquad 9b \cdot 3b^2 = 27b^3 \text{ gives a perfect cube in the denominator}$$

$$= \frac{\sqrt[3]{5a}\,\sqrt[3]{3b^2}}{\sqrt[3]{3^2b}\,\sqrt[3]{3b^2}}$$

$$= \frac{\sqrt[3]{(5a)(3b^2)}}{\sqrt[3]{(3^2b)(3b^2)}}$$

$$= \frac{\sqrt[3]{15ab^2}}{\sqrt[3]{(3b)^3}}$$

$$= \frac{\sqrt[3]{15ab^2}}{3b}$$

When a fraction has a denominator of the form $(c\sqrt{x} + d\sqrt{y})$ or $(c\sqrt{x} - d\sqrt{y})$, we can use the rule $(a - b)(a + b) = a^2 - b^2$ to rationalize it.

**TO RATIONALIZE THE DENOMINATOR OF A FRACTION WITH TWO TERMS IN THE DENOMINATOR**

1. Simplify the fraction as much as possible.
2. If the denominator is of the form $(c\sqrt{x} + d\sqrt{y})$, multiply numerator and denominator by $(c\sqrt{x} - d\sqrt{y})$ and simplify.
3. If the denominator is of the form $(c\sqrt{x} - d\sqrt{y})$, multiply numerator and denominator by $(c\sqrt{x} + d\sqrt{y})$ and simplify.

**EXAMPLE 6** Rationalize the denominator.

$$\frac{3}{\sqrt{2}+\sqrt{3}} = \frac{3}{(\sqrt{2}+\sqrt{3})} \cdot \frac{(\sqrt{2}-\sqrt{3})}{(\sqrt{2}-\sqrt{3})} \qquad \text{Multiply by } 1 = \frac{\sqrt{2}-\sqrt{3}}{\sqrt{2}-\sqrt{3}}$$

$$= \frac{3(\sqrt{2}-\sqrt{3})}{(\sqrt{2})^2-(\sqrt{3})^2} \qquad \text{Use } (a+b)(a-b) = a^2 - b^2$$

$$= \frac{3(\sqrt{2}-\sqrt{3})}{2-3}$$

$$= \frac{3(\sqrt{2}-\sqrt{3})}{-1}$$

$$= -3(\sqrt{2}-\sqrt{3}) = 3(\sqrt{3}-\sqrt{2})$$

Before we work similar examples, we consider several products of binomial radical expressions.

**EXAMPLE 7** Simplify.

$$\sqrt{3}\,(\sqrt{15}+\sqrt{60}) = \sqrt{3}\sqrt{15}+\sqrt{3}\sqrt{60} \qquad \text{Distributive law}$$

$$= \sqrt{3 \cdot 15} + \sqrt{3 \cdot 60}$$

$$= \sqrt{3^2 \cdot 5} + \sqrt{3^2 \cdot 2^2 \cdot 5} \qquad \text{Factor out perfect squares}$$

$$= 3\sqrt{5} + 3 \cdot 2\sqrt{5}$$

$$= 3\sqrt{5} + 6\sqrt{5}$$

$$= 9\sqrt{5}$$

**EXAMPLE 8** Simplify.

$$(\sqrt{3}-\sqrt{2})(\sqrt{3}+5\sqrt{2}) = \overset{F}{\sqrt{3}\sqrt{3}} + \overset{O}{\sqrt{3}(5\sqrt{2})} - \overset{I}{\sqrt{2}\sqrt{3}} - \overset{L}{\sqrt{2}(5\sqrt{2})}$$

Use FOIL to multiply

$$= 3 + 5\sqrt{6} - \sqrt{6} - 5(2)$$

$$= 3 - 10 + 5\sqrt{6} - \sqrt{6}$$

$$= -7 + 4\sqrt{6}$$

**EXAMPLE 9** Simplify.

$$(\sqrt{5}+\sqrt{7})(\sqrt{5}-\sqrt{7}) = (\sqrt{5})^2 - (\sqrt{7})^2 \qquad (a+b)(a-b) = a^2 - b^2$$
$$= 5 - 7$$
$$= -2$$

**EXAMPLE 10** Rationalize the denominator.

$$\frac{\sqrt{5}}{2\sqrt{7}-\sqrt{5}} = \frac{\sqrt{5}}{(2\sqrt{7}-\sqrt{5})} \cdot \frac{(2\sqrt{7}+\sqrt{5})}{(2\sqrt{7}+\sqrt{5})} \qquad \text{Use the rule}$$
$$= \frac{\sqrt{5}\,(2\sqrt{7}+\sqrt{5})}{(2\sqrt{7})^2 - (\sqrt{5})^2}$$
$$= \frac{\sqrt{5}\,(2\sqrt{7}) + \sqrt{5}\sqrt{5}}{4 \cdot 7 - 5} \qquad (2\sqrt{7})^2 = 2^2(\sqrt{7})^2 = 4 \cdot 7$$
$$= \frac{2\sqrt{35}+5}{28-5}$$
$$= \frac{2\sqrt{35}+5}{23}$$

**EXAMPLE 11** Rationalize the denominator.

$$\frac{2\sqrt{3}+\sqrt{11}}{\sqrt{3}-\sqrt{11}} = \frac{(2\sqrt{3}+\sqrt{11})}{(\sqrt{3}-\sqrt{11})} \cdot \frac{(\sqrt{3}+\sqrt{11})}{(\sqrt{3}+\sqrt{11})} \qquad \text{Use the rule}$$
$$= \frac{2\sqrt{3}\sqrt{3} + 2\sqrt{3}\sqrt{11} + \sqrt{11}\sqrt{3} + \sqrt{11}\sqrt{11}}{(\sqrt{3})^2 - (\sqrt{11})^2}$$
$$= \frac{2 \cdot 3 + 2\sqrt{33} + \sqrt{33} + 11}{3 - 11}$$
$$= \frac{6 + 11 + 2\sqrt{33} + \sqrt{33}}{-8}$$
$$= \frac{17 + 3\sqrt{33}}{-8}$$

**EXAMPLE 12** Rationalize the denominator.

$$\frac{3-\sqrt{x}}{\sqrt{x}+5} = \frac{(3-\sqrt{x})}{(\sqrt{x}+5)} \cdot \frac{(\sqrt{x}-5)}{(\sqrt{x}-5)} \qquad \text{Use the rule}$$
$$= \frac{3\sqrt{x} - 3 \cdot 5 - \sqrt{x}\sqrt{x} + 5\sqrt{x}}{(\sqrt{x})^2 - 5^2}$$
$$= \frac{3\sqrt{x} + 5\sqrt{x} - 15 - x}{x - 25} \qquad x \geq 0 \text{ so } -\sqrt{x}\sqrt{x} = -\sqrt{x^2} = -|x| = -x$$
$$= \frac{8\sqrt{x} - 15 - x}{x - 25}$$

## EXERCISES 4.5

*Simplify. Rationalize all denominators.*

1. $\frac{\sqrt{3}}{\sqrt{2}}$

2. $\sqrt{\frac{12}{10}}$

3. $\frac{\sqrt{24}}{\sqrt{50}}$

4. $\frac{\sqrt{99}}{\sqrt{63}}$

5. $\frac{\sqrt{147}}{\sqrt{75}}$

6. $\frac{\sqrt{250}}{\sqrt{27}}$

7. $\frac{\sqrt[3]{3}}{\sqrt[3]{5}}$

8. $\sqrt[3]{\frac{24}{81}}$

9. $\frac{\sqrt[3]{40}}{\sqrt[3]{250}}$

10. $\frac{\sqrt{7x}}{\sqrt{5y}}$

11. $\frac{\sqrt{8a}}{\sqrt{27ab}}$

12. $\frac{\sqrt{3x^3y^2}}{\sqrt{150xy^3}}$

13. $\frac{\sqrt[3]{x}}{\sqrt[3]{2y}}$

14. $\frac{\sqrt[3]{54a}}{\sqrt[3]{7b}}$

15. $\frac{\sqrt[3]{125x^2}}{\sqrt[3]{3y^2}}$

16. $\sqrt{3}\,(\sqrt{3} - 2\sqrt{5})$

17. $\sqrt{8}\,(\sqrt{2} + 3\sqrt{12})$

18. $\sqrt{6}\,(2\sqrt{27} + \sqrt{24})$

19. $(\sqrt{3} + \sqrt{7})(\sqrt{3} - \sqrt{7})$

20. $(2\sqrt{5} - \sqrt{2})(3\sqrt{5} + 4\sqrt{2})$

21. $\frac{\sqrt{5}}{\sqrt{11} + \sqrt{5}}$

22. $\frac{\sqrt{8} - \sqrt{3}}{\sqrt{2} + \sqrt{3}}$

23. $\frac{2\sqrt{2} - 3}{3\sqrt{2} - 2}$

24. $\frac{\sqrt{x} + 1}{\sqrt{x} - 1}$

**25.** $5\sqrt{3}+\dfrac{7}{\sqrt{3}}$

**26.** $2\sqrt{45y}-\dfrac{3\sqrt{y}}{\sqrt{5}}$

**27.** $\dfrac{3\sqrt{18}-7\sqrt{50}}{\sqrt{2}}$

**28.** $\dfrac{\sqrt{3}}{\sqrt{5}}+\dfrac{\sqrt{5}}{\sqrt{3}}$

**29.** $\dfrac{\sqrt{3}-\dfrac{1}{\sqrt{3}}}{1+\sqrt{3}\cdot\dfrac{1}{\sqrt{3}}}$

**30.** $\dfrac{3+\dfrac{1}{2}}{1-\dfrac{1}{2}\sqrt{3}}$

---

ANSWERS: **1.** $\frac{\sqrt{6}}{2}$ **2.** $\frac{\sqrt{30}}{5}$ **3.** $\frac{2\sqrt{3}}{5}$ **4.** $\frac{\sqrt{77}}{7}$ **5.** $\frac{7}{5}$ **6.** $\frac{5\sqrt{30}}{9}$ **7.** $\frac{\sqrt[3]{75}}{5}$ **8.** $\frac{2}{3}$ **9.** $\frac{\sqrt[3]{20}}{5}$ **10.** $\frac{\sqrt{35xy}}{5y}$ **11.** $\frac{2\sqrt{6b}}{9b}$ **12.** $\frac{x\sqrt{2y}}{10y}$ **13.** $\frac{\sqrt[3]{4xy^2}}{2y}$ **14.** $\frac{3\sqrt[3]{98ab^2}}{7b}$ **15.** $\frac{5\sqrt[3]{9x^2y}}{3y}$ **16.** $3-2\sqrt{15}$ **17.** $4+12\sqrt{6}$ **18.** $18\sqrt{2}+12$ **19.** $-4$ **20.** $22+5\sqrt{10}$ **21.** $\frac{\sqrt{55}-5}{6}$ **22.** $3\sqrt{6}-7$ **23.** $\frac{6-5\sqrt{2}}{14}$ **24.** $\frac{x+2\sqrt{x}+1}{x-1}$ **25.** $\frac{22\sqrt{3}}{3}$ **26.** $\frac{27\sqrt{5y}}{5}$ **27.** $-26$ **28.** $\frac{8\sqrt{15}}{15}$ **29.** $\frac{\sqrt{3}}{3}$ **30.** $14+7\sqrt{3}$

## 4.6 Rational Number Exponents

In this section we give meaning to expressions such as $8^{2/3}$, $(81)^{-3/4}$, and $x^{7/5}$. If we assume that the rules for integer exponents also hold for rational exponents, we have the following results.

$$a^{1/2}\cdot a^{1/2}=a^{1/2+1/2}=a$$
$$a^{1/3}\cdot a^{1/3}\cdot a^{1/3}=a^{1/3+1/3+1/3}=a$$

Thus, $a^{1/2}=\sqrt{a}$ and $a^{1/3}=\sqrt[3]{a}$. In fact, for any root $k$,

$$a^{1/k}=\sqrt[k]{a}.$$

Furthermore, note that $(a^{1/2})^3=a^{3/2}$ and, therefore,

$$a^{3/2}=(\sqrt{a})^3=(a^3)^{1/2}=\sqrt{a^3}.$$

This leads to the following definition. (The restriction $a\geq 0$ is not necessary if the radical has an odd index, for example $(\sqrt[3]{-8})^2=(-8)^{2/3}$, but we will not consider these cases.) Also, the definition follows from

$$a^{m/n}=(a^{1/n})^m=(a^m)^{1/n}.$$

If $a \geq 0$,

$$a^{m/n} = (\sqrt[n]{a})^m = \sqrt[n]{a^m}.$$

All the rules of integer exponents hold for fractional exponents. We list these rules with rational exponents $p/q$ and $r/s$.

$$a^{p/q}\, a^{r/s} = a^{p/q+r/s}$$

$$\frac{a^{p/q}}{a^{r/s}} = a^{p/q-r/s} \qquad (a \neq 0)$$

$$(a^{p/q})^{r/s} = a^{(p/q)\cdot(r/s)}$$

$$a^{-p/q} = \frac{1}{a^{p/q}} \qquad (a \neq 0)$$

$$(ab)^{p/q} = a^{p/q}\, b^{p/q}$$

$$\left(\frac{a}{b}\right)^{p/q} = \frac{a^{p/q}}{b^{p/q}} \qquad (b \neq 0)$$

We can now simplify exponential and radical expressions.

**EXAMPLE 1** Simplify.

**(a)** $9^{1/2} = \sqrt{9} = 3$

**(b)** $8^{2/3} = (8^{1/3})^2 = (\sqrt[3]{8})^2 = (2)^2 = 4$

**(c)** $81^{-3/4} = (81^{1/4})^{-3} = (\sqrt[4]{81})^{-3} = 3^{-3} = \dfrac{1}{3^3} = \dfrac{1}{27}$

In the work which follows, all variables represent positive numbers.

**EXAMPLE 2** Simplify.

$$\begin{aligned} \sqrt{25x^8y^6} &= (25x^8y^6)^{1/2} && \sqrt{a} = a^{1/2} \\ &= (5^2)^{1/2}\,(x^8)^{1/2}\,(y^6)^{1/2} && (ab)^{p/q} = a^{p/q}\, b^{p/q} \\ &= 5^{2\cdot 1/2}\, x^{8\cdot 1/2}\, y^{6\cdot 1/2} && (a^{p/q})^{r/s} = a^{(p/q)\cdot(r/s)} \\ &= 5x^4y^3 \end{aligned}$$

**EXAMPLE 3** Simplify.

$$\begin{aligned} \sqrt[3]{\frac{8x^{10}}{27y^9}} &= \left(\frac{2^3\, x^{10}}{3^3\, y^9}\right)^{1/3} && \sqrt[3]{a} = a^{1/3} \\ &= \frac{(2^3)^{1/3}\,(x^{10})^{1/3}}{(3^3)^{1/3}\,(y^9)^{1/3}} && (ab)^{p/q} = a^{p/q}\, b^{p/q} \text{ and } \left(\frac{a}{b}\right)^{p/q} = \frac{a^{p/q}}{b^{p/q}} \\ &= \frac{2x^{10/3}}{3y^{9/3}} && (a^{p/q})^{r/s} = a^{(p/q)\cdot(r/s)} \\ &= \frac{2x^{9/3}\, x^{1/3}}{3y^3} && a^{10/3} = a^{9/3+1/3} = a^{9/3}\cdot a^{1/3} \\ &= \frac{2x^3\sqrt[3]{x}}{3y^3} \end{aligned}$$

**EXAMPLE 4** Simplify.

$$\left(\frac{16a^4b^{-10}}{81a^{-4}b^2}\right)^{3/4} = \left(\frac{2^4a^8}{3^4b^{12}}\right)^{3/4} \qquad \frac{a^m}{a^n} = a^{m-n}$$

$$= \frac{(2^4)^{3/4}\,(a^8)^{3/4}}{(3^4)^{3/4}\,(b^{12})^{3/4}}$$

$$= \frac{2^{4\cdot 3/4}\,a^{8\cdot 3/4}}{3^{4\cdot 3/4}\,b^{12\cdot 3/4}}$$

$$= \frac{2^3\,a^6}{3^3\,b^9}$$

$$= \frac{8a^6}{27b^9}$$

**EXAMPLE 5** Simplify.

$$\left(\frac{9x^6y^{-4}}{z^{-8}}\right)^{-3/2} = \frac{(3^2)^{-3/2}(x^6)^{-3/2}(y^{-4})^{-3/2}}{(z^{-8})^{-3/2}} \qquad (ab)^{-3/2} = a^{-3/2}b^{-3/2} \text{ and } \left(\frac{a}{b}\right)^{-3/2} = \frac{a^{-3/2}}{b^{-3/2}}$$

$$= \frac{3^{-3}x^{-9}y^6}{z^{12}} \qquad (a^{p/q})^{r/s} = a^{(p/q)\cdot(r/s)}$$

$$= \frac{y^6}{27x^9z^{12}}$$

**EXAMPLE 6** Rationalize the denominator.

$$\left(\frac{1}{a^2b}\right)^{2/3} = \frac{1^{2/3}}{(a^2)^{2/3}b^{2/3}}$$

$$= \frac{1}{a^{4/3}b^{2/3}} \qquad 1^{2/3} = 1$$

$$= \frac{1}{a^{1+1/3}b^{2/3}} \qquad a^{4/3} = a^{1+1/3}$$

$$= \frac{1}{a\cdot a^{1/3}b^{2/3}}\cdot\frac{a^{2/3}b^{1/3}}{a^{2/3}b^{1/3}} \qquad a^{1/3}a^{2/3} \text{ and } b^{2/3}b^{1/3} \text{ gives perfect cubes in the denominator}$$

$$= \frac{a^{2/3}b^{1/3}}{aa^{1/3+2/3}b^{2/3+1/3}}$$

$$= \frac{(a^2b)^{1/3}}{a\cdot a\cdot b}$$

$$= \frac{\sqrt[3]{a^2b}}{a^2b}$$

## EXERCISES 4.6

*Write in exponential notation.*

**1.** $\sqrt{7}$ **2.** $\sqrt[3]{7^2}$ **3.** $\sqrt[4]{2x}$

**4.** $\sqrt[5]{y^2}$ **5.** $\sqrt[8]{a^6}$ **6.** $\sqrt[3]{x^2y^3}$

*Write in radical notation.*

7. $7^{1/2}$

8. $3^{2/3}$

9. $3^{4/3}$

10. $x^{5/6}$

11. $a^{2/3}b^{1/3}$

12. $y^{-1/3}$

*Simplify. Write all radicals in exponential notation to simplify, but give answers in radical notation. Rationalize all denominators. (Assume all variables positive.)*

13. $\sqrt{25x^2}$

14. $\sqrt[3]{125y^6}$

15. $\sqrt[4]{a^8b^{12}}$

16. $\sqrt{8x^4y^8}$

17. $\sqrt[6]{a^{12}b^{18}}$

18. $\sqrt[3]{8x^4}$

19. $\sqrt{\dfrac{25a^4}{9b^6}}$

20. $\sqrt[3]{\dfrac{16x^9}{y^{12}}}$

21. $\sqrt[5]{\dfrac{a^{20}}{625b^{15}}}$

22. $(4x^2y^4)^{3/2}$

23. $(27a^6b^9)^{2/3}$

24. $(8x^{-3}y^6)^{4/3}$

25. $\left(\dfrac{96a^3b^{-2}}{3a^{-2}b^8}\right)^{1/5}$

26. $(x^2y)^{-1/2}$

27. $\left(\dfrac{8a^2}{b^{-3}}\right)^{-1/3}$

28. $(8^{2/3}x^{-1/3}y^4)^3$

29. $(3a^2b)^{1/2}b^{1/2}$

30. $(xy^{-1})^{-1/3}$

*Simplify. Rationalize denominators.*

31. $3\sqrt{5x} + \dfrac{2\sqrt{x}}{\sqrt{5}}$

32. $(\sqrt{6} - 2\sqrt{5})(4\sqrt{6} - \sqrt{5})$

33. $\dfrac{\sqrt{11} + \sqrt{2}}{\sqrt{11} - \sqrt{2}}$

34. $\dfrac{2\sqrt{x} - 3}{\sqrt{x} + 3}$

---

ANSWERS: 1. $7^{1/2}$ 2. $7^{2/3}$ 3. $(2x)^{1/4}$ 4. $y^{2/5}$ 5. $a^{3/4}$ 6. $x^{2/3}y$ 7. $\sqrt{7}$ 8. $\sqrt[3]{3^2}$ 9. $\sqrt[3]{3^4} = 3\sqrt[3]{3}$ 10. $\sqrt[6]{x^5}$ 11. $\sqrt[3]{a^2b}$ 12. $\dfrac{1}{\sqrt[3]{y}}$ or $\dfrac{\sqrt[3]{y^2}}{y}$ (rationalized) 13. $5x$ 14. $5y^2$ 15. $a^2b^3$ 16. $2x^2y^4\sqrt{2}$ 17. $a^2b^3$ 18. $2x\sqrt[3]{x}$ 19. $\dfrac{5a^2}{3b^3}$ 20. $\dfrac{2x^3\sqrt[3]{2}}{y^4}$ 21. $\dfrac{a^4\sqrt[5]{5}}{5b^3}$ 22. $8x^3y^6$ 23. $9a^4b^6$ 24. $\dfrac{16y^8}{x^4}$ 25. $\dfrac{2a}{b^2}$ 26. $\dfrac{\sqrt{y}}{xy}$ 27. $\dfrac{\sqrt[3]{a}}{2ab}$ 28. $\dfrac{64y^{12}}{x}$ 29. $ab\sqrt{3}$ 30. $\dfrac{\sqrt[3]{x^2y}}{x}$ 31. $\dfrac{17\sqrt{5x}}{5}$ 32. $34 - 9\sqrt{30}$ 33. $\dfrac{13 + 2\sqrt{22}}{9}$ 34. $\dfrac{2x - 9\sqrt{x} + 9}{x - 9}$

## 4.7 Radical Equations that Result in Linear Equations

In this section we solve radical equations which result in linear equations. The key to solving radical equations is the following rule.

**RULE OF SQUARING**

If $a$ and $b$ are numbers or expressions and if $a = b$, then $a^2 = b^2$.

When this rule is applied to an equation involving a variable, the resulting equation may have more solutions than the original. For example, the equation

$$x = 5$$

has only one solution, 5, but

$$x^2 = 25$$

has two solutions, 5 and $-5$. Thus, when the rule of squaring is used, any possible solutions must always be checked in the original equation since $a^2 = b^2$ and $a = b$ may not be equivalent. Solutions which do not check in the original equation are called **extraneous roots.**

**TO SOLVE AN EQUATION INVOLVING RADICALS**

1. If only one radical is present, isolate this radical on one side of the equation, simplify, and proceed to Step 3.
2. If two radicals are present, isolate one of the radicals on one side of the equation.
3. Square both sides.
4. Solve the resulting equation. If a radical remains in the equation, isolate it and square once more.
5. Check all possible solutions in the *original* equation since the equations involved may not be equivalent.

CAUTION: Isolating the radicals, as indicated in Step 1 of the rule, is necessary since the entire quantity on each side of the equation must be squared. *Do not square term by term.* For example, $\sqrt{4} + \sqrt{9} = 2 + 3 = 5$, but $(\sqrt{4})^2 + (\sqrt{9})^2 = 4 + 9$, not $5^2$ or 25.

**EXAMPLE 1** Solve.

$$\sqrt{2x+3} + 5 = 8$$

$$\sqrt{2x+3} = 3 \quad \text{Subtract 5 from both sides to isolate the radical}$$

$$(\sqrt{2x+3})^2 = (3)^2 \quad \text{Use rule of squaring}$$

$$2x + 3 = 9$$

$$2x = 6$$

$$x = 3$$

Check:
$$\sqrt{2(3)+3}+5 \stackrel{?}{=} 8$$
$$\sqrt{9}+5 \stackrel{?}{=} 8$$
$$3+5 \stackrel{?}{=} 8$$
$$8=8$$

The solution is 3.

**EXAMPLE 2** Solve.

$$3\sqrt{2y-5}-\sqrt{y+23}=0$$
$$3\sqrt{2y-5}=\sqrt{y+23} \quad \text{Isolate the radicals}$$
$$(3\sqrt{2y-5})^2=(\sqrt{y+23})^2 \quad \text{Use rule of squaring}$$
$$9(2y-5)=y+23 \quad \text{Square 3 also}$$
$$18y-45=y+23$$
$$17y=68$$
$$y=4$$

Check:
$$3\sqrt{2(4)-5}-\sqrt{4+23} \stackrel{?}{=} 0$$
$$3\sqrt{3}-\sqrt{27} \stackrel{?}{=} 0$$
$$3\sqrt{3}-3\sqrt{3}=0$$

The solution is 4.

**EXAMPLE 3** Solve.

$$\sqrt{x^2+5}-x+5=0$$
$$\sqrt{x^2+5}=x-5 \quad \text{Isolate the radical}$$
$$(\sqrt{x^2+5})^2=(x-5)^2 \quad \text{Use rule of squaring}$$
$$x^2+5=x^2-\mathbf{10x}+25$$

$(x-5)^2$ is *not* $x^2-25$; do not forget the middle term $-10x$

$$5=-10x+25 \quad \text{Subtract } x^2$$
$$-20=-10x$$
$$2=x$$

Check:
$$\sqrt{2^2+5}-2+5 \stackrel{?}{=} 0$$
$$\sqrt{9}-2+5 \stackrel{?}{=} 0$$
$$3-2+5 \stackrel{?}{=} 0$$
$$6 \neq 0$$

The equation has no solution.

**EXAMPLE 4** Solve.

| | |
|---|---|
| $\sqrt{y+3}-\sqrt{y-2}=1$ | |
| $\sqrt{y+3}=1+\sqrt{y-2}$ | Isolate $\sqrt{y+3}$ |
| $(\sqrt{y+3})^2=(1+\sqrt{y-2})^2$ | Use rule of squaring |
| $y+3=1+2\sqrt{y-2}+y-2$ | Don't forget the middle term $2\sqrt{y-2}$ |
| $y+3=2\sqrt{y-2}+y-1$ | |
| $3=2\sqrt{y-2}-1$ | A radical remains which must be isolated |
| $4=2\sqrt{y-2}$ | $\sqrt{y-2}$ is isolated by adding 1 to both sides |
| $2=\sqrt{y-2}$ | Always simplify |
| $2^2=(\sqrt{y-2})^2$ | Use rule of squaring again |
| $4=y-2$ | |
| $6=y$ | |

Check: $\sqrt{6+3}-\sqrt{6-2}\stackrel{?}{=}1$

$$\sqrt{9}-\sqrt{4}\stackrel{?}{=}1$$
$$3-2\stackrel{?}{=}1$$
$$1=1$$

The solution is 6.

## EXERCISES 4.7

*Solve.*

**1.** $\sqrt{x-2}=3$

**2.** $5\sqrt{3x}-20=0$

**3.** $4\sqrt{2y-1}-2=0$

**4.** $\sqrt{4y-3}-2=3$

**5.** $5\sqrt{a-3}+10=0$

**6.** $6\sqrt{a}=2\sqrt{a+16}$

**7.** $\sqrt{x+7} = \sqrt{x} + 2$

**8.** $3\sqrt{x+1} = \sqrt{x+33}$

**9.** $\sqrt{3y+2} - \sqrt{y+12} = 0$

**10.** $\sqrt{5y-3} - \sqrt{2y+1} = 0$

**11.** $\dfrac{8}{\sqrt{a}} = 2$
[*Hint:* First clear the fraction.]

**12.** $\dfrac{15}{\sqrt{a}} = \dfrac{20}{\sqrt{a+7}}$

**13.** $\sqrt{x^2-5} + x - 5 = 0$

**14.** $\sqrt{x^2+2} - x - 2 = 0$

**15.** $\sqrt{y^2-13} - y + 1 = 0$

**16.** $\sqrt{y^2+9} + y + 1 = 0$

**17.** $\sqrt{a+6} + \sqrt{a+11} = 5$

**18.** $\sqrt{a+20} - \sqrt{a+4} = 2$

**19.** $\sqrt{x+2} + \sqrt{x+6} = 4$

**20.** $\sqrt{x+10} - \sqrt{x-6} = 2$

**21.** Twice the square root of a number is the same as the square root of 9 more than three times the number. Find the number.

**22.** If the square root of 1 less than a number is subtracted from the square root of 15 more than the number, the result is 2. Find the number.

*Simplify.*

**23.** $\sqrt{75x^4y^8}$

**24.** $\sqrt[3]{54x^9y^7}$

**25.** $\sqrt[3]{\dfrac{24a^{-2}b^3}{a^4b^{-2}}}$

**26.** $\sqrt{\dfrac{8a^2b^3}{a^5b^{-5}}}$

**27.** $(81x^{12}y^{-16})^{1/4}$

**28.** $(4x^{-4}y^5)^{3/2}$

**29.** $\left(\dfrac{8a^4}{b^3}\right)^{-1/2}$

**30.** $\left(\dfrac{a^5b^{-15}}{32}\right)^{2/5}$

---

ANSWERS: 1. 11 2. $\frac{16}{3}$ 3. $\frac{5}{8}$ 4. 7 5. no solution 6. 2 7. no solution 8. 3 9. 5 10. $\frac{4}{3}$ 11. 16 12. 9 13. 3 14. $-\frac{1}{2}$ 15. 7 16. no solution 17. $-2$ 18. 5 19. $\frac{1}{4}$ 20. 15 21. 9 22. 10 23. $5x^2y^4\sqrt{3}$ 24. $3x^3y^2\sqrt[3]{2y}$ 25. $\frac{2b\sqrt[3]{3b^2}}{a^2}$ 26. $\frac{2b^4\sqrt{2a}}{a^2}$ 27. $\frac{3x^3}{y^4}$ 28. $\frac{8y^7\sqrt{y}}{x^6}$ 29. $\frac{b\sqrt{2b}}{4a^2}$ 30. $\frac{a^2}{4b^6}$

## 4.8 Solving Literal Equations

In Section 2.9 we solved literal equations that were linear in the variables. In this section we solve literal equations which involve powers and radicals. Recall that literal equations are equations, such as formulas, which involve several variables. For example

$$r = \sqrt{\frac{A}{\pi}}$$

is a literal equation that becomes $A = \pi r^2$ when solved for $A$. We first look at review examples from Chapter 2, giving similar numerical examples to help explain the solution procedures.

**EXAMPLE 1** Solve $P = 2 \cdot l + 2 \cdot w$ for $w$.

*Similar numerical equation*

$$P = 2 \cdot l + 2 \cdot w \qquad 12 = 2 \cdot 3 + 2 \cdot w$$

$$P - 2l = 2 \cdot l - 2 \cdot l + 2 \cdot w \qquad 12 - 2 \cdot 3 = 2 \cdot 3 - 2 \cdot 3 + 2 \cdot w$$

$$P - 2l = 2 \cdot w \qquad 6 = 2 \cdot w$$

$$\frac{1}{2}(P - 2l) = \frac{1}{2} \cdot 2 \cdot w \qquad \frac{1}{2} \cdot 6 = \frac{1}{2} \cdot 2 \cdot w$$

$$\frac{P - 2l}{2} = w \qquad 3 = w$$

**EXAMPLE 2** Solve $r = \sqrt{\frac{A}{\pi}}$ for $A$.

*Similar numerical equation*

$$r = \sqrt{\frac{A}{\pi}} \qquad 5 = \sqrt{\frac{A}{3}}$$

$$r^2 = \frac{A}{\pi} \quad \text{Square both sides} \quad 5^2 = \frac{A}{3}$$

$$\pi r^2 = \frac{A}{\pi} \cdot \pi \quad \text{Multiply by } \pi \quad 3 \cdot 5^2 = \frac{A}{3} \cdot 3$$

$$\pi r^2 = A \qquad 75 = A$$

**EXAMPLE 3** Solve $a = \sqrt{b - c}$ for $c$.

*Similar numerical equation*

$$a = \sqrt{b - c} \qquad 3 = \sqrt{5 - c}$$

$$a^2 = (\sqrt{b - c})^2 \qquad (3)^2 = (\sqrt{5 - c})^2$$

$$a^2 = b - c \qquad 9 = 5 - c$$

$$a^2 - b = b - b - c \qquad 9 - 5 = 5 - 5 - c$$

$$a^2 - b = -c \qquad 9 - 5 = -c$$

$$(-1)(a^2 - b) = (-1)(-c) \qquad (-1)(4) = (-1)(-c)$$

$$b - a^2 = c \qquad -4 = c$$

Although we have not formally discussed the **rule of cubing,** it is similar to the rule of squaring. That is, if $a = b$, then $a^3 = b^3$.

**EXAMPLE 4** Solve $\left(\frac{x}{y}\right)^{1/3} = z^2$ for $y$.

| | *Similar numerical equation* |
|---|---|
| $\left(\frac{x}{y}\right)^{1/3} = z^2$ | $\left(\frac{3}{y}\right)^{1/3} = 2^2$ |
| $\left[\left(\frac{x}{y}\right)^{1/3}\right]^3 = (z^2)^3$ | $\left[\left(\frac{3}{y}\right)^{1/3}\right]^3 = (2^2)^3$ |
| $\frac{x}{y} = z^6$ | $\frac{3}{y} = 64$ |
| $\frac{x}{y} \cdot y = z^6 \cdot y$ | $\frac{3}{y} \cdot y = 64 \cdot y$ |
| $x = z^6 \cdot y$ | $3 = 64 \cdot y$ |
| $\frac{1}{z^6} \cdot x = \frac{1}{z^6} \cdot z^6 \cdot y$ | $\frac{1}{64} \cdot 3 = \frac{1}{64} \cdot 64y$ |
| $\frac{x}{z^6} = y$ | $\frac{3}{64} = y$ |

## EXERCISES 4.8

*Solve. Assume all variables are suitably chosen so that all radicals are meaningful.*

**1.** $2a + b = c$ for $a$

**2.** $2ab = c$ for $b$

**3.** $x = \frac{1}{2}yz$ for $z$

**4.** $xy + z = w$ for $x$

**5.** $A = \frac{1}{2}(b_1 + b_2)h$ for $b_2$

**6.** $a = \frac{3}{b + c}$ for $c$

**7.** $z = yx^2$ for $y$

**8.** $x^2 + y = z^2$ for $y$

**9.** $b^2 = 5 - a^2c$ for $c$

**10.** $a = \frac{1}{2}gt^2$ for $g$

**11.** $\sqrt{\frac{x}{y}} = z$ for $y$

**12.** $z + x^2z = y^2$ for $z$
[*Hint:* Factor out $z$ first.]

**13.** $a = b\sqrt{1 + x}$ for $x$

**14.** $a = \sqrt{1 + \frac{b}{x}}$ for $x$

**15.** $\sqrt{xy + 3} = z$ for $y$

**16.** $z^{1/3} + x = y$ for $z$

**17.** $\sqrt{2x + 3} - \sqrt{3x - 2} = 0$

**18.** $\sqrt{x - 6} = \sqrt{x + 7}$

**19.** $3\sqrt{3y + 7} - 5\sqrt{y + 3} = 0$

**20.** $\sqrt{y^2 + 9} - y - 1 = 0$

**21.** $\sqrt{x + 1} - \sqrt{x - 6} = 1$

**22.** $\sqrt{x + 19} - \sqrt{x - 14} = 3$

---

ANSWERS: **1.** $\frac{c - b}{2}$ **2.** $\frac{c}{2a}$ **3.** $\frac{2x}{y}$ **4.** $\frac{w - z}{y}$ **5.** $\frac{2A}{h} - b_1 = \frac{2A - b_1h}{h}$ **6.** $\frac{3}{a} - b = \frac{3 - ba}{a}$ **7.** $\frac{z}{x^2}$
**8.** $z^2 - x^2$ **9.** $\frac{5 - b^2}{a^2}$ **10.** $\frac{2a}{t^2}$ **11.** $\frac{x}{z^2}$ **12.** $\frac{y^2}{1 + x^2}$ **13.** $\frac{a^2 - b^2}{b^2}$ **14.** $\frac{b}{a^2 - 1}$ **15.** $\frac{z^2 - 3}{x}$ **16.** $(y - x)^3$
**17.** 5 **18.** no solution **19.** 6 **20.** 4 **21.** 15 **22.** 30

## Chapter 4 Summary

### Words And Phrases For Review

[4.1] integer exponent
scientific notation
[4.2] square root
radical
absolute value
cube root
$k$th root
principal $k$th root
radicand
index
[4.7] rule of squaring
[4.8] rule of cubing

### Brief Reminders

[4.1] For real numbers $a$ and $b$ and integers $m$ and $n$:

$$a^m \cdot a^n = a^{m+n}. \qquad \frac{a^m}{a^n} = a^{m-n} \text{ if } a \neq 0.$$

$$(a^m)^n = a^{m \cdot n}. \qquad a^0 = 1 \text{ if } a \neq 0.$$

$$a^{-n} = \frac{1}{a^n} \text{ if } a \neq 0. \qquad \frac{1}{a^{-n}} = a^n \text{ if } a \neq 0.$$

$$(a \cdot b)^n = a^n \cdot b^n. \qquad \left(\frac{a}{b}\right)^n = \frac{a^n}{b^n} \text{ if } b \neq 0.$$

[4.2] **1.** For any real number $a$, $\sqrt{a^2} = |a| \geq 0$.

**2.** $\sqrt[k]{a^k} = |a|$, if $k$ is even. (Use absolute value when $k$ is even.)

**3.** $\sqrt[k]{a^k} = a$, if $k$ is odd. (Do not use absolute value when $k$ is odd.)

[4.3] **1.** If $\sqrt[k]{a}$ and $\sqrt[k]{b}$ are defined, then $\sqrt[k]{a}\,\sqrt[k]{b} = \sqrt[k]{ab}$.

**2.** If $\sqrt[k]{a}$ and $\sqrt[k]{b}$ are defined and $b \neq 0$, then $\dfrac{\sqrt[k]{a}}{\sqrt[k]{b}} = \sqrt[k]{\dfrac{a}{b}}$.

[4.4] **1.** $\sqrt{a+b}$ is *not* $\sqrt{a} + \sqrt{b}$.

**2.** $\sqrt{a-b}$ is *not* $\sqrt{a} - \sqrt{b}$.

**3.** When "removing" a perfect square from under a square root radical, multiply it by the coefficient.

[4.5] To rationalize the denominator of an expression whose denominator is of the form $a\sqrt{x} - b\sqrt{y}$ (or $a\sqrt{x} + b\sqrt{y}$), multiply numerator and denominator by $a\sqrt{x} + b\sqrt{y}$ (or $a\sqrt{x} - b\sqrt{y}$). For example,

$$\frac{3}{\sqrt{3}-\sqrt{2}} = \frac{3}{(\sqrt{3}-\sqrt{2})} \cdot \frac{(\sqrt{3}+\sqrt{2})}{(\sqrt{3}+\sqrt{2})} = \frac{3(\sqrt{3}+\sqrt{2})}{3-2} = 3(\sqrt{3}+\sqrt{2}).$$

[4.6] **1.** If $a \geq 0$, $a^{m/n} = (\sqrt[n]{a})^m = \sqrt[n]{a^m}$.

**2.** All rules of exponents listed in Section 4.1 apply for rational number exponents as well as integer exponents.

[4.7] To solve a radical equation, isolate a radical on one side of the equation and square both sides. (Do *not* simply square term by term.) Check all solutions in the *original* equation.

[4.8] When solving a literal equation, it may help to solve a similar numerical equation first.

## CHAPTER 4 REVIEW EXERCISES

[4.1] *Simplify and write without negative exponents.*

1. $\dfrac{x^7y^3}{xy^2}$

2. $\dfrac{3a^{-2}b^3}{12a^{-3}b}$

3. $\left(\dfrac{5^0x^2y}{x^{-2}y^2}\right)^{-2}$

*Simplify and give answers in scientific notation.*

4. $(300{,}000{,}000)(.00005)$

5. $\dfrac{(.00000003)(4{,}000{,}000)}{200{,}000}$

[4.2–4.6] *Simplify. Rationalize all denominators. Assume all variables are positive.*

6. $\sqrt{121x^6y^{10}}$

7. $\sqrt[3]{-27a^3b^{12}}$

8. $\sqrt[4]{-16x^2y^2}$

9. $\sqrt{27a^3b^5}$

10. $\sqrt[3]{81x^4y^6}$

11. $\sqrt{3ab}\,\sqrt{12a^3b^4}$

12. $\dfrac{\sqrt{48a^{-2}b^3}}{\sqrt{3a^2b^{-4}}}$

13. $\dfrac{\sqrt[3]{54x^7y^{-3}}}{\sqrt[3]{x^{-1}y^{-8}}}$

14. $\sqrt[5]{\dfrac{5^2(a-b)^2}{5^{-3}(a-b)^{-7}}}$ $(a-b \geq 0)$

15. $5\sqrt{147} - 4\sqrt{48}$

16. $-6\sqrt[3]{-54} + 8\sqrt[3]{250}$

17. $3\sqrt{\dfrac{98}{25}} + 6\dfrac{\sqrt{54}}{\sqrt{75}}$

18. $5a\sqrt{63a} - 4\sqrt{175a^3}$

19. $\sqrt[3]{27x^5y^9} - 3y\sqrt[3]{125x^5y^6}$

20. $8\sqrt{75} - 3\sqrt{24} + 6\sqrt{12}$

21. $\dfrac{\sqrt{x^2y^3}}{\sqrt{2xy^4}}$

22. $\dfrac{\sqrt[3]{27a^3}}{\sqrt[3]{2a^2b^2}}$

23. $\dfrac{\sqrt{3} - 2\sqrt{2}}{\sqrt{3} + \sqrt{2}}$

24. $\dfrac{\sqrt{5} + \sqrt{7}}{2\sqrt{5} - \sqrt{7}}$

25. $\left(\dfrac{25x^5y^{-3}}{x^{-5}y^{-8}}\right)^{3/2}$

26. $\left(\dfrac{54a^6b^7}{2a^{-3}b^2}\right)^{-1/3}$

[4.6] *Rewrite* without *fractional exponents.*

**27.** $(x^3y)^{3/5}$

**28.** $(a^2b^2)^{3/7}$

*Rewrite* with *fractional exponents.*

**29.** $\sqrt[4]{x^3}$

**30.** $\sqrt[7]{a}$

[4.7] *Solve.*

**31.** $\sqrt{4x-3}-\sqrt{x+9}=0$

**32.** $\sqrt{x+5}-\sqrt{x-4}=1$

**33.** $\sqrt{y^2+4}+y+2=0$

**34.** $\sqrt{y+2}-\sqrt{y-6}=2$

[4.8] **35.** Solve $a^2b-1=c^2$ for $b$

**36.** Solve $\sqrt{ab+c}=5$ for $a$

**37.** Solve $z^{1/3}\,x^{1/3}=y$ for $z$

**38.** Solve $x=z\sqrt{1-y}$ for $y$

---

**ANSWERS:** **1.** $x^6y$ **2.** $\frac{ab^2}{4}$ **3.** $\frac{y^2}{x^8}$ **4.** $1.5\times 10^4$ **5.** $6\times 10^{-7}$ **6.** $11\,x^3y^5$ **7.** $-3ab^4$ **8.** not a real number **9.** $3ab^2\sqrt{3ab}$ **10.** $3xy^2\sqrt[3]{3x}$ **11.** $6a^2b^2\sqrt{b}$ **12.** $\frac{4b^3\sqrt{b}}{a^2}$ **13.** $3x^2y\sqrt[3]{2x^2y^2}$ **14.** $5(a-b)\sqrt[5]{(a-b)^4}$ **15.** $19\sqrt{3}$ **16.** $58\sqrt[3]{2}$ **17.** $\frac{39\sqrt{2}}{5}$ **18.** $-5a\sqrt{7a}$ **19.** $-12xy^3\sqrt[3]{x^2}$ **20.** $52\sqrt{3}-6\sqrt{6}$ **21.** $\frac{\sqrt{2xy}}{2y}$ **22.** $\frac{3\sqrt[3]{4ab}}{2b}$ **23.** $7-3\sqrt{6}$ **24.** $\frac{17+3\sqrt{35}}{13}$ **25.** $125x^{15}y^7\sqrt{y}$ **26.** $\frac{\sqrt[3]{b}}{3a^3b^2}$ **27.** $\sqrt[5]{(x^3y)^3}=x\sqrt[5]{x^4y^3}$ **28.** $\sqrt[7]{(a^2b^2)^3}=\sqrt[7]{a^6b^6}$ **29.** $x^{3/4}$ **30.** $a^{1/7}$ **31.** 4 **32.** 20 **33.** no solution **34.** 7 **35.** $\frac{c^2+1}{a^2}$ **36.** $\frac{25-c}{b}$ **37.** $\frac{y^3}{x}$ **38.** $\frac{z^2-x^2}{z^2}$

# 5

# Quadratic Equations and Complex Numbers

## 5.1 Solving Quadratic Equations Using the Method of Factoring

An important property of our number system that can be used to solve certain types of equations involves zero products. For example, $2 \cdot 0 = 0$ and $0 \cdot 1/2 = 0$. In fact, for any number $a$, $a \cdot 0 = 0$. Also, if a product of two or more factors is zero, then at least one of the factors must be zero. This property is stated in the following rule.

**ZERO-PRODUCT RULE**

If $a$ and $b$ are numbers or expressions and

$$\text{if } a \cdot b = 0, \quad \text{then } a = 0 \quad \text{or} \quad b = 0.$$

The solutions (or roots) of an equation like

$$(x - 3)(x + 5) = 0$$

are found by setting each factor equal to zero and solving for $x$.

**EXAMPLE 1** Solve $(x - 3)(x + 5) = 0$.

By the zero-product rule,

$$x - 3 = 0 \quad \text{or} \quad x + 5 = 0$$
$$x = 3 \qquad\qquad x = -5$$

The solutions are 3 and $-5$.

**EXAMPLE 2** Solve $\frac{1}{2}x(2x + 1) = 0$.

$$\frac{1}{2} = 0 \quad \text{or} \quad x = 0 \quad \text{or} \quad 2x + 1 = 0$$

no solution

$$2x = -1$$

$$x = -\frac{1}{2}$$

Zero-product rule extends to products of three or more factors

The solutions are 0 and $-\frac{1}{2}$. (Check.)

When a constant factor is set equal to zero, we obtain no solution. In Example 2, we could have multiplied both sides by 2 and solved the equivalent equation $x(2x+1)=0$.

**EXAMPLE 3** Solve $x(x-3)(2x+3)=0$.

$$x=0 \quad \text{or} \quad x-3=0 \quad \text{or} \quad 2x+3=0 \qquad \text{Zero-product rule}$$
$$x=3 \qquad\qquad 2x=-3$$
$$x=-\frac{3}{2}$$

The solutions are 0, 3, and $-\frac{3}{2}$. (Check.)

**CAUTION:** Do not apply the zero-product rule when the given equation does not involve a zero product.

**EXAMPLE 4** Solve $(x-2)-(2x-1)=0$.

We first clear parentheses and then combine like terms. The zero-product rule does not apply to a zero difference or a zero sum.

$$(x-2) - (2x-1)=0$$
$$x-2 - 2x + 1=0 \qquad \text{Watch signs}$$
$$-x-1=0$$
$$-x=1$$
$$x=-1$$

Check the solution $-1$.

In Example 1, if we had started with

$$x^2+2x-15=0$$

we could have factored to get

$$(x-3)(x+5)=0$$

and then continued as in the example. We have, in effect, solved a *quadratic* equation. A **quadratic equation** or **second-degree equation** is an equation which can be written in the form

$$ax^2+bx+c=0$$

where $x$ is the variable and $a$, $b$, and $c$ are constants with $a \neq 0$. We insist that $a \neq 0$ since a quadratic equation must have an $x^2$ term. We call $ax^2+bx+c=0$ the **general form** of the quadratic equation and usually write a quadratic equation in this form before we attempt to solve it.

**EXAMPLE 5** Write $3x^2=-5x+4$ in general form.

Adding $5x$ and $-4$ to both sides of the equation, we obtain

$$3x^2+5x-4=0.$$

This is in general form with $a=3$, $b=5$, and $c=-4$ (not 4).

**EXAMPLE 6** Write $2x^2 + 9 = 7x^2 - 4x$ in general form.

Adding $-7x^2$ and $4x$ to both sides, we have

$$-5x^2 + 4x + 9 = 0.$$

Thus $a = -5$, $b = 4$, and $c = 9$. If we multiply both sides of the equation by $(-1)$, we obtain

$$5x^2 - 4x - 9 = 0$$

where $a = 5$, $b = -4$, and $c = -9$. These equations are equivalent (have the same solutions), but the second one is easier to factor.

If an equation in general form has no $x$ term, then $b = 0$. If there is no constant term, then $c = 0$. Thus, for

$$4x^2 - 9 = 0$$

$a = 4$, $b = 0$, and $c = -9$, and for

$$4x^2 - 7x = 0$$

$a = 4$, $b = -7$, and $c = 0$.

**TO SOLVE A QUADRATIC EQUATION BY FACTORING**

1. Write the equation in general form.
2. Clear any fractions, divide out any constant factors, and make the coefficient of $x^2$ positive by multiplying the equation by the proper number.
3. Factor the left side into a product of two factors.
4. Use the zero-product rule to solve the resulting equation.
5. Check your answers in the original equation.

**EXAMPLE 7** Solve the following quadratic equation by factoring.

$$x^2 - x - 12 = 0$$

$$(x - 4)(x + 3) = 0 \qquad \text{Factor}$$

$$x - 4 = 0 \quad \text{or} \quad x + 3 = 0 \qquad \text{Zero-product rule}$$

$$x = 4 \qquad\qquad x = -3$$

Check: $(4)^2 - 4 - 12 \stackrel{?}{=} 0$ $\qquad (-3)^2 - (-3) - 12 \stackrel{?}{=} 0$

$16 - 4 - 12 \stackrel{?}{=} 0$ $\qquad 9 + 3 - 12 \stackrel{?}{=} 0$

$16 - 16 = 0$ $\qquad 12 - 12 = 0$

The solutions are 4 and $-3$.

**EXAMPLE 8** Solve $5x^2 - 2x + \frac{1}{5} = 0$.

$$25x^2 - 10x + 1 = 0 \quad \text{Multiply by 5 to clear fractions}$$
$$(5x - 1)^2 = 0$$
$$5x - 1 = 0 \quad \text{or} \quad 5x - 1 = 0 \quad \text{Zero-product rule}$$
$$5x = 1 \qquad 5x = 1$$
$$x = \frac{1}{5} \qquad x = \frac{1}{5}$$

There is only one solution, $\frac{1}{5}$. (Check.)

**EXAMPLE 9** Solve $2x^2 - 2x = 12$.

$$2x^2 - 2x - 12 = 0 \quad \text{Write in general form}$$
$$2(x^2 - x - 6) = 0 \quad \text{Factor out 2 and divide by 2}$$
$$x^2 - x - 6 = 0$$
$$(x - 3)(x + 2) = 0$$
$$x - 3 = 0 \quad \text{or} \quad x + 2 = 0 \quad \text{Zero-product rule}$$
$$x = 3 \qquad x = -2$$

The solutions are 3 and $-2$. (Check.)

**EXAMPLE 10** Solve $5x^2 = 3x$.

$$5x^2 - 3x = 0 \quad \text{Note that } c = 0$$
$$x(5x - 3) = 0$$
$$x = 0 \quad \text{or} \quad 5x - 3 = 0 \quad \text{When } c = 0 \text{ one solution is always } 0$$
$$5x = 3$$
$$x = \frac{3}{5}$$

The solutions are 0 and $\frac{3}{5}$. (Check.)

**EXAMPLE 11** Solve $3(y^2 - 6) = y(y + 7) - 3$.

$$3y^2 - 18 = y^2 + 7y - 3 \quad \text{First clear parentheses}$$
$$2y^2 - 18 = 7y - 3$$
$$2y^2 - 7y - 18 = -3$$
$$2y^2 - 7y - 15 = 0 \quad \text{Write in general form}$$
$$(2y + 3)(y - 5) = 0$$
$$2y + 3 = 0 \quad \text{or} \quad y - 5 = 0 \quad \text{Zero-product rule}$$
$$2y = -3 \qquad y = 5$$
$$y = -\frac{3}{2}.$$

The solutions are $-\frac{3}{2}$ and 5. (Check.)

## EXERCISES 5.1

*Solve.*

**1.** $(x-1)(x+2)=0$

**2.** $(2y+3)(y-4)=0$

**3.** $(4z+1)(4z-1)=0$

**4.** $x(x+8)=0$

**5.** $(y-7)y=0$

**6.** $5(z+3)z=0$

**7.** $\frac{1}{3}x\left(x-\frac{2}{3}\right)=0$

**8.** $(8.2z-24.6)(7.2z-14.4)=0$

**9.** $(x-1)(x+2)(x-5)=0$

**10.** $4z(z+3)(z+3)=0$

**11.** $(x-1)-(2x+3)=0$

**12.** $(z+3)+(z-3)=0$

*Solve the following quadratic equations. (Check your answers.)*

**13.** $x^2-5x+6=0$

**14.** $2y^2-8y-24=0$

**15.** $4u^2 - 8u = 0$

**16.** $4v^2 + 12v + 9 = 0$

**17.** $x^2 = x + 72$

**18.** $y^2 - 10y = -25$

**19.** $u^2 = 6 + u$

**20.** $6v = -12v^2$

**21.** $3x^2 = 7 - 20x$

**22.** $\frac{1}{2}y^2 - 3y + 4 = 0$

**23.** $5u^2 - 4u = 3u^2 + 9u + 7$

**24.** $4v^2 + 4 = 7v^2 + 11v + 14$

**25.** $5x(x + 2) = 7x$

**26.** $y(8 - 2y) = 6$

**27.** $4u^2 - 5 = 31$

**28.** $(v + 1)(v - 5) = -5v + 7$

**29.** $(3x - 1)(2x + 1) = 3(2x + 1)$

**30.** $3y^2 - 8y = 147 - 8y$

**31.** $2u(u + 1) = 1 + 3u$

**32.** $9(v^2 + 1) = 24v - 7$

**33.** $3x^2 - 6x + 2 = 3x(x - 2)$

**34.** $2y^2 + 6y = 2y(y + 3)$

**35.** $(2u - 3)(u + 1) = 4(2u - 3)$

**36.** $40v^2 + 12v + 2 = 4(v + 2)$

---

ANSWERS: **1.** $1, -2$ **2.** $-\frac{3}{2}, 4$ **3.** $\frac{1}{4}, -\frac{1}{4}$ **4.** $0, -8$ **5.** $0, 7$ **6.** $0, -3$ **7.** $0, \frac{2}{3}$ **8.** $3, 2$ **9.** $1, -2, 5$ **10.** $0, -3$ **11.** $-4$ **12.** $0$ **13.** $2, 3$ **14.** $6, -2$ **15.** $0, 2$ **16.** $-\frac{3}{2}$ **17.** $9, -8$ **18.** $5$ **19.** $3, -2$ **20.** $0, -\frac{1}{2}$ **21.** $\frac{1}{3}, -7$ **22.** $2, 4$ **23.** $7, -\frac{1}{2}$ **24.** $-\frac{5}{3}, -2$ **25.** $0, -\frac{3}{5}$ **26.** $1, 3$ **27.** $3, -3$ **28.** $3, -4$ **29.** $\frac{4}{3}, -\frac{1}{2}$ **30.** $7, -7$ **31.** $1, -\frac{1}{2}$ **32.** $\frac{4}{3}$ **33.** no solution **34.** any number is a solution **35.** $3, \frac{3}{2}$ **36.** $\frac{3}{10}, -\frac{1}{2}$

## 5.2 Solving Quadratic Equations Using the Quadratic Formula

Factoring is the easiest and best way to solve certain quadratic equations, but many quadratic equations cannot be solved using this technique. For such equations we use the *quadratic formula,* which is derived by a process called **completing the square.** We illustrate this process in the following examples before using it to derive the formula.

First, consider the square of the binomial $x + d$.

$$(x + d)^2 = x^2 + 2dx + d^2$$

In the expression $x^2 + 2dx + d^2$, the coefficient of $x$ is $2d$. If we had only $x^2 + 2dx$, we could **complete the square** by adding the square of one half the coefficient of $x$,

$$\left[\frac{1}{2} \cdot (2d)\right]^2 = d^2,$$

to $x^2 + 2dx$. Further examples are shown in the following table.

| *To complete square on* | *Add (one half the coefficient of x)²* | *Obtaining the perfect square* |
|---|---|---|
| $x^2 + 6x$ | $9 \quad \left[9 = \left(\frac{1}{2} \cdot 6\right)^2\right]$ | $x^2 + 6x + 9 = (x + 3)^2$ |
| $x^2 - 10x$ | $25 \quad \left[25 = \left(\frac{1}{2} \cdot (-10)\right)^2\right]$ | $x^2 - 10x + 25 = (x - 5)^2$ |
| $x^2 - x$ | $\frac{1}{4} \quad \left[\frac{1}{4} = \left(\frac{1}{2} \cdot (-1)\right)^2\right]$ | $x^2 - x + \frac{1}{4} = \left(x - \frac{1}{2}\right)^2$ |

We now use the process of completing the square to solve the equation $x^2 - 6x + 8 = 0$. Note that this equation can be solved by the method of factoring: $(x - 2)(x - 4) = 0$, and 2 and 4 are solutions.

First we isolate the constant term on the right side of the equation, leaving space as indicated.

$$x^2 - 6x \qquad = -8$$

Then we complete the square on the left side by adding $[\frac{1}{2} \cdot (-6)]^2 = (-3)^2 = 9$. But if 9 is added on the left side, to keep both sides equal, 9 must be added to the right side.

$$x^2 - 6x + 9 = -8 + 9$$

$$(x - 3)^2 = 1 \qquad \text{Check by squaring the left side}$$

To complete the solution, we must take the square root of both sides of this equation.

$x - 3 = \pm\sqrt{1}$ — $\pm\sqrt{1}$ represents two numbers, $\sqrt{1}$ and $-\sqrt{1}$

$x - 3 = \pm 1$ — $\sqrt{1} = 1$

$x = 3 \pm 1$ — Solve for $x$ by adding 3 to both sides

$x = 3 + 1$ or $x = 3 - 1$ — $x = 3 \pm 1$ means $x = 3 + 1$ or $x = 3 - 1$

$x = 4 \qquad x = 2$

We obtain the same two solutions as before. (Of course, the factoring method was much easier and would be preferred in this case.)

**EXAMPLE 1** Solve $x^2 + 3x - 10 = 0$ by completing the square.

$$x^2 + 3x = 10 \qquad \text{Isolate constant on right side}$$

$$x^2 + 3x + \frac{9}{4} = 10 + \frac{9}{4} \qquad \left(\frac{1}{2} \cdot 3\right)^2 = \left(\frac{3}{2}\right)^2 = \frac{9}{4}$$

$$\left(x + \frac{3}{2}\right)^2 = \frac{49}{4} \qquad \text{Check this}$$

$$x + \frac{3}{2} = \pm\sqrt{\frac{49}{4}} \qquad \text{Take square root of both sides}$$

$$x + \frac{3}{2} = \pm\frac{7}{2}$$

$$x = -\frac{3}{2} \pm \frac{7}{2}$$

$$x = -\frac{3}{2} + \frac{7}{2} \quad \text{or} \quad x = -\frac{3}{2} - \frac{7}{2}$$

$$= \frac{4}{2} = 2 \qquad\qquad = -\frac{10}{2} = -5$$

The solutions are 2 and $-5$.

If the coefficient of $x^2$ in a quadratic equation is not 1, we *cannot* complete the square by taking one half the coefficient of $x$ and squaring it. For example, $(2x + 3)^2 = 4x^2 + 12x + 9$ but $(\frac{1}{2} \cdot 12)^2 = 6^2 \neq 9$. The procedure in such a case is illustrated in the following example.

**EXAMPLE 2** Solve $2x^2 + 2x - 3 = 0$ by completing the square.

$$2x^2 + 2x = 3 \qquad \text{Isolate the constant}$$

$$x^2 + x = \frac{3}{2} \qquad \text{Divide by 2 to make the coefficient of } x^2 \text{ equal to 1}$$

$$x^2 + x + \frac{1}{4} = \frac{3}{2} + \frac{1}{4} \qquad \left(\frac{1}{2} \cdot 1\right)^2 = \frac{1}{4}$$

$$\left(x + \frac{1}{2}\right)^2 = \frac{7}{4} \qquad \frac{3}{2} + \frac{1}{4} = \frac{6}{4} + \frac{1}{4} = \frac{7}{4}$$

$$x + \frac{1}{2} = \pm\sqrt{\frac{7}{4}}$$

$$x + \frac{1}{2} = \pm\frac{\sqrt{7}}{\sqrt{4}} \qquad \sqrt{\frac{a}{b}} = \frac{\sqrt{a}}{\sqrt{b}}$$

$$x + \frac{1}{2} = \frac{\pm\sqrt{7}}{2}$$

$$x = -\frac{1}{2} \pm \frac{\sqrt{7}}{2} = \frac{-1 \pm \sqrt{7}}{2}$$

$$x = \frac{-1 + \sqrt{7}}{2} \quad \text{or} \quad x = \frac{-1 - \sqrt{7}}{2}$$

The solutions are $\frac{-1 + \sqrt{7}}{2}$ and $\frac{-1 - \sqrt{7}}{2}$.

Rather than continuing to solve particular quadratic equations by completing the square, we now use completing the square to solve a general quadratic equation and in the process derive the quadratic formula. Recall the general form of a quadratic equation.

| | |
|---|---|
| $\mathbf{ax^2 + bx + c = 0}$ | $a \neq 0$ |
| $ax^2 + bx = -c$ | Isolate the constant |
| $\dfrac{ax^2}{a} + \dfrac{b}{a}x = -\dfrac{c}{a}$ | Divide by $a$ to make the coefficient of $x^2$ equal to 1 |
| $x^2 + \dfrac{b}{a}x + \left(\dfrac{b}{2a}\right)^2 = -\dfrac{c}{a} + \left(\dfrac{b}{2a}\right)^2$ | $\left(\dfrac{1}{2} \cdot \dfrac{b}{a}\right)^2 = \left(\dfrac{b}{2a}\right)^2$ |
| $\left(x + \dfrac{b}{2a}\right)^2 = -\dfrac{c}{a} + \dfrac{b^2}{4a^2}$ | |
| $\left(x + \dfrac{b}{2a}\right)^2 = -\dfrac{4ac}{4a^2} + \dfrac{b^2}{4a^2}$ | $4a^2$ is LCD |
| $\left(x + \dfrac{b}{2a}\right)^2 = \dfrac{b^2 - 4ac}{4a^2}$ | Subtract |
| $x + \dfrac{b}{2a} = \pm\sqrt{\dfrac{b^2 - 4ac}{4a^2}}$ | |
| $x + \dfrac{b}{2a} = \dfrac{\pm\sqrt{b^2 - 4ac}}{\sqrt{4a^2}}$ | $\sqrt{\dfrac{a}{b}} = \dfrac{\sqrt{a}}{\sqrt{b}}$ |
| $x + \dfrac{b}{2a} = \dfrac{\pm\sqrt{b^2 - 4ac}}{2a}$ | |
| $x = -\dfrac{b}{2a} \pm \dfrac{\sqrt{b^2 - 4ac}}{2a}$ | Subtract $\dfrac{b}{2a}$ |
| $\mathbf{x = \dfrac{-b \pm \sqrt{b^2 - 4ac}}{2a}}$ | $2a$ is the denominator of the entire expression |

This last formula is called the **quadratic formula** and it must be memorized. To solve a quadratic equation, we identify the constants $a$, $b$, and $c$ and substitute into the quadratic formula. Note that there are two solutions in general.

$$x = \frac{-b + \sqrt{b^2 - 4ac}}{2a} \quad \text{and} \quad x = \frac{-b - \sqrt{b^2 - 4ac}}{2a}$$

**TO SOLVE A QUADRATIC EQUATION USING THE QUADRATIC FORMULA**

1. Write the equation in general form, $ax^2 + bx + c = 0$. Be sure that all common factors have been divided out and all fractions have been cleared.
2. Identify the constants $a$, $b$, and $c$.
3. Substitute the values of $a$, $b$, and $c$ into the quadratic formula,
$$x = \frac{-b \pm \sqrt{b^2 - 4ac}}{2a}.$$
4. Simplify the numerical expression to obtain the solutions.

**EXAMPLE 3** Solve $x^2 - 6x + 8 = 0$ using the quadratic formula.

The equation is in the simplest general form, and therefore $a = 1$, $b = -6$ (not 6), and $c = 8$.

$$x = \frac{-b \pm \sqrt{b^2 - 4ac}}{2a}$$

$$= \frac{-(-6) \pm \sqrt{(-6)^2 - 4(1)(8)}}{2(1)}$$

Multiply 4(1)(8) and then subtract the product from $(-6)^2$

$$= \frac{6 \pm \sqrt{36 - 32}}{2}$$

$$= \frac{6 \pm \sqrt{4}}{2} = \frac{6 \pm 2}{2}$$

$$x = \frac{6+2}{2} = \frac{8}{2} = 4 \quad \text{or} \quad x = \frac{6-2}{2} = \frac{4}{2} = 2$$

The solutions are 4 and 2.

**EXAMPLE 4** Solve $6x^2 - 10 = -8x$ using the quadratic formula.

First write the equation in general form and simplify.

$$6x^2 + 8x - 10 = 0$$ Add $8x$ to both sides

$$2(3x^2 + 4x - 5) = 0$$ Factor out 2

$$3x^2 + 4x - 5 = 0$$ Divide by 2 to simplify

Thus $a = 3$, $b = 4$, and $c = -5$. $c = -5$, not 5

$$x = \frac{-b \pm \sqrt{b^2 - 4ac}}{2a}$$

$$= \frac{-4 \pm \sqrt{(4)^2 - 4(3)(-5)}}{2(3)}$$

$$= \frac{-4 \pm \sqrt{16 + 60}}{6}$$ Watch the signs

$$= \frac{-4 \pm \sqrt{76}}{6}$$

$$= \frac{-4 \pm \sqrt{4 \cdot 19}}{6}$$ 4 is a perfect square

$$= \frac{-4 \pm \sqrt{4}\sqrt{19}}{6}$$ $\sqrt{ab} = \sqrt{a}\sqrt{b}$

$$= \frac{-4 \pm 2\sqrt{19}}{6}$$ $\sqrt{4} = 2$

$$= \frac{\cancel{2}(-2 \pm \sqrt{19})}{\cancel{2} \cdot 3}$$ Factor out 2 and cancel

$$= \frac{-2 \pm \sqrt{19}}{3}$$ This is simplified

The solutions are $\frac{-2 + \sqrt{19}}{3}$ and $\frac{-2 - \sqrt{19}}{3}$.

If you are asked to solve a quadratic equation and no particular method is specified, first try the method of factoring. If this fails, use the quadratic formula.

## EXERCISES 5.2

*What must be added to complete the square?*

**1.** $x^2 - 10x +$ ________ **2.** $y^2 - 6y +$ ________ **3.** $z^2 + 3z +$ ________

**4.** $x^2 + x +$ ________ **5.** $y^2 - \frac{1}{3}y +$ ________ **6.** $z^2 + \frac{8}{5}z +$ ________

*Solve by completing the square.*

**7.** $x^2 + 6x + 5 = 0$

**8.** $y^2 - 10y + 21 = 0$

**9.** $u^2 - 2u - 8 = 0$

**10.** $3v^2 + 12v = 135$

**11.** $2x^2 + 3x = 2$

**12.** $y^2 = 12 - 10y$

*Write in general form and identify a, b, and c.*

**13.** $2x^2 = 7 - 3x$

**14.** $u(u - 3) = 5 - 3u$

**15.** $2(y + 3) = 6 - y^2$

**16.** $3x^2 + 1 = 2x(x - 1)$

*Solve by using the quadratic formula.*

**17.** $u^2 + 5u - 14 = 0$

**18.** $v^2 - 8v + 15 = 0$

**19.** $x^2 - 16 = 0$

**20.** $y^2 = 4y + 21$

**21.** $3u^2 + 10u = -3u - 4$

**22.** $4v^2 = 20v - 25$

**23.** $4x(x + 3) = 9x$

**24.** $y(y + 1) + 2y = 4$

**25.** $5u^2 - u = 1$

**26.** $4v^2 + 1 = 2v(v + 2)$

**27.** $6x^2 - 9x - 12 = 0$

**28.** $\frac{1}{2}y^2 - \frac{5}{4}y + \frac{1}{4} = 0$

**29.** $(2u - 1)(u - 2) = 2(u + 4) + 3$

**30.** $10v^2 = 5(v^2 + 25)$

*Solve. (First try factoring and if that fails use the quadratic formula.)*

**31.** $x^2 - 6 = 1 - 6x$

**32.** $5y^2 + 3y = 2$

**33.** $3u^2 - 3u + 2 = 1 + 2u$

**34.** $(v - 1)(v + 1) = -2v$

**35.** $x(x + 5) = -4$

**36.** $10y^2 = 12y + 6$

**37.** $3u(u - 2) = 12 - 6u$

**38.** $\frac{2}{3}v^2 - \frac{1}{3}v - 1 = 0$

**39.** $\frac{1}{2}x^2 + x - 1 = 0$

**40.** $2(y^2 - 3) = y(5y + 9)$

---

ANSWERS: **1.** 25 **2.** 9 **3.** $\frac{9}{4}$ **4.** $\frac{1}{4}$ **5.** $\frac{1}{36}$ **6.** $\frac{16}{25}$ **7.** $-1, -5$ **8.** 3, 7 **9.** $4, -2$ **10.** $5, -9$ **11.** $\frac{1}{2}, -2$ **12.** $-5 \pm \sqrt{37}$ **13.** $a = 2, b = 3, c = -7$ **14.** $a = 1, b = 0, c = -5$ **15.** $a = 1, b = 2, c = 0$ **16.** $a = 1, b = 2, c = 1$ **17.** $2, -7$ **18.** 3, 5 **19.** $4, -4$ **20.** $7, -3$ **21.** $-\frac{1}{3}, -4$ **22.** $\frac{5}{2}$ **23.** $0, -\frac{3}{4}$ **24.** $1, -4$ **25.** $\frac{1 \pm \sqrt{21}}{10}$ **26.** $\frac{2 \pm \sqrt{2}}{2}$ **27.** $\frac{3 \pm \sqrt{41}}{4}$ **28.** $\frac{5 \pm \sqrt{17}}{4}$ **29.** $\frac{9}{2}, -1$ **30.** $5, -5$ **31.** $1, -7$ **32.** $\frac{2}{5}, -1$ **33.** $\frac{5 \pm \sqrt{13}}{6}$ **34.** $-1 \pm \sqrt{2}$ **35.** $-1, -4$ **36.** $\frac{3 \pm 2\sqrt{6}}{5}$ **37.** $2, -2$ **38.** $\frac{3}{2}, -1$ **39.** $-1 \pm \sqrt{3}$ **40.** $-1, -2$

## 5.3 Solving Equations Quadratic in Form

Some equations which are not quadratic can be solved using the techniques of Sections 5.1 and 5.2 by making an appropriate substitution. For example,

$$x^4 - 13x^2 + 36 = 0$$

is not a quadratic equation (it contains $x^4$), but if we substitute $u$ for $x^2$, we obtain the equation

$$u^2 - 13u + 36 = 0$$

which is quadratic in $u$. This is done in the following example

**EXAMPLE 1** Solve.

$$x^4 - 13x^2 + 36 = 0$$
$$(x^2)^2 - 13x^2 + 36 = 0 \qquad x^4 = (x^2)^2$$
$$u^2 - 13u + 36 = 0 \qquad \text{Let } u = x^2$$
$$(u - 4)(u - 9) = 0 \qquad \text{Use the zero-product rule}$$
$$u - 4 = 0 \quad \text{or} \quad u - 9 = 0$$
$$u = 4 \qquad\qquad u = 9$$

Since $u = x^2$,

$$x^2 = 4 \qquad\qquad x^2 = 9 \qquad \text{Solve for the original variable}$$
$$x = \pm 2 \qquad\qquad x = \pm 3$$

Check: $x^4 - 13x^2 + 36 = 0$

$$(2)^4 - 13(2)^2 + 36 \stackrel{?}{=} 0$$
$$16 - 52 + 36 \stackrel{?}{=} 0$$
$$0 = 0$$

2 checks and $-2$ will check also.

$$(3)^4 - 13(3)^2 + 36 \stackrel{?}{=} 0$$
$$81 - 117 + 36 = 0$$

3 checks and $-3$ will check also.

The solutions are 2, $-2$, 3, and $-3$.

CAUTION: 4 and 9 are solutions to the equation in the variable $u$ and are not solutions to the original equation. Do not forget to find the solutions relative to the original variable.

Equations like the one in Example 1 are called **quadratic in form.** More problems of this type are presented in the following examples.

**EXAMPLE 2** Solve.

$$(y+4)^2 - 5(y+4) + 6 = 0$$
$$u^2 - 5u + 6 = 0 \qquad \text{Let } u = y + 4$$
$$(u-2)(u-3) = 0$$
$$u - 2 = 0 \quad \text{or} \quad u - 3 = 0$$
$$u = 2 \qquad\qquad u = 3$$

Since $u = y + 4$,

$$y + 4 = 2 \quad \text{or} \quad y + 4 = 3 \qquad \text{Solve for the original variable}$$
$$y = -2 \qquad\qquad y = -1$$

Check: $(y+4)^2 - 5(y+4) + 6 = 0$

$$(-2+4)^2 - 5(-2+4) + 6 \stackrel{?}{=} 0$$
$$2^2 - 5(2) + 6 \stackrel{?}{=} 0$$
$$4 - 10 + 6 = 0$$

$-2$ checks. (Check $-1$).

The solutions are $-2$ and $-1$.

**EXAMPLE 3** Solve.

$$x - 3 - 4\sqrt{x-3} + 3 = 0$$
$$u^2 - 4u + 3 = 0 \qquad \text{Let } u = \sqrt{x-3}; \text{ then } u^2 = x - 3$$
$$(u-1)(u-3) = 0$$
$$u - 1 = 0 \quad \text{or} \quad u - 3 = 0$$
$$u = 1 \qquad\qquad u = 3$$

Since $u = \sqrt{x-3}$,

$$\sqrt{x-3} = 1 \qquad\qquad \sqrt{x-3} = 3 \qquad \text{Solve for the original variable}$$
$$(\sqrt{x-3})^2 = 1^2 \qquad\qquad (\sqrt{x-3})^2 = 3^2$$
$$x - 3 = 1 \qquad\qquad x - 3 = 9$$
$$x = 4 \qquad\qquad x = 12$$

Check: $4 - 3 - 4\sqrt{4-3} + 3 \stackrel{?}{=} 0 \qquad 12 - 3 - 4\sqrt{12-3} + 3 \stackrel{?}{=} 0$

$$1 - 4\sqrt{1} + 3 \stackrel{?}{=} 0 \qquad\qquad 9 - 4\sqrt{9} + 3 \stackrel{?}{=} 0$$
$$1 - 4 + 3 = 0 \qquad\qquad 9 - 12 + 3 = 0$$

The solutions are 4 and 12.

In the next example we encounter the equation $y^{1/3} = 6$. This is the same as $\sqrt[3]{y} = 6$, and is solved by cubing both sides.

$$(y^{1/3})^3 = 6^3$$
$$y = 216$$

**EXAMPLE 4** Solve.

$$y^{2/3} - 5y^{1/3} - 6 = 0$$

$$(y^{1/3})^2 - 5y^{1/3} - 6 = 0 \qquad \text{Let } u = y^{1/3}\text{; then } u^2 = y^{2/3}$$

$$u^2 - 5u - 6 = 0$$

$$(u + 1)(u - 6) = 0$$

$$u + 1 = 0 \quad \text{or} \quad u - 6 = 0$$

$$u = -1 \qquad\qquad u = 6$$

Since $u = y^{1/3}$,

$$y^{1/3} = -1 \qquad\qquad y^{1/3} = 6$$

$$(y^{1/3})^3 = (-1)^3 \qquad\qquad (y^{1/3})^3 = 6^3$$

$$y = -1 \qquad\qquad y = 216$$

Check: $(-1)^{2/3} - 5(-1)^{1/3} - 6 \stackrel{?}{=} 0 \qquad (216)^{2/3} - 5(216)^{1/3} - 6 \stackrel{?}{=} 0$

$$(-1^{1/3})^2 - 5(-1) - 6 \stackrel{?}{=} 0 \qquad (216^{1/3})^2 - 5(6) - 6 \stackrel{?}{=} 0$$

$$(-1)^2 + 5 - 6 \stackrel{?}{=} 0 \qquad 6^2 - 30 - 6 \stackrel{?}{=} 0$$

$$1 + 5 - 6 = 0 \qquad 36 - 30 - 6 = 0$$

The solutions are $-1$ and 216.

## EXERCISES 5.3

*Solve.*

**1.** $x^4 - 5x^2 + 4 = 0$ (let $u = x^2$)

**2.** $x - 5\sqrt{x} + 4 = 0$ (let $u = \sqrt{x}$)

**3.** $y^4 - 5y^2 + 4 = 0$ (let $u = y^2$)

**4.** $(y + 3)^2 - 6(y + 3) + 8 = 0$ (let $u = y + 3$)

**5.** $x - 6 - 3\sqrt{x - 6} + 2 = 0$ (let $u = \sqrt{x - 6}$)

**6.** $x^{2/3} - 5x^{1/3} + 6 = 0$ (let $u = x^{1/3}$)

**7.** $2(y - 5)^2 + 5(y - 5) = 3$

**8.** $y^4 - 16 = 0$

**9.** $x^4 + 20 = 9x^2$

**10.** $x - 11\sqrt{x} + 28 = 0$

**11.** $y^{2/3} + 2y^{1/3} = 15$

**12.** $(y - 7)^4 - 13(y - 7)^2 + 42 = 0$

---

ANSWERS: **1.** $\pm 1, \pm 2$ **2.** 1, 16 **3.** $\pm 1, \pm 2$ **4.** $-1, 1$ **5.** 7, 10 **6.** 8, 27 **7.** $2, \frac{11}{2}$ **8.** $2, -2$ **9.** $\pm 2, \pm\sqrt{5}$ **10.** 16, 49 **11.** $27, -125$ **12.** $7 \pm \sqrt{6}, 7 \pm \sqrt{7}$

## 5.4 Radical Equations That Result in Quadratic Equations

Recall from Section 4.7 that the fundamental rule in solving radical equations involving square roots is the rule of squaring, which states that

$$\text{if } a = b \quad \text{then } a^2 = b^2.$$

We now repeat the procedure for solving radical equations.

**TO SOLVE AN EQUATION INVOLVING RADICALS**

1. If only one radical is present, isolate this radical on one side of the equation, simplify, and proceed to (3).
2. If two radicals are present, isolate one of the radicals on one side of the equation.
3. Square both sides.
4. Solve the resulting equation. If a radical remains, isolate it and square once more.
5. Check all possible solutions in the original equation.

EXAMPLE 1 Solve the following radical equation.

$$\sqrt{x^2 - 13} + 5 = 11$$

$$\sqrt{x^2 - 13} = 6 \qquad \text{Isolate the radical by subtracting 5 from both sides}$$

$$x^2 - 13 = 36 \qquad \text{Square both sides}$$

$$x^2 - 49 = 0$$

$$(x - 7)(x + 7) = 0 \qquad \text{Use the zero-product rule}$$

$$x - 7 = 0 \quad \text{or} \quad x + 7 = 0$$

$$x = 7 \qquad\qquad x = -7$$

We must check to make sure that squaring has not introduced extraneous roots.

Check:
$$\sqrt{(7)^2 - 13} + 5 \stackrel{?}{=} 11 \qquad \sqrt{(-7)^2 - 13} + 5 \stackrel{?}{=} 11$$
$$\sqrt{49 - 13} + 5 \stackrel{?}{=} 11 \qquad \sqrt{49 - 13} + 5 \stackrel{?}{=} 11$$
$$\sqrt{36} + 5 \stackrel{?}{=} 11 \qquad \sqrt{36} + 5 \stackrel{?}{=} 11$$
$$6 + 5 = 11 \qquad 6 + 5 = 11$$

The solutions are 7 and $-7$.

**EXAMPLE 2** $\sqrt{2y + 11} - y - 4 = 0$

$$\sqrt{2y + 11} = y + 4$$ Isolate the radical

$$(\sqrt{2y + 11})^2 = (y + 4)^2$$ Square both sides

$$2y + 11 = y^2 + 8y + 16$$ $(y + 4)^2 \neq y^2 + 16$; remember the middle term $8y$

$$y^2 + 6y + 5 = 0$$
$$(y + 1)(y + 5) = 0$$

$$y + 1 = 0 \quad \text{or} \quad y + 5 = 0$$
$$y = -1 \qquad\qquad y = -5$$

Check:
$$\sqrt{2(-1) + 11} - (-1) - 4 \stackrel{?}{=} 0 \qquad \sqrt{2(-5) + 11} - (-5) - 4 \stackrel{?}{=} 0$$
$$\sqrt{-2 + 11} + 1 - 4 \stackrel{?}{=} 0 \qquad \sqrt{-10 + 11} + 5 - 4 \stackrel{?}{=} 0$$
$$\sqrt{9} + 1 - 4 \stackrel{?}{=} 0 \qquad \sqrt{1} + 5 - 4 \stackrel{?}{=} 0$$
$$3 + 1 - 4 \stackrel{?}{=} 0 \qquad 1 + 5 - 4 \stackrel{?}{=} 0$$
$$0 = 0 \qquad 2 \neq 0$$

5 is an extraneous root.

The only solution is $-1$.

**EXAMPLE 3** $\sqrt{x^2 - 7x + 15} - \sqrt{4x - 13} = 0$

$$\sqrt{x^2 - 7x + 15} = \sqrt{4x - 13}$$ Isolate the radicals

$$x^2 - 7x + 15 = 4x - 13$$ Square both sides

$$x^2 - 11x + 28 = 0$$
$$(x - 7)(x - 4) = 0$$

$$x - 7 = 0 \quad \text{or} \quad x - 4 = 0$$
$$x = 7 \qquad\qquad x = 4$$

Check:
$$\sqrt{(7)^2 - 7(7) + 15} - \sqrt{4(7) - 13} \stackrel{?}{=} 0$$
$$\sqrt{49 - 49 + 15} - \sqrt{28 - 13} \stackrel{?}{=} 0$$
$$\sqrt{15} - \sqrt{15} = 0$$

$$\sqrt{(4)^2 - 7(4) + 15} - \sqrt{4(4) - 13} \stackrel{?}{=} 0$$
$$\sqrt{16 - 28 + 15} - \sqrt{16 - 13} \stackrel{?}{=} 0$$
$$\sqrt{3} - \sqrt{3} = 0$$

The solutions are 7 and 4.

**EXAMPLE 4** $\sqrt{3x+1} - \sqrt{x+9} = 2$

$\sqrt{3x+1} = 2 + \sqrt{x+9}$ Isolate one radical

$(\sqrt{3x+1})^2 = (2 + \sqrt{x+9})^2$ Square both sides

$3x + 1 = 4 + 4\sqrt{x+9} + x + 9$ Use $(a+b)^2 = a^2 + 2ab + b^2$: $4\sqrt{x+9}$ is $2ab$

$2x - 12 = 4\sqrt{x+9}$ Isolate the radical

$x - 6 = 2\sqrt{x+9}$ Factor out 2 and divide by 2

$(x-6)^2 = (2\sqrt{x+9})^2$ Square both sides

$x^2 - 12x + 36 = 4(x+9)$ Square 2 on the right

$x^2 - 12x + 36 = 4x + 36$

$x^2 - 16x = 0$

$x(x - 16) = 0$

$x = 0$ or $x - 16 = 0$

$x = 16$

Check: $\sqrt{3(0)+1} - \sqrt{0+9} \stackrel{?}{=} 2$ $\sqrt{3(16)+1} - \sqrt{16+9} \stackrel{?}{=} 2$

$\sqrt{1} - \sqrt{9} \stackrel{?}{=} 2$ $\sqrt{49} - \sqrt{25} \stackrel{?}{=} 2$

$1 - 3 \neq 2$ $7 - 5 = 2$

The only solution is 16.

## EXERCISES 5.4

*Solve. (Be sure to check all answers.)*

**1.** $\sqrt{x^2+3} - 2 = 0$

**2.** $\sqrt{x^2-7} - 3 = 0$

**3.** $\sqrt{y^2+1} - 2 = 0$

**4.** $\sqrt{3y^2-50} - y = 0$

**5.** $\sqrt{u+4} - u + 8 = 0$

**6.** $\sqrt{2u+11} - u + 2 = 0$

**7.** $\sqrt{5x-1}+x-3=0$

**8.** $\sqrt{3x+1}-2x+6=0$

**9.** $\sqrt{y^2-2y+10}-\sqrt{3y+4}=0$

**10.** $\sqrt{y^2+6y}+\sqrt{2y+21}=0$

**11.** $\sqrt{2x+5}-\sqrt{x+2}=1$

**12.** $\sqrt{3x+1}-\sqrt{x-4}=3$

**13.** $\sqrt{3y+3}+\sqrt{y-1}=4$

**14.** $\sqrt{5y+4}-\sqrt{3y-2}=2$

**15.** $x^4-5x^2=-6$

**16.** $x-6\sqrt{x}+5=0$

**17.** $y^{2/3}=y^{1/3}+12$

**18.** $(y-8)^2+5(y-8)-6=0$

---

ANSWERS: **1.** 1, −1 **2.** 4, −4 **3.** $\sqrt{3}, -\sqrt{3}$ **4.** 5 **5.** 12 **6.** 7 **7.** 1 **8.** 5 **9.** 2, 3 **10.** no solution **11.** 2, −2 **12.** 5, 8 **13.** 2 **14.** 1, 9 **15.** $\pm\sqrt{2}, \pm\sqrt{3}$ **16.** 1, 25 **17.** 64, −27 **18.** 2, 9

## 5.5 Word Problems that Result in Quadratic Equations

Many number, age, and geometry problems result in quadratic equations. When solving word problems, write out in detail the definitions of the variables and construct and label any geometric figure. Show all details as in the following examples and in Chapter 2.

**EXAMPLE 1** The product of 1 less than a number and 7 more than the number is 153. Find the number.

Let $x =$ the desired number
$x - 1 = 1$ less than the number
$x + 7 = 7$ more than the number

Since the product is 153, the equation to be solved is

$$(x-1)(x+7) = 153. \quad \text{Do not try to use the zero-product rule here since } 153 \neq 0$$

$$x^2 + 6x - 7 = 153$$

$$x^2 + 6x - 160 = 0$$

$$(x-10)(x+16) = 0 \quad \text{Now use zero-product rule}$$

$$x - 10 = 0 \quad \text{or} \quad x + 16 = 0$$

$$x = 10 \qquad x = -16 \quad \text{Check in the original equation}$$

The solutions are 10 and $-16$.

**EXAMPLE 2** One third the product of two consecutive positive odd integers is 85. Find the integers.

Let $x =$ the first positive odd integer
$x + 2 =$ the next consecutive odd integer
$\frac{1}{3}x(x+2) = \frac{x(x+2)}{3} =$ one third their product

The equation we must solve is

$$\frac{x(x+2)}{3} = 85.$$

$$x(x+2) = 255 \quad \text{Clear the fraction}$$

$$x^2 + 2x = 255$$

$$x^2 + 2x - 255 = 0$$

$$(x-15)(x+17) = 0$$

$$x - 15 = 0 \quad \text{or} \quad x + 17 = 0$$

$$x = 15 \qquad x = -17$$

Since $x$ must be a positive integer, we rule out $-17$. Thus, 15 is the first integer, and the next odd integer is 17 ($x + 2 = 17$). (Check.)

**EXAMPLE 3** If the square of Arnie's age is decreased by 300 the result is twenty times his age. How old is Arnie?

Let $x =$ Arnie's present age
$x^2 - 300 =$ the square of Arnie's age decreased by 300
$20x =$ twenty times Arnie's age

The equation we must solve is

$$x^2 - 300 = 20x.$$
$$x^2 - 20x - 300 = 0$$
$$(x - 30)(x + 10) = 0$$
$$x - 30 = 0 \quad \text{or} \quad x + 10 = 0$$
$$x = 30 \qquad\qquad x = -10$$

Since $-10$ cannot be an age, the answer is 30 (check). Arnie is 30 years old.

**EXAMPLE 4** Find the length and width of a rectangle if the length is 5 cm more than the width and the area is 84 cm². [Area = (length) · (width)] See the sketch in Figure 5.1.

Let $\quad x =$ width of the rectangle
$x + 5 =$ length of the rectangle

$$x(x + 5) = 84$$
$$x^2 + 5x = 84$$
$$x^2 + 5x - 84 = 0$$
$$(x - 7)(x + 12) = 0$$
$$x - 7 = 0 \quad \text{or} \quad x + 12 = 0$$
$$x = 7 \qquad\qquad x = -12$$

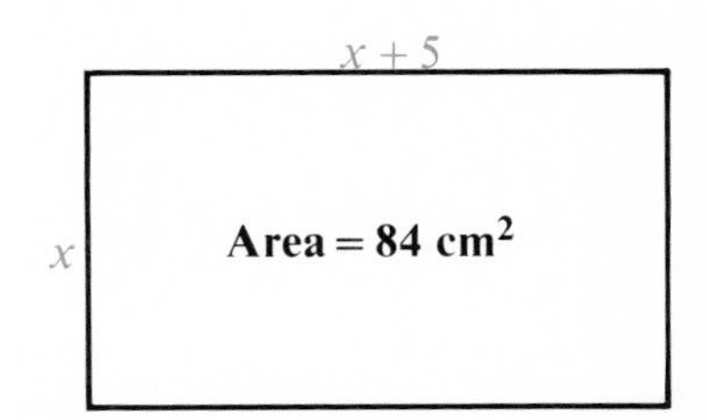

Figure 5.1

Since $-12$ cannot be a width of a rectangle, the width is 7 cm and the length is $x + 5 = 7 + 5 = 12$ cm.

**EXAMPLE 5** If the hypotenuse of a right triangle is 7 in and one leg is 3 in more than the other, find the measure of each leg.

The triangle is shown in Figure 5.2. Recall that the Pythagorean theorem states that the sum of the squares of the legs of a right triangle is equal to the square of the hypotenuse.

Let $\quad x =$ the measure of one leg
$x + 3 =$ the measure of the other leg

Use the Pythagorean theorem.

$$x^2 + (x + 3)^2 = 7^2$$
$$x^2 + x^2 + 6x + 9 = 49$$
$$2x^2 + 6x + 9 = 49$$
$$2x^2 + 6x - 40 = 0$$
$$x^2 + 3x - 20 = 0$$

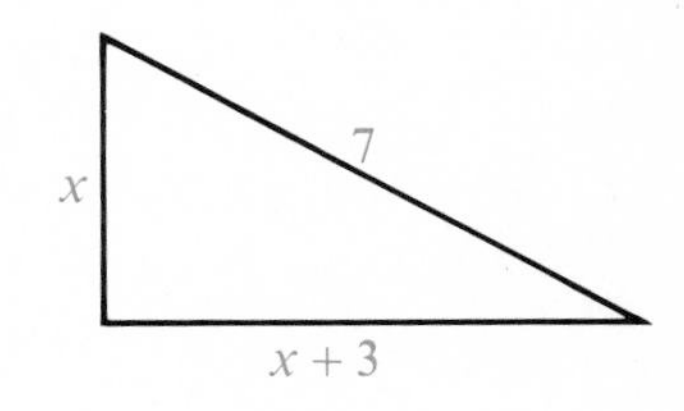

Figure 5.2

Since we cannot factor, we use the quadratic formula.

$$x = \frac{-b \pm \sqrt{b^2 - 4ac}}{2a}$$

$$x = \frac{-3 \pm \sqrt{(3)^2 - 4(1)(-20)}}{2(1)} \qquad a = 1,\ b = 3,\ c = -20$$

$$= \frac{-3 \pm \sqrt{9 + 80}}{2}$$

$$= \frac{-3 \pm \sqrt{89}}{2}$$

We obtain $\frac{-3+\sqrt{89}}{2}$ and $\frac{-3-\sqrt{89}}{2}$ as solutions to the quadratic equation.
But since $\frac{-3-\sqrt{89}}{2}$ is negative, it is discarded so that

$$x = \frac{-3+\sqrt{89}}{2} = \frac{\sqrt{89}-3}{2} \quad \text{and} \quad x+3 = \frac{\sqrt{89}-3}{2} + 3 = \frac{\sqrt{89}-3}{2} + \frac{6}{2} = \frac{\sqrt{89}+3}{2}.$$

The legs are $\frac{\sqrt{89}-3}{2}$ in and $\frac{\sqrt{89}+3}{2}$ in.

## EXERCISES 5.5

*Solve.*

1. The square of a number is 5 more than four times the number. Find the number.

   Let $x =$ the number
   $x^2 =$ the square of the number
   $x^2 - 5 = 4x$

2. The product of Samantha's present age and her age 7 years ago is 170. How old is Sam?

3. The product of two consecutive positive even integers is 440. Find the integers.

   Let $x =$ first positive even integer
   $x + 2 =$ second positive even integer

**4.** One more than a number times 1 less than the number is 35. Find the number.

**5.** Jose is 12 years older than Mike. The product of their ages is 220. How old is each?

Let $x =$ Mike's age
$x + 12 =$ Jose's age

**6.** The hypotenuse of a right triangle is 5 cm, and one leg is 1 cm longer than the other. Find the length of each leg.

Let $x =$ length of one leg
$x + 1 =$ length of the other leg

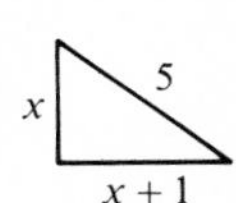


**7.** The number of square inches in the area of a square is 12 more than the number of inches in its perimeter. Find the length of each side.

$[A = l \cdot w = x \cdot x = x^2$ and $P = 2l + 2w = 2x + 2x = 4x]$

**8.** Wanda's age 4 years ago times her age in 6 years is 144. How old is Wanda?

**9.** Five times the product of two positive consecutive integers is 550. Find the integers.

**10.** Adding 4 to the square of Zeke's age is the same as subtracting 3 from eight times his age. How old is Zeke?

**11.** The length of a rectangle is 10 m more than the width. Find the length and width if the area is 264 $m^2$.

**12.** The sum of the squares of two positive consecutive even integers is 100. Find the integers.

**13.** If the sides of a square are lengthened by 3 cm the area will be 196 $cm^2$. Find the length of a side.

**14.** The area of a triangle is 114 $ft^2$ and the height is 7 ft more than the base. Find the height and base.

**15.** The square of a number is 48 more than 8 times the number. Find the number.

**16.** The sum of the squares of a number and 2 more than the number is 6. Find the number.

**17.** If twice Alma's age is squared, the result is the same as the square of her age 3 years from now. How old is Alma?

**18.** Find the length of a side of an equilateral triangle if the height is 12 m.

**19.** The area of a parallelogram is 55 $ft^2$. If the base is 1 ft greater than twice the altitude, find the base and altitude.

**20.** A box is 12 cm high. The length is 7 cm more than the width. If the volume is 1728 $cm^3$, find the length and width.

**21.** $\sqrt{u^2 - 9} - 4 = 0$

**22.** $\sqrt{u + 8} - u + 4 = 0$

**23.** $\sqrt{3x^2 - x} - \sqrt{9x - 8} = 0$

**24.** $\sqrt{6x + 4} - \sqrt{2x + 5} = 3$

---

ANSWERS: **1.** 5, −1 **2.** 17 **3.** 20, 22 **4.** 6 or −6 **5.** Jose: 22 yr, Mike: 10 yr **6.** 3 cm, 4 cm **7.** 6 in **8.** 12 **9.** 10, 11 **10.** 1 or 7 **11.** 22 m, 12 m **12.** 6, 8 **13.** 11 cm **14.** 19 ft, 12 ft **15.** 12, −4 **16.** $-1 \pm \sqrt{2}$ **17.** 3 **18.** $8\sqrt{3}$ m **19.** 11 ft, 5 ft **20.** 16 cm, 9 cm **21.** 5, −5 **22.** 8 **23.** 2, $\frac{4}{3}$ **24.** 10

## 5.6 Complex Numbers

Suppose that we attempt to use the quadratic formula to solve

$$x^2 - 2x + 2 = 0. \qquad a = 1,\ b = -2,\ c = 2$$

$$x = \frac{-b \pm \sqrt{b^2 - 4ac}}{2a}$$

$$= \frac{-(-2) \pm \sqrt{4 - 4(1)(2)}}{2(1)}$$

$$= \frac{2 \pm \sqrt{-4}}{2}$$

What number squared equals −4? We know that for any real number $a$, $a^2 \geq 0$, so there is no real number equal to $\sqrt{-4}$. As a result, certain very simple equations (such as the one above) have no solution in the system of real numbers. In order to provide solutions for these equations and square roots for negative numbers, we introduce a new kind of number called an **imaginary number.**

By agreeing to follow rules of radicals similar to those for real numbers, only one new numeral is needed. We use $i$ as the numeral corresponding to $\sqrt{-1}$.

$$\sqrt{-1} = i \qquad \text{and} \qquad i^2 = -1$$

Every other imaginary number is expressible as a product of a real number and the number $i$. For example,

$$\sqrt{-4} = \sqrt{4(-1)} = \sqrt{4}\sqrt{-1} = 2i.$$

As a result, the solutions to the equation

$$x^2 - 2x + 2 = 0$$

are

$$x = \frac{2 \pm \sqrt{-4}}{2} = \frac{2 \pm 2i}{2} = \frac{2(1 \pm i)}{2} = 1 \pm i.$$

A number of the form

$$a + bi, \qquad a \text{ and } b \text{ real numbers,}$$

is called a **complex number.** The number $a$ is called the **real part** and $b$ is called the **imaginary part** of the complex number $a + bi$. If $b = 0$ in $a + bi$, the complex number is simply the real number $a$, while if $a = 0$, it is the **pure imaginary number** $bi$. Thus, the complex numbers include both the real numbers and the imaginary numbers.

**EXAMPLE 1**

**(a)** $\sqrt{-9} = \sqrt{9 \cdot (-1)} = \sqrt{9}\sqrt{-1} = 3i$

**(b)** $\sqrt{-2}\sqrt{-5} = \sqrt{2 \cdot (-1)}\sqrt{5(-1)} = \sqrt{2} \cdot i \cdot \sqrt{5} \cdot i$

$= \sqrt{2}\sqrt{5}\, i^2 = \sqrt{10}\,(-1) = -\sqrt{10}$ Remember $i^2 = -1$

CAUTION: Do not write $\sqrt{-2}\sqrt{-5} = \sqrt{(-2)(-5)} = \sqrt{10}$. The rule $\sqrt{a}\sqrt{b} = \sqrt{ab}$ does not apply when $a$ and $b$ are both negative. We can use this rule when $a \geq 0$ and $b = -1$ to express square roots of negative numbers in terms of the imaginary number $i$.

**EXAMPLE 2**

**(a)** $\dfrac{\sqrt{-30}}{\sqrt{-6}} = \dfrac{\sqrt{30(-1)}}{\sqrt{6(-1)}} = \dfrac{\sqrt{30}\,i}{\sqrt{6}\,i} = \sqrt{\dfrac{30}{6}} = \sqrt{5}$

**(b)** $\sqrt{-16} + \sqrt{-49} = \sqrt{16(-1)} + \sqrt{49(-1)} = \sqrt{16}\,i + \sqrt{49}\,i = 4i + 7i = 11i$

**(c)** $i^3 = i^2 \cdot i = (-1)i = -i$

**(d)** $i^4 = i^2 \cdot i^2 = (-1)(-1) = 1$

**(e)** $i^5 = i \cdot i^4 = i \cdot 1 = i$

**(f)** $i^6 = i^2 \cdot i^4 = (-1) \cdot 1 = -1$

**(g)** $i^7 = i^3 \cdot i^4 = (-i)(1) = -i$

Observe that the first four powers of $i$ are $i^1 = i$, $i^2 = -1$, $i^3 = -i$, $i^4 = 1$, and thereafter, powers of $i$ occur in the same pattern, repeating in cycles of four.

Equality of complex numbers is defined as follows.

**DEFINITION** Two complex numbers $a + bi$ and $c + di$ are **equal,**

$$a + bi = c + di,$$

if and only if their real parts are equal ($a = c$) and their imaginary parts are equal ($b = d$).

**EXAMPLE 3** (a) If $a + bi = 3 + \sqrt{2}i$, then $a = 3$ and $b = \sqrt{2}$.

(b) If $c + di = 5$, then $c = 5$ and $d = 0$ since $5 = 5 + 0i$.

(c) If $u + vi = -\sqrt{3}i$, then $u = 0$ and $v = -\sqrt{3}$ since $-\sqrt{3}i = 0 - \sqrt{3}i$.

(d) If $2x + yi = 3x - 2 + 7i$, then

$$\begin{aligned} 2x &= 3x - 2 \qquad \text{and} \qquad y = 7 \\ -x &= -2 \\ x &= 2 \end{aligned}$$

The basic operations on complex numbers are easy to learn since the number $i$ acts the same as any other literal number in calculations. The following examples illustrate the various computational techniques.

**EXAMPLE 4**

$$\begin{aligned} (3 + 2i) + (-4 - i) &= 3 - 4 + 2i - i && \text{Use commutative rule} \\ &= (3 - 4) + (2 - 1)i && \text{Add real parts and imaginary parts} \\ &= -1 + i \end{aligned}$$

**EXAMPLE 5**

$$\begin{aligned} (2 - \sqrt{3}i) - (4 + 3\sqrt{3}i) &= 2 - 4 - \sqrt{3}i - 3\sqrt{3}i \\ &= (2 - 4) - (\sqrt{3} + 3\sqrt{3})i \\ &= -2 - 4\sqrt{3}i \end{aligned}$$

Notice that for addition and subtraction, we add (subtract) the real parts and then add (subtract) the imaginary parts. Multiplication of complex numbers is accomplished by taking products just as if they were products of binomials and then substituting $-1$ for $i^2$.

**EXAMPLE 6**



$$\begin{aligned} (4 + 2i)(3 - 5i) &= \overset{F}{4 \cdot 3} - \overset{O}{4 \cdot 5i} + \overset{I}{2 \cdot 3i} - \overset{L}{2 \cdot 5 \cdot i^2} \\ &= 12 - 20i + 6i - 10i^2 \\ &= 12 - 14i - 10(-1) \\ &= 12 + 10 - 14i \\ &= 22 - 14i \end{aligned}$$

**EXAMPLE 7**

$$\begin{aligned} (5 + \sqrt{2}i)(5 - \sqrt{2}i) &= 5 \cdot 5 - 5\sqrt{2}i + 5\sqrt{2}i - \sqrt{2}\sqrt{2}i^2 \\ &= 25 - 2i^2 \\ &= 25 - 2(-1) \\ &= 25 + 2 = 27 \end{aligned}$$

In the preceding example, a special product was found. The numbers $5 + \sqrt{2}i$ and $5 - \sqrt{2}i$ are called *conjugates,* and whenever two conjugates are multiplied, the result is a real number. The **conjugate** of the complex number $a + bi$ is $a - bi$, and the **conjugate** of $a - bi$ is $a + bi$. Thus, $a + bi$ and $a - bi$ are called **conjugates** of each other.

The notion of a conjugate is an important one in the study of complex numbers. One application of this relationship is found in the division process. To divide one complex number by another, for example

$$\frac{a+bi}{c+di},$$

we multiply both numerator and denominator by the conjugate of $c+di$, which is $c-di$. The next example illustrates this technique.

**EXAMPLE 8** Divide $(3+7i)$ by $(2-5i)$.

$$\begin{aligned}\frac{3+7i}{2-5i} &= \frac{(3+7i)(2+5i)}{(2-5i)(2+5i)}\\ &= \frac{6+15i+14i+35i^2}{4+10i-10i-25i^2}\\ &= \frac{6+29i-35}{4+25} \qquad i^2=-1\\ &= \frac{-29+29i}{29}\\ &= \frac{-29}{29}+\frac{29i}{29} = -1+i\end{aligned}$$

**EXAMPLE 9** Find the reciprocal of $3+2i$ and express it in the form $a+bi$.

The reciprocal of $3+2i$ is $\frac{1}{3+2i}$. We must divide the complex number $1 = 1+0i$ by the complex number $3+2i$.

$$\begin{aligned}\frac{1}{3+2i} &= \frac{1(3-2i)}{(3+2i)(3-2i)} = \frac{3-2i}{9-6i+6i-4i^2}\\ &= \frac{3-2i}{9+4} = \frac{3-2i}{13}\\ &= \frac{3}{13}-\frac{2}{13}i\end{aligned}$$

**EXAMPLE 10** Solve $x^2-2x+3=0$.

Use the quadratic formula with $a=1$, $b=-2$, and $c=3$.

$$\begin{aligned}x &= \frac{-(-2)\pm\sqrt{(-2)^2-4(1)(3)}}{2(1)}\\ &= \frac{2\pm\sqrt{4-12}}{2} = \frac{2\pm\sqrt{-8}}{2} = \frac{2\pm2\sqrt{2}i}{2}\\ &= 1\pm\sqrt{2}i\end{aligned}$$

Check:
$$(1 + \sqrt{2}i)^2 - 2(1 + \sqrt{2}i) + 3 \stackrel{?}{=} 0$$
$$1 + 2\sqrt{2}i + 2i^2 - 2 - 2\sqrt{2}i + 3 \stackrel{?}{=} 0$$
$$1 + 2\sqrt{2}i - 2 - 2 - 2\sqrt{2}i + 3 \stackrel{?}{=} 0$$
$$0 = 0$$

$1 - \sqrt{2}i$ also checks.

The solutions are $1 + \sqrt{2}i$ and $1 - \sqrt{2}i$.

## EXERCISES 5.6

*Express in terms of the imaginary number i.*

**1.** $\sqrt{-9}$ **2.** $\sqrt{-4}$ **3.** $\sqrt{-36}$ **4.** $-\sqrt{-36}$

**5.** $\sqrt{-5}$ **6.** $\sqrt{-8}$ **7.** $-\sqrt{-3}$ **8.** $\sqrt{-\pi^2}$

*Simplify and express each answer as a real number or in terms of i.*

**9.** $\sqrt{-4} + \sqrt{-9}$ **10.** $\sqrt{-4} - \sqrt{-9}$ **11.** $\sqrt{-7} + 2\sqrt{-7}$

**12.** $\sqrt{-3} - \sqrt{-7}$ **13.** $\sqrt{-4}\sqrt{-9}$ **14.** $\sqrt{-3}\sqrt{-3}$

**15.** $\sqrt{-7}\sqrt{-3}$ **16.** $\dfrac{\sqrt{-4}}{\sqrt{-9}}$ **17.** $\dfrac{\sqrt{-7}}{\sqrt{-3}}$

*Using the definition of equality of complex numbers, determine x and y.*

**18.** $x + yi = 2 + 3i$ **19.** $x + yi = 2 - 3i$ **20.** $x + yi = -5 + 7i$

**21.** $3x + 2yi = 4i$ **22.** $7x + 3yi = 21$ **23.** $(4x + 1) + 7yi = 5 - i$

*Perform the indicated operations.*

**24.** $(2 + 3i) + (5 + 8i)$ **25.** $(2 + 3i) - (5 + 8i)$ **26.** $3 + (5 - 2i)$

**27.** $-6i + (8 + 2i)$

**28.** $(-3 - 2i) - (6 - 7i)$

**29.** $(-5 + 4i) + (5 - 4i)$

**30.** $7(2 - 3i)$

**31.** $7i(2 - 3i)$

**32.** $(1 - 2i)(3 + 4i)$

**33.** $(8 + 4i)(-7 + 3i)$

**34.** $(2 + 3i)(2 - 3i)$

**35.** $(2 + 3i)^2$

**36.** $i^9$

**37.** $i^{15}$

**38.** $i^{40}$

*Give the conjugate of each of the following complex numbers.*

**39.** $4 + 6i$

**40.** $4 - 6i$

**41.** $-4 + 6i$

**42.** $-4 - 6i$

**43.** $8$

**44.** $6i$

**45.** $-7i$

**46.** $2 + \sqrt{2}i$

*Find the following quotients.*

**47.** $\frac{3 + 5i}{1 + i}$

**48.** $\frac{3 + 5i}{1 - i}$

**49.** $\frac{7}{4 + 3i}$

**50.** $\frac{7}{4 - 3i}$

**51.** $\frac{3i}{5 - 4i}$

**52.** $\frac{8 - 7i}{5 + 4i}$

*Find the reciprocals of the following numbers.*

**53.** $1 + i$

**54.** $1 - i$

**55.** $3 + 4i$

**56.** $-2i$

**57.** $3$

**58.** $-4 + 8i$

*Solve.*

**59.** $x^2 + 2x + 2 = 0$

**60.** $x^2 + 9 = 0$

**61.** $2x^2 - 3x + 4 = 0$

**62.** $x^4 - x^2 - 12 = 0$
[*Hint:* Let $u = x^2$ and factor.]

**63.** The sum of the squares of two positive consecutive odd integers is 74. Find the integers.

**64.** The area of a triangle is 65 cm² and the height is 3 cm less than the base. Find the height and base.

---

ANSWERS: **1.** $3i$ **2.** $2i$ **3.** $6i$ **4.** $-6i$ **5.** $\sqrt{5}i$ **6.** $2\sqrt{2}i$ **7.** $-\sqrt{3}i$ **8.** $\pi i$ **9.** $5i$ **10.** $-i$ **11.** $3\sqrt{7}i$ **12.** $\sqrt{3}i - \sqrt{7}i$ **13.** $-6$ **14.** $-3$ **15.** $-\sqrt{21}$ **16.** $\frac{2}{3}$ **17.** $\frac{\sqrt{7}}{\sqrt{3}}$ **18.** $x = 2, y = 3$ **19.** $x = 2, y = -3$ **20.** $x = -5, y = 7$ **21.** $x = 0, y = 2$ **22.** $x = 3, y = 0$ **23.** $x = 1, y = -\frac{1}{7}$ **24.** $7 + 11i$ **25.** $-3 - 5i$ **26.** $8 - 2i$ **27.** $8 - 4i$ **28.** $-9 + 5i$ **29.** $0$ **30.** $14 - 21i$ **31.** $21 + 14i$ **32.** $11 - 2i$ **33.** $-68 - 4i$ **34.** $13$ **35.** $-5 + 12i$ **36.** $i$ **37.** $-i$ **38.** $1$ **39.** $4 - 6i$ **40.** $4 + 6i$ **41.** $-4 - 6i$ **42.** $-4 + 6i$ **43.** $8$ **44.** $-6i$ **45.** $7i$ **46.** $2 - \sqrt{2}i$ **47.** $4 + i$ **48.** $-1 + 4i$ **49.** $\frac{28}{25} - \frac{21}{25}i$ **50.** $\frac{28}{25} + \frac{21}{25}i$ **51.** $-\frac{12}{41} + \frac{15}{41}i$ **52.** $\frac{12}{41} - \frac{67}{41}i$ **53.** $\frac{1}{2} - \frac{1}{2}i$ **54.** $\frac{1}{2} + \frac{1}{2}i$ **55.** $\frac{3}{25} - \frac{4}{25}i$ **56.** $\frac{1}{2}i$ **57.** $\frac{1}{3}$ **58.** $-\frac{1}{20} - \frac{1}{10}i$ **59.** $-1 \pm i$ **60.** $\pm 3i$ **61.** $\frac{3}{4} \pm \frac{\sqrt{23}}{4}i$ **62.** $\pm\sqrt{3}i, \pm 2$ **63.** 5, 7 **64.** 10, 13

## 5.7 The Discriminant and Properties of Solutions to Quadratic Equations

In the following example we use the quadratic formula to solve three different quadratic equations with three different types of solutions. (The equations in the first two parts would normally be solved by the method of factoring, but we are using the formula to illustrate a point.)

**EXAMPLE 1** **(a)** $x^2 - 2x - 3 = 0$

$$x = \frac{-b \pm \sqrt{b^2 - 4ac}}{2a}$$

$$= \frac{-(-2) \pm \sqrt{(-2)^2 - 4(1)(-3)}}{2(1)} \qquad a = 1,\ b = -2,\ c = -3$$

$$= \frac{2 \pm \sqrt{4 + 12}}{2} = \frac{2 \pm \sqrt{16}}{2}$$

$$= \frac{2 \pm 4}{2} = 1 \pm 2$$

In this case we obtain exactly two real solutions, $x = 3$ and $x = -1$.

**(b)** $x^2 - 2x + 1 = 0$

$$x = \frac{-(-2) \pm \sqrt{(-2)^2 - 4(1)(1)}}{2(1)} \qquad a = 1, b = -2, c = 1$$

$$= \frac{2 \pm \sqrt{4 - 4}}{2} = \frac{2 \pm \sqrt{0}}{2}$$

$$= \frac{2}{2} = 1$$

In this case we obtain exactly one real solution, $x = 1$.

**(c)** $x^2 - 2x + 2 = 0$

$$x = \frac{-(-2) \pm \sqrt{(-2)^2 - 4(1)(2)}}{2(1)} \qquad a = 1, b = -2, c = 2$$

$$= \frac{2 \pm \sqrt{4 - 8}}{2} = \frac{2 \pm \sqrt{-4}}{2}$$

$$= \frac{2 \pm 2i}{2} = \frac{2(1 \pm i)}{2} = 1 \pm i$$

In this case we obtain two complex solutions which are conjugates, $x = 1 + i$ and $x = 1 - i$.

If we examine the solutions given in each case by the quadratic formula,

$$x = \frac{-b \pm \sqrt{b^2 - 4ac}}{2a},$$

we see that the kind of solution is determined by the number under the radical,

$$b^2 - 4ac.$$

This number is called the **discriminant** of the quadratic equation $ax^2 + bx + c = 0$.
If $b^2 - 4ac > 0$, then there are two real solutions,

$$x_1 = \frac{-b + \sqrt{b^2 - 4ac}}{2a} \quad \text{and} \quad x_2 = \frac{-b - \sqrt{b^2 - 4ac}}{2a}.$$

In Example 1 (a), $x^2 - 2x - 3 = 0$ has two real solutions and the discriminant is 16, which is greater than zero. In (b), the discriminant of $x^2 - 2x + 1 = 0$ is zero, and there is exactly one real solution. In (c), the discriminant of $x^2 - 2x + 2 = 0$ is $-4$ which is less than zero, and there are two complex solutions. These results are generalized in the following rule.

**NUMBER OF SOLUTIONS TO A QUADRATIC EQUATION**

The quadratic equation $ax^2 + bx + c = 0$ has

**1.** Two real number solutions if $b^2 - 4ac > 0$.

**2.** One real number solution if $b^2 - 4ac = 0$.

**3.** Two complex number solutions if $b^2 - 4ac < 0$.

**EXAMPLE 2** Determine the nature of the solutions to the following quadratic equations by calculating the discriminants.

**(a)** $x^2 - 6x + 7 = 0 \qquad (a = 1, b = -6, c = 7)$

$$\begin{aligned} b^2 - 4ac &= (-6)^2 - 4(1)(7) \\ &= 36 - 28 \\ &= 8 > 0 \end{aligned}$$

There are two real-number solutions.

**(b)** $2x^2 + 9 = 0 \qquad (a = 2, b = 0, c = 9)$

$$\begin{aligned} b^2 - 4ac &= (0)^2 - 4(2)(9) \\ &= 0 - 72 \\ &= -72 < 0 \end{aligned}$$

There are two complex-number solutions.

**(c)** $-x^2 + 10x - 25 = 0 \qquad (a = -1, b = 10, c = -25)$

$$\begin{aligned} b^2 - 4ac &= (10)^2 - 4(-1)(-25) \\ &= 100 - 100 \\ &= 0 \end{aligned}$$

There is only one real-number solution.

We now turn to the problem of finding a quadratic equation when we are given its solutions (roots). To see how this is done, solve the following equation by factoring.

$$\begin{gathered} x^2 - 2x - 3 = 0 \\ (x + 1)(x - 3) = 0 \\ x + 1 = 0 \quad \text{or} \quad x - 3 = 0 \\ x = -1 \qquad\qquad x = 3 \end{gathered}$$

We need only reverse the steps in the problem above to construct the quadratic equation from the solutions. Suppose that we are given two solutions $-1$ and $3$.

$$\begin{gathered} x = -1 \quad \text{or} \quad x = 3 \\ x - (-1) = 0 \quad \text{or} \quad x - 3 = 0 \\ x + 1 = 0 \quad \text{or} \quad x - 3 = 0 \\ (x + 1)(x - 3) = 0 \\ x^2 - 2x - 3 = 0 \end{gathered}$$

If $x_1$ and $x_2$ are solutions to a quadratic equation, the equation is

$$(x - x_1)(x - x_2) = 0.$$

**EXAMPLE 3** Find the quadratic equation which has the given solutions.

**(a)** $x_1 = -2$ and $x_2 = 5$

$$(x - x_1)(x - x_2) = 0$$

$$(x - (-2))(x - 5) = 0$$ Substitute the given solutions

$$(x + 2)(x - 5) = 0$$

$$x^2 - 5x + 2x - 10 = 0$$ Multiply the binomial factors

$$x^2 - 3x - 10 = 0$$

**(b)** $x_1 = 1 + 3i$ and $x_2 = 1 - 3i$

$$(x - x_1)(x - x_2) = 0$$

$$[x - (1 + 3i)][x - (1 - 3i)] = 0$$ Substitute the solutions

$$x^2 - (1 - 3i)x - (1 + 3i)x + (1 + 3i)(1 - 3i) = 0$$ Multiply the binomial factors

$$x^2 - 2x + 10 = 0$$ Simplify

Other interesting properties of the solutions to a quadratic equation involve their sum and product. Let us add the solutions.

$$x_1 + x_2 = \frac{-b + \sqrt{b^2 - 4ac}}{2a} + \frac{-b - \sqrt{b^2 - 4ac}}{2a}$$

$$= \frac{-b + \sqrt{b^2 - 4ac} - b - \sqrt{b^2 - 4ac}}{2a}$$

$$= \frac{-2b}{2a} = -\frac{b}{a}$$

Similarly, we may multiply the solutions.

$$x_1 \cdot x_2 = \left(\frac{-b + \sqrt{b^2 - 4ac}}{2a}\right) \cdot \left(\frac{-b - \sqrt{b^2 - 4ac}}{2a}\right)$$

$$= \frac{(-b)^2 - (\sqrt{b^2 - 4ac})^2}{(2a)^2}$$ Use $(a + b)(a - b) = a^2 - b^2$ in the numerator

$$= \frac{b^2 - (b^2 - 4ac)}{4a^2}$$

$$= \frac{b^2 - b^2 + 4ac}{4a^2}$$

$$= \frac{4ac}{4a^2} = \frac{c}{a}$$

**SUM AND PRODUCT OF SOLUTIONS TO QUADRATIC EQUATIONS**

If $x_1$ and $x_2$ are solutions to $ax^2 + bx + c = 0$, then

$$x_1 + x_2 = -\frac{b}{a} \quad \text{and} \quad x_1 \cdot x_2 = \frac{c}{a}.$$

**EXAMPLE 4** Find the sum and product of the solutions of $3x^2 - 4x + 2 = 0$ $(a = 3, b = -4, c = 2)$.

$$x_1 + x_2 = -\frac{b}{a} = -\left(\frac{-4}{3}\right) = \frac{4}{3}$$

$$x_1 \cdot x_2 = \frac{c}{a} = \frac{2}{3}$$

It is also true that if

$$x_1 + x_2 = -\frac{b}{a} \quad \text{and} \quad x_1 \cdot x_2 = \frac{c}{a}$$

then $x_1$ and $x_2$ are solutions to $ax^2 + bx + c = 0$.

**EXAMPLE 5** Are $3 - i$ and $3 + i$ solutions of $2x^2 - 12x + 20 = 0$?

$$x_1 + x_2 = 3 - i + 3 + i = 6$$

and

$$-\frac{b}{a} = -\frac{(-12)}{2} = 6. \qquad a = 2,\ b = -12 \text{ in } 2x^2 - 12x + 20 = 0$$

Also,

$$x_1 \cdot x_2 = (3 - i)(3 + i) = 9 - i^2 = 9 - (-1) = 10$$

and

$$\frac{c}{a} = \frac{20}{2} = 10. \qquad a = 2,\ c = 20$$

Since $x_1 + x_2 = -\frac{b}{a} = 6$ and $x_1 \cdot x_2 = \frac{c}{a} = 10$, $3 - i$ and $3 + i$ are solutions to $2x^2 - 12x + 20 = 0$.

## EXERCISES 5.7

*Use the discriminant to determine the nature of the solutions.*

**1.** $x^2 - 3x + 2 = 0$

**2.** $x^2 + 3x + 2 = 0$

**3.** $-x^2 + 3x - 2 = 0$

**4.** $-x^2 - 3x - 2 = 0$

**5.** $x^2 + 4x + 5 = 0$

**6.** $x^2 + 6x + 9 = 0$

**7.** $3x^2 - 2x + 5 = 0$

**8.** $-2x^2 - 3x + 1 = 0$

**9.** $5x^2 + 7 = 0$

**10.** $-3x^2 + 2x - 4 = 0$

**11.** $4x^2 - 4x + 1 = 0$

**12.** $7x^2 - 3 = 0$

*Find the quadratic equation which has the following solutions.*

**13.** 2, 3

**14.** −2, 3

**15.** $2i, -2i$

**16.** 7

**17.** $2 + i, 2 - i$

**18.** $\frac{2}{3}, -1$

**19.** $0, 4$

**20.** $2 + 3i, 2 - 3i$

*Find the sum and product of the solutions.*

**21.** $x^2 - 7x + 6 = 0$

**22.** $x^2 + 7x + 6 = 0$

**23.** $3x^2 - 2x + 5 = 0$

**24.** $4x^2 - 9 = 0$

**25.** $4x^2 + 9 = 0$

**26.** $4x^2 - 5x = 0$

*Use the formulas for the sum and product of solutions to determine if $x_1$ and $x_2$ are solutions to the given quadratic equation.*

**27.** $x^2 - 5x + 6 = 0$; $x_1 = 2$, $x_2 = 3$

**28.** $x^2 + 4x - 5 = 0$; $x_1 = -5$, $x_2 = 1$

**29.** $x^2 + 3x + 2 = 0$; $x_1 = 2$, $x_2 = 1$

**30.** $x^2 - 2x + 2 = 0$; $x_1 = 1 - i$, $x_2 = 1 + i$

**31.** $3x^2 + 5x - 2 = 0;\ x_1 = \frac{1}{3},\ x_2 = -2$

**32.** $2x^2 + 8 = 0;\ x_1 = 4i,\ x_2 = -4i$

---

ANSWERS: **1.** two real solutions **2.** two real solutions **3.** two real solutions **4.** two real solutions **5.** two complex solutions **6.** one real solution **7.** two complex solutions **8.** two real solutions **9.** two complex solutions **10.** two complex solutions **11.** one real solution **12.** two real solutions **13.** $x^2 - 5x + 6 = 0$ **14.** $x^2 - x - 6 = 0$ **15.** $x^2 + 4 = 0$ **16.** $x^2 - 14x + 49 = 0$ **17.** $x^2 - 4x + 5 = 0$ **18.** $3x^2 + x - 2 = 0$ **19.** $x^2 - 4x = 0$ **20.** $x^2 - 4x + 13 = 0$ **21.** $x_1 + x_2 = 7,\ x_1 \cdot x_2 = 6$ **22.** $x_1 + x_2 = -7,\ x_1 \cdot x_2 = 6$ **23.** $x_1 + x_2 = \frac{2}{3},\ x_1 \cdot x_2 = \frac{5}{3}$ **24.** $x_1 + x_2 = 0,\ x_1 \cdot x_2 = -\frac{9}{4}$ **25.** $x_1 + x_2 = 0,\ x_1 \cdot x_2 = \frac{9}{4}$ **26.** $x_1 + x_2 = \frac{5}{4},\ x_1 \cdot x_2 = 0$ **27.** yes **28.** yes **29.** no **30.** yes **31.** yes **32.** no

## 5.8 Solving Quadratic Inequalities

Inequalities such as

$$2x^2 + 5x - 3 < 0 \qquad \text{and} \qquad x^2 - 3x + 7 \geq 0$$

are called **quadratic inequalities.** In both instances, if the inequality symbol were replaced by an equal sign, the result would be a quadratic equation. There are numerous techniques for solving such inequalities, but the method we present is perhaps the simplest. The following examples outline the procedure which is then summarized in a rule.

EXAMPLE 1 Solve $2x^2 + 5x - 3 < 0$.

First replace the inequality symbol with an equal sign, then solve and graph the solution as in Figure 5.3.

$$(2x - 1)(x + 3) = 0 \qquad \text{Factor}$$

$$2x - 1 = 0 \quad \text{or} \quad x + 3 = 0 \qquad \text{Zero-product rule}$$

$$2x = 1 \qquad x = -3$$

$$x = \frac{1}{2}$$

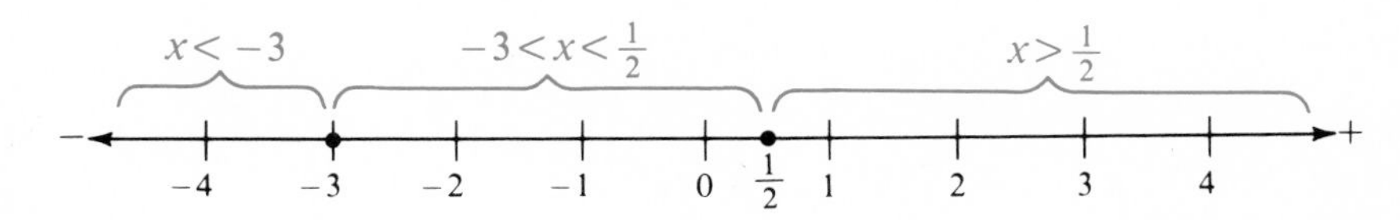


**Figure 5.3**

The solutions to the equation separate the number line into three segments:

$$x < -3, \qquad -3 < x < \frac{1}{2}, \qquad x > \frac{1}{2}.$$

The numbers $-3$ and $1/2$ are *not* solutions to $2x^2 + 5x - 3 < 0$ since they *are* solutions to the equation $2x^2 + 5x - 3 = 0$. All the numbers in a segment behave the same way in the inequality. For example, any number in the segment $x < -3$, when substituted for $x$ in $2x^2 + 5x - 3$, results in a number greater than 0. Thus, we may choose any point in each of the three segments and evaluate the quadratic expression.

Let $x = -4$: $2x^2 + 5x - 3 = 2(-4)^2 + 5(-4) - 3 = 32 - 20 - 3 = 9 > 0$.
Let $x = 0$: $2x^2 + 5x - 3 = 2(0)^2 + 5(0) - 3 = 0 + 0 - 3 = -3 < 0$.
Let $x = 1$: $2x^2 + 5x - 3 = 2(1)^2 + 5(1) - 3 = 2 + 5 - 3 = 4 > 0$.

Then, for any $x$ satisfying $x < -3$, $2x^2 + 5x - 3 > 0$. Similarly, if $-3 < x < 1/2$, $2x^2 + 5x - 3 < 0$, and if $x > 1/2$, $2x^2 + 5x - 3 > 0$. We often summarize this by writing the appropriate inequality above each segment on a number line,

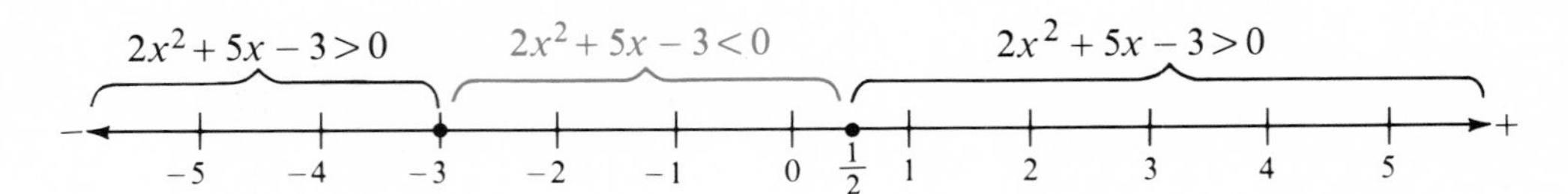


**Figure 5.4**

as in Figure 5.4. For every $x$ satisfying $x < -3$ or $x > 1/2$, $2x^2 + 5x - 3 > 0$, and for every $x$ satisfying $-3 < x < 1/2$, $2x^2 + 5x - 3 < 0$. The solution to the original inequality is

$$-3 < x < \frac{1}{2}.$$

If we had solved the inequality

$$2x^2 + 5x - 3 > 0,$$

in view of the above remarks the solution would be

$$x < -3 \quad \text{or} \quad x > \frac{1}{2}.$$

Similarly, the solution to

$$2x^2 + 5x - 3 \leq 0$$

is $-3 \leq x \leq \frac{1}{2}$ (equality does hold this time), and the solution to

$$2x^2 + 5x - 3 \geq 0$$

is $x \leq -3$ or $x \geq \frac{1}{2}$ (again equality holds).

EXAMPLE 2 Solve $x^2 - 4x + 4 > 0$.

We first solve $x^2 - 4x + 4 = 0$.

$$(x - 2)(x - 2) = 0$$
$$x - 2 = 0 \quad \text{or} \quad x - 2 = 0$$
$$x = 2 \qquad\qquad x = 2$$

There is only one solution so the number line is divided into two segments,

$$x < 2 \quad \text{and} \quad x > 2.$$

Evaluate at 0 and 3.

$$\text{Let } x = 0: \quad 0^2 - 4(0) + 4 = 4 > 0.$$
$$\text{Let } x = 3: \quad 3^2 - 4(3) + 4 = 9 - 12 + 4 = 1 > 0.$$

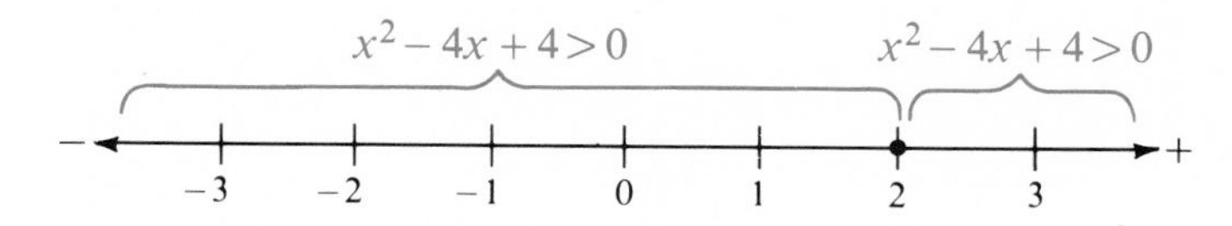


**Figure 5.5**

The results are shown in Figure 5.5. The solution to $x^2 - 4x + 4 > 0$ is

$$x < 2 \quad \text{or} \quad x > 2.$$

In view of the above, the solution to the inequality

$$x^2 - 4x + 4 \geq 0$$

is the entire collection of real numbers. Also, the solution to

$$x^2 - 4x + 4 \leq 0$$

is $x = 2$, and the inequality

$$x^2 - 4x + 4 < 0$$

has no solution.

EXAMPLE 3 Solve $x^2 + x + 2 \geq 0$.

We first solve $x^2 + x + 2 = 0$.

$$x = \frac{-1 \pm \sqrt{(1)^2 - 4(1)(2)}}{2(1)} \qquad \text{This time we cannot factor}$$
$$= \frac{-1 \pm \sqrt{-7}}{2} = \frac{-1 \pm \sqrt{7}i}{2}$$

Since only real (not imaginary) numbers are identified with points on a number line, there are no points on the line associated with the solutions $(-1 \pm \sqrt{7}i)/2$. As a result, the solutions in this case do not divide the number line into separate parts, so $x^2 + x + 2$ has the same sign regardless of the choice of value for the variable $x$. Since

$$0^2 + 0 + 2 = 2 > 0$$

every real number is a solution to the given inequality, as shown in Figure 5.6.

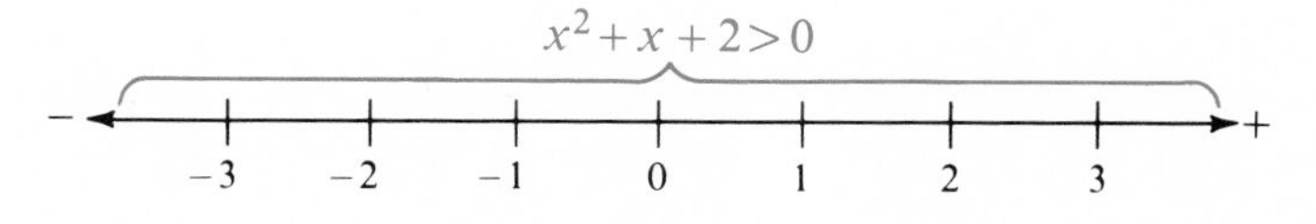


**Figure 5.6**

In view of the above, the solution to

$$x^2 + x + 2 > 0$$

is also the entire collection of real numbers, and the inequalities

$$x^2 + x + 2 < 0 \quad \text{and} \quad x^2 + x + 2 \le 0$$

have no solution.

**TO SOLVE A QUADRATIC INEQUALITY**

1. Replace the inequality symbol with an equal sign and solve the resulting quadratic equation.
2. If there are two real solutions, they separate the number line into three distinct segments. Evaluate the quadratic expression at an arbitrary value chosen from each segment and, using these values, specify the solution.
3. If there is one real solution, it separates the number line into two distinct segments. Evaluate at an arbitrary value in each segment and specify the solution.
4. If there are no real solutions, only complex ones, either every number is a solution or else there is no solution.

**EXAMPLE 4** Solve $(x + 3)(x - 1) \le -2$.

Simplify.

$$x^2 + 2x - 3 \le -2$$
$$x^2 + 2x - 1 \le 0$$

Thus, we must solve the following equation.

$$x^2 + 2x - 1 = 0 \qquad \text{Writing in general form}$$

$$x = \frac{-2 \pm \sqrt{2^2 - 4(1)(-1)}}{2(1)}$$

$$= \frac{-2 \pm \sqrt{8}}{2} = \frac{-2 \pm 2\sqrt{2}}{2}$$

$$= \frac{\not{2}(-1 \pm \sqrt{2})}{\not{2}} = -1 \pm \sqrt{2}$$

Since $\sqrt{2} \approx 1.4$, $-1 \pm \sqrt{2} \approx -1 \pm 1.4$ so that the solutions are, approximately, 0.4 and $-2.4$.

Let $x = -3$: $(-3)^2 + 2(-3) - 1 = 9 - 6 - 1 = 2 > 0$
Let $x = 0$: $0^2 + 2(0) - 1 = -1 < 0$
Let $x = 1$: $1^2 + 2(1) - 1 = 2 > 0$

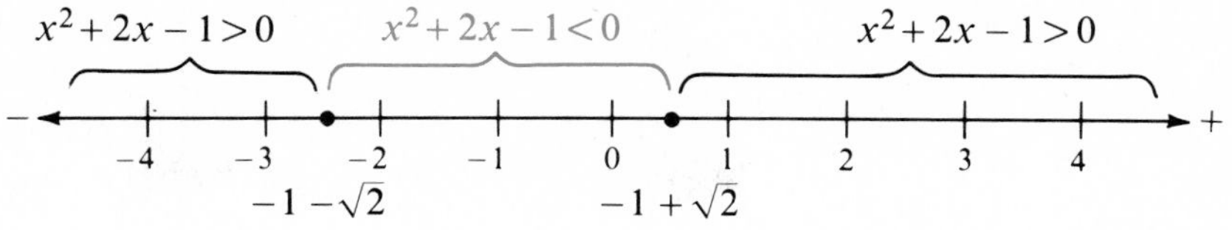

$x^2 + 2x - 1 > 0$ | $x^2 + 2x - 1 < 0$ | $x^2 + 2x - 1 > 0$

−4 −3 −2 −1 0 1 2 3 4

$-1 - \sqrt{2}$ $-1 + \sqrt{2}$

**Figure 5.7**

The solution, $-1 - \sqrt{2} \le x \le -1 + \sqrt{2}$ (equality holds), is shown in Figure 5.7.

## EXERCISES 5.8

*Solve.*

**1.** $x^2 - 5x + 6 > 0$

**2.** $x^2 - x < 6$

**3.** $2x^2 \le 1 - x$

**4.** $2x^2 + x \ge 15$

**5.** $x^2 - 2x - 2 > 0$

**6.** $x^2 + 2x + 2 > 0$

**7.** $4x^2 - 12x + 9 \le 0$

**8.** $4x^2 - 12x + 9 < 0$

**9.** $4x^2 - 12x + 9 > 0$

**10.** $4x^2 - 12x + 9 \geq 0$

**11.** $x(x - 1) \leq -1$

**12.** $3x^2 - 2x \leq 2$

*Use the discriminant to determine the nature of the solutions.*

**13.** $x^2 - 3x - 4 = 0$

**14.** $9x^2 + 6x + 1 = 0$

**15.** $2x^2 - 2x + 5 = 0$

**16.** $x^2 - 5x = -6$

---

ANSWERS: **1.** $x < 2$ or $x > 3$ **2.** $-2 < x < 3$ **3.** $-1 \leq x \leq \frac{1}{2}$ **4.** $x \geq \frac{5}{2}$ or $x \leq -3$ **5.** $x > 1 + \sqrt{3}$ or $x < 1 - \sqrt{3}$ **6.** Every real number is a solution. **7.** $x = \frac{3}{2}$ **8.** no solution **9.** $x < \frac{3}{2}$ or $x > \frac{3}{2}$ **10.** Every real number is a solution. **11.** no solution **12.** $\frac{1 - \sqrt{7}}{3} \leq x \leq \frac{1 + \sqrt{7}}{3}$ **13.** two real solutions **14.** one real solution **15.** two complex solutions **16.** two real solutions

## Chapter 5 Summary

### Words and Phrases for Review

[5.1] quadratic (second-degree) equation
[5.2] complete the square
[5.3] equations quadratic in form
[5.4] radical equations
[5.6] imaginary number
complex number
real part
imaginary part
pure imaginary number
conjugate
[5.7] discriminant
[5.8] quadratic inequality

### Brief Reminders

[5.1] **1.** Zero-product rule: If $a \cdot b = 0$, then $a = 0$ or $b = 0$. This rule only applies when one side of the equation is zero. For example, if $a \cdot b = 7$, *do not* conclude that $a = 7$ or $b = 7$.

**2.** Do not use the zero-product rule on a zero-sum or zero-difference equation. For example, $(2x + 1) - (x + 5) = 0$ is *not* a zero-product equation. To solve it, clear parentheses and combine like terms. *Do not* set $2x + 1$ and $x + 5$ equal to zero.

[5.2] Quadratic formula: The solutions of $ax^2 + bx + c = 0$ are given by

$$x = \frac{-b \pm \sqrt{b^2 - 4ac}}{2a}.$$

[5.3] When solving an equation using a substitution, be sure to give solutions for the original variable and not for the substituted variable.

[5.6] **1.** $\sqrt{-1} = i$, that is, $i^2 = -1$.

**2.** To divide one complex number by another, for example,

$$\frac{a + bi}{c + di},$$

multiply numerator and denominator by the conjugate of the denominator, $c - di$.

[5.7] **1.** The discriminant of the quadratic equation $ax^2 + bx + c = 0$ is the number $b^2 - 4ac$.

If $b^2 - 4ac > 0$, the quadratic equation has two real number solutions.

If $b^2 - 4ac = 0$, the quadratic equation has one real number solution.

If $b^2 - 4ac < 0$, the quadratic equation has two complex number solutions.

**2.** If $x_1$ and $x_2$ are solutions to $ax^2 + bx + c = 0$, then

$$x_1 + x_2 = -\frac{b}{a} \quad \text{and} \quad x_1 \cdot x_2 = \frac{c}{a}.$$

[5.8] To solve $ax^2 + bx + c > 0$ or $ax^2 + bx + c < 0$, first solve $ax^2 + bx + c = 0$ and then determine the value of $ax^2 + bx + c$ in the intervals determined by the solutions of $ax^2 + bx + c = 0$.

## CHAPTER 5 REVIEW EXERCISES

[5.1] **1.** The zero-product rule states that if $a \cdot b = 0$, then ____________.

**2.** $ax^2 + bx + c = 0$ is called the ________________ form of the quadratic equation.

[5.2] **3.** The quadratic formula is ________________.

**4.** When solving a quadratic equation always try **(a)** ______________ first, then if this fails try **(b)** ________________________.

[5.6] **5.** In the complex number $a + bi$, $a$ is called the **(a)** ________________ and $b$ is called the **(b)** ________________.

**6.** $a + bi = c + di$, if and only if ________________.

**7.** The complex number $a - bi$ is called the ________________ of $a + bi$.

[5.7] **8.** The discriminant of $ax^2 + bx + c = 0$ is ________________.

**9.** If the discriminant is negative, the quadratic equation has ______ ________________ solutions.

[5.8] **10.** An inequality of the form $ax^2 + bx + c > 0$ is called a(n) ______ ________________________.

[5.1–5.2] *Solve.*

**11.** $y^2 - 2y - 48 = 0$

**12.** $y^2 = 4y - 2$

**13.** $3x(x + 5) = -5x - 25$

**14.** $5x^2 - 125 = 0$

**15.** $y^2 + 100 = 20y$

**16.** $y^2 - \frac{15}{2}y + 14 = 0$

**17.** $3x^2 = 2x + 2$

**18.** $(x - 5)(3x + 7) = -30$

[5.3] *Solve.*

**19.** $x^4 - 11x^2 + 28 = 0$

**20.** $(x - 7)^2 - 10(x - 7) + 25 = 0$

**21.** $y + 4 - 5\sqrt{y + 4} + 6 = 0$

**22.** $y^{2/3} - y^{1/3} - 6 = 0$

[5.4] *Solve.*

**23.** $\sqrt{3x^2 - 5} - 4 = 0$

**24.** $\sqrt{4x + 1} - x + 5 = 0$

**25.** $\sqrt{2y^2 - 5y + 4} - \sqrt{y^2 + 6y - 20} = 0$

**26.** $\sqrt{4x + 1} - \sqrt{2x + 4} = 3$

[5.5] *Solve.*

**27.** Twice the product of two positive consecutive odd integers is 390. Find the integers.

**28.** Ten less than the square of Ezra's age is the same as 8 more than three times his age. How old is Ezra?

**29.** Five times the number of square centimeters in the area of a square is the same as the number of centimeters in the perimeter increased by 3. Find the length of a side.

[5.6] *Simplify.*

**30.** $\sqrt{-16} + 4\sqrt{-9}$

**31.** $\sqrt{-16} - 4\sqrt{-9}$

**32.** $\sqrt{-5} - 2\sqrt{-3}$

**33.** $\sqrt{-16} \cdot \sqrt{-9}$

**34.** $\dfrac{\sqrt{-16}}{\sqrt{-9}}$

**35.** $\sqrt{-5}\,\sqrt{-5}$

*Perform the indicated operations.*

**36.** $(7 - 4i) + (3 + 6i)$

**37.** $(7 - 4i) - (3 + 6i)$

**38.** $2i - (3 - 6i)$

**39.** $-6i(5 - 2i)$

**40.** $(-2 + 3i)(5 - 7i)$

**41.** $(4 - 3i)(4 + 3i)$

**42.** $(4 - 3i)^2$

**43.** $i^{10}$

**44.** $i^{25}$

**45.** $\dfrac{1}{2 + i}$

**46.** $\dfrac{i}{2 + i}$

**47.** $\dfrac{3 - 2i}{4 + 5i}$

*Solve.*

**48.** $x^2 + 2x + 2 = 0$

**49.** $x^2 - 4x = -7$

[5.7] *Use the discriminant to determine the nature of the solutions.*

**50.** $9x^2 + 6x + 1 = 0$

**51.** $2x^2 - 3x = 7$

**52.** $2x^2 = 3x - 7$

*Determine the quadratic equation which has the following solutions.*

**53.** 4, 1

**54.** $1 + i,\ 1 - i$

**55.** $-1, \frac{3}{2}$

*Find the sum and product of the solutions.*

**56.** $3x^2 + 5x + 6 = 0$

**57.** $-2x^2 + 7x + 10 = 0$

*Using the formulas for the sum and product of solutions, determine if $x_1$ and $x_2$ are solutions to the given quadratic equation.*

**58.** $3x^2 + 2x - 5 = 0;\ x_1 = -\frac{5}{3},\ x_2 = 1$

**59.** $2x^2 + x - 5 = 0;\ x_1 = \frac{5}{2},\ x_2 = -1$

[5.8] *Solve.*

**60.** $9x^2 - 6x + 1 > 0$

**61.** $3x^2 + 4x - 4 \leq 0$

---

ANSWERS: **1.** $a = 0$ or $b = 0$ **2.** general **3.** $x = \frac{-b \pm \sqrt{b^2 - 4ac}}{2a}$ **4.** (a) factoring (b) the quadratic formula **5.** (a) real part (b) imaginary part **6.** $a = c$ and $b = d$ **7.** conjugate **8.** $b^2 - 4ac$ **9.** two complex **10.** quadratic inequality **11.** 8, −6 **12.** $2 \pm \sqrt{2}$ **13.** $-\frac{5}{3}, -5$ **14.** 5, −5 **15.** 10 **16.** $\frac{7}{2}, 4$ **17.** $\frac{1 \pm \sqrt{7}}{3}$ **18.** $\frac{4 \pm \sqrt{31}}{3}$ **19.** $\pm 2, \pm\sqrt{7}$ **20.** 12 **21.** 0, 5 **22.** 27, −8 **23.** $\pm\sqrt{7}$ **24.** 12 **25.** 3, 8 **26.** 30 **27.** 13, 15 **28.** 6 **29.** $\frac{2 + \sqrt{19}}{5}$ **30.** $16i$ **31.** $-8i$ **32.** $(\sqrt{5} - 2\sqrt{3})i$ **33.** −12 **34.** $\frac{4}{3}$ **35.** −5 **36.** $10 + 2i$ **37.** $4 - 10i$ **38.** $-3 + 8i$ **39.** $-12 - 30i$ **40.** $11 + 29i$ **41.** 25 **42.** $7 - 24i$ **43.** −1 **44.** $i$ **45.** $\frac{2 - i}{5}$ **46.** $\frac{1 + 2i}{5}$ **47.** $\frac{2 - 23i}{41}$ **48.** $-1 \pm i$ **49.** $2 \pm \sqrt{3}i$ **50.** one real **51.** two real **52.** two complex **53.** $x^2 - 5x + 4 = 0$ **54.** $x^2 - 2x + 2 = 0$ **55.** $2x^2 - x - 3 = 0$ **56.** $x_1 + x_2 = -\frac{5}{3},\ x_1 \cdot x_2 = 2$ **57.** $x_1 + x_2 = \frac{7}{2},\ x_1 \cdot x_2 = -5$ **58.** yes **59.** no **60.** $x > \frac{1}{3}$ or $x < \frac{1}{3}$ **61.** $-2 \leq x \leq \frac{2}{3}$

# 6

# Fractional Expressions

## 6.1 Algebraic Fractions: Basic Concepts

In Chapter 1 we saw that a fraction is a ratio of integers. In this chapter we define an **algebraic fraction** as a fraction containing a variable in either the numerator or denominator (or both). The following are algebraic fractions.

$$\frac{2x}{x-1}, \quad \frac{\sqrt{y}+1}{2y+1}, \quad \frac{(a+b)^2}{2}, \quad \frac{7}{(x-y)(y+1)}$$

Most of the algebraic fractions we consider are ratios of polynomials.

Any value of the variable which makes the denominator zero is called a **meaningless replacement** for the variable since division by zero is not defined.

**TO FIND MEANINGLESS REPLACEMENTS IN AN ALGEBRAIC FRACTION**

1. Set the denominator equal to zero.
2. Solve the resulting equation; any solution is a meaningless replacement.

EXAMPLE 1 What are the meaningless replacements for $x$ in $\frac{3x}{x-5}$?

Setting $x - 5 = 0$ and solving we obtain $x = 5$. Therefore, the only meaningless replacement is 5.

EXAMPLE 2 What are the meaningless replacements for $y$ in $\frac{3y}{y^2 - y - 6}$?

$$y^2 - y - 6 = 0$$
$$(y-3)(y+2) = 0 \qquad \text{Factor}$$
$$y - 3 = 0 \quad \text{or} \quad y + 2 = 0 \qquad \text{Zero-product rule}$$
$$y = 3 \qquad\qquad y = -2$$

The meaningless replacements are 3 and $-2$.

**EXAMPLE 3** What are the meaningless replacements in $\frac{x^2 - y^2}{15}$?

Set the denominator equal to zero and solve.

$$15 = 0$$

Since there is no solution to this equation, there are no meaningless replacements.

**EXAMPLE 4** What are the meaningless replacements for the variables in $\frac{2a - 3b + 1}{a^2 - b^2}$?

Set $a^2 - b^2 = 0$ and factor.

$$(a - b)(a + b) = 0$$

$$a - b = 0 \quad \text{or} \quad a + b = 0$$

$$a = b \qquad\qquad a = -b$$

The fraction is meaningless if $a$ and $b$ are equal or if $a$ and $b$ are negatives of each other.

In Chapter 1, we saw that multiplying or dividing the numerator and denominator of a fraction by the same nonzero number gives us equivalent fractions.

$$\frac{4}{10} = \frac{4 \cdot 5}{10 \cdot 5} = \frac{20}{50} \qquad \frac{4}{10} \text{ is equivalent to } \frac{20}{50}$$

$$\frac{4}{10} = \frac{4 \div 2}{10 \div 2} = \frac{2}{5} \qquad \frac{4}{10} \text{ is equivalent to } \frac{2}{5}$$

If the numerator and denominator of an algebraic fraction are multiplied or divided by the same nonzero expression or number, the resulting algebraic fraction is **equivalent** to the original.

**EXAMPLE 5** The fractions

$$\frac{x}{x - y} \quad \text{and} \quad \frac{x(x + y)}{(x - y)(x + y)}$$

are equivalent (if $x + y \neq 0$). Note that $\frac{x(x + y)}{(x - y)(x + y)}$ has been built up from $\frac{x}{x - y}$ by multiplying numerator and denominator by $(x + y)$, while $\frac{x}{x - y}$ has been reduced from $\frac{x(x + y)}{(x - y)(x + y)}$ by dividing numerator and denominator by $(x + y)$.

**EXAMPLE 6** The fractions $\frac{3x}{x+2}$ and $\frac{9}{x-3}$ are *not* equivalent since there is no expression that one can be multiplied by to give the other.

A fraction is **reduced to lowest terms** when 1 (or $-1$) is the only number or expression that divides both numerator and denominator. In Chapter 1 we reduced $\frac{-12}{42}$ to lowest terms as follows.

$$\frac{-12}{42} = \frac{(-1)\cdot\cancel{2}\cdot 2\cdot\cancel{3}}{\cancel{2}\cdot\cancel{3}\cdot 7} = \frac{-2}{7}$$ Factor completely and cancel common factors

When we factor completely, we factor into **primes,** numbers or expressions that can only be divided by themselves and 1.

**CAUTION:** Cancel factors only, never cancel terms. Expression which are canceled must be multiplied (never added or subtracted) by every other expression in that numerator or denominator. For example,

$$\frac{2+4}{2} = \frac{6}{2} = 3$$

but $\frac{\cancel{2}+4}{\cancel{2}}$ is certainly not 3.

**TO REDUCE AN ALGEBRAIC FRACTION TO LOWEST TERMS**

1. Factor numerator and denominator completely.
2. Divide out all common factors.
3. Multiply the remaining factors.

**EXAMPLE 7** Reduce $\frac{20x^2y}{50x^3y^4}$ to lowest terms.

Factoring, we have

$$\frac{20x^2y}{50x^3y^4} = \frac{\cancel{2}\cdot 2\cdot\cancel{5}\cdot\cancel{x}\cdot\cancel{x}\cdot\cancel{y}}{\cancel{2}\cdot\cancel{5}\cdot 5\cdot\cancel{x}\cdot\cancel{x}\cdot x\cdot\cancel{y}\cdot y\cdot y\cdot y} = \frac{2}{5x\cdot y\cdot y\cdot y} = \frac{2}{5xy^3}.$$

Compare this with the result obtained by applying the rules of exponents to $x$ and $y$.

When canceling (dividing) factors, remember that the quotient is 1 (not zero). For example,

$$\frac{x}{3x^2} = \frac{\cancel{x}}{3\cdot\cancel{x}\cdot x} = \frac{1}{3\cdot 1\cdot x} = \frac{1}{3x} \qquad \frac{x}{x} = \frac{1}{1} = 1$$

and *not* $\frac{0}{3x}$, a common error.

**EXAMPLE 8** Reduce $\frac{6x^3 + 2x}{2x}$ to lowest terms.

$$\frac{6x^3 + 2x}{2x} = \frac{\cancel{2x}(3x^2 + 1)}{\cancel{2x}} = \frac{3x^2 + 1}{1} = 3x^2 + 1$$

*Never cancel terms* as in $\frac{6x^3 + \cancel{2x}}{\cancel{2x}}$, which is not equal to $\frac{6x^3 + 1}{1}$.

**EXAMPLE 9** Reduce $\frac{x + xy}{y}$ to lowest terms.

$$\frac{x + xy}{y} = \frac{x(1 + y)}{y} \qquad \text{Cannot be reduced}$$

**EXAMPLE 10** Reduce $\frac{x^2 + x - 2}{x^2 - 1}$ to lowest terms.

$$\frac{x^2 + x - 2}{x^2 - 1} = \frac{\cancel{(x - 1)}(x + 2)}{\cancel{(x - 1)}(x + 1)} = \frac{x + 2}{x + 1}$$

(Why *not* cancel the $x$ in numerator and denominator?)

**EXAMPLE 11** Reduce $\frac{2a^3 + a^2b - ab^2}{a^2 + ab}$ to lowest terms.

$$\frac{2a^3 + a^2b - ab^2}{a^2 + ab} = \frac{a(2a^2 + ab - b^2)}{a(a + b)} = \frac{\cancel{a(a + b)}(2a - b)}{\cancel{a(a + b)}} = \frac{2a - b}{1} = 2a - b$$

## EXERCISES 6.1

*What are the meaningless replacements for the variable or variables?*

1. $\frac{2}{a - 1}$
2. $\frac{x}{3x + 1}$
3. $\frac{y^2 + 3}{y + 3}$
4. $\frac{a - 1}{2}$
5. $\frac{3x + 1}{x}$
6. $\frac{y + 3}{y^2 + 3}$
7. $\frac{a + 1}{(a - 1)(a + 2)}$
8. $\frac{x}{x^2 - 1}$
9. $\frac{5}{y^2 - 2y + 1}$

10. $\dfrac{a+b}{2a^3+7a^2+6a}$

11. $\dfrac{4x+5}{x^2-xy}$

12. $\dfrac{u-w}{u^2+w^2}$

*Are the given fractions equivalent? Explain.*

13. $\dfrac{2}{3}, \dfrac{6}{9}$

14. $\dfrac{4}{x}, \dfrac{x}{4}$

15. $\dfrac{5}{a}, \dfrac{-5}{-a}$

16. $\dfrac{-7}{x}, \dfrac{7}{-x}$

17. $\dfrac{x+1}{y}, \dfrac{2(x+1)}{2y}$

18. $\dfrac{a+b}{3}, \dfrac{5a+5b}{15}$

19. $\dfrac{2}{x^3}, \dfrac{2+y}{x^3+y}$

20. $\dfrac{a+b}{a^2-b^2}, \dfrac{1}{a-b}$

21. $\dfrac{a^2-b^2+c^2}{b-c}, \dfrac{-a^2+b^2-c^2}{c-b}$

*Reduce to lowest terms.*

22. $\dfrac{45}{105}$

23. $\dfrac{91}{39}$

24. $\dfrac{9x^7y^4}{18x^3y^5}$

25. $\dfrac{7a^2b^3z^5}{21ab^5z^2}$

26. $\dfrac{2x}{2x^2+2x}$

27. $\dfrac{y+4}{y^2+4}$

28. $\dfrac{a^2-b^2}{a^2-2ab+b^2}$

29. $\dfrac{x^3-y^3}{x-y}$

30. $\dfrac{a-z}{z-a}$
[*Hint:* $z-a=-1(a-z)$.]

31. $\dfrac{xz+xw+yz+yw}{z^2+zw}$

32. $\dfrac{a^4-b^4}{a^2+b^2}$

33. $\dfrac{15x+7x^2-2x^3}{x^2-8x+15}$

---

**ANSWERS:** **1.** 1 **2.** $-\frac{1}{3}$ **3.** $-3$ **4.** none **5.** 0 **6.** none **7.** 1, $-2$ **8.** 1, $-1$ **9.** 1 **10.** $a=0, -2, -\frac{3}{2}$; $b$ has none **11.** $x=0$; $x=y$ **12.** $u$ and $w$ both zero **13.** yes **14.** no **15.** yes **16.** yes **17.** yes **18.** yes **19.** no **20.** yes, if $a+b\neq 0$ **21.** yes **22.** $\frac{3}{7}$ **23.** $\frac{7}{3}$ **24.** $\frac{x^4}{2y}$ **25.** $\frac{az^3}{3b^2}$ **26.** $\frac{1}{x+1}$ **27.** $\frac{y+4}{y^2+4}$ **28.** $\frac{a+b}{a-b}$ **29.** $x^2+xy+y^2$ **30.** $-1$ **31.** $\frac{x+y}{z}$ **32.** $a^2-b^2$ **33.** $\frac{x(2x+3)}{3-x}$

## 6.2 Multiplication of Fractions

The product of two or more arithmetic fractions is equal to the product of all numerators divided by the product of all denominators. The same is true for algebraic fractions. In both cases, the resulting fraction should be reduced to lowest terms. Reducing a fraction to lowest terms is easier if all common factors in the numerator and denominator are canceled before multiplying, as in the following example.

EXAMPLE 1 $\frac{6}{35}\cdot\frac{7}{3}=\frac{6\cdot 7}{35\cdot 3}=\frac{2\cdot\cancel{3}\cdot\cancel{7}}{5\cdot\cancel{7}\cdot\cancel{3}}=\frac{2}{5}$

If we had multiplied first to obtain $\frac{42}{105}$, we would have had to factor (undo this multiplication) to reduce the product to $\frac{2}{5}$.

**TO MULTIPLY TWO OR MORE FRACTIONS**

1. Factor all numerators and denominators completely.
2. Place the indicated product of all numerator factors over all denominator factors.
3. Cancel (divide out) common factors to reduce to lowest terms.

EXAMPLE 2 $\frac{3a}{8}\cdot\frac{2}{9a^2}=\frac{3\cdot a}{2\cdot 2\cdot 2}\cdot\frac{2}{3\cdot 3\cdot a\cdot a}=\frac{\cancel{3}\cdot\cancel{a}\cdot\cancel{2}}{\cancel{2}\cdot 2\cdot 2\cdot\cancel{3}\cdot 3\cdot\cancel{a}\cdot a}=\frac{1}{2\cdot 2\cdot 3\cdot a}=\frac{1}{12a}$

EXAMPLE 3 $\frac{x^2-5x+6}{x^2+x-6}\cdot\frac{x+3}{x-2}=\frac{(x-2)(x-3)}{(x+3)(x-2)}\cdot\frac{(x+3)}{(x-2)}=\frac{\cancel{(x-2)}(x-3)\cancel{(x+3)}}{\cancel{(x+3)}\cancel{(x-2)}(x-2)}=\frac{x-3}{x-2}$

EXAMPLE 4 $\frac{a^2-4b^2}{49a^3-4ab^2}\cdot\frac{7a^2-2ab}{a+2b}=\frac{(a-2b)(a+2b)}{a(7a-2b)(7a+2b)}\cdot\frac{a(7a-2b)}{a+2b}$

$$=\frac{(a-2b)\cancel{(a+2b)}\cdot\cancel{a(7a-2b)}}{\cancel{a(7a-2b)}(7a+2b)\cancel{(a+2b)}}=\frac{a-2b}{7a+2b}$$

### EXERCISES 6.2

*Multiply.*

1. $\frac{3}{a^3}\cdot\frac{2a}{9}$

2. $\frac{xy^2}{3}\cdot\frac{6}{x^3y}$

3. $\frac{2ab}{5}\cdot\frac{25}{4a^3}$

**4.** $\dfrac{2y}{(y+1)^2} \cdot \dfrac{y+1}{y}$

**5.** $\dfrac{3x^2}{x+2} \cdot \dfrac{x^2-4}{3x}$

**6.** $\dfrac{ab}{a-3b} \cdot \dfrac{(a-3b)^2}{a^2b-3ab^2}$

**7.** $\dfrac{4x+8}{9x+9} \cdot \dfrac{x^2-1}{2x+4}$

**8.** $\dfrac{a^3+a^2b}{5a} \cdot \dfrac{25}{3a+3b}$

**9.** $\dfrac{y}{y^2-yz-12z^2} \cdot \dfrac{y^2+3yz}{y-4z}$

**10.** $\dfrac{x^3-y^3}{2xy} \cdot \dfrac{4x^2y^2}{x^2+xy+y^2}$

**11.** $\dfrac{a^2+ab}{a-5b} \cdot \dfrac{5b-a}{b^2-a^2}$

**12.** $\dfrac{x^2-4}{x^2-4x+4} \cdot \dfrac{x^2-9x+14}{x^3+2x^2}$

**13.** $\dfrac{9a^2-6a+1}{28+7a} \cdot \dfrac{a^2-16}{3a^2-13a+4}$

**14.** $\dfrac{2x^2-5xy-3y^2}{4x^2-y^2} \cdot \dfrac{8x^2+10xy+3y^2}{4x^2-9xy-9y^2}$

*Are the given fractions equivalent?*

**15.** $\dfrac{x+y}{x^2-y^2}, \dfrac{1}{x-y}$

**16.** $\dfrac{x^2}{4}, \dfrac{x^2+5}{9}$

**17.** $\dfrac{1}{a+b}, \dfrac{a+b}{1}$

*Reduce to lowest terms.*

**18.** $\dfrac{a^3+b^3}{a+b}$

**19.** $\dfrac{x^2+3x}{x^2+5x+6}$

**20.** $\dfrac{a^3-ab^2}{a^2-ab}$

*What are the meaningless replacements for the variable or variables?*

**21.** $\dfrac{x^2 + y^2 + z^2}{5}$ **22.** $\dfrac{x + 7}{x^2 + 5x + 6}$ **23.** $\dfrac{3}{xy}$

---

ANSWERS: **1.** $\dfrac{2}{3a^2}$ **2.** $\dfrac{2y}{x^2}$ **3.** $\dfrac{5b}{2a^2}$ **4.** $\dfrac{2}{y+1}$ **5.** $x(x-2)$ **6.** 1 **7.** $\dfrac{2(x-1)}{9}$ **8.** $\dfrac{5a}{3}$ **9.** $\dfrac{y^2}{(y-4z)^2}$
**10.** $2xy(x-y)$ **11.** $\dfrac{-a}{b-a}$ or $\dfrac{a}{a-b}$ **12.** $\dfrac{x-7}{x^2}$ **13.** $\dfrac{3a-1}{7}$ **14.** $\dfrac{2x+y}{2x-y}$ **15.** yes, if $x + y \neq 0$ **16.** no
**17.** no **18.** $a^2 - ab + b^2$ **19.** $\dfrac{x}{x+2}$ **20.** $a + b$ **21.** none **22.** $x = -2, -3$ **23.** $x = 0$ and $y = 0$

## 6.3 Division of Fractions

When one arithmetic fraction is divided by another, for example,

$$\frac{2}{3} \div \frac{2}{15} \qquad \text{or} \qquad \frac{\frac{2}{3}}{\frac{2}{15}}$$

the fraction $\frac{2}{15}$ is called the **divisor.** Recall that to calculate the **quotient** (the result when the division is performed), we multiply $\frac{2}{3}$ by the reciprocal of $\frac{2}{15}$.

$$\frac{2}{3} \div \frac{2}{15} = \frac{2}{3} \cdot \frac{15}{2} = \frac{2}{3} \cdot \frac{3 \cdot 5}{2} = \frac{\cancel{2} \cdot \cancel{3} \cdot 5}{\cancel{3} \cdot \cancel{2}} = 5$$

The **reciprocal** of a fraction is the fraction formed by interchanging the numerator and denominator. Some fractions and their reciprocals are given in the following table.

| *Fraction* | $\frac{3}{y}$ | $\frac{x+y}{(x-y)^2}$ | $5\ \left(\frac{5}{1}\right)$ | $y\ \left(\frac{y}{1}\right)$ | $\frac{1}{2}$ | $a + b$ |
|---|---|---|---|---|---|---|
| *Reciprocal* | $\frac{y}{3}$ | $\frac{(x-y)^2}{x+y}$ | $\frac{1}{5}$ | $\frac{1}{y}$ | $2\ \left(\frac{2}{1}\right)$ | $\frac{1}{a+b}$ $\left(\text{not } \frac{1}{a} + \frac{1}{b}\right)$ |

Notice that in all cases the product of a fraction and its reciprocal is 1. Division of algebraic fractions is similar to division of arithmetic fractions.

**TO DIVIDE ONE ALGEBRAIC FRACTION BY ANOTHER**

1. Find the reciprocal of the divisor and multiply.
2. Follow the same procedure as when multiplying fractions.

**EXAMPLE 1** $\frac{a}{a+2} \div \frac{a}{5} = \frac{a}{a+2} \cdot \frac{5}{a}$ The reciprocal of $\frac{a}{5}$ is $\frac{5}{a}$

$= \frac{\cancel{a} \cdot 5}{(a+2) \cdot \cancel{a}} = \frac{5}{a+2}$

**EXAMPLE 2** $\frac{x+y}{x} \div x = \frac{x+y}{x} \cdot \frac{1}{x}$ The reciprocal of $x$ is $\frac{1}{x}$

$= \frac{x+y}{x^2}$

**EXAMPLE 3** $\frac{x^2-6x+9}{x^2-4} \div \frac{x-3}{x+2} = \frac{x^2-6x+9}{x^2-4} \cdot \frac{x+2}{x-3} = \frac{(x-3)(x-3)}{(x-2)(x+2)} \cdot \frac{(x+2)}{(x-3)}$

$= \frac{\cancel{(x-3)}(x-3)\cancel{(x+2)}}{(x-2)\cancel{(x+2)}\cancel{(x-3)}} = \frac{x-3}{x-2}$

## EXERCISES 6.3

*Find the reciprocal of each.*

1. $\frac{4}{5}$
2. $\frac{1}{7}$
3. $8$
4. $\frac{7}{y}$
5. $\frac{x}{3}$
6. $y$
7. $2z$
8. $\frac{1}{y+z}$
9. $\frac{u}{v+u}$
10. $\frac{a-b}{b}$
11. $x+3$
12. $\frac{3x-5}{2y+1}$
13. $4x+3$
14. $2\frac{2}{3}$
15. $.27$

*Divide.*

16. $\frac{a}{a+3} \div \frac{a}{5}$
17. $\frac{3x}{x-1} \div \frac{6x}{x-1}$
18. $\frac{3(a+b)}{7} \div \frac{9(a+b)}{28}$
19. $\frac{x^2-x-2}{y^2} \div \frac{x^2-1}{y^2}$
20. $\frac{a^2-ab}{a+b} \div \frac{a-b}{a^2+2ab+b^2}$

**21.** $\frac{4}{a^2 - 9} \div \frac{4a^2 - 4a - 24}{a^2 - 6a + 9}$

**22.** $\frac{uv - uw + xv - xw}{v - w} \div \frac{v^2 - 2vw + w^2}{xv - xw}$

**23.** $\frac{a^3 - 27}{a^2 - 3a} \div \frac{a^2 + 3a + 9}{a}$

**24.** $\frac{x - y}{4x + 4y} \div \frac{x^2 - 2xy + y^2}{x^2 - y^2}$

**25.** $\frac{2a^2 - 2a}{2a + 4} \div \frac{a^2 - 1}{a^2 + 3a + 2}$

**26.** $\frac{z^2 - z - 6}{z^2 - 81} \div \frac{z^2 - 9z + 18}{4z + 36}$

**27.** $\frac{a^2 - 4a + 3}{a^2 - 2a - 15} \div \frac{2a^2 - 6a + 4}{a^2 + 3a - 10}$

**28.** $\frac{x^2 - y^2}{x^3 - y^3} \div \frac{x^2 + 2xy + y^2}{x^2 + xy + y^2}$

*Multiply.*

**29.** $\frac{a^2 - 6a - 16}{2a^2 - 128} \cdot \frac{a^2 + 16a + 64}{3a^2 + 30a + 48}$

**30.** $\frac{x^2 + 4x - 5}{2x^2 + 16x} \cdot \frac{x^3 - 64x}{x^2 - 9x + 8}$

**31.** Convert $\frac{5}{x}$ to an equivalent fraction with each of the following as denominator.

(a) $3x$ (b) $x^2$ (c) $x^4$

(d) $x(x + 2)$ (e) $x(1 - x)$ (f) $x(x^2 + 1)$

---

ANSWERS: 1. $\frac{5}{4}$ 2. 7 3. $\frac{1}{8}$ 4. $\frac{y}{7}$ 5. $\frac{3}{x}$ 6. $\frac{1}{y}$ 7. $\frac{1}{2z}$ 8. $y + z$ 9. $\frac{v + u}{u}$ 10. $\frac{b}{a - b}$ 11. $\frac{1}{x + 3}$ 12. $\frac{2y + 1}{3x - 5}$ 13. $\frac{1}{4x + 3}$ (not $\frac{1}{4x} + \frac{1}{3}$! Why?) 14. $\frac{3}{8}$ 15. $\frac{100}{27}$ 16. $\frac{5}{a + 3}$ 17. $\frac{1}{2}$ 18. $\frac{4}{3}$ 19. $\frac{x - 2}{x - 1}$ 20. $a(a + b)$ 21. $\frac{1}{(a + 3)(a + 2)}$ 22. $\frac{x(u + x)}{v - w}$ 23. 1 24. $\frac{1}{4}$ 25. $a$ 26. $\frac{4(z + 2)}{(z - 9)(z - 6)}$ 27. $\frac{(a - 3)(a + 5)}{2(a - 5)(a + 3)}$ 28. $\frac{1}{x + y}$ 29. $\frac{1}{6}$ 30. $\frac{x + 5}{2}$ 31. (a) $\frac{15}{3x}$ (b) $\frac{5x}{x^2}$ (c) $\frac{5x^3}{x^4}$ (d) $\frac{5(x + 2)}{x(x + 2)}$ (e) $\frac{5(1 - x)}{x(1 - x)}$ (f) $\frac{5(x^2 + 1)}{x(x^2 + 1)}$

## 6.4 Addition and Subtraction of Fractions: Same Denominators

Fractions must have the same denominator before they can be added or subtracted. When the denominators are the same, we add (or subtract) the numerators and their sum (or difference) is the numerator of the answer. The common denominator is the denominator of the answer.

**EXAMPLE 1** $\frac{4}{9} + \frac{2}{9} = \frac{4 + 2}{9} = \frac{6}{9} = \frac{\not{3} \cdot 2}{\not{3} \cdot 3} = \frac{2}{3}$

Always reduce any sum or difference of fractions to lowest terms.

**EXAMPLE 2** $\frac{a + b}{5a} + \frac{a^2 - b}{5a} = \frac{(a + b) + (a^2 - b)}{5a} = \frac{a + a^2}{5a} = \frac{\not{a}(1 + a)}{5\not{a}} = \frac{1 + a}{5}$

**EXAMPLE 3** $\frac{4}{5} - \frac{2}{5} = \frac{4 - 2}{5} = \frac{2}{5}$

**EXAMPLE 4** $\frac{2x + y}{x - 1} - \frac{x - y}{x - 1} = \frac{(2x + y) - (x - y)}{x - 1}$ Use parentheses

$= \frac{2x + y - x + y}{x - 1} = \frac{x + 2y}{x - 1}$

CAUTION: It is extremely important to use parentheses, especially in a subtraction problem. In the above example, $-(x - y)$ becomes $-x + y$ when the parentheses are cleared. If you do not enclose the numerator in parentheses, it is easy to make a sign error.

Often, two fractions have denominators which are negatives of each other. When this occurs, multiply both the numerator and denominator of one of the fractions by $-1$. The resulting fraction is equivalent to the original (why?), and it has the same denominator as the other fraction.

EXAMPLE 5 $\dfrac{x}{3}+\dfrac{x-y}{-3}=\dfrac{x}{3}+\dfrac{(-1)(x-y)}{(-1)(-3)}=\dfrac{x}{3}+\dfrac{-x+y}{3}=\dfrac{x+(-x+y)}{3}=\dfrac{y}{3}$

EXAMPLE 6

$$\begin{aligned}\frac{4y}{y-2}-\frac{y+4}{2-y}&=\frac{4y}{y-2}-\frac{(-1)(y+4)}{(-1)(2-y)}\\&=\frac{4y}{y-2}-\frac{-y-4}{-2+y}\\&=\frac{4y}{y-2}-\frac{(-y-4)}{y-2}\\&=\frac{4y-(-y-4)}{y-2}=\frac{4y+y+4}{y-2}\\&=\frac{5y+4}{y-2}\end{aligned}$$

Watch signs

## EXERCISES 6.4

*Perform the indicated operation.*

1. $\dfrac{1}{4}+\dfrac{5}{4}$
2. $\dfrac{y}{11}+\dfrac{7}{11}$
3. $\dfrac{4}{7}-\dfrac{x}{7}$
4. $\dfrac{3}{x}+\dfrac{8}{x}$
5. $\dfrac{4}{2y}-\dfrac{5}{2y}$
6. $\dfrac{3}{x+y}+\dfrac{5}{x+y}$
7. $\dfrac{4}{2a-1}-\dfrac{a}{2a-1}$
8. $\dfrac{3y+2}{y-4}+\dfrac{-2-3y}{y-4}$
9. $\dfrac{3b}{b+1}-\dfrac{2b-1}{b+1}$
10. $\dfrac{a+b}{a+b}+\dfrac{a^2+b}{a+b}$
11. $\dfrac{8y^3}{2y^3-1}-\dfrac{6y^3}{2y^3-1}$
12. $\dfrac{1}{(x+1)^2}+\dfrac{x}{(x+1)^2}$

**13.** $\frac{a}{2} + \frac{2a - 1}{-2}$

**14.** $\frac{4}{y} + \frac{3y - 1}{-y}$

**15.** $\frac{3}{4z} - \frac{1 + z}{-4z}$

**16.** $\frac{2a}{a - b} + \frac{3a}{b - a}$

**17.** $\frac{4x}{3 - x} - \frac{x + 1}{x - 3}$

**18.** $\frac{3}{a^2 - ab + b^2} - \frac{3 - a}{a^2 - ab + b^2}$

**19.** $\frac{a}{a^2 - ab} + \frac{b}{a(a - b)}$

**20.** $\frac{x}{8x - 6} - \frac{5}{2(3 - 4x)}$

**21.** $\frac{z}{1 - z^2} - \frac{1 - z}{(z - 1)(z + 1)}$

**22.** $\frac{a^3 - b^3}{a^2 - b^2} \cdot \frac{a^2 + 2ab + b^2}{a^2 + ab + b^2}$

**23.** $\frac{x^2 - 2xy + y^2}{xy - y^2} \div \frac{x^3 - xy^2}{x^2 + 2xy + y^2}$

**24.** In order to add or subtract two fractions, their denominators must be ______ ______________________.

---

ANSWERS: **1.** $\frac{3}{2}$ **2.** $\frac{y + 7}{11}$ **3.** $\frac{4 - x}{7}$ **4.** $\frac{11}{x}$ **5.** $\frac{-1}{2y}$ **6.** $\frac{8}{x + y}$ **7.** $\frac{4 - a}{2a - 1}$ **8.** 0 **9.** 1 **10.** $\frac{a^2 + a + 2b}{a + b}$ **11.** $\frac{2y^3}{2y^3 - 1}$ **12.** $\frac{1}{x + 1}$ **13.** $\frac{1 - a}{2}$ **14.** $\frac{5 - 3y}{y}$ **15.** $\frac{4 + z}{4z}$ **16.** $\frac{-a}{a - b}$ or $\frac{a}{b - a}$ **17.** $\frac{5x + 1}{3 - x}$ **18.** $\frac{a}{a^2 - ab + b^2}$ **19.** $\frac{a + b}{a^2 - ab}$ **20.** $\frac{x + 5}{8x - 6}$ **21.** $\frac{1}{1 - z^2}$ **22.** $a + b$ **23.** $\frac{x + y}{xy}$ **24.** the same or equal

## 6.5 Finding Least Common Denominators

In order to add or subtract fractions that have different denominators, we must convert them to equivalent fractions that have a common denominator. Recall that to add $\frac{4}{15}$ and $\frac{5}{6}$, we first must find a **common denominator.** One such denominator is $15 \cdot 6 = 90$. Since

$$\frac{4}{15} = \frac{4 \cdot 6}{15 \cdot 6} = \frac{24}{90} \quad \text{and} \quad \frac{5}{6} = \frac{5 \cdot 15}{6 \cdot 15} = \frac{75}{90},$$

we have

$$\frac{4}{15} + \frac{5}{6} = \frac{4 \cdot 6}{15 \cdot 6} + \frac{5 \cdot 15}{6 \cdot 15} = \frac{24}{90} + \frac{75}{90} = \frac{99}{90} = \frac{\cancel{9} \cdot 11}{\cancel{9} \cdot 10} = \frac{11}{10}.$$

It is often wise to find a common denominator smaller than the product of all denominators. If we use the **least common denominator,** we shorten the computation and decrease the effort needed to reduce the final sum or difference to lowest terms. For example, we could have used 30 as the common denominator above.

$$\frac{4}{15} + \frac{5}{6} = \frac{4 \cdot 2}{15 \cdot 2} + \frac{5 \cdot 5}{6 \cdot 5} = \frac{8}{30} + \frac{25}{30} = \frac{33}{30} = \frac{\cancel{3} \cdot 11}{\cancel{3} \cdot 10} = \frac{11}{10}$$

Exactly the same procedures should be followed when working with algebraic fractions. The method used for finding the least common denominator (LCD) of two or more fractions is important.

**TO FIND THE LCD OF TWO OR MORE FRACTIONS**

1. Factor all denominators and reduce fractions to lowest terms.
2. When there are no common factors in any two denominators, the LCD is the product of *all* denominators.
3. When there are common factors in two or more denominators, each factor appears in the LCD as many times as it appears in the denominator where it is found the greatest number of times.

**EXAMPLE 1** Find the LCD of $\frac{7}{150}$ and $\frac{2}{315}$.

Factor the denominators.

$$150 = 2 \cdot 3 \cdot 5 \cdot 5 \quad \text{and} \quad 315 = 3 \cdot 3 \cdot 5 \cdot 7$$

The LCD must consist of one 2, two 3s, two 5s and one 7. Thus the LCD is

$$2 \cdot 3 \cdot 3 \cdot 5 \cdot 5 \cdot 7 = 3150.$$

**EXAMPLE 2** Find the LCD of $\frac{3}{x}$ and $\frac{5}{x+y}$.

Since $x$ and $x + y$ are already completely factored, and since there are no common factors in the two denominators, the LCD is

$$x(x+y).$$

**EXAMPLE 3** Find the LCD of $\frac{3a+1}{a^2-b^2}$ and $\frac{7b}{2a-2b}$.

Factor the denominators.

$$a^2 - b^2 = (a-b)\ (a+b) \qquad \text{and} \qquad 2a - 2b = 2\ (a-b)$$

The LCD must consist of one $(a - b)$, one $(a + b)$ and one 2. Thus the LCD is

$$2(a-b)(a+b).$$

**EXAMPLE 4** Find the LCD of $\frac{5}{9x^4-36x^3+36x^2}$ and $\frac{2x+7}{15x^5-45x^4+30x^3}$.

$$9x^4 - 36x^3 + 36x^2 = 9x^2(x^2 - 4x + 4) = 3 \cdot 3\ xx\ (x-2)(x-2)$$
$$15x^5 - 45x^4 + 30x^3 = 15x^3(x^2 - 3x + 2) = 3 \cdot 5\ xxx\ (x-2)\ (x-1)$$

The LCD must consist of two 3s, one 5, three $x$s, two $(x - 2)$s and one $(x - 1)$. Therefore, the LCD is

$$3 \cdot 3 \cdot 5 \cdot x^3(x-2)^2(x-1) = 45x^3(x-2)^2(x-1).$$

**EXAMPLE 5** Find the LCD of $\frac{a+1}{a^2+2a+1}$ and $\frac{5-a}{a^2-4a-5}$.

Factor the denominators.

$$a^2 + 2a + 1 = (a+1)(a+1) \qquad \text{and} \qquad a^2 - 4a - 5 = (a+1)(a-5)$$

We are tempted to conclude that the LCD is $(a + 1)^2(a - 5)$. However, we can reduce the given fractions to lowest terms as follows.

$$\frac{a+1}{a^2+2a+1} = \frac{\cancel{(a+1)}}{(a+1)\cancel{(a+1)}} = \frac{1}{a+1}$$

$$\frac{5-a}{a^2-4a-5} = \frac{5-a}{(a-5)(a+1)} = \frac{(-1)\cancel{(a-5)}}{\cancel{(a-5)}(a+1)} = \frac{-1}{a+1}$$

$5 - a$ and $a - 5$ are negatives

Thus, to find the LCD of the given fractions, all we need to do is find the LCD of $\frac{1}{a+1}$ and $\frac{-1}{a+1}$, which is simply $a + 1$.

Always make sure that the given fractions are reduced to lowest terms before attempting to determine the LCD.

## EXERCISES 6.5

*Find the LCD of the given fractions.*

**1.** $\frac{2}{15}$ and $\frac{5}{9}$

**2.** $\frac{3}{5}$ and $\frac{2}{13}$

**3.** $\frac{1}{39}$ and $\frac{5}{33}$

**4.** $\frac{2}{y}$ and $\frac{3}{5y}$

**5.** $\frac{x+1}{x}$ and $\frac{2}{x^2}$

**6.** $\frac{3}{z}$ and $\frac{2}{z+1}$

**7.** $\frac{a+b}{a}$ and $\frac{3}{4a^5}$

**8.** $\frac{6}{5b}$ and $\frac{3}{5b+5}$

**9.** $\frac{2}{x-y}$ and $\frac{3}{x+y}$

**10.** $\frac{a}{3a+3b}$ and $\frac{b}{a^2-b^2}$

**11.** $\frac{2}{x^2-y^2}$ and $\frac{3}{x^2-2xy+y^2}$

**12.** $\frac{3}{b^2+5b+6}$ and $\frac{b}{b^2-5b-6}$

**13.** $\dfrac{2}{a-b}$ and $\dfrac{a}{b-a}$

**14.** $\dfrac{z+2}{z^2+4z+4}$ and $\dfrac{z-1}{z^2+z-2}$

**15.** $\dfrac{a+3}{a^3-27}$ and $\dfrac{a}{a^2+3a+9}$

**16.** $\dfrac{x+3}{5x^4-15x^3-50x^2}$ and $\dfrac{x-7}{10x^3-100x^2+250x}$

*Perform the indicated operations.*

**17.** $\dfrac{x}{x-5}+\dfrac{2x+1}{5-x}$

**18.** $\dfrac{1}{a^2-4a+4}-\dfrac{3-a}{a^2-4a+4}$

**19.** $\dfrac{2x^2+9x-5}{2x^2-3x+1}\div\dfrac{x+5}{x-1}$

**20.** $\dfrac{3a+6b}{a^2+ab-2b^2}\cdot\dfrac{a^3-b^3}{a^2+ab+b^2}$

---

ANSWERS: **1.** 45 **2.** 65 **3.** 429 **4.** $5y$ **5.** $x^2$ **6.** $z(z+1)$ **7.** $4a^5$ **8.** $5b(b+1)$ **9.** $(x-y)(x+y)$ **10.** $3(a-b)(a+b)$ **11.** $(x+y)(x-y)^2$ **12.** $(b+2)(b+3)(b-6)(b+1)$ **13.** $a-b$ **14.** $(z+2)$ **15.** $(a-3)(a^2+3a+9)$ **16.** $10x^2(x-5)^2(x+2)$ **17.** $\dfrac{-x-1}{x-5}$ or $-\dfrac{x+1}{x-5}$ or $\dfrac{x+1}{5-x}$ **18.** $\dfrac{1}{a-2}$ **19.** 1 **20.** 3

## 6.6 Addition and Subtraction of Fractions: Different Denominators

To add or subtract fractions with different denominators, we first convert them to equivalent fractions having the LCD as denominator.

**EXAMPLE 1** $\dfrac{2}{21}+\dfrac{3}{35}$

$=\dfrac{2}{3\cdot 7}+\dfrac{3}{5\cdot 7}$ — $21=3\cdot 7$ and $35=5\cdot 7$; the LCD is $3\cdot 5\cdot 7 = 105$

$=\dfrac{2\cdot(5)}{3\cdot 7\cdot(5)}+\dfrac{3\cdot(3)}{5\cdot 7\cdot(3)}$ — Multiply numerator and denominator of $\dfrac{2}{21}$ by 5 and numerator and denominator of $\dfrac{3}{35}$ by 3.

$=\dfrac{2\cdot 5+3\cdot 3}{3\cdot 5\cdot 7}$ — Denominators are now the same; add numerators and place the sum over the LCD

$=\dfrac{10+9}{3\cdot 5\cdot 7}$ — Leave denominator factored and simplify the numerator.

$=\dfrac{19}{3\cdot 5\cdot 7}=\dfrac{19}{105}$ — Since 19 has no factor of 3, 5, or 7, the resulting fraction is in lowest terms

The above example illustrates the technique of adding (subtracting is similar) arithmetic fractions. Exactly the same procedure applies to algebraic fractions.

**TO ADD OR SUBTRACT TWO (OR MORE) ALGEBRAIC FRACTIONS**

1. Express denominators in completely factored form and reduce all fractions.
2. Determine the LCD of the fractions.
3. Multiply the numerator and denominator of each fraction by all factors present in the LCD but missing in the denominator of the particular fraction, so the fractions have the same denominator.
4. Indicate the sum or difference of all numerators, using parentheses (if applicable).
5. Simplify the resulting fraction.

**EXAMPLE 2** $\dfrac{5}{6a}+\dfrac{a+1}{15a^2}=\dfrac{5}{2\cdot 3a}+\dfrac{a+1}{3\cdot 5\cdot a\cdot a}$ — Factor denominators

$=\dfrac{5\cdot(5\cdot a)}{2\cdot 3\cdot a\cdot(5\cdot a)}+\dfrac{(a+1)(2)}{3\cdot 5\cdot a\cdot a\cdot(2)}$ — Supply missing factors: LCD is $2\cdot 3\cdot 5\cdot a\cdot a$.

$=\dfrac{5\cdot 5\cdot a+(a+1)\cdot 2}{2\cdot 3\cdot a\cdot 5\cdot a}$ — Add numerators over the LCD

$=\dfrac{25a+2a+2}{2\cdot 3\cdot 5\cdot a\cdot a}$

$=\dfrac{27a+2}{30a^2}$ — No common factors; the sum is in lowest terms

**EXAMPLE 3** $\frac{3}{a+b} - \frac{2}{a-b}$ — Denominators already factored: LCD $= (a+b)(a-b)$

$= \frac{3(a-b)}{(a+b)(a-b)} - \frac{2(a+b)}{(a+b)(a-b)}$ — Supply missing factors

$= \frac{3(a-b) - 2(a+b)}{(a+b)(a-b)}$ — Subtract numerators over LCD

$= \frac{3a - 3b - 2a - 2b}{(a+b)(a-b)}$ — Watch signs

$= \frac{a - 5b}{(a+b)(a-b)}$ — Since no common factors exist, this is the difference in lowest terms

**EXAMPLE 4** $\frac{y^2 - 1}{y^3 - 1} - \frac{y}{2y^2 + 2y + 2}$

$= \frac{\cancel{(y-1)}(y+1)}{\cancel{(y-1)}(y^2+y+1)} - \frac{y}{2(y^2+y+1)}$ — Factor and reduce fractions

$= \frac{y+1}{y^2+y+1} - \frac{y}{2(y^2+y+1)}$ — LCD $= 2(y^2+y+1)$

$= \frac{2(y+1)}{2(y^2+y+1)} - \frac{y}{2(y^2+y+1)}$ — Supply missing factors

$= \frac{2(y+1) - y}{2(y^2+y+1)}$ — Subtract numerators over LCD

$= \frac{2y + 2 - y}{2(y^2+y+1)} = \frac{y+2}{2(y^2+y+1)}$ — Already reduced to lowest terms

**EXAMPLE 5** $\frac{2xy}{x^2 - y^2} - \frac{y}{x+y} = \frac{2xy}{(x-y)(x+y)} - \frac{y}{x+y}$ — LCD $= (x-y)(x+y)$

$= \frac{2xy}{(x-y)(x+y)} - \frac{y(x-y)}{(x-y)(x+y)}$

$= \frac{2xy - y(x-y)}{(x-y)(x+y)}$

$= \frac{2xy - yx + y^2}{(x-y)(x+y)}$

$= \frac{xy + y^2}{(x-y)(x+y)}$ — This one can be simplified

$= \frac{y\cancel{(x+y)}}{(x-y)\cancel{(x+y)}} = \frac{y}{x-y}$

**EXAMPLE 6** $a + 1 + \frac{a}{a-1} = \frac{a+1}{1} + \frac{a}{a-1}$ — $a + 1 = \frac{a+1}{1}$

$= \frac{(a+1)(a-1)}{1 \cdot (a-1)} + \frac{a}{a-1}$ — LCD $= 1(a-1) = a - 1$

$= \frac{(a+1)(a-1) + a}{a-1}$

$= \frac{a^2 - 1 + a}{a-1} = \frac{a^2 + a - 1}{a-1}$

**EXAMPLE 7**

$$\frac{2}{3x-21}+\frac{x}{49-x^2}=\frac{2}{3(x-7)}+\frac{x}{(7-x)(7+x)}$$

LCD = $3(x-7)(x+7)$

$$=\frac{2}{3(x-7)}+\frac{(-1)(x)}{(-1)(7-x)(7+x)}$$

$x-7$ and $7-x$ are negatives

$$=\frac{2}{3(x-7)}+\frac{-x}{(x-7)(7+x)}$$

$$=\frac{2(x+7)}{3(x-7)(x+7)}+\frac{3(-x)}{3(x-7)(7+x)}$$

Obtain equivalent fractions with the same LCD

$$=\frac{2(x+7)-3x}{3(x-7)(x+7)}$$

$$=\frac{2x+14-3x}{3(x-7)(x+7)}$$

$$=\frac{-x+14}{3(x-7)(x+7)}$$

## EXERCISES 6.6

*Perform the indicated operation.*

**1.** $\frac{3}{5x}+\frac{5-6x}{10x^2}$

**2.** $\frac{3}{a+4}-\frac{2}{a-4}$

**3.** $\frac{7}{a^2-25}+\frac{2}{a-5}$

**4.** $\frac{y^2+1}{y^2-1}-\frac{y-1}{y+1}$

**5.** $\frac{4a}{a^2-36}-\frac{4}{a+6}$

**6.** $\frac{2x}{9-x^2}-\frac{1}{3-x}$

7. $\frac{5}{3y-3}+\frac{3}{1-y}$

8. $\frac{x-2}{x-3}+\frac{x-x^2}{x^2-9}$

9. $\frac{xy}{3x-3y}+\frac{xy}{5y-5x}$

10. $\frac{3}{2a^2-2a}+\frac{5}{2-2a}$

11. $\frac{2x+1}{x+2}+x$

12. $a-b-\frac{-ab^2}{a^2+ab}$

13. $\frac{y}{y^2-y-20}+\frac{2}{y+4}$

14. $\frac{2a-1}{a^2-4a+4}+\frac{2a+3}{4-a^2}$

15. $\frac{x+2}{x^2+5x+6}-\frac{x+1}{x^2+4x+3}$

16. $\frac{4-3a}{3a^2+6a+12}+\frac{a^2-4}{a^3-8}$

17. $\dfrac{1 - 8a^3}{2a - 4a^2} \cdot \dfrac{2a^3}{4a^2 + 2a + 1}$

18. $\dfrac{x^2y - x^3}{y} \div \dfrac{x^4 - x^2y^2}{y^2 + xy}$

---

ANSWERS: 1. $\dfrac{1}{2x^2}$ 2. $\dfrac{a - 20}{(a + 4)(a - 4)}$ 3. $\dfrac{2a + 17}{(a - 5)(a + 5)}$ 4. $\dfrac{2y}{(y - 1)(y + 1)}$ 5. $\dfrac{24}{(a - 6)(a + 6)}$ 6. $\dfrac{-1}{x + 3}$ 7. $\dfrac{-4}{3(y - 1)}$ 8. $\dfrac{2}{x + 3}$ 9. $\dfrac{2xy}{15(x - y)}$ 10. $\dfrac{3 - 5a}{2a(a - 1)}$ 11. $\dfrac{x^2 + 4x + 1}{x + 2}$ 12. $\dfrac{a^2}{a + b}$ 13. $\dfrac{3y - 10}{(y - 5)(y + 4)}$ 14. $\dfrac{4(a + 1)}{(a - 2)^2 (a + 2)}$ 15. 0 16. $\dfrac{10}{3(a^2 + 2a + 4)}$ 17. $a^2$ 18. $-1$

## 6.7 Simplifying Complex Fractions

A fractional expression which contains at least one other fraction within it is called a **complex fraction.** The following are examples of complex fractions.

$$\frac{\frac{1}{5}}{\frac{2}{3}}, \quad \frac{a}{\frac{3}{4}}, \quad \frac{\frac{x}{y}}{7}, \quad \frac{2 + \frac{1}{y}}{3}, \quad \frac{2 + \frac{a}{b}}{2 - \frac{a}{b}}$$

A complex fraction is **simplified** when all of its component fractions have been eliminated and a simple fraction obtained. There are two basic methods for simplifying a complex fraction. The first involves changing each of the numerators and denominators to a single fraction.

**TO SIMPLIFY A COMPLEX FRACTION**

1. Change the numerator and denominator to single fractions.
2. Divide the two fractions.
3. Reduce to lowest terms.

**EXAMPLE 1** Simplify $\dfrac{\frac{x}{y} + 2}{2 - \frac{x}{y}}$.

$$\frac{x}{y} + 2 = \frac{x}{y} + \frac{2}{1} = \frac{x}{y} + \frac{2 \cdot y}{1 \cdot y} = \frac{x + 2y}{y}$$ **Add the numerator fractions**

$$2 - \frac{x}{y} = \frac{2}{1} - \frac{x}{y} = \frac{2 \cdot y}{1 \cdot y} - \frac{x}{y} = \frac{2y - x}{y}$$ **Subtract the denominator fractions**

Thus

$$\frac{\frac{x}{y}+2}{2-\frac{x}{y}} = \frac{\frac{x+2y}{y}}{\frac{2y-x}{y}} = \frac{x+2y}{y} \div \frac{2y-x}{y} = \frac{x+2y}{y} \cdot \frac{y}{2y-x}$$

$$= \frac{(x+2y)\cdot \cancel{y}}{\cancel{y}\cdot(2y-x)} = \frac{x+2y}{2y-x}.$$

The second method involves clearing all fractions within the expression by multiplying numerator and denominator by the LCD of all the internal fractions.

**TO SIMPLIFY A COMPLEX FRACTION**

1. Find the LCD of all fractions within the complex fraction.
2. Multiply numerator and denominator of the complex fraction by the LCD to obtain an equivalent fraction.
3. Reduce to lowest terms.

**EXAMPLE 2**

$$\frac{\frac{x}{y}+2}{2-\frac{x}{y}} = \frac{\left(\frac{x}{y}+2\right)\cdot y}{\left(2-\frac{x}{y}\right)\cdot y}$$ The LCD of $\frac{x}{y}$ and $2 = \frac{2}{1}$ is $y$

$$= \frac{\frac{x}{\cancel{y}}\cdot \cancel{y} + 2\cdot y}{2\cdot y - \frac{x}{\cancel{y}}\cdot \cancel{y}}$$ Distributive laws

$$= \frac{x+2y}{2y-x}$$ The result is already reduced to lowest terms

**EXAMPLE 3** Simplify $\dfrac{a-\frac{4}{a}}{\frac{16}{a^3}-a}$. The LCD of $\frac{a}{1}$, $\frac{4}{a}$, and $\frac{16}{a^3}$ is $a^3$

$$\frac{\left(a-\frac{4}{a}\right)\cdot a^3}{\left(\frac{16}{a^3}-a\right)\cdot a^3} = \frac{a\cdot a^3 - \frac{4}{a}\cdot a^3}{\frac{16}{a^3}\cdot a^3 - a\cdot a^3}$$ Multiply numerator and denominator by LCD $= a^3$ and distribute

$$= \frac{a^4-4a^2}{16-a^4}$$ Simplify

$$= \frac{a^2(a^2-4)}{(4-a^2)(4+a^2)}$$ Factor

$$= \frac{-a^2}{4+a^2}$$ $\frac{a^2-4}{4-a^2} = -1$ (Why?)

**EXAMPLE 4** Simplify $\dfrac{1 + \dfrac{y}{x-y}}{\dfrac{y}{x+y} - 1}$.

$$\frac{\left(1 + \dfrac{y}{x-y}\right)(x-y)(x+y)}{\left(\dfrac{y}{x+y} - 1\right)(x-y)(x+y)}$$

$$= \frac{1 \cdot (x-y)(x+y) + \dfrac{y}{\cancel{(x-y)}}\cancel{(x-y)}(x+y)}{\dfrac{y}{\cancel{(x+y)}} \cdot (x-y)\cancel{(x+y)} - 1 \cdot (x-y)(x+y)}$$ Multiply by LCD, $(x-y)(x+y)$, and distribute

$$= \frac{(x^2 - y^2) + y(x+y)}{y(x-y) - (x^2 - y^2)}$$ Use parentheses

$$= \frac{x^2 - y^2 + yx + y^2}{yx - y^2 - x^2 + y^2}$$ Watch signs

$$= \frac{x^2 + yx}{yx - x^2} = \frac{\cancel{x}(x+y)}{\cancel{x}(y-x)}$$

$$= \frac{x+y}{y-x}$$

When we evaluate an algebraic expression, the result is often a complex fraction which must be simplified.

**EXAMPLE 5** Evaluate $\dfrac{a-2}{a+3}$ when $a = \dfrac{x-y}{y}$.

Substitute $\dfrac{x-y}{y}$ for $a$.

$$\frac{\dfrac{x-y}{y} - 2}{\dfrac{x-y}{y} + 3} = \frac{\left(\dfrac{x-y}{y} - 2\right) \cdot y}{\left(\dfrac{x-y}{y} + 3\right) \cdot y}$$ Multiply by the LCD, $y$

$$= \frac{\dfrac{x-y}{\cancel{y}} \cdot \cancel{y} - 2 \cdot y}{\dfrac{x-y}{\cancel{y}} \cdot \cancel{y} + 3 \cdot y}$$ Distribute

$$= \frac{x - y - 2y}{x - y + 3y} = \frac{x - 3y}{x + 2y}$$

## EXERCISES 6.7

*Simplify.*

1. $\dfrac{\dfrac{1}{a} + 1}{\dfrac{1}{a} - 1}$

2. $\dfrac{1 - \dfrac{2}{3y}}{y - \dfrac{4}{9y}}$

3. $\dfrac{\dfrac{2}{a} - \dfrac{2}{b}}{\dfrac{1}{a^2} - \dfrac{1}{b^2}}$

4. $\dfrac{1 - \dfrac{x}{y}}{y - \dfrac{x^2}{y}}$

5. $\dfrac{\dfrac{1}{a} - \dfrac{1}{b}}{\dfrac{a}{b} - \dfrac{b}{a}}$

6. $\dfrac{x - \dfrac{4}{x}}{1 + \dfrac{2}{x}}$

7. $\dfrac{1 - \dfrac{a^2}{b^2}}{1 - \dfrac{a}{b}}$

8. $\dfrac{\dfrac{2}{x} - \dfrac{5}{y}}{\dfrac{4}{x^2} - \dfrac{25}{y^2}}$

9. $\dfrac{\dfrac{2}{a-2} + 1}{\dfrac{2}{a+2} - 1}$

10. $\dfrac{1 - \dfrac{2}{x} - \dfrac{3}{x^2}}{1 + \dfrac{1}{x}}$

11. $\dfrac{a - 3 + \dfrac{2}{a}}{a - 4 + \dfrac{3}{a}}$

12. $\dfrac{\dfrac{1}{x+y} - \dfrac{1}{x-y}}{\dfrac{-2}{x-y}}$

13. Evaluate $\dfrac{a+1}{a-1}$ for $a = \dfrac{x+y}{y}$.

14. Evaluate $\dfrac{1-x}{x}$ for $x = \dfrac{a}{b}$.

*Perform the indicated operation.*

15. $\dfrac{2a-1}{a^2-4a+4} + \dfrac{2a+1}{4-a^2}$

16. $\dfrac{x^3-x^2}{x^2-1} \div \dfrac{3-3x}{x^2-2x+1}$

---

ANSWERS: 1. $\dfrac{1+a}{1-a}$ 2. $\dfrac{3}{3y+2}$ 3. $\dfrac{2ab}{b+a}$ 4. $\dfrac{1}{y+x}$ 5. $\dfrac{-1}{a+b}$ 6. $x-2$ 7. $\dfrac{b+a}{b}$ 8. $\dfrac{xy}{2y+5x}$ 9. $-\dfrac{a+2}{a-2}$ or $\dfrac{a+2}{2-a}$ 10. $\dfrac{x-3}{x}$ 11. $\dfrac{a-2}{a-3}$ 12. $\dfrac{y}{x+y}$ 13. $\dfrac{x+2y}{x}$ 14. $\dfrac{b-a}{a}$ 15. $\dfrac{6a}{(a-2)^2(a+2)}$ 16. $\dfrac{x^2(1-x)}{3(x+1)}$

## 6.8 Division of Polynomials

We first consider the problem of dividing a polynomial by a monomial (a polynomial with only one term).

> To divide a polynomial by a monomial, divide each term of the polynomial by the monomial.

EXAMPLE 1

$$\frac{5x^3 + 10x^2 - 20x}{5x} = \frac{5x^3}{5x} + \frac{10x^2}{5x} - \frac{20x}{5x} = x^2 + 2x - 4$$

Notice that the first step is the reverse of adding fractions with the same denominators.

EXAMPLE 2

$$\frac{3a^3b^4 + 6a^2b^3 - 9ab^5 + 3ab}{3a^2b^2} = \frac{3a^3b^4}{3a^2b^2} + \frac{6a^2b^3}{3a^2b^2} - \frac{9ab^5}{3a^2b^2} + \frac{3ab}{3a^2b^2}$$

$$= ab^2 + 2b - \frac{3b^3}{a} + \frac{1}{ab}$$

Remember the rules of exponents

EXAMPLE 3

$$(7y^5 - 14y^3 + 21y) \div 7y^2 = \frac{7y^5}{7y^2} - \frac{14y^3}{7y^2} + \frac{21y}{7y^2}$$

$$= y^3 - 2y + \frac{3}{y}$$

When dividing, with practice we should be able to proceed directly to the final answer.

Dividing a polynomial by a binomial (a polynomial with two terms) is somewhat more difficult than dividing by a monomial. The rule is illustrated in the examples that follow.

**TO DIVIDE A POLYNOMIAL BY A BINOMIAL**

1. Arrange the terms of both in descending order.
2. Divide the first term of the binomial (the divisor) into the first term of the polynomial (the dividend) to obtain the first term of the quotient.
3. Multiply the binomial by the first (new) term of the quotient and subtract the result from the dividend. Bring down the next term to obtain a new polynomial which becomes the new dividend.
4. Divide the new dividend polynomial by the binomial and continue this process until the variable in the first term of the remainder dividend is raised to a lower power than the variable in the first term of the divisor.

**EXAMPLE 4** Divide $(x^2 - 15 + 2x)$ by $(x - 3)$.

1. Arrange terms in descending order.

$$x - 3\overline{)x^2 + 2x - 15}$$

2. Divide the first term of the binomial into the first term of the polynomial.

equals

$$\begin{array}{r} x \\ x - 3\overline{)x^2 + 2x - 15} \end{array} \qquad x^2 \div x = x$$

divided by

3. Multiply the binomial by the first term of the quotient and subtract the results from the dividend. Bring down the next terms to obtain a new dividend.

$$\begin{array}{r} x \\ x - 3\overline{)x^2 + 2x - 15} \\ \underline{x^2 - 3x} \\ 5x - 15 \end{array}$$

$x(x - 3) = x^2 - 3x$

Subtract $x^2 - 3x$ from $x^2 + 2x$ and bring down $-15$

4. Divide the new dividend polynomial by the binomial using Steps 2 and 3.

$$\begin{array}{r} x + 5 \\ x - 3\overline{)x^2 + 2x - 15} \\ \underline{x^2 - 3x} \\ 5x - 15 \\ \underline{5x - 15} \\ 0 \end{array}$$

Divide $5x - 15$ by $x - 3$

Multiply $x - 3$ by 5

No variable in the new dividend; the process terminates

**EXAMPLE 5** Divide $(21y + 18y^2 + 40 + 20y^3)$ by $(7 + 5y)$.

$$\begin{array}{r} 4y^2 - 2y + 7 \\ 5y + 7\overline{)20y^3 + 18y^2 + 21y + 40} \\ \underline{20y^3 + 28y^2} \\ -10y^2 + 21y \\ \underline{-10y^2 - 14y} \\ 35y + 40 \\ \underline{35y + 49} \\ -9 \end{array}$$

The quotient of $20y^3$ and $5y$

The quotient of $-10y^2$ and $5y$

The quotient of $35y$ and $5y$

Arrange in descending order

The quotient $4y^2$ times the divisor $5y + 7$

Subtract $20y^3 + 28y^2$ from $20y^3 + 18y^2$ and bring down $21y$

The quotient $-2y$ times $5y + 7$

Subtract $-10y^2 - 14y$ from $-10y^2 + 21y$ and bring down 40

The quotient 7 times $5y + 7$

Subtract; no variable in new dividend

The answer is $4y^2 - 2y + 7$ with a remainder of $-9$, or $4y^2 - 2y + 7 - \dfrac{9}{5y + 7}$.

There are two forms for the answer, as for a numerical division problem:

$11 \div 4 = 2$ with a remainder of 3, or $11 \div 4 = 2 + \dfrac{3}{4}$.

**EXAMPLE 6** Divide $(a^3 + 8)$ by $(a + 2)$.

When terms are missing, either leave space for them or write them with zero coefficients.

$$
\begin{array}{rl}
 & a^2 - 2a + 4 \\
a + 2 & \overline{)\, a^3 \qquad\qquad\quad + 8} \\
 & \underline{a^3 + 2a^2} \\
 & -2a^2 \\
 & \underline{-2a^2 - 4a} \\
 & \qquad 4a + 8 \\
 & \qquad \underline{4a + 8} \\
 & \qquad\qquad 0
\end{array}
\qquad\qquad
\begin{array}{rl}
 & a^2 - 2a + 4 \\
a + 2 & \overline{)\, a^3 + 0a^2 + 0 \cdot a + 8} \\
 & \underline{a^3 + 2a^2} \\
 & -2a^2 + 0 \cdot a \\
 & \underline{-2a^2 - \quad 4a} \\
 & \qquad 4a + 8 \\
 & \qquad \underline{4a + 8} \\
 & \qquad\qquad 0
\end{array}
$$

To divide polynomials with two variables, we arrange the terms of each in order of descending powers of the same variable and divide as before.

**EXAMPLE 7** $(2x^2 + 5xy - 3y^2) \div (2x - y)$

Here the terms are arranged in descending order relative to the variable $x$.

$$
\begin{array}{rl}
 & x + 3y \\
2x - y & \overline{)\, 2x^2 + 5xy - 3y^2} \\
 & \underline{2x^2 - \quad xy} \\
 & \qquad 6xy - 3y^2 \\
 & \qquad \underline{6xy - 3y^2} \\
 & \qquad\qquad 0
\end{array}
$$

We could also arrange in descending order relative to $y$ and then divide.

$$
\begin{array}{rl}
 & 3y + x \\
-y + 2x & \overline{)\, -3y^2 + 5xy + 2x^2} \\
 & \underline{-3y^2 + 6xy} \\
 & \qquad - xy + 2x^2 \\
 & \qquad \underline{- xy + 2x^2} \\
 & \qquad\qquad 0
\end{array}
$$

In either case, the quotient is $x + 3y$.

## EXERCISES 6.8

*Divide.*

1. $\dfrac{11a^3 - 22a^2 + 11a}{11a}$

2. $\dfrac{5x^5 - 30x^4 + 25x^3 - 5}{5x^2}$

3. $\dfrac{28a^5 + 44a^4 - 8a^3}{-4a}$

4. $\dfrac{66x^6 - 33x^3 + 11x}{11x^2}$

**5.** $\dfrac{-72y^3 + 24y^2 + 88y}{-8y}$

**6.** $(-16x^3 + 28x^2 - 8x) \div (-4x)$

**7.** $\dfrac{15a^4b^3 - 3a^2b^2 + 9a^3b}{-3a^2b^2}$

**8.** $\dfrac{35x^5y^3 - 49x^4y^5 + 28x^2y}{-7x^3y^2}$

**9.** $a - 2\overline{)3a^2 - 8a + 4}$

**10.** $(x^3 - 27) \div (x - 3)$

**11.** $(4a - 12 + a^2) \div (a + 6)$
[*Hint:* Arrange in descending order.]

**12.** $x - 1\overline{)x^3 \qquad + x - 3}$

**13.** $2x - 3\overline{)4x^4 \qquad - 7x^2 + x + 3}$

**14.** $2x - 5\overline{)12x^3 + 4x^2 - 41x + 140}$

**15.** $x + 2\overline{)x^3 + 2x^2 - x - 2}$

**16.** $3a + 2\overline{)6a^3 + 13a^2 + 24a + 12}$

**17.** $(5xy + x^2 + 6y^2) \div (x + 2y)$

**18.** $3a + b\overline{)3a^3 + a^2b - 6a - 2b}$

19. $a + b\overline{)a^3 \qquad + b^3}$

20. Evaluate $\dfrac{1-a}{a-1}$ for $a = \dfrac{x+y}{x-y}$.

*Simplify the complex fractions.*

21. $\dfrac{y - \dfrac{9}{y}}{1 + \dfrac{3}{y}}$

22. $\dfrac{\dfrac{3}{a} + \dfrac{7}{b}}{\dfrac{9b}{a} - \dfrac{49a}{b}}$

---

ANSWERS: 1. $a^2 - 2a + 1$ 2. $x^3 - 6x^2 + 5x - \dfrac{1}{x^2}$ 3. $-7a^4 - 11a^3 + 2a^2$ 4. $6x^4 - 3x + \dfrac{1}{x}$ 5. $9y^2 - 3y - 11$ 6. $4x^2 - 7x + 2$ 7. $-5a^2b + 1 - \dfrac{3a}{b}$ 8. $-5x^2y + 7xy^3 - \dfrac{4}{xy}$ 9. $3a - 2$ 10. $x^2 + 3x + 9$ 11. $a - 2$ 12. $x^2 + x + 2 - \dfrac{1}{x-1}$ 13. $2x^3 + 3x^2 + x + 2 + \dfrac{9}{2x-3}$ 14. $6x^2 + 17x + 22 + \dfrac{250}{2x-5}$ 15. $x^2 - 1$ 16. $2a^2 + 3a + 6$ 17. $x + 3y$ 18. $a^2 - 2$ 19. $a^2 - ab + b^2$ 20. $-1$ 21. $y - 3$ 22. $\dfrac{1}{3b - 7a}$

## 6.9 Solving Fractional Equations

An equation that contains one or more fractional expressions is called a **fractional equation.** We eliminate the fractions by multiplying both sides of the equation by the LCD of all fractions (this process is called **clearing the fractions**). The resulting equation can be solved by using previous techniques.

**TO SOLVE A FRACTIONAL EQUATION**

1. Find the LCD of all fractions in the equation.
2. Multiply both sides of the equation by the LCD to clear fractions. Make sure that *all* terms are multiplied.
3. Solve the resulting equation (which will be free of fractions).
4. Be sure to check your solutions in the original equation.

**EXAMPLE 1** Solve $\frac{5}{x} - \frac{7}{6} = \frac{3}{2x}$.

$$6x \cdot \left[\frac{5}{x} - \frac{7}{6}\right] = 6x \cdot \left[\frac{3}{2x}\right] \qquad \text{Multiply both sides by the LCD, } 6x$$

$$6x \cdot \frac{5}{x} - 6x \cdot \frac{7}{6} = 6x \cdot \frac{3}{2x} \qquad \text{Distribute}$$

$$30 - 7x = 9 \qquad \text{Clear fractions}$$
$$-7x = -21 \qquad \text{Subtract 30 from both sides}$$
$$x = 3 \qquad \text{Divide by } -7$$

Check:

$$\frac{5}{3} - \frac{7}{6} \stackrel{?}{=} \frac{3}{2 \cdot 3}$$

$$\frac{2 \cdot 5}{2 \cdot 3} - \frac{7}{6} \stackrel{?}{=} \frac{3}{6}$$

$$\frac{10}{6} - \frac{7}{6} \stackrel{?}{=} \frac{3}{6}$$

$$\frac{3}{6} = \frac{3}{6}$$

The solution is 3.

**EXAMPLE 2** Solve $\frac{1}{y} = \frac{1}{4 - y}$.

$$y(4 - y)\left[\frac{1}{y}\right] = y(4 - y)\left[\frac{1}{4 - y}\right] \qquad \text{Multiply both sides by the LCD, } y(4 - y)$$

$$4 - y = y \qquad \text{Clear fractions}$$
$$4 = 2y \qquad \text{Add } y$$
$$2 = y \qquad \text{Divide by 2}$$

Check:

$$\frac{1}{2} \stackrel{?}{=} \frac{1}{4 - 2}$$

$$\frac{1}{2} = \frac{1}{2}$$

The solution is 2.

**EXAMPLE 3** Solve $\frac{a^2 + 9}{a^2 - 9} - \frac{3}{a + 3} = \frac{-a}{3 - a}$.

We first observe that $a^2 - 9 = (a - 3)(a + 3)$. If we multiply the numerator and denominator of $\frac{-a}{3 - a}$ by $-1$, we obtain the equivalent fraction $\frac{a}{a - 3}$. Then the LCD is $(a - 3)(a + 3)$, and we multiply both sides by it.

$$(a - 3)(a + 3)\left[\frac{a^2 + 9}{(a - 3)(a + 3)} - \frac{3}{(a + 3)}\right] = (a - 3)(a + 3)\left[\frac{a}{(a - 3)}\right]$$

$$(a - 3)(a + 3) \cdot \frac{a^2 + 9}{(a - 3)(a + 3)} - (a - 3)(a + 3) \cdot \frac{3}{(a + 3)} = (a - 3)(a + 3) \cdot \frac{a}{(a - 3)}$$

$$a^2 + 9 - (a - 3) \cdot 3 = (a + 3) \cdot a$$
$$a^2 + 9 - (3a - 9) = a^2 + 3a$$
$$a^2 + 9 - 3a + 9 = a^2 + 3a$$
$$-3a + 18 = 3a$$
$$18 = 6a$$
$$3 = a$$

Check: $\dfrac{3^2 + 9}{3^2 - 9} - \dfrac{3}{3 + 3} \stackrel{?}{=} \dfrac{-3}{3 - 3}$

$$\frac{18}{0} - \frac{3}{6} = \frac{-3}{0}$$

Since division by 0 is undefined, there is no solution.

**EXAMPLE 4** Solve $\dfrac{y}{3} - \dfrac{6}{y} = 1$.

$$3y\left[\frac{y}{3} - \frac{6}{y}\right] = 3y\,[1] \qquad \text{Multiply by LCD, } 3y$$

$$\cancel{3}y \cdot \frac{y}{\cancel{3}} - 3\cancel{y} \cdot \frac{6}{\cancel{y}} = 3y \qquad \text{Distribute}$$

$$y^2 - 18 = 3y \qquad \text{This time we have a quadratic equation}$$
$$y^2 - 3y - 18 = 0$$
$$(y + 3)(y - 6) = 0 \qquad \text{Factor}$$
$$y + 3 = 0 \quad \text{or} \quad y - 6 = 0$$
$$y = -3 \qquad\qquad y = 6$$

Check: $\dfrac{-3}{3} - \dfrac{6}{-3} \stackrel{?}{=} 1$ $\qquad$ $\dfrac{6}{3} - \dfrac{6}{6} \stackrel{?}{=} 1$

$-1 + 2 \stackrel{?}{=} 1$ $\qquad$ $2 - 1 \stackrel{?}{=} 1$

$1 = 1$ $\qquad$ $1 = 1$

$-3$ checks $\qquad$ 6 checks

The solutions are $-3$ and 6.

## EXERCISES 6.9

*Solve.*

1. $\dfrac{2}{x} - \dfrac{1}{x} = -6$

2. $\dfrac{3}{a + 1} = \dfrac{5}{a}$

3. $\frac{1}{y-1}+\frac{2}{y+1}=0$

4. $\frac{x+1}{x+2}=\frac{x+2}{x-1}$

5. $\frac{1}{a+2}+\frac{1}{a-2}=\frac{1}{a^2-4}$

6. $\frac{6}{x+3}+\frac{2}{x-3}=\frac{20}{x^2-9}$

7. $\frac{x}{x+4}=\frac{4}{x-4}+\frac{x^2+16}{x^2-16}$

8. $\frac{9+y}{16+2y}=\frac{-7}{16+2y}$

9. $\frac{1}{a}=\frac{-6}{a^2+5}$

10. $\frac{10}{y-1}=\frac{16}{y-1}-3y$

11. $\frac{x^2}{x+12}=1$

12. $\frac{a+2}{a-1}+\frac{3}{a}=1$

13. $\frac{25}{y-5}-\frac{25}{y}=\frac{1}{4}$

14. $\frac{3x}{2x+3}=\frac{8x^2+12}{4x^2-9}-\frac{x}{2x-3}$

**15.** Simplify $\dfrac{\frac{2}{a}+\frac{3}{a}}{\frac{7}{a}-\frac{2}{a}}$.

**16.** $2x+3\overline{)4x^3+6x^2+10x+8}$

---

ANSWERS: **1.** $-\frac{1}{6}$ **2.** $-\frac{5}{2}$ **3.** $\frac{1}{3}$ **4.** $-\frac{5}{4}$ **5.** $\frac{1}{2}$ **6.** 4 **7.** no solution **8.** $-16$ **9.** $-5, -1$ **10.** $2, -1$ **11.** $4, -3$ **12.** $\frac{1}{2}$ **13.** $25, -20$ **14.** $-2$ **15.** 1 **16.** $2x^2+5$ with a remainder of $-7$

## 6.10 Word Problems: Number, Age, and Proportion Problems

Many number and age problems translate into fractional equations.

**EXAMPLE 1** If the numerator of a fraction exceeds the denominator by 6 and the value of the fraction is $\frac{5}{3}$, find the fraction.

Let $x =$ the denominator of the fraction
$x + 6 =$ the numerator of the fraction (Why?)

The fraction is $\frac{x+6}{x}$, so the equation we must solve is

$$\frac{x+6}{x}=\frac{5}{3}. \qquad \text{The LCD is } 3x$$

$$3x\cdot\frac{x+6}{x}=3x\cdot\frac{5}{3}$$

$$3(x+6)=x\cdot 5 \qquad \text{Use parentheses}$$

$$3x+18=5x$$

$$18=2x$$

$$9=x$$

$$x+6=15$$

The fraction is $\frac{15}{9}$. The numerator 15 is 6 more than the denominator 9, and $\frac{15}{9}$ reduces to $\frac{5}{3}$.

**EXAMPLE 2** The sum of Bill's and Jan's ages is 65 years, and Bill's age divided by Jan's is $\frac{7}{6}$. How old is each?

Let $x =$ Bill's age
$65 - x =$ Jan's age (The sum of $x$ and $65 - x$ is 65.)

The equation we must solve is

$$\frac{x}{65-x} = \frac{7}{6}. \qquad \text{The LCD is } 6(65-x)$$

$$6(65-x) \cdot \frac{x}{(65-x)} = 6(65-x) \cdot \frac{7}{6}$$

$$6x = (65-x)7$$
$$6x = 455 - 7x$$
$$13x = 455$$
$$x = 35$$
$$65 - x = 30$$

Bill is 35 years old and Jan is 30 years old. (Check this answer.)

**EXAMPLE 3** The reciprocal of 3 more than a number is twice the reciprocal of the number. Find the number.

Let $x =$ the number
$x + 3 =$ 3 more than the number
$\frac{1}{x+3} =$ the reciprocal of 3 more than the number
$\frac{1}{x} =$ the reciprocal of the number
$2 \cdot \frac{1}{x} = \frac{2}{x} =$ twice the reciprocal of the number

The equation we must solve is

$$\frac{1}{x+3} = \frac{2}{x}. \qquad \text{The LCD is } x(x+3)$$

$$x(x+3) \cdot \frac{1}{x+3} = x(x+3) \cdot \frac{2}{x}$$

$$x = (x+3)2$$
$$x = 2x + 6$$
$$-x = 6$$
$$x = -6$$

The number is $-6$. (Check.)

**EXAMPLE 4** A student made 75, 80, and 82 on the first three of four tests in a course. What must she make on the fourth test for an average score of 83?

Let $x =$ score on the fourth test

To find the average of four numbers, we add them and divide by 4. Thus we must solve

$$\frac{75 + 80 + 82 + x}{4} = 83.$$

$$\not{4} \cdot \frac{237 + x}{\not{4}} = 4 \cdot 83 \qquad \text{Multiply by the LCD, 4}$$

$$237 + x = 332$$

$$x = 95$$

She must make a 95 on the fourth test to average 83. (Check.)

The **ratio** of one number $a$ to another number $b$ is the quotient $\frac{a}{b}$. An equation which states that two ratios are equal is called a **proportion.** For example,

$$\frac{3}{4} = \frac{15}{20} \quad \text{and} \quad \frac{x}{2} = \frac{9}{12}$$

are proportions. When a variable is present in one or more **terms of a proportion** (numerator or denominator) as in

$$\frac{y}{y + 1} = \frac{2}{5},$$

the proportion is simply a fractional equation. Equations like this are solved by multiplying both sides by the LCD: in our example, $5(y + 1)$.

$$5\cancel{(y + 1)} \cdot \frac{y}{\cancel{(y + 1)}} = \cancel{5}(y + 1) \cdot \frac{2}{\cancel{5}}$$

**(1)** $$5y = (y + 1) \cdot 2$$

When working with proportions, we can shorten our work by forming the **cross-product equation.** If we multiply $y$, the numerator of the first fraction, by 5, the denominator of the second, and set this equal to the product of the numerator of the second fraction, 2, by the denominator of the first, $(y + 1)$, we obtain equation (1). The cross-product is indicated by the arrows in the following equation.

$$\frac{y}{y + 1} \times \frac{2}{5}$$

**CAUTION:** Never attempt to find the cross-product equation when the original equation is *not* a proportion. For example, the notion of the cross-product equation for an equation such as

$$\frac{2}{x} + \frac{3}{x - 1} = 7$$

is meaningless.

Proportions are useful in many word problems. Consider the problem: If 2 hours are required to type 10 pages, how many hours would be required to type 25 pages? Letting $x$ represent the number of hours required to type 25 pages, the proportion

$$\begin{matrix} \textbf{first time} \longrightarrow \\ \textbf{first number of pages} \longrightarrow \end{matrix} \frac{2}{10} = \frac{x}{25} \begin{matrix} \longleftarrow \textbf{second time} \\ \longleftarrow \textbf{second number of pages} \end{matrix}$$

describes the problem. That is, the ratio of the first number of hours to the first number of pages must equal the ratio of the second number of hours to the second number of pages. Forming the cross-product equation, we obtain

$$10x = 50$$
$$x = 5.$$

Thus it would take 5 hours to type 25 pages. (Does this time seem reasonable?)

**EXAMPLE 5** If a boat uses 14 gallons of gas to go 102 miles, how many gallons would be needed to go 510 miles?

Let $x =$ the number of gallons required to go 510 miles.

The equation we must solve is

$$\frac{14}{102} = \frac{x}{510}$$

miles on 14 gallons → 102 ; 510 ← miles on $x$ gallons

$$102x = 510 \cdot 14 \qquad \text{Cross-product equation}$$

$$x = \frac{510 \cdot 14}{102}$$

$$x = \frac{\cancel{102} \cdot 5 \cdot 14}{\cancel{102}} = 70$$

It would take 70 gallons to go 510 miles.

## EXERCISES 6.10

*Solve.*

**1.** Find two numbers whose sum is 45 and whose quotient is $\frac{2}{3}$.

Let $x =$ the first number
$45 - x =$ the second number

**2.** Harriet is three-fourths as old as Lorraine, and the difference of their ages is 10 years. How old is each?

Let $x =$ Lorraine's age
$\frac{3}{4}x =$ Harriet's age

**3.** The reciprocal of 5 less than a number is twice the reciprocal of the number. Find the number.

Let $x =$ the desired number

$x - 5 =$ 5 less than the number

$\frac{1}{x-5} =$ the reciprocal of 5 less than the number

$\frac{1}{x} =$

**4.** In an election the winning candidate won by a 5 to 3 margin. If he received 820 votes, how many votes did the loser receive?

Let $x =$ the number of votes for the loser

$$\frac{820}{x} = \frac{5}{3}$$

**5.** The numerator of a fraction is 4 less than the denominator. If both the numerator and denominator are increased by 2, the value of the fraction is 2/3. Find the fraction.

**6.** If Brenda is 5 years older than Jane and the quotient of their ages is $\frac{5}{4}$, how old is each?

**7.** If the average weight of the four linemen for the Rams is 243 lb, and the weights of three of them are 220 lb, 240 lb, and 260 lb, how much does the fourth one weigh?

**8.** If 150 ft of wire weighs 45 lb, what will 240 ft of the same wire weigh?

**9.** If on a map $\frac{3}{4}$ inch represents 20 miles, how many miles will be represented by 9 inches?

**10.** In a sample of 92 tires, 3 were defective. How many defective tires would you expect in a sample of 644 tires?

**11.** If five times the reciprocal of a number is equal to the reciprocal of 4 more than the number, find the number.

**12.** The three starting defensive linemen for the Denver Broncos in Super Bowl XII had an average weight of 252 lb. If Carter weighed 256 lb and Chavous weighed 250 lb, how much did Alzado weigh?

*Solve the fractional equations.*

**13.** $\frac{1}{x^2 - 1} = \frac{1}{x - 1}$

**14.** $\frac{a^2}{12 - a} = 1$

**15.** Subtract. $\frac{2}{a+b} - \frac{2a}{a^2 + 2ab + b^2}$

**16.** Divide. $(x^4 - 1) \div (x - 1)$

ANSWERS: 1. 18, 27 2. Harriet is 30, Lorraine is 40 3. 10 4. 492 5. $\frac{6}{10}$
6. Brenda is 25, Jane is 20 7. 252 lb 8. 72 lb 9. 240 mi 10. 21 11. $-5$ 12. 250 lb 13. 0
14. 3, $-4$ 15. $\frac{2b}{(a+b)^2}$ 16. $x^3 + x^2 + x + 1$

## 6.11 Word Problems: Motion Problems

Some motion problems, involving the formula

$$d = rt \qquad \text{(distance} = \text{rate} \cdot \text{time),}$$

translate into fractional equations. When solving motion problems that involve two distances, two times and two rates, it is especially helpful if we precisely describe each of these six quantities. Study the following examples carefully.

**EXAMPLE 1** The speed of a freight train is 10 mph slower than the speed of a passenger train. If the freight train travels 275 miles in the same time that the passenger train travels 325 miles, find the speed of each train.

Let $x$ = speed of the freight train
$x + 10$ = speed of the passenger train (Why?)
$275$ = distance the freight train travels
$325$ = distance the passenger train travels

Since $d = r \cdot t$, we have that $t = \frac{d}{r}$. Thus,

$$\frac{275}{x} = \text{time the freight train travels}$$

$$\frac{325}{x+10} = \text{time the passenger train travels.}$$

Since the two times are equal (why?), we must solve

$$\frac{275}{x} = \frac{325}{x+10}.$$

$$275(x + 10) = 325x \qquad \text{Cross-product equation}$$
$$275x + 2750 = 325x$$
$$2750 = 50x$$
$$55 = x$$
$$65 = x + 10$$

The freight train travels 55 mph, and the passenger train travels 65 mph. (Check.)

**EXAMPLE 2** The speed of a stream is 4 mph. A boat travels 48 mi upstream in the same time it takes to travel 72 mi downstream. What is the speed of the boat in still water?

Let $x$ = speed of the boat in still water

When the boat travels downstream, its speed relative to the bank of the stream is its speed in still water *increased* by the speed of the stream. Thus,

$$x + 4 = \text{speed of the boat when traveling downstream.}$$

When the boat travels upstream, its speed relative to the bank of the stream is its speed in still water *decreased* by the speed of the stream. Thus,

$$\begin{aligned} x - 4 &= \text{speed of the boat when traveling upstream.} \\ 72 &= \text{distance traveled downstream} \\ 48 &= \text{distance traveled upstream} \end{aligned}$$

Again we calculate the two times using $t = \frac{d}{r}$.

$$\frac{72}{x+4} = \text{time of travel downstream}$$

$$\frac{48}{x-4} = \text{time of travel upstream}$$

Since the two times are equal (why?), we must solve the following equation.

$$\begin{aligned} \frac{72}{x+4} &= \frac{48}{x-4} \\ 72(x-4) &= 48(x+4) && \text{Cross product} \\ 72x - 288 &= 48x + 192 \\ 24x - 288 &= 192 && \text{Subtract } 48x \\ 24x &= 480 && \text{Add } 288 \\ x &= 20 \end{aligned}$$

The speed of the boat in still water is 20 mph.

CAUTION: In the above example, a common mistake is to conclude that the speed upstream is $4 - x$ rather than $x - 4$. Which number, $x$ or 4, must be larger if the boat actually makes progress upstream? Do you see why $x - 4$ is the correct rate?

## EXERCISES 6.11

*Solve.*

1. A freight train travels 5 mph slower than a passenger train. If the freight train travels 260 miles in the same time that the passenger train travels 280 miles, find the speed of each.

Let $x =$ speed of the passenger train

$x - 5 =$ speed of the freight train

$260 =$

$280 =$

$\frac{280}{x} =$

$\frac{260}{x-5} =$

**2.** The speed of a stream is 5 mph. If a boat travels 50 miles downstream in the same time that it takes to travel 25 miles upstream, what is the speed of the boat in still water?

Let $x =$ speed of the boat in still water

$x - 5 =$ speed of the boat going upstream

$x + 5 =$

$50 =$

$25 =$

$\frac{50}{x+5} =$

$\frac{25}{x-5} =$

**3.** A plane flies 480 miles with the wind and 330 miles against the wind in the same length of time. If the speed of the wind is 25 mph, what is the speed of the plane in still air?

Let $x =$ speed of the plane in still air

$x + 25 =$

$x - 25 =$

$480 =$

$330 =$

**4.** The speed of a boat in still water is 24 mph. If the boat travels 54 miles upstream in the same time that it takes to travel 90 miles downstream, what is the speed of the stream?

**5.** A man walks a distance of 12 miles at a rate 8 mph slower than the rate he rides a bicycle for a distance of 24 miles. If the total time of the trip is 5 hours, how fast does he walk? How fast does he ride?

$\left[ \textit{Hint: } \frac{12}{x} + \frac{24}{x+8} = 5 \right]$

**6.** A woman flies from Phoenix to Denver (a distance of 800 miles) at a rate 40 mph faster than on the return trip. If the total time of the trip is 9 hours, what was her rate going to Denver, and what was her rate returning to Phoenix?

**7.** Find the number which when added to both the numerator and denominator of $\frac{3}{11}$ will produce a fraction equivalent to $\frac{1}{2}$.

**8.** If a car used 16 gallons of gas for a trip of 264 miles, how much gas will be used for a trip of 1100 miles?

---

ANSWERS: **1.** freight train: 65 mph, passenger train: 70 mph **2.** 15 mph **3.** 135 mph **4.** 6 mph **5.** walk: 4 mph, ride: 12 mph **6.** 200 mph going, 160 mph returning **7.** 5 **8.** $66\frac{2}{3}$ gal

## 6.12 Word Problems: Work Problems

Consider the following problem: Jim can do a job in 3 hr and Dave can do the same job in 7 hr. How long would it take them to do the job if they worked together?

This type of problem is usually called a *work problem* and three important principles must be kept in mind.

1. The time required to do a job when the individuals work together must be less than the time required for the fastest worker to complete the job alone. Thus the time together is *not* the average of the two times (which would be 5 hours in this case). Since Jim can do the job alone in 3 hr, with help the time must clearly be less than 3 hr.

2. If a job can be done in $t$ hr, in 1 hour $\frac{1}{t}$ of the job will be completed. For example, since Jim can do the job in 3 hr, in 1 hour he would do $\frac{1}{3}$ of the job. Similarly, in 1 hour Dave would do $\frac{1}{7}$ of the job since he can do it all in 7 hours.

3. The work done by Jim in 1 hour added to the work done by Dave in 1 hour equals the amount of work done together in 1 hour. Thus, if $t$ is the time required to complete the job working together,

$$\begin{aligned}&(\text{Amount Jim does in 1 hr}) + (\text{Amount Dave does in 1 hr})\\ &\qquad = (\text{Amount done together in 1 hr})\end{aligned}$$

translates to

$$\frac{1}{3} + \frac{1}{7} = \frac{1}{t}.$$

Thus, such a word problem translates to a fractional equation. Multiply both sides by the LCD, $21t$.

$$21t\left[\frac{1}{3} + \frac{1}{7}\right] = 21t \cdot \frac{1}{t}$$

$$21t \cdot \frac{1}{3} + 21t \cdot \frac{1}{7} = 21$$

$$7t + 3t = 21$$

$$10t = 21$$

$$t = \frac{21}{10}$$

It would take Jim and Dave $\frac{21}{10}$ hr (2 hr 6 min) to complete the job working together. (Does this seem reasonable in view of (1) above? Check.)

A simple variation of this type of problem is to request the time for one individual and to give the time required when working together.

**EXAMPLE 1** When Sue and Dee work together, it takes 5 minutes to do a job. If Dee can do the job alone in 7 minutes, how long would it take Sue to complete the job if she works alone?

$5 =$ the number of minutes to do the job together

$\frac{1}{5} =$ the amount done together in 1 minute

$7 =$ the number of minutes for Dee to do the job alone

$\frac{1}{7} =$ the amount done by Dee in 1 minute

Let $t =$ the number of minutes required for Sue to do the job alone

$\frac{1}{t} =$ the amount done by Sue in 1 minute

$$(\text{Amount by Sue}) + (\text{Amount by Dee}) = (\text{Amount together})$$

$$\frac{1}{t} + \frac{1}{7} = \frac{1}{5}$$

$$35t\left(\frac{1}{t} + \frac{1}{7}\right) = 35t \cdot \frac{1}{5} \qquad \text{The LCD is } 35t$$

$$35t \cdot \frac{1}{t} + 35t \cdot \frac{1}{7} = 35t \cdot \frac{1}{5}$$

$$35 + 5t = 7t$$

$$35 = 2t$$

$$\frac{35}{2} = t$$

It would take Sue $17\frac{1}{2}$ minutes to do the job alone. (Check.)

It is important when solving work problems (as with any word problem) to be neat and complete. Do not take shortcuts, especially when writing down the pertinent information. Writing detailed descriptions of the variables can eliminate errors. Your work should be patterned after the examples.

**EXAMPLE 2** It takes pipe $A$ 9 days longer to fill a reservoir than pipe $B$. If the two pipes are turned on together, they can fill the reservoir in 20 days. How long would it take each pipe to fill the reservoir alone?

Let $t =$ the number of days required for $B$ to fill the reservoir
$t + 9 =$ the number of days required for $A$ to fill the reservoir
$20 =$ the number of days required to fill the reservoir working together

$\frac{1}{t} =$ the amount filled by $B$ in 1 day

$\frac{1}{t+9} =$ the amount filled by $A$ in 1 day

$\frac{1}{20} =$ the amount filled by $A$ and $B$ working together in 1 day

We must solve

$$\frac{1}{t} + \frac{1}{t+9} = \frac{1}{20}. \qquad \text{The LCD} = 20t(t+9)$$

$$20\cancel{t}(t+9) \cdot \frac{1}{\cancel{t}} + 20t\cancel{(t+9)} \cdot \frac{1}{\cancel{t+9}} = \cancel{20}t(t+9) \cdot \frac{1}{\cancel{20}}$$

$$\begin{aligned} 20(t+9) \cdot 1 + 20t \cdot 1 &= t(t+9) \cdot 1 \\ 20t + 180 + 20t &= t^2 + 9t \\ 40t + 180 &= t^2 + 9t \\ 0 &= t^2 - 31t - 180 \end{aligned}$$

This time we obtain a quadratic equation to solve. Factoring,

$$(t+5)(t-36) = 0.$$

$$t + 5 = 0 \quad \text{or} \quad t - 36 = 0$$

$$t = -5 \qquad\qquad t = 36$$

$$t + 9 = 45$$

Since $-5$ is meaningless for a time in a work problem, we have $t = 36$ as the only solution. Therefore it takes pipe $B$ 36 days to fill the reservoir alone, and pipe $A$ requires 45 days to fill it. (Check.)

## EXERCISES 6.12

*Solve.*

**1.** If Clyde can do a job in 8 days and Irv can do the same job in 3 days, how long would it take them to do the job together?

$8 =$ the number of days for Clyde to do the job

$\frac{1}{8} =$ the amount Clyde does in 1 day

$3 =$

$\frac{1}{3} =$

Let $t =$ the number of days to do the job together

$\frac{1}{t} =$ the amount done in 1 day working together

2. If pipe $A$ can fill a tank in 20 hr and pipe $B$ can fill the same tank in 15 hr, how long would it take to fill the tank if both pipes fill together?

$20 =$

$\frac{1}{20} =$

$15 =$

$\frac{1}{15} =$

Let $t =$ the number of hr to fill tank working together

$\frac{1}{t} =$

3. On the day that the Wisconsin cheese shipment arrives at Perko's Delicatessen, it takes Perko 3 hours to slice and display the cheese. When Perko and his assistant Graydon work together, it only takes 2 hours to process the same shipment. How long would it take Graydon to slice and display the cheese if he worked alone?

Let $t =$ time required for Graydon to cut the cheese

$\frac{1}{t} =$

$3 =$

$\frac{1}{3} =$

$2 =$

$\frac{1}{2} =$

**4.** When Barb does a job by herself, it takes her 7 hours. If Barb and Wilma work together it takes 6 hours to do the same job. How long does it take Wilma to do the job if she works alone?

Let $t =$ time required for Wilma to do the job

**5.** When each works alone, Burford can mow a lawn in 3 hr less time than Ernie. When they work together, it takes 2 hours. How long does it take each to do the job by himself?

**6.** It takes Sybil 6 hr longer to paint a room than it takes Joanne. Working together, they can paint the room in 4 hr. How long would it take each working alone to paint the room?

**7.** John drives 12 mph faster than his mother. If John travels 310 miles in the same time that his mother travels 250 miles, find the speed of each.

**8.** A plane flies 1160 km with the wind and 840 km against the wind in the same length of time. If the speed of the plane in still air is 250 km/hr, what is the wind speed?

---

ANSWERS: **1.** $\frac{24}{11}$ days **2.** $\frac{60}{7}$ hr **3.** 6 hr **4.** 42 hr **5.** Burford: 3 hr, Ernie: 6 hr **6.** Sybil: 12 hr, Joanne: 6 hr **7.** John: 62 mph, his mother: 50 mph **8.** 40 km/hr

## Chapter 6 Summary

### Words and Phrases for Review

[6.1] algebraic fraction
meaningless replacement
equivalent fractions
prime factor

[6.5] common denominator
least common denominator (LCD)

[6.7] complex fraction

[6.9] fractional equation
clearing fractions

[6.10] ratio
proportion

### Brief Reminders

[6.1] When reducing fractions, cancel or divide out factors only, never cancel terms. For example,

$$\frac{\not{3}(x+2)}{\not{3}} = x+2 \quad \text{but} \quad \frac{3+(x+2)}{3} \text{ is } not\ \frac{\not{3}+(x+2)}{\not{3}}.$$

[6.2–6.3] **1.** When multiplying or dividing fractions, factor numerators and denominators and cancel common factors. Do *not* find the LCD.

**2.** The reciprocal of $a+b$ is $1/(a+b)$ and *not* $1/a+1/b$.

[6.4] When subtracting fractions, use parentheses to avoid sign errors.

[6.5] Before finding the LCD, reduce all fractions to lowest terms.

[6.6] Do not cancel out terms when adding or subtracting fractions, cancel factors only.

[6.7] To simplify a complex fraction, multiply all terms by the LCD of all fractions involved.

[6.9] To solve a fractional equation, multiply both sides by the LCD of all fractions. Check all answers in the *original* equation and omit any which result in division by zero.

[6.10] Form the cross-product equation only for proportions.

## CHAPTER 6 REVIEW EXERCISES

[6.1] *What are the meaningless replacements for the variable or variables?*

1. $\dfrac{x+2}{x(x-5)}$

2. $\dfrac{2x+1}{x-y}$

3. $\dfrac{a+8}{a^3+a^2-2a}$

*Are the given fractions equivalent?*

4. $\dfrac{2}{x-y}, \dfrac{-2}{y-x}$

5. $\dfrac{a+b}{5}, \dfrac{3a+3b}{15}$

6. $\dfrac{x}{a}, \dfrac{x+7}{a+7}$

[6.2] *Multiply.*

7. $\dfrac{x^2+2xy+y^2}{x-2} \cdot \dfrac{x^2-4}{(x+y)^2}$

8. $\dfrac{a^2+2ab+b^2}{a^2-b^2} \cdot \dfrac{b-a}{4b+4a}$

[6.3] *Divide.*

9. $\dfrac{x^2-x-6}{x^2-36} \div \dfrac{2x-6}{x^2-4x-12}$

10. $\dfrac{a^2-ab-2b^2}{a^2-3ab+2b^2} \div \dfrac{2a+b}{a-b}$

[6.6] *Add.*

11. $\dfrac{6}{x+y} + \dfrac{3x+y}{x^2-y^2}$

12. $\dfrac{b}{a^2+ab+b^2} + \dfrac{a^2-ab}{a^3-b^3}$

*Subtract.*

**13.** $\dfrac{7x}{x^2 - 2x + 1} - \dfrac{3}{x - 1}$

**14.** $\dfrac{a^2 - 3ab}{a^2 - b^2} - \dfrac{-2ab - b^2}{2a^2 - ab - b^2}$

[6.7] *Simplify.*

**15.** $\dfrac{\dfrac{x}{y} - \dfrac{y}{x}}{\dfrac{1}{y} - \dfrac{1}{x}}$

**16.** $\dfrac{x - 1 - \dfrac{6}{x}}{1 + \dfrac{2}{x}}$

[6.8] *Divide.*

**17.** $x - 4\overline{)5x^3 - 18x^2 - 14x + 24}$

**18.** $2x + 1\overline{)4x^3 - 4x^2 - x + 5}$

[6.9] *Solve.*

**19.** $1 - \dfrac{a + 2}{a - 1} = \dfrac{3}{a}$

**20.** $3y = \dfrac{16}{y - 1} - \dfrac{10}{y - 1}$

[6.10] **21.** The reciprocal of 8 more than a number is three times the reciprocal of the number. Find the number.

**22.** Larry is 6 years older than Amy and the ratio of their ages is $\frac{7}{6}$. How old is each?

**23.** If 30 ft of wire weighs 9 lb, what will 48 ft of the same wire weigh?

[6.11] **24.** The speed of a river is 2 mph. If a boat can travel 104 miles upstream in the same time it can travel 120 miles downstream, what is the speed of the boat in still water?

[6.12] **25.** If pipe $A$ can fill a tank in 5 hr and pipe $B$ requires 2 hr to fill the same tank, how long would it take to fill the tank if the pipes work together?

**26.** If Lizette can do a job in 12 hours less time than Bo, and together they can do the job in 8 hours, how long would it take each to do it alone?

---

**ANSWERS:** **1.** 0, 5 **2.** $x = y$ **3.** 0, 1, −2 **4.** yes **5.** yes **6.** no **7.** $x + 2$ **8.** $-\frac{1}{4}$ **9.** $\frac{(x+2)^2}{2(x+6)}$ **10.** $\frac{a+b}{2a+b}$ **11.** $\frac{9x-5y}{(x+y)(x-y)}$ **12.** $\frac{a+b}{a^2+ab+b^2}$ **13.** $\frac{4x+3}{(x-1)^2}$ **14.** $\frac{a-b}{a+b}$ **15.** $x + y$ **16.** $x - 3$ **17.** $5x^2 + 2x - 6$ **18.** $2x^2 - 3x + 1 + \frac{4}{2x+1}$ **19.** $\frac{1}{2}$ **20.** 2, −1 **21.** −12 **22.** Larry is 42, Amy is 36 **23.** $\frac{72}{5}$ lb **24.** 28 mph **25.** $\frac{10}{7}$ hr **26.** Lizette: 12 hr, Bo: 24 hr

# 7

# Graphs and Functions

## 7.1 The Cartesian Coordinate System

In the second chapter we learned to graph equations and inequalities in one variable. At that time we plotted on a number line the points that corresponded to the solutions of the given equation or inequality. Recall that a number line is a line that has been marked off in unit lengths with each point on the line identified with a real number, as in Figure 7.1. We identify the number zero with the origin. Points to the right of the origin are associated with the positive numbers, and points to the left of the origin correspond to the negative numbers. We think of the line as extending infinitely far in both directions.

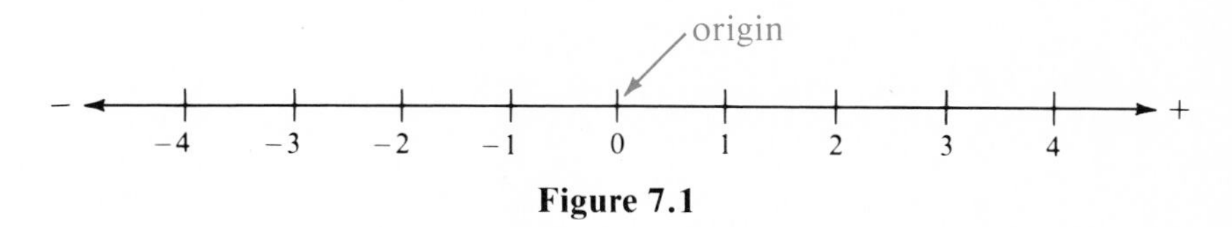


**Figure 7.1**

Equivalently, we could have a *vertical* number line on which, traditionally, the positive numbers correspond to points above the origin, and the negative numbers correspond to points below the origin. A horizontal and a vertical number line are shown in Figure 7.2

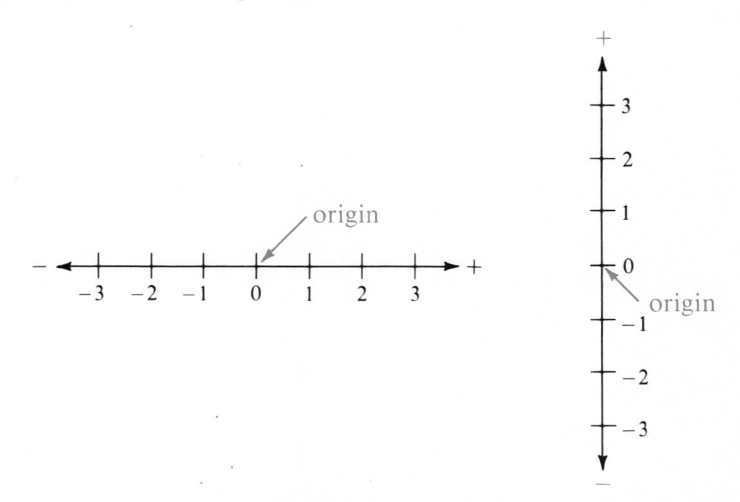


**Figure 7.2**

When a horizontal number line and a vertical number line are placed together, as in Figure 7.3 on p. 316 so that the two origins coincide and the lines are perpendicular, the result is called a **rectangular** or **Cartesian coordinate system** (named after French mathematician René Descartes) or a **coordinate plane** (a plane is a flat surface that extends infinitely far).

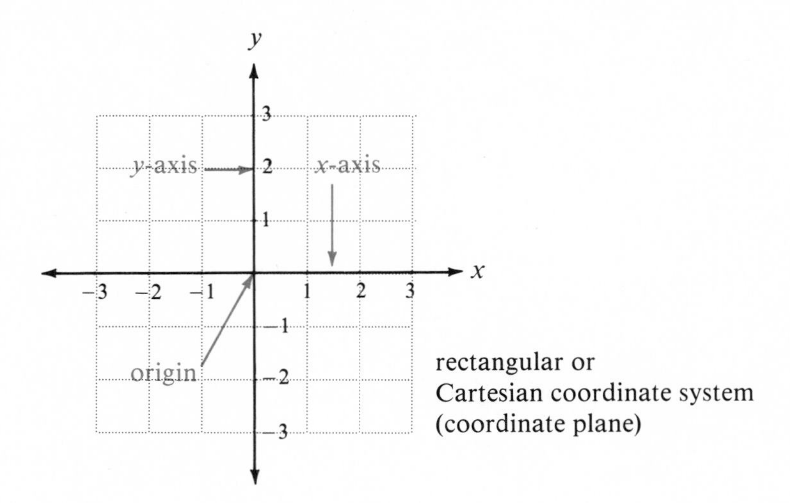


**Figure 7.3**

The horizontal number line is called the **horizontal axis** or ***x*-axis,** and the vertical number line is called the **vertical axis** or ***y*-axis.** The point of intersection of the axes is called the **origin,** and a Cartesian coordinate system is sometimes called an ***x, y*-coordinate system.**

Recall that there is one and only one point on a number line associated with each real number. A similar situation exists for points in a plane and **ordered pairs** of numbers. For example, the ordered pair (2, 3) can be identified with a point in a coordinate plane as follows:

The first number, 2, called the **first-coordinate** or ***x*-coordinate** of the point, is associated with a point on the horizontal or $x$-axis.

The second number, 3, called the **second coordinate** or ***y*-coordinate** of the point, is associated with a point on the vertical or $y$-axis.

The pair (2, 3) is associated with the point where the vertical line through 2 on the $x$-axis and the horizontal line through 3 on the $y$-axis intersect. See Figure 7.4.

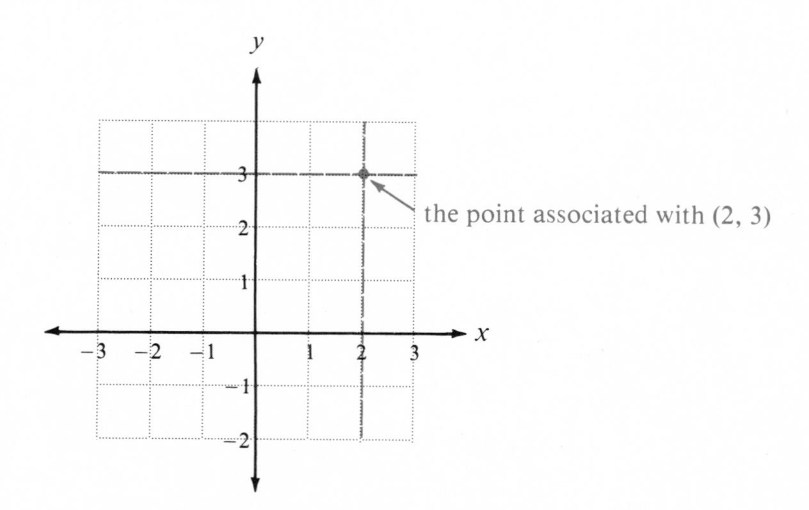


**Figure 7.4**

EXAMPLE 1 The points associated with the ordered pairs (3, 2), (−1, 3), (−3, −2), and (2, −2) are *plotted* in the coordinate plane in Figure 7.5.

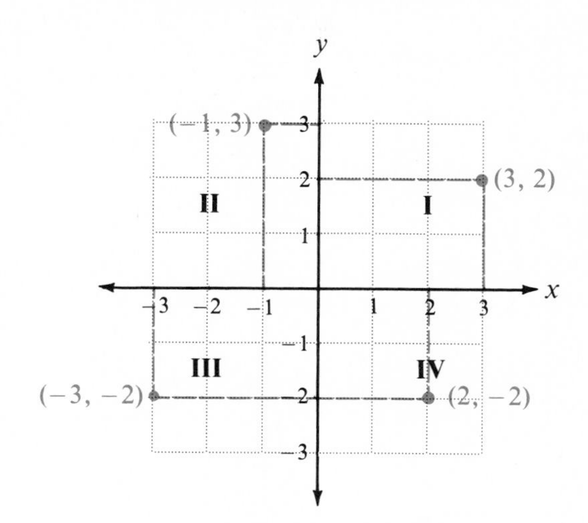


**Figure 7.5**

The axes in a coordinate system separate the plane into four sections called **quadrants.** The first, second, third, and fourth quadrants are identified by the Roman numerals I, II, III, and IV, respectively, in the coordinate plane in Figure 7.5. The signs of the $x$-coordinate (first) and $y$-coordinate (second) in the various quadrants are as follows.

I: (+, +) II: (−, +), III: (−, −), IV: (+, −).

We often use $(x, y)$ to refer to a general ordered pair of numbers. Thus, the point in the plane associated with the pair $(x, y)$ has $x$-coordinate or **abscissa** $x$ and $y$-coordinate or **ordinate** $y$. When we identify the point in a plane associated with the given ordered pair, we say that we **plot** the point.

## EXERCISES 7.1

1. Plot the points associated with the given pairs $A(1, 4)$, $B(4, -2)$, $C(-3, 2)$, $D(-3, 0)$, $E(3, 0)$, $F(0, 0)$, $G(-3, -3)$, and $H(0, -2)$.

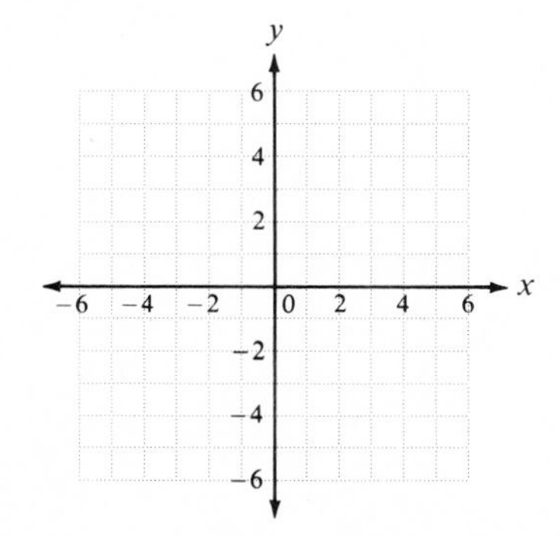

**2.** Give the coordinates of the points $A$, $B$, $C$, $D$, $E$, $F$, $G$, and $H$.

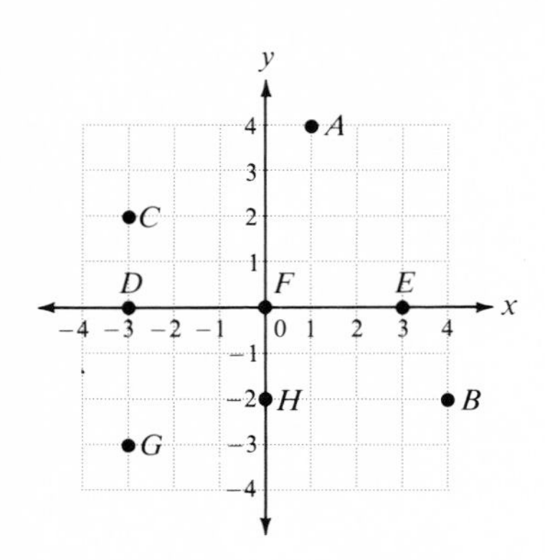


**3.** Plot the points associated with the pairs $J\left(\frac{1}{2}, 2\right)$, $K\left(-\frac{5}{2}, 1\right)$, $L\left(-2, -\frac{7}{4}\right)$, and $M\left(3, -\frac{3}{4}\right)$.

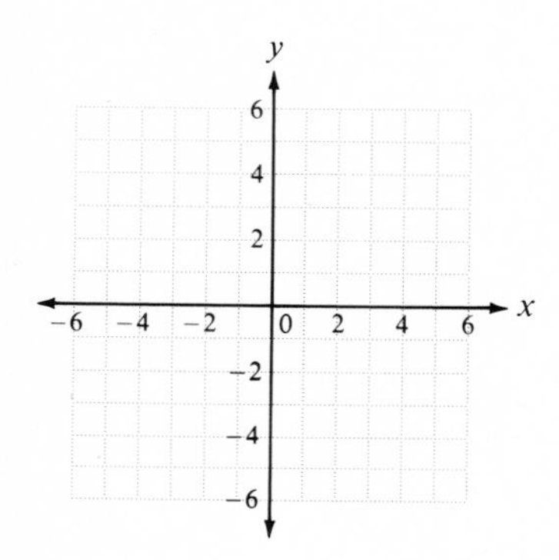


**4.** In which quadrants are the points $J$, $K$, $L$, and $M$ of Exercise 3 located?

**5.** A configuration in which points are plotted in a plane is called ________________ ________________.

**6.** The ordered pair (3, −4) determines a point in quadrant **(a)** ________________ with $x$-coordinate **(b)** ________________ and $y$-coordinate **(c)** ________________ ________________.

**7.** In quadrant II the first coordinate of a point is always **(a)** ________________ and the second coordinate is always **(b)** ________________.

**8.** In quadrant IV the first coordinate of a point is always **(a)** ________________ and the second coordinate is always **(b)** ________________.

**9.** In quadrant I the first coordinate of a point is always **(a)** ________________ and the second coordinate is always **(b)** ________________.

**10.** In quadrant III the first coordinate of a point is always **(a)** ________________ and the second coordinate is always **(b)** ________________.

**11.** The horizontal axis is called the ________________.

**12.** The vertical axis is called the ________________.

**13.** The perpendicular lines in a Cartesian coordinate system are called collectively the ______________.

**14.** The coordinates of the origin are ______________.

**15.** An $x$, $y$-coordinate system separates the plane into four regions called ______________.

**16.** When we identify the point associated with a given ordered pair of numbers, we say that we ______________ the point.

---

ANSWERS: 1. See the coordinate system in Exercise 2. 2. See Exercise 1.
3.

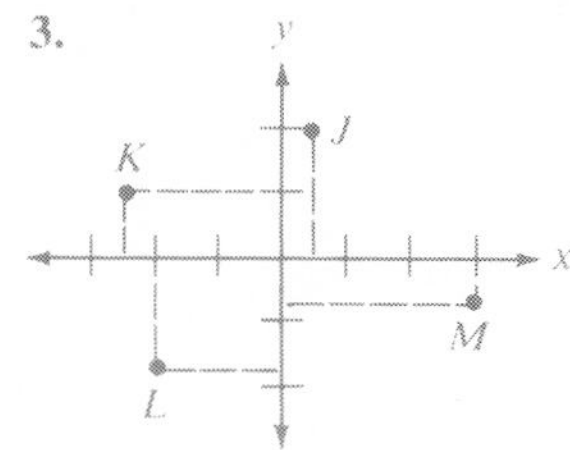


4. $J$: I; $K$: II; $L$: III; $M$: IV 5. Cartesian coordinate system or $x$, $y$-coordinate system or coordinate plane 6. (a) IV (b) 3 (c) −4 7. (a) − (b) + 8. (a) + (b) − 9. (a) + (b) + 10. (a) − (b) − 11. $x$-axis 12. $y$-axis 13. axes 14. (0, 0) 15. quadrants 16. plot

## 7.2 Graphing Linear Equations

All of the equations we graphed in the second chapter involved only one variable. Graphing such equations required only a single number line. Graphing equations that contain two variables requires a Cartesian coordinate system since solutions to equations in two variables are ordered pairs of numbers.

An equation like

$$y = 3x - 2$$

has infinitely many solutions. One solution is (2, 4) since if $x$ is replaced with 2 and $y$ is replaced with 4, the resulting equation is true. Carrying out this replacement we have

$$\begin{aligned} 4 &= 3(2) - 2 \\ 4 &= 6 - 2 \\ 4 &= 4. \end{aligned}$$

Also, since

$$\begin{aligned} -2 &= 3(0) - 2 \\ -2 &= 0 - 2 \\ -2 &= -2 \end{aligned} \quad \text{and} \quad \begin{aligned} -5 &= 3(-1) - 2 \\ -5 &= -3 - 2 \\ -5 &= -5 \end{aligned}$$

(0, −2) and (−1, −5) are also solutions to the equation $y = 3x - 2$.

The **graph** of an equation with two variables $x$ and $y$ is the set of points in a Cartesian coordinate system that correspond to **solutions** (ordered pairs of numbers that make the equation true) of the equation. Since there are usually infinitely many solutions, we cannot determine and plot each possible pair. Generally we plot enough points to enable us to see a pattern, and then connect these points with a line or curve to graph the equation.

One way to find and store a collection of solutions is to make a **table of values.** We choose several values for $x$, substitute these values into the equation, and compute the corresponding values for $y$. We begin by making a table such as the one below.

| $x$ | 0 | 1 | −1 | 2 | −2 | 3 | −3 |
|---|---|---|---|---|---|---|---|
| $y$ | | | | | | | |

| Substitution in $y = 3x - 2$ | Result |
|---|---|
| $x = 0$ | $y = 3(0) - 2 = -2$ |
| $x = 1$ | $y = 3(1) - 2 = 1$ |
| $x = -1$ | $y = 3(-1) - 2 = -5$ |
| $x = 2$ | $y = 3(2) - 2 = 4$ |
| $x = -2$ | $y = 3(-2) - 2 = -8$ |
| $x = 3$ | $y = 3(3) - 2 = 7$ |
| $x = -3$ | $y = 3(-3) - 2 = -11$ |

If we place the computed $y$-value below the $x$-value used to calculate it, the table has the following form.

| $x$ | 0 | 1 | −1 | 2 | −2 | 3 | −3 |
|---|---|---|---|---|---|---|---|
| $y$ | −2 | 1 | −5 | 4 | −8 | 7 | −11 |

We usually make our calculations mentally or as scratch work and place the values in the table. The table above lists seven solutions to the equation $y = 3x - 2$;

$$(0, -2),\quad (1, 1),\quad (-1, -5),\quad (2, 4),\quad (-2, -8),\quad (3, 7),\quad (-3, -11).$$

Next, we plot the points that correspond to these solutions in a Cartesian coordinate system, as in Figure 7.6.

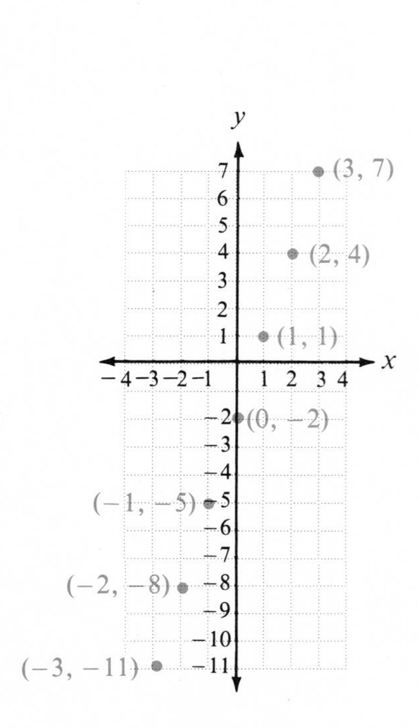


Figure 7.6

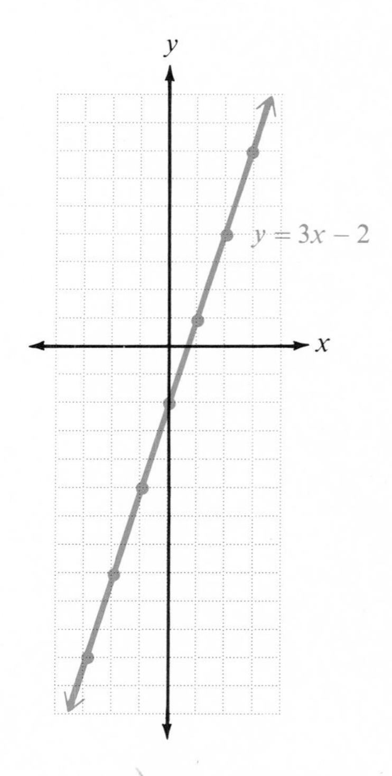


Figure 7.7

It appears that all seven points lie on a straight line. Thus, it is reasonable to assume that the graph of this equation is the straight line in Figure 7.7 which passes through these seven points.

**TO GRAPH AN EQUATION IN THE TWO VARIABLES $x$ AND $y$**

1. Make a table of values.
2. Plot the points which correspond to these pairs of numbers in a Cartesian coordinate system.
3. Connect the points with a line or curve.

Earlier, we graphed equations in one variable on a number line. Generally, for graphing purposes, an equation in one variable such as

$$y = 3 \qquad \text{or} \qquad x + 2 = 0$$

is thought of as an equation in two variables with the coefficient of the missing variable equal to zero. That is,

$$y = 3 \quad \text{is the same as} \quad y = 0 \cdot x + 3$$

and

$$x + 2 = 0 \quad \text{is the same as} \quad x + 0 \cdot y + 2 = 0$$

With this in mind, such equations can be graphed in a Cartesian coordinate system.

**EXAMPLE 1** Graph $y = 3$ in a Cartesian coordinate system.

Solutions to this equation always have $y$-coordinate 3, with $x$-coordinate any number at all. For example, (5, 3), (0, 3), and (−1, 3) are all solutions since in $y = 3$ or $y = 0 \cdot x + 3$, $0 \cdot (\textit{any number}) + 3 = 3$. In the table of values for $y = 3$, if we let $x$ be 0, −1, 1, 2, −2, 3, or −3, in all cases, $y$ is 3.

| $x$ | 0 | 1 | −1 | 2 | −2 | 3 | −3 |
|---|---|---|---|---|---|---|---|
| $y$ | 3 | 3 | 3 | 3 | 3 | 3 | 3 |

Plotting the seven points in the table and drawing the line between them, we see that the graph of $y = 3$ in Figure 7.8 is a straight line parallel to the $x$-axis and 3 units above it.

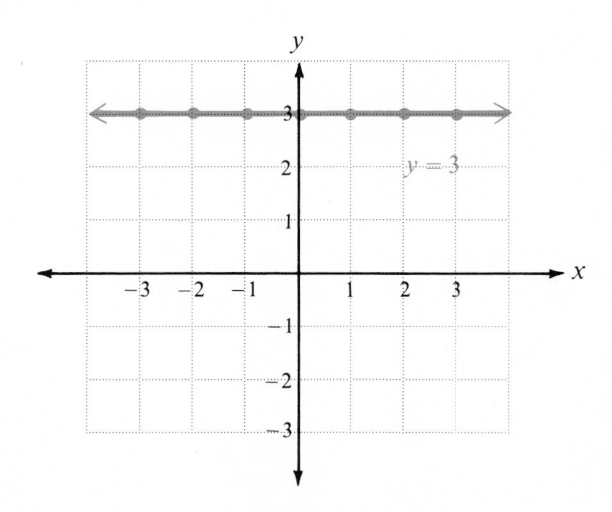


**Figure 7.8**

**EXAMPLE 2** Graph $x + 2 = 0$ in a Cartesian coordinate system.

Solutions to this equation (which is equivalent to $x = -2$) always have $x$-coordinate $-2$, with $y$-coordinate any number at all. For example, $(-2, 0)$, $(-2, 1)$, and $(-2, -1)$ are all solutions since in $x + 2 = 0$ or $x + 0 \cdot y + 2 = 0$, $-2 + 0 \cdot (\textit{any number}) + 2 = -2 + 2 = 0$.

| $x$ | $-2$ | $-2$ | $-2$ | $-2$ | $-2$ | $-2$ | $-2$ |
|---|---|---|---|---|---|---|---|
| $y$ | $0$ | $1$ | $-1$ | $2$ | $-2$ | $3$ | $-3$ |

Plotting the seven points in the table and drawing the line between them as in Figure 7.9, we see that the graph of $x + 2 = 0$ ($x = -2$) is a straight line parallel to the $y$-axis, 2 units to the left of the $y$-axis.

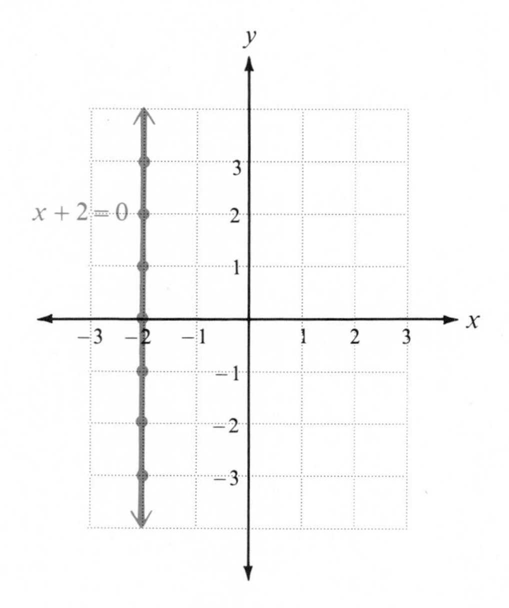


**Figure 7.9**

From now on, given an equation such as $x = 7$ or $y = -3$ and the instruction to find its graph, assume that the equation involves two variables with the coefficient of the missing variable equal to zero. Thus the graph will be plotted in a Cartesian coordinate system and not simply on a number line.

A **linear equation** in two variables has as its graph a straight line. It is easy to recognize a linear equation since the variables are raised to the first power only and it can always be written in the **general form**

$$ax + by + c = 0$$

where $a$, $b$, and $c$ are real number constants, $a$ and $b$ not both zero. (If $a = 0$ and $b = 0$, the equation contains no variable, and it will be true if $c$ is zero and false otherwise.) Linear equations are also called **first-degree equations** since the variables occur to the first power only. All of the equations we have considered thus far have been linear.

EXAMPLE 3

| *Equation* | *Nature* | *General form* | *Constants* |
|---|---|---|---|
| $3x - y = 4$ | linear | $3x - y - 4 = 0$ | $a = 3, b = -1, c = -4$ |
| $x = y + 3$ | linear | $x - y - 3 = 0$ | $a = 1, b = -1, c = -3$ |
| $x + 2 = 0$ | linear | $x + 0 \cdot y + 2 = 0$ | $a = 1, b = 0, c = 2$ |
| $3y - 8 = 0$ | linear | $0 \cdot x + 3y - 8 = 0$ | $a = 0, b = 3, c = -8$ |
| $y = x^2 + 1$ | not linear ($x$ to second power) | | |
| $2x + 3y^3 + 1 = 0$ | not linear ($y$ to third power) | | |
| $xy = -1$ | not linear (cannot be written in form $ax + by + c = 0$) | | |
| $y = 3 - \frac{2}{x}$ | not linear (cannot be written in form $ax + by + c = 0$) | | |

Knowing that the graph of a linear equation is a straight line, we need only two solutions to graph it. In most cases, the two pairs which are easiest to determine are the **intercepts.** The points at which a line crosses the $x$-axis and the $y$-axis are called, respectively, the ***x*-intercept** and the ***y*-intercept.** To find the intercepts, we fill in the following table.

| $x$ | 0 | |
|---|---|---|
| $y$ | | 0 |

Since the $x$-intercept is a point on the $x$-axis, it will have $y$-coordinate 0. Similarly, the $y$-intercept is a point on the $y$-axis, and it will have $x$-coordinate 0. For example, to find the intercepts for the linear equation

$$3x - 5y - 15 = 0,$$

we first substitute 0 for $x$ and solve for $y$.

$$\begin{aligned} 3 \cdot 0 - 5y - 15 &= 0 \\ -5y - 15 &= 0 \\ -5y &= 15 \\ y &= -3 \end{aligned}$$

Then substitute 0 for $y$ and solve for $x$.

$$\begin{aligned} 3x - 5 \cdot 0 - 15 &= 0 \\ 3x - 15 &= 0 \\ 3x &= 15 \\ x &= 5 \end{aligned}$$

The completed table

| $x$ | 0 | 5 |
|---|---|---|
| $y$ | $-3$ | 0 |

displays the $y$-intercept $(0, -3)$ and the $x$-intercept $(5, 0)$. If we plot these two points and draw the line between them, we obtain the graph of $3x - 5y - 15 = 0$ in Figure 7.10.

Clearly, if we know in advance what the graph looks like, it is much easier to plot.

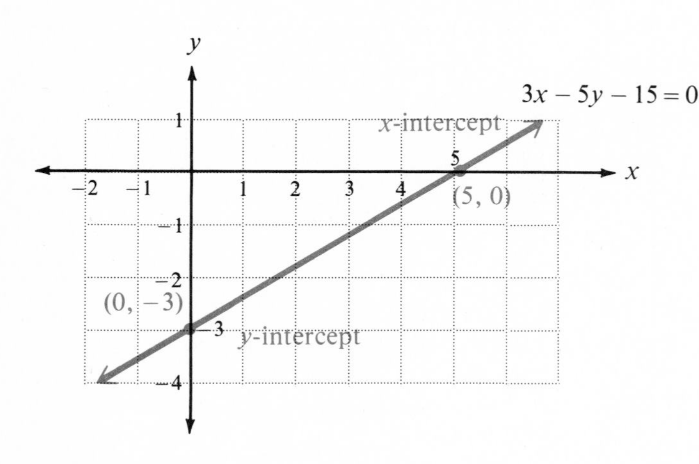


Figure 7.10

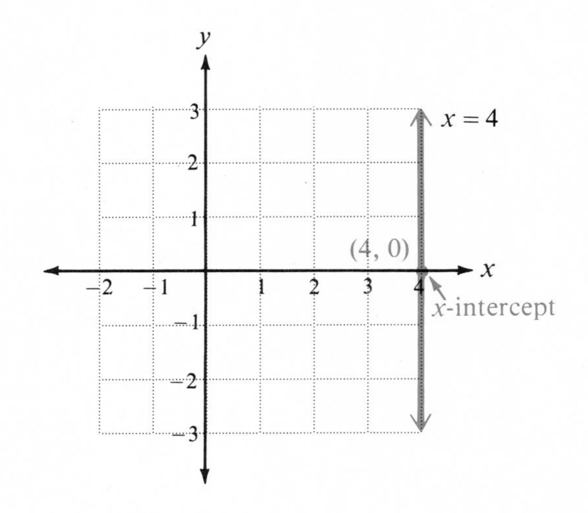


Figure 7.11

**EXAMPLE 4** Graph $x = 4$.

In this case, with no $y$ term in the equation, solutions are of the form $(4, y)$, for any real number $y$. We cannot complete the table

| $x$ | 0 | |
|---|---|---|
| $y$ | | 0 |

since $x$ cannot be zero (why?). However, if $y$ is 0, $x$ is 4 (why?). Thus the line has *no* $y$-intercept and has $x$-intercept $(4, 0)$. An equation of the form $x = 4$ has as its graph the line passing through the point $(4, 0)$ parallel to the $y$-axis, as shown in Figure 7.11.

**EXAMPLE 5** Graph $3y + 5 = 0$.

Since there is no $x$ term, the equation can be simplified to $y = -\frac{5}{3}$, and the solutions are of the form $(x, -\frac{5}{3})$ for any real number $x$. Thus $y$ can never be zero (why?) so there is no $x$-intercept. Also, when $x = 0$, $y = -\frac{5}{3}$ so that $(0, -\frac{5}{3})$ is the $y$-intercept. As before, we should recognize that the graph of this equation is the line passing through the point $(0, -\frac{5}{3})$ parallel to the $x$-axis, which is shown in Figure 7.12.

An equation $ax + by + c = 0$ in which either $a = 0$ (the $x$-term is missing) or $b = 0$ (the $y$-term is missing) will always have only one intercept and its graph will be a straight line parallel to the $x$-axis or $y$-axis, respectively.

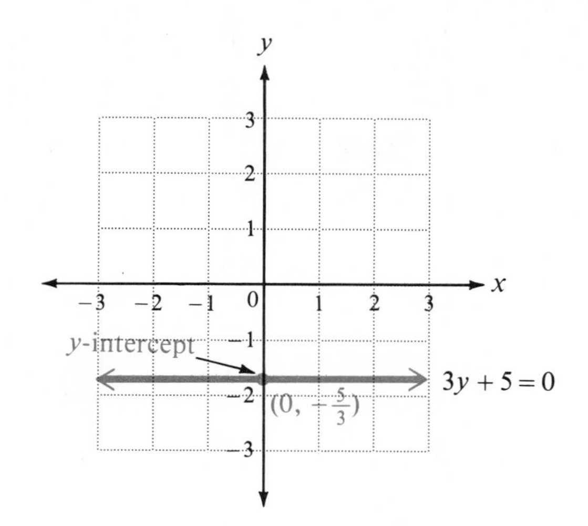


Figure 7.12

Figure 7.13

**EXAMPLE 6** Graph $3x + y = 0$.

In this case, with no constant term, the table of intercepts is the following.

| $x$ | 0 | 0 |
|---|---|---|
| $y$ | 0 | 0 |

That is, the $x$-intercept and $y$-intercept are the same point, the origin (0, 0). When this occurs, we must determine one additional solution. If $x = 1$, $y = -3$; therefore, $(1, -3)$ is a second point (in addition to the origin). See Figure 7.13.

**TO GRAPH A LINEAR EQUATION OF THE FORM $ax + by + c = 0$**

1. If $a \neq 0$ and $b \neq 0$, find the intercepts, plot them, and draw the line through them. If both intercepts are (0, 0) (that is, $c = 0$), find one other point and draw the line between (0, 0) and the additional point.
2. If $a = 0$, the equation becomes $by + c = 0$ or $y = -\frac{c}{b}$. The graph is a line through $y$-intercept $\left(0, -\frac{c}{b}\right)$ parallel to the $x$-axis.
3. If $b = 0$, the equation becomes $ax + c = 0$ or $x = -\frac{c}{a}$. The graph is a line through $x$-intercept $\left(-\frac{c}{a}, 0\right)$ parallel to the $y$-axis.

## EXERCISES 7.2

**1.** *Which of the following are linear equations? Explain.*

**(a)** $x + y + 1 = 0$ **(b)** $x^2 + xy + y^2 = 7$ **(c)** $x - xy = 0$

**(d)** $\frac{2}{x} + \frac{3}{y} = 1$ **(e)** $x = 5$ **(f)** $3y - 7 = 0$

2. The graph of a linear equation is always a(n) ______________________.

3. Explain why linear equations are also called first-degree equations.

4. The general form of a linear equation is ______________________.

5. An equation of the type $ax + c = 0$ has as its graph a straight line parallel to ______________________.

6. An equation of the type $by + c = 0$ has as its graph a straight line parallel to ______________________.

7. An equation of the type $ax + by = 0$ ($a \neq 0$ and $b \neq 0$) has $x$-intercept and $y$-intercept ______________________.

8. A point at which a graph crosses the $x$-axis is called the ________________.

9. A point at which a graph crosses the $y$-axis is called the ________________.

10. To graph a general linear equation, only **(a)** ______________ points are necessary and often the best points to use are the **(b)** ______________.

11. The coordinates of the origin are ______________.

12. Every point on the $x$-axis has __________-coordinate equal to zero.

13. Every point on the $y$-axis has __________-coordinate equal to zero.

14. Every point above the $x$-axis has __________-coordinate equal to a positive number.

15. Every point to the left of the $y$-axis has __________-coordinate equal to a negative number.

16. Points that have both coordinates positive are in quadrant __________.

17. Points that have both coordinates negative are in quadrant __________.

18. Points which have $x$-coordinate negative and $y$-coordinate positive are in quadrant __________.

19. Points which have $x$-coordinate positive and $y$-coordinate negative are in quadrant __________.

20. A configuration in which graphs of equations are plotted is called ______________________.

*Determine the intercepts and graph the following:*

21. $3x + y = 6$

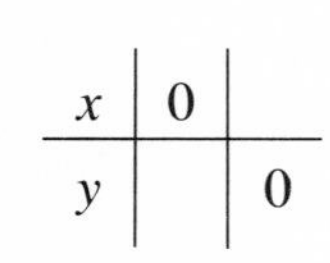

| $x$ | 0 | |
|---|---|---|
| $y$ | | 0 |

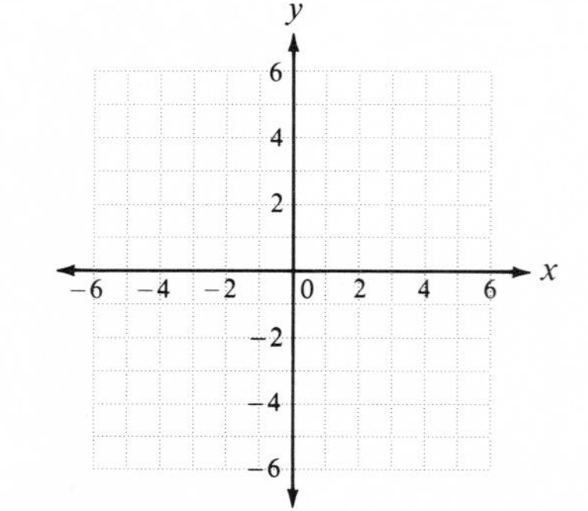


22. $x - y = 2$

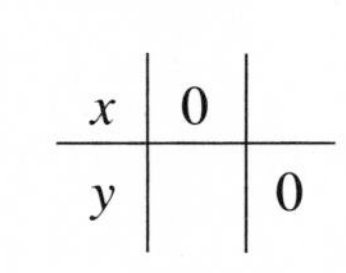

| $x$ | 0 | |
|---|---|---|
| $y$ | | 0 |

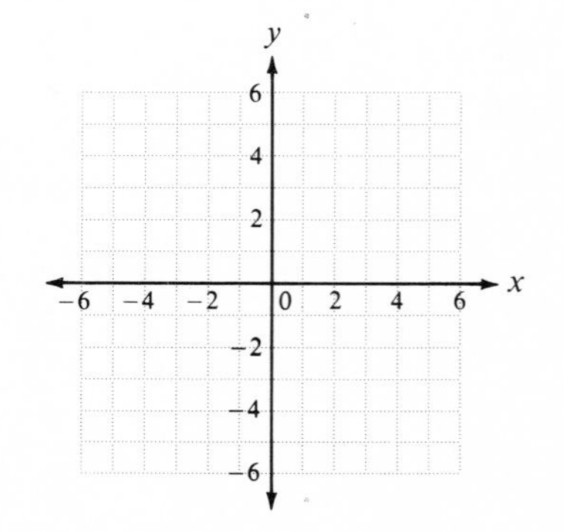

**23.** $x - y = 0$

| $x$ | 0 | |
|---|---|---|
| $y$ | | 0 |

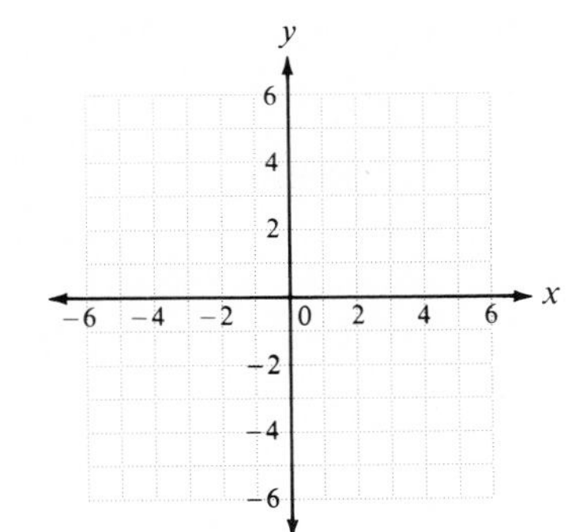

**24.** $3x - 7 = 0$

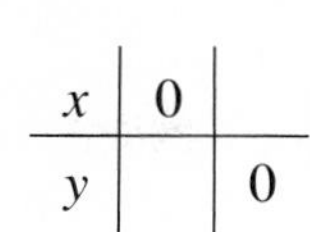

| $x$ | 0 | |
|---|---|---|
| $y$ | | 0 |

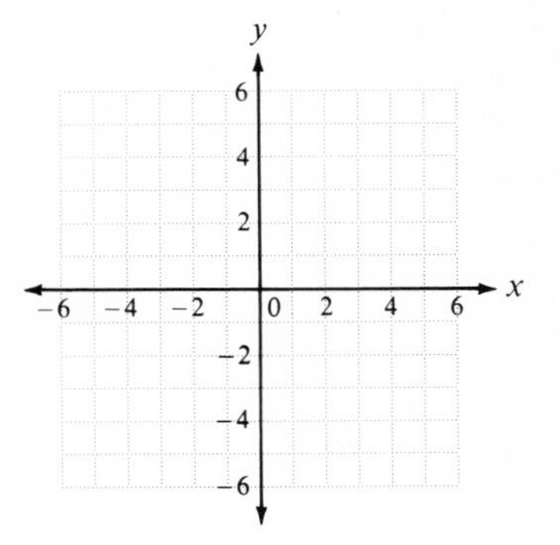

**25.** $y = -1$

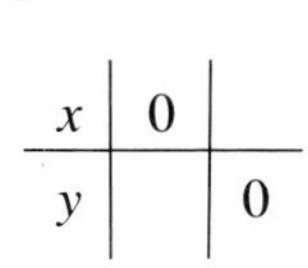

| $x$ | 0 | |
|---|---|---|
| $y$ | | 0 |

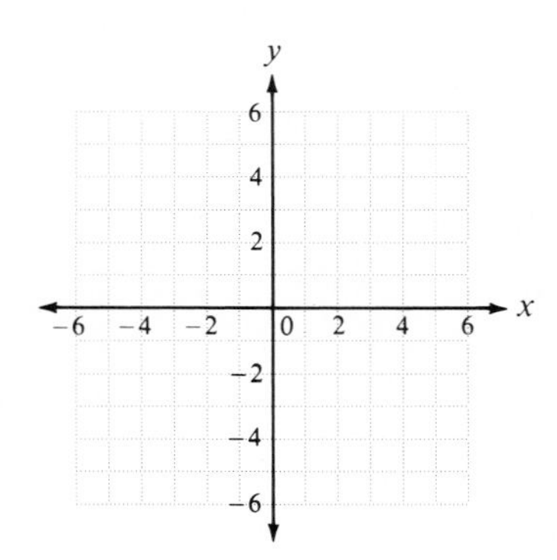

**26.** $3x - 2y = 0$

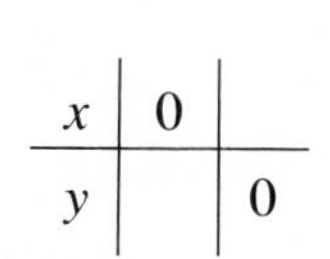

| $x$ | 0 | |
|---|---|---|
| $y$ | | 0 |

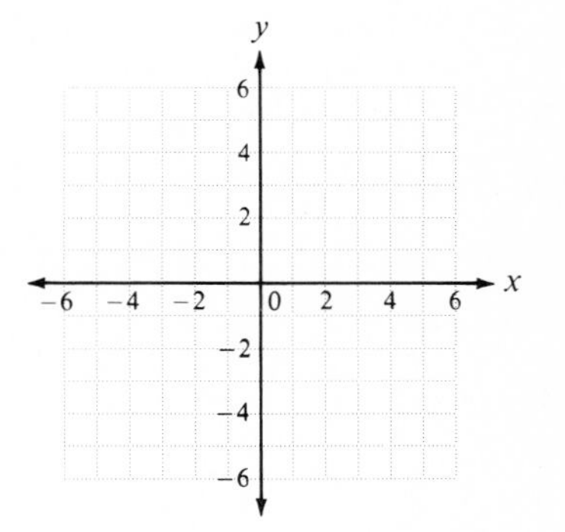

**27.** What are the intercepts of the line $x = 0$? What is its graph?

**28.** What are the intercepts of the line $y = 0$? What is its graph?

---

ANSWERS: **1.** Only (a), (e), and (f) are linear equations. **2.** straight line **3.** The variables occur to the first power only. **4.** $ax + by + c = 0$ $(a \neq 0$ or $b \neq 0)$ **5.** $y$-axis **6.** $x$-axis **7.** (0, 0) **8.** $x$-intercept **9.** $y$-intercept **10 (a)** two **(b)** intercepts **11.** (0, 0) **12.** $y$ **13.** $x$ **14.** $y$ **15.** $x$ **16.** I **17.** III **18.** II **19.** IV **20.** a Cartesian coordinate system or an $x$, $y$-coordinate system

**21.** $x$-intercept (2, 0); $y$-intercept (0, 6)

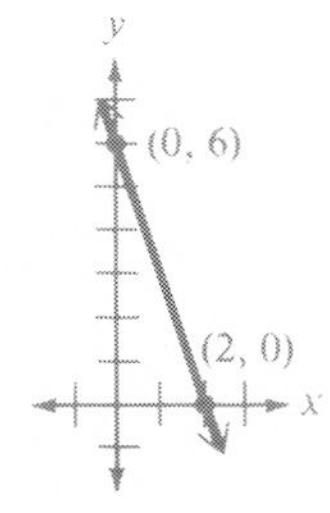


**22.** $x$-intercept (2, 0); $y$-intercept (0, −2)

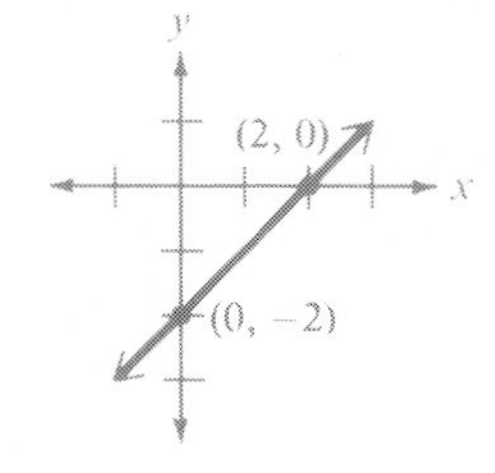


**23.** $x$-intercept (0, 0); $y$-intercept (0, 0)

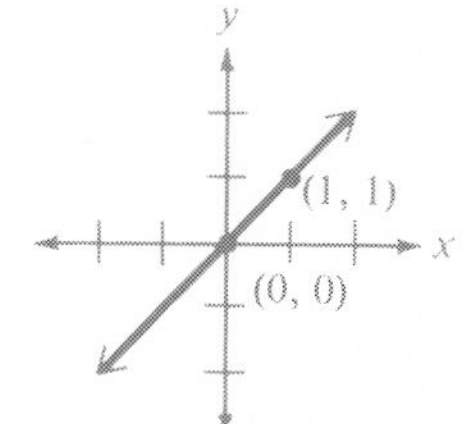


**24.** $x$-intercept $\left(\frac{7}{3}, 0\right)$; no $y$-intercept

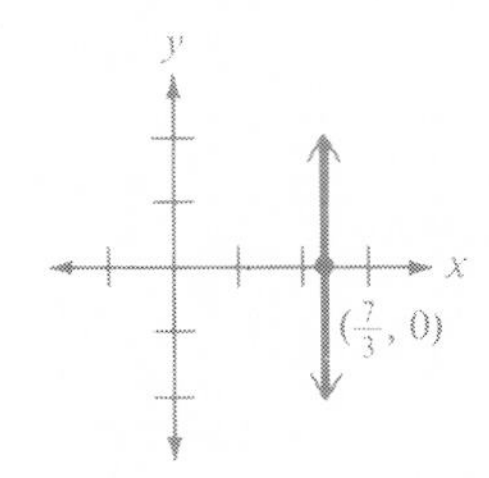


**25.** no $x$-intercept; $y$-intercept (0, −1)

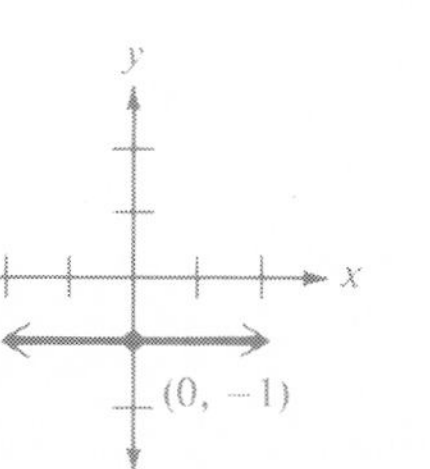


**26.** $x$-intercept (0, 0); $y$-intercept (0, 0)

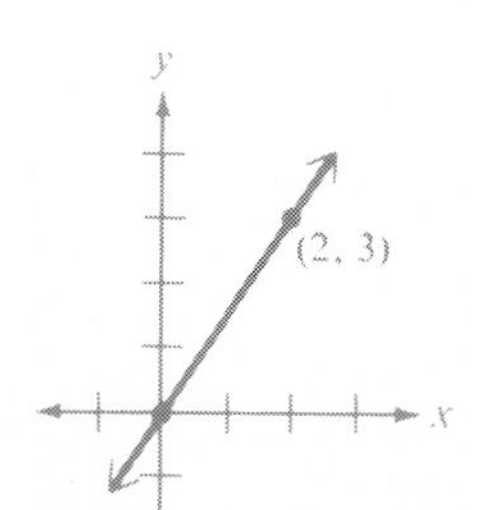


**27.** $x$-intercept (0, 0); every point on the $y$-axis is a $y$-intercept; the graph is the $y$-axis.

**28.** Every point on the $x$-axis is an $x$-intercept; $y$-intercept (0, 0); the graph is the $x$-axis.

## 7.3 Slope of a Line

We have seen that the graph of a given linear equation may be horizontal (parallel to the $x$-axis), vertical (parallel to the $y$-axis), may "slope" upward from lower left to upper right, or may "slope" downward from upper left to lower right. The *slope* of a line (or lack of slope) can be precisely defined.

Suppose that two points $P$ and $Q$ with coordinates $(x_1, y_1)$ and $(x_2, y_2)$, respectively, lie on the same straight line. (See Figure 7.14.) The difference $y_2 - y_1$ is the change in $y$-coordinates while $x_2 - x_1$ is the change in $x$-coordinates of the two points. The **slope** of the line (often denoted by the letter $m$) is given by

$$m = \frac{y_2 - y_1}{x_2 - x_1} = \frac{\text{change in } y\text{-coordinates}}{\text{change in } x\text{-coordinates}}.$$

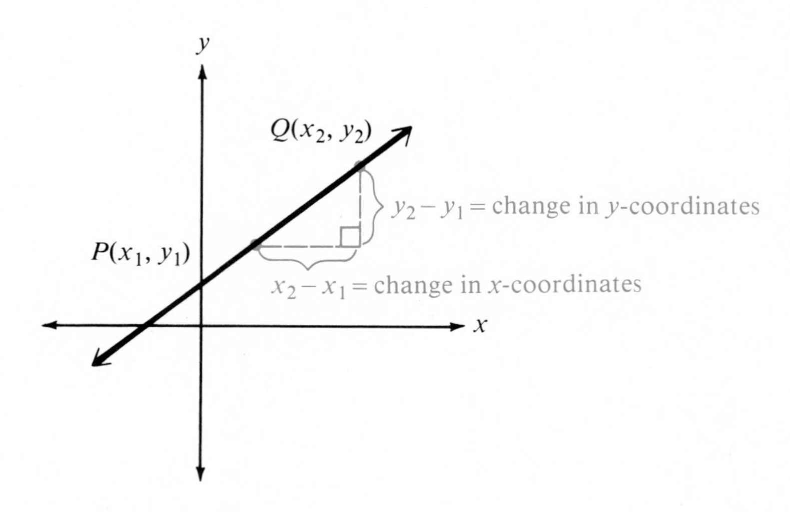


**Figure 7.14**

EXAMPLE 1 Find the slope of the line passing through the two points (4, 3) and (1, 2).

See the graph in Figure 7.15. Suppose we identify point $P(x_1, y_1)$ with (1, 2) and point $Q(x_2, y_2)$ with (4, 3). The slope will then be given by

$$m = \frac{y_2 - y_1}{x_2 - x_1} = \frac{3 - 2}{4 - 1} = \frac{1}{3}.$$

What happens if we identify $P(x_1, y_1)$ with (4, 3) and, $Q(x_2, y_2)$ with (1, 2)? In this case we have

$$m = \frac{y_2 - y_1}{x_2 - x_1} = \frac{2 - 3}{1 - 4} = \frac{-1}{-3} = \frac{1}{3}.$$

Thus we see that the slope is the same regardless of how the two points are identified.

CAUTION: It is important to make sure that the coordinates are subtracted in the same order. That is, *do not compute*

$$\frac{y_2 - y_1}{x_1 - x_2}. \quad \text{This is wrong}$$

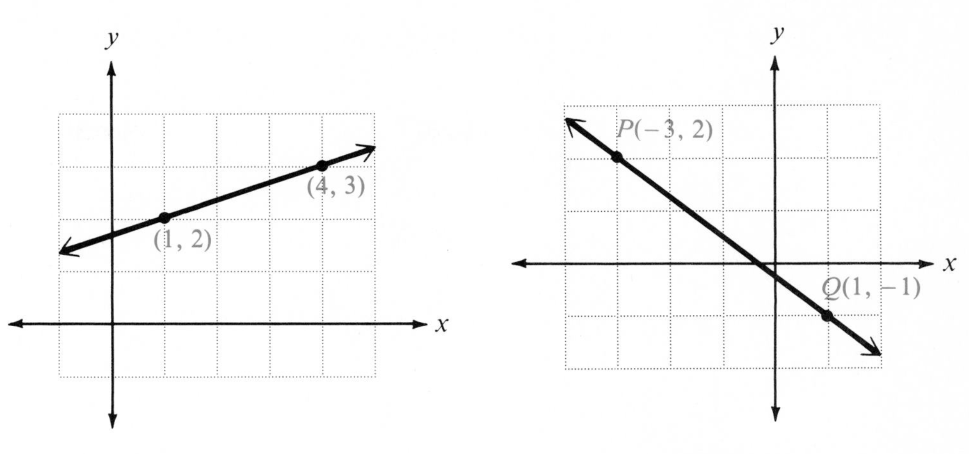


Figure 7.15

Figure 7.16

EXAMPLE 2 Find the slope of the line passing through the two points (−3, 2) and (1, −1).

Let us identify $P(x_1, y_1)$ with (−3, 2) and $Q(x_2, y_2)$ with (1, −1) as in Figure 7.16. The slope is given by

$$m = \frac{y_2 - y_1}{x_2 - x_1} = \frac{(-1) - (2)}{(1) - (-3)} \qquad \text{Watch signs}$$

$$= \frac{-3}{1 + 3} = -\frac{3}{4}.$$

EXAMPLE 3 Find the slope of the line passing through the two points (3, 2) and (−1, 2).

Identifying $P(x_1, y_1)$ with (3, 2) and $Q(x_2, y_2)$ with (−1, 2) in Figure 7.17, we obtain

$$m = \frac{y_2 - y_1}{x_2 - x_1} = \frac{2 - 2}{-1 - 3} = \frac{0}{-4} = 0.$$

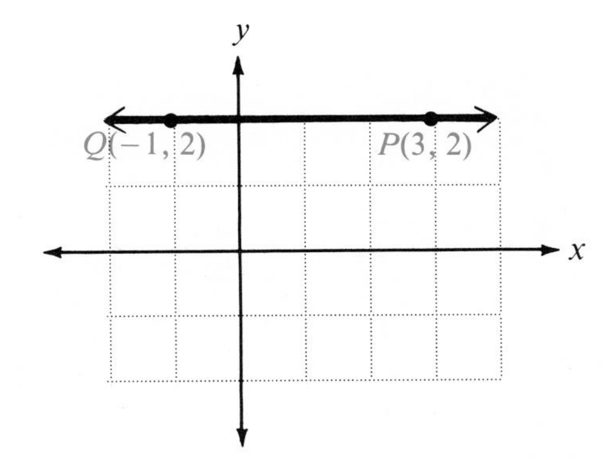


Figure 7.17

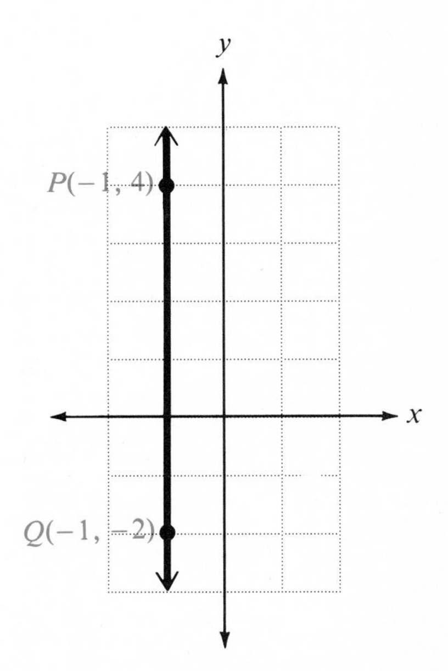


Figure 7.18

**EXAMPLE 4** Find the slope of the line passing through the two points (−1, 4) and (−1, −2).

Identifying $P(x_1, y_1)$ with (−1, 4) and $Q(x_2, y_2)$ with (−1, −2) in Figure 7.18 (on p. 329), we obtain

$$m = \frac{y_2 - y_1}{x_2 - x_1} = \frac{-2 - 4}{-1 - (-1)} = \frac{-6}{-1 + 1}$$

$$= \frac{-6}{0} \quad \text{which is undefined.}$$

In view of the above examples, the next rule seems reasonable.

**THE SLOPE OF A LINE**

1. A line which "slopes" from lower left to upper right has **slope** a positive number.
2. A line which "slopes" from upper left to lower right has **slope** a negative number.
3. A horizontal line (parallel to the $x$-axis) has **slope** zero.
4. A vertical line (parallel to the $y$-axis) has **no slope.**

The graphs in Figure 7.19 show the four possibilities listed in the rule.

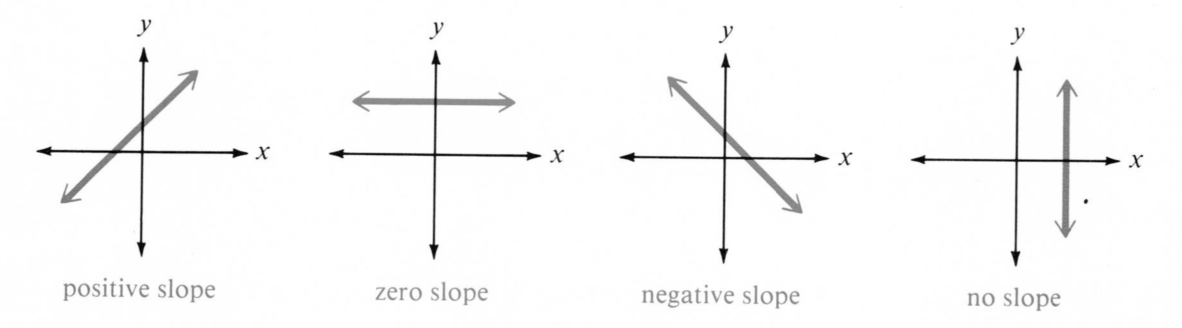


**Figure 7.19**

The notion of slope can be useful in solving a variety of problems. For example, suppose that we are given three points $P$, $Q$, and $R$ and we wish to know if the three points are **collinear** (lie on the same straight line). We first determine the slope of the line through $P$ and $Q$ and then determine the slope of the line through $Q$ and $R$. The points are on the same straight line through $P$, $Q$, and $R$ provided the two slopes are the same. (Why?)

**EXAMPLE 5** Determine whether $P(5, 3)$, $Q(1, 1)$, and $R(-3, -1)$ are collinear.

The slope of the line through $P$ and $Q$ is

$$m_1 = \frac{1 - 3}{1 - 5} = \frac{-2}{-4} = \frac{1}{2}.$$

The slope of the line through $Q$ and $R$ is

$$m_2 = \frac{-1 - 1}{-3 - 1} = \frac{-2}{-4} = \frac{1}{2}.$$

Since $m_1 = m_2$, the three points must all lie on the same straight line (that is, they are collinear).

The notion of the slope of a line can also be used to determine when two lines are **parallel** (never intersect) or **perpendicular** (intersect at right angles).

Given two distinct lines with slopes $m_1$ and $m_2$:

**1.** If $m_1 = m_2$, the lines are parallel (equal slopes determine parallel lines).

**2.** If $m_1 = -\frac{1}{m_2}$, the lines are perpendicular (slopes which are negative reciprocals determine perpendicular lines).

EXAMPLE 6 Verify that the line $l_1$ between (1, 8) and (−2, −1) and the line $l_2$ between (2, 4) and (−1, −5) in Figure 7.20 are parallel.

$$\text{The slope of } l_1 \text{ is } m_1 = \frac{8 - (-1)}{1 - (-2)} = \frac{8 + 1}{1 + 2}$$
$$= \frac{9}{3} = 3.$$
$$\text{The slope of } l_2 \text{ is } m_2 = \frac{4 - (-5)}{2 - (-1)} = \frac{4 + 5}{2 + 1}$$
$$= \frac{9}{3} = 3.$$

Since $m_1 = m_2$, the lines are parallel.

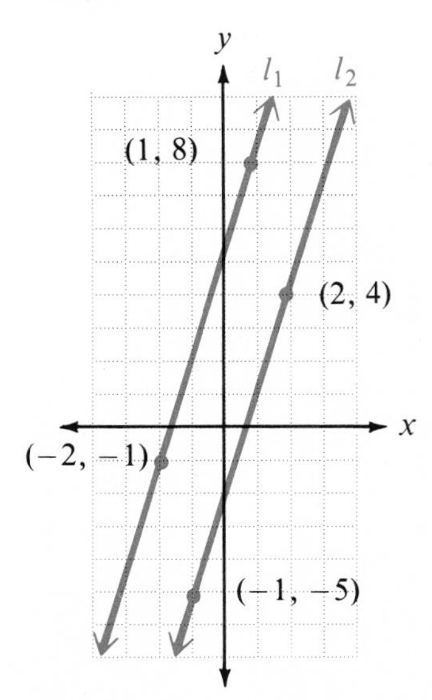


Figure 7.20

**EXAMPLE 7** Verify that the line $l_1$ between (0, 2) and (5, 6) and the line $l_2$ between (4, −4) and (−4, 6) in Figure 7.21 are perpendicular.

$$\text{The slope of } l_1 \text{ is } m_1 = \frac{2-6}{0-5} = \frac{-4}{-5} = \frac{4}{5}.$$

$$\text{The slope of } l_2 \text{ is } m_2 = \frac{-4-6}{4-(-4)} = \frac{-10}{8}$$

$$= -\frac{5}{4}.$$

Since

$$m_1 = \frac{4}{5} \quad \text{and} \quad -\frac{1}{m_2} = -\frac{1}{-\frac{5}{4}} = (-1)\cdot\left(-\frac{4}{5}\right) = \frac{4}{5},$$

$$m_1 = -\frac{1}{m_2}$$

so that the lines are perpendicular.

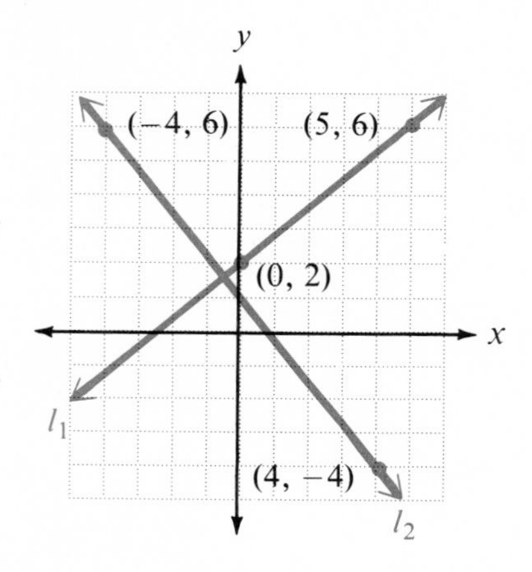


**Figure 7.21**

**EXAMPLE 8** Find the slope of the line with equation $x - 3y + 6 = 0$.

We must determine two points on the line. When $x = 0$,

$$\begin{aligned} 0 - 3y + 6 &= 0 \\ -3y &= -6 \\ y &= 2. \end{aligned}$$

Thus (0, 2) is one point on the line. When $x = 3$,

$$\begin{aligned} 3 - 3y + 6 &= 0 \\ -3y + 9 &= 0 \\ -3y &= -9 \\ y &= 3. \end{aligned}$$

Thus (3, 3) is another point on the line. The slope of the line is

$$m = \frac{2-3}{0-3} = \frac{-1}{-3} = \frac{1}{3}.$$

## EXERCISES 7.3

**1.** The slope of a line passing through the two points $P(x_1, y_1)$ and $Q(x_2, y_2)$ is

$m =$ ______________________.

*Answer Exercises 2–7 with one of the phrases* **(a)** positive slope **(b)** negative slope **(c)** zero slope **(d)** no slope.

**2.** A line parallel to the $y$-axis has ______________________.

**3.** A line which "slopes" from lower left to upper right has ______________

______________________.

**4.** A line which "slopes" from upper left to lower right has ______________

______________________.

**5.** A line parallel to the $x$-axis has ______________________.

**6.** The $x$-axis has ______________________.

**7.** The $y$-axis has ______________________.

*Find the slope of the line passing through the given pair of points.*

**8.** $(7, -1)$ and $(3, 3)$

**9.** $(-5, 2)$ and $(-1, -6)$

**10.** $(-4, 1)$ and $(-1, 3)$

**11.** $(4, 2)$ and $(4, -2)$

**12.** $(1, 7)$ and $(3, 3)$

**13.** $(-2, 7)$ and $(-5, 7)$

**14.** A line which is parallel to the line through $(5, -1)$ and $(3, 3)$ must have slope

______________________. Explain.

**15.** A line which is perpendicular to the line through $(5, -1)$ and $(3, 3)$ must

have slope ______________________. Explain.

**16.** Verify that the line $l_1$ between $(1, 5)$ and $(-2, -1)$ and the line $l_2$ between $(-1, -7)$ and $(4, 3)$ are parallel.

**17.** Verify that the line $l_1$ between (0, 2) and (1, −1) and the line $l_2$ between (0, −1) and (3, 0) are perpendicular.

**18.** Determine the slope of the line with equation $5x - 2y = 8$.

**19.** Are the points $P(-2, 1)$, $Q(2, 3)$, and $R(0, 2)$ collinear? Explain.

**20.** Is the triangle with vertices $A(-2, 7)$, $B(-3, 3)$, and $C(-6, 8)$ a right triangle? Explain.
[*Hint:* Determine if the lines passing through two vertices are perpendicular.]

**21.** Find the intercepts and graph the line with equation $3x - 2y - 6 = 0$. What is the slope of this line?

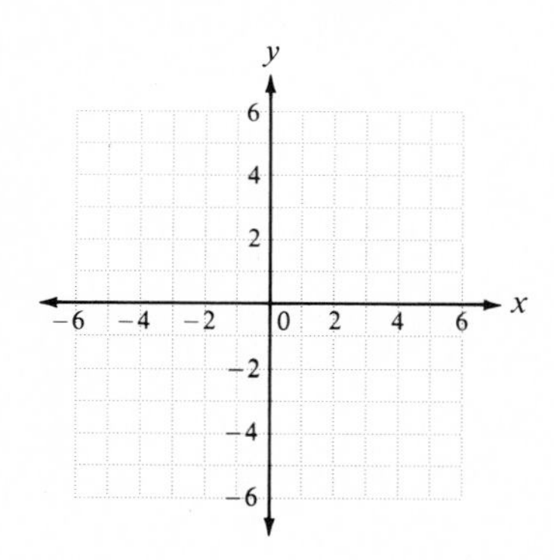


**22.** Find the intercepts and graph the line with equation $3y - 4 = 0$. What is the slope of this line?

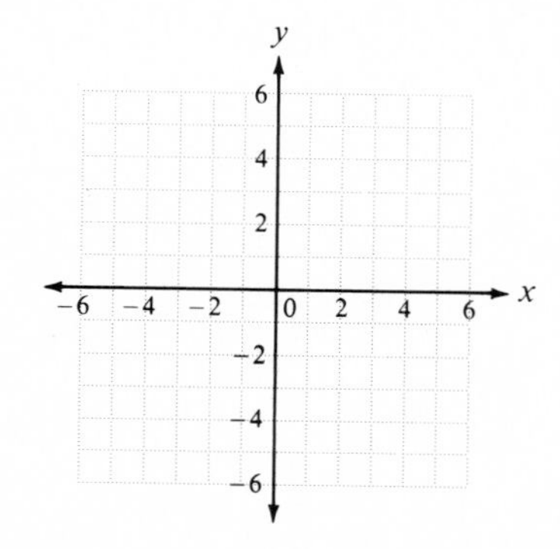

ANSWERS: **1.** $m = \frac{y_2 - y_1}{x_2 - x_1}$ **2.** no slope **3.** positive slope **4.** negative slope **5.** zero slope **6.** zero slope **7.** no slope **8.** $-1$ **9.** $-2$ **10.** $\frac{2}{3}$ **11.** no slope **12.** $-2$ **13.** 0 **14.** $-2$; Parallel lines have the same slope and the given line has slope $-2$. **15.** $\frac{1}{2}$; Perpendicular lines must have slopes which are negative reciprocals and the given line has slope $-2$, the negative reciprocal of which is $\frac{1}{2}$. **16.** both have slope 2 **17.** $m_1 = -3$ and $m_2 = \frac{1}{3}$ (negative reciprocals) **18.** $\frac{5}{2}$ **19.** yes. The slope of the line through $P$ and $Q$ is $\frac{1}{2}$, and the slope of the line through $Q$ and $R$ is $\frac{1}{2}$. **20.** Yes. Line through $A$ and $B$ has slope 4, and the line through $A$ and $C$ has slope $-\frac{1}{4}$.

**21.** $x$-intercept (2, 0);
$y$-intercept (0, $-3$);
slope 3/2

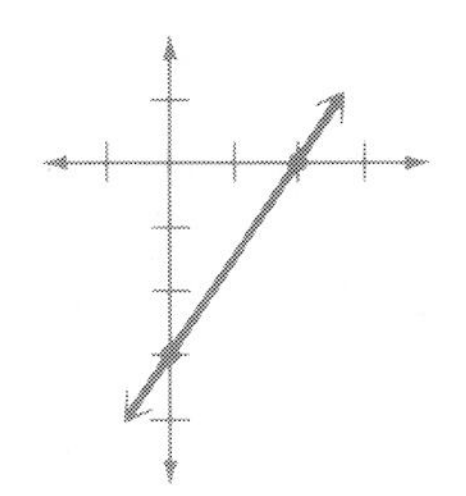

**22.** no $x$-intercept;
$y$-intercept (0, 4/3);
slope 0

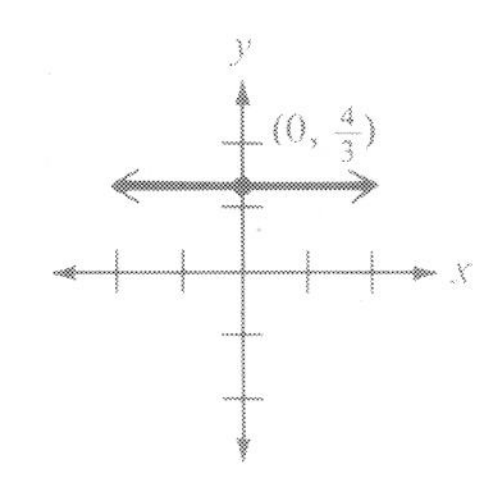


## 7.4 Forms of the Equation of a Line

The **general form** of an equation of a line is

$$\mathbf{ax + by + c = 0} \quad (a \neq 0 \text{ or } b \neq 0).$$

In this section, we will consider three additional forms, the *point-slope form,* the *two-point form,* and the *slope-intercept form.* All of these forms depend directly or indirectly on the notion of slope of a line.

**POINT-SLOPE FORM**

To find the equation of a line which has slope $m$ and passes through the point $(x_1, y_1)$, substitute into the formula

$$\mathbf{y - y_1 = m(x - x_1).}$$

If we think of a general point on the line through $(x_1, y_1)$ with slope $m$ as having coordinates $(x, y)$, the slope $m$ is given by

$$m = \frac{y - y_1}{x - x_1},$$

where $(x, y)$ plays the role of the point $(x_2, y_2)$ in the formula for $m$. Multiplying both sides by $(x - x_1)$, we obtain

$$(x - x_1) \cdot m = (x - x_1) \cdot \frac{(y - y_1)}{(x - x_1)},$$

which is equivalent to the point-slope form $y - y_1 = m(x - x_1)$.

**EXAMPLE 1** Find the point-slope form of the equation of the line with slope $\frac{1}{2}$ passing through the point $(-3, 1)$. Also write the general form of the equation of this line.

We have $(x_1, y_1) = (-3, 1)$ and $m = \dfrac{1}{2}$ so substituting in $y - y_1 = m(x - x_1)$ we obtain the point-slope form

$$y - 1 = \frac{1}{2}(x - (-3)). \qquad \text{Watch the sign}$$

To obtain the general form of the equation, we eliminate fractions, clear parentheses, and then collect all terms on the left side of the equation.

$$\begin{aligned}
y - 1 &= \frac{1}{2}(x + 3) & \\
2(y - 1) &= 2 \cdot \frac{1}{2}(x + 3) & \text{Multiply by LCD} = 2 \\
2y - 2 &= x + 3 & \text{Clear parentheses} \\
2y - 2 - x - 3 &= x + 3 - x - 3 & \text{Subtract } x \text{ and } 3 \\
-x + 2y - 5 &= 0 & \\
x - 2y + 5 &= 0 & \text{Multiply by } -1
\end{aligned}$$

We usually write the general form with the coefficient of $x$, which is $a$, positive instead of negative.

**EXAMPLE 2** Find the general form of the equation of a line with $y$-intercept $(0, -2)$ and slope 3.

We are given a point $(0, -2)$ and slope 3. Substitute in the point-slope form.

$$\begin{aligned}
y - (-2) &= 3(x - 0) & \text{Watch signs} \\
y + 2 &= 3x & \text{Clear parentheses} \\
0 &= 3x - y - 2 & \text{Subtract } y \text{ and } 2
\end{aligned}$$

The general form is $3x - y - 2 = 0$.

**EXAMPLE 3** Find the general form of the equation of a line passing through the point $(-1, 2)$ and perpendicular to a line with slope 3.

Since the line is perpendicular to a line with slope 3, it must have slope $-\frac{1}{3}$ (negative reciprocal of 3). We substitute in the point-slope form.

$$y - 2 = -\frac{1}{3}(x - (-1))$$

$$y - 2 = -\frac{1}{3}(x + 1) \qquad \text{Watch signs}$$

$$3(y - 2) = 3\left(-\frac{1}{3}\right)(x + 1) \qquad \text{Clear fractions}$$

$$3(y - 2) = -(x + 1)$$

$$3y - 6 = -x - 1 \qquad \text{Clear parentheses}$$

$$x + 3y - 5 = 0 \qquad \text{Add } x \text{ and } 1$$

The general form is $x + 3y - 5 = 0$.

The point-slope form enables us to find the equation of a line when we are given a point on the line and the slope of the line. A special case occurs when the given point is the $y$-intercept.

**SLOPE-INTERCEPT FORM**

To find the equation of a line that has slope $m$ and $y$-intercept $(0, b)$, substitute into the formula

$$y = mx + b.$$

Use the $y$-intercept $(0, b)$ as the given point in the point-slope form, and substitute

$$y - b = m(x - 0)$$
$$y - b = mx$$
$$y = mx + b$$

to obtain the slope-intercept form.

**EXAMPLE 4** Find the slope-intercept form of the equation of the line with slope $-2$ and $y$-intercept $(0, 7)$.

Substituting $-2$ for $m$ and 7 for $b$ we obtain

$$y = -2x + 7.$$

The importance of the slope-intercept form is that once this form is obtained, the slope (the coefficient of $x$) can be read directly as can the $y$-coordinate of the $y$-intercept (the constant term). Given any form of the equation of a line, if the equation is solved for $y$, the coefficient of $x$ is the slope and the constant term is the $y$-coordinate of the $y$-intercept.

**EXAMPLE 5** What is the slope and $y$-intercept of the line with the following equation?

$$y = \frac{1}{2}x - 3$$

Since the equation is already solved for $y$ (that is, already in slope-intercept form), the slope is 1/2 and the $y$-intercept is $(0, -3)$. (Note that $b = -3$ not 3.) The following diagram might be helpful for remembering these facts.

$$y = \frac{1}{2}x - 3 = \frac{1}{2}x + (-3) = mx + b$$

$$m = \frac{1}{2} = \text{slope} \qquad (0, -3) = y\text{-intercept } [-3 = b]$$

**EXAMPLE 6** Find the slope and $y$-intercept of the line with equation

$$2x - 3y + 5 = 0.$$

We first solve the equation for $y$; that is, we write the equation in slope-intercept form.

$$-3y = -2x - 5 \qquad \text{Subtract } 2x \text{ and } 5$$

$$\left(-\frac{1}{3}\right)(-3y) = \left(-\frac{1}{3}\right)(-2x - 5) \qquad \text{Multiply by } -\frac{1}{3}$$

$$y = \frac{2}{3}x + \frac{5}{3} \qquad \text{Clear parentheses}$$

$$m = \frac{2}{3} = \text{slope} \qquad \left(0, \frac{5}{3}\right) = y\text{-intercept } \left[\frac{5}{3} = b\right]$$

The slope is $\frac{2}{3}$ and the $y$-intercept is $\left(0, \frac{5}{3}\right)$.

In the last section we found the slope of the line with equation $x - 3y + 6 = 0$ by first determining two points on the line and then using the definition of the slope. We used the two points $(0, 2)$ and $(3, 3)$, and substituted to obtain

$$m = \frac{y_2 - y_1}{x_2 - x_1} = \frac{2 - 3}{0 - 3} = \frac{-1}{-3} = \frac{1}{3}.$$

With what we have just learned, we can solve the same problem in a much simpler way by solving for $y$.

$$-3y = -x - 6 \qquad \text{Subtract } x \text{ and } 6$$

$$y = \frac{1}{3}x + 2 \qquad \text{Divide through by } -3$$

$$\downarrow$$

$$m = \text{slope} = \frac{1}{3}$$

The ability to determine the slope of a line more quickly by solving for $y$ is also helpful when considering parallel and perpendicular lines.

**EXAMPLE 7** Determine whether the lines with equations $3x - 2y + 7 = 0$ and $2x + 3y - 6 = 0$ are parallel.

We solve each equation for $y$ (write each in slope-intercept form) to determine the slope.

$$\begin{aligned} 3x - 2y + 7 &= 0 \\ -2y &= -3x - 7 \\ y &= \frac{3}{2}x + \frac{7}{2} \\ &\downarrow \\ m_1 &= \frac{3}{2} \end{aligned} \qquad \begin{aligned} 2x + 3y - 6 &= 0 \\ 3y &= -2x + 6 \\ y &= -\frac{2}{3}x + 2 \\ &\downarrow \\ m_2 &= -\frac{2}{3} \end{aligned}$$

Since $m_1 = \frac{3}{2}$ and $m_2 = -\frac{2}{3}$ are negative reciprocals, the two lines are perpendicular, not parallel.

**EXAMPLE 8** Determine whether the lines with equations $x - 4y + 2 = 0$ and $3x - 12y + 6 = 0$ are parallel.

Solve each equation for $y$.

$$\begin{aligned} x - 4y + 2 &= 0 \\ -4y &= -x - 2 \\ y &= \frac{1}{4}x + \frac{2}{4} \\ y &= \frac{1}{4}x + \frac{1}{2} \\ &\downarrow \\ &m_1 \end{aligned} \qquad \begin{aligned} 3x - 12y + 6 &= 0 \\ -12y &= -3x - 6 \\ y &= \frac{3}{12}x + \frac{6}{12} \\ y &= \frac{1}{4}x + \frac{1}{2} \\ &\downarrow \\ &m_2 \end{aligned}$$

Since $m_1 = m_2$ we are tempted to conclude that the two lines are parallel. However, the $y$-intercepts are also equal (both are $(0, \frac{1}{2})$) so the lines coincide (the equations determine the same line) and are really not parallel.

**EXAMPLE 9** Find the equation of the line passing through the point $(-2, 5)$ parallel to the line with equation $4x + 2y - 9 = 0$.

Since we want a line parallel to the line with equation $4x + 2y - 9 = 0$, we must find the slope of this line. Solving for $y$,

$$\begin{aligned} 2y &= -4x + 9 \\ y &= -2x + \frac{9}{2}. \end{aligned}$$

The slope is $-2$, and this is also the slope of the desired line (parallel lines have equal slopes) which must pass through $(-2, 5)$. To find the desired line we substitute in the point-slope form.

$$\begin{aligned} y - y_1 &= m(x - x_1) \\ y - 5 &= -2(x - (-2)) \\ y - 5 &= -2(x + 2) \\ y - 5 &= -2x - 4 \\ 2x + y - 1 &= 0 \end{aligned}$$

The general form of the equation of the line through $(-2, 5)$ parallel to the line $4x + 2y - 9 = 0$ is $2x + y - 1 = 0$.

The third form that we consider involves finding the line when we are given two points on that line.

**TWO-POINT FORM**

To find the equation of a line that passes through the two distinct points $P(x_1, y_1)$ and $Q(x_2, y_2)[x_1 \neq x_2]$, substitute into the formula

$$y - y_1 = \frac{y_2 - y_1}{x_2 - x_1}(x - x_1).$$

The two-point form of the equation of a line follows directly from the point-slope form and the definition of the slope of a line,

$$m = \frac{y_2 - y_1}{x_2 - x_1}.$$

Substituting into $y - y_1 = m(x - x_1)$,

$$y - y_1 = \frac{y_2 - y_1}{x_2 - x_1}(x - x_1).$$

EXAMPLE 10 Find the general form of the equation of a line passing through the points $(4, -1)$ and $(2, 7)$.

We let $(x_1, y_1) = (4, -1)$ and $(x_2, y_2) = (2, 7)$ and substitute into the two-point form.

$$\begin{aligned}
y - (-1) &= \frac{7 - (-1)}{2 - 4}(x - 4)\\
y + 1 &= \frac{8}{-2}(x - 4)\\
y + 1 &= -4(x - 4)\\
y + 1 &= -4x + 16\\
4x + y - 15 &= 0
\end{aligned}$$

The general form of the equation of the line passing through $(4, -1)$ and $(2, 7)$ is $4x + y - 15 = 0$. If we identify the points in the other order, that is $(x_1, y_1) = (2, 7)$ and $(x_2, y_2) = (4, -1)$, we still obtain the same equation.

$$\begin{aligned}
y - 7 &= \frac{-1 - 7}{4 - 2}(x - 2)\\
y - 7 &= \frac{-8}{2}(x - 2)\\
y - 7 &= -4(x - 2)\\
y - 7 &= -4x + 8\\
4x + y - 15 &= 0
\end{aligned}$$

The names of the forms of the equations identify the quantities necessary for determining each equation.

The *point-slope form* requires knowing a *point* and the *slope*.
The *slope-intercept form* requires knowing (actually identifies) the *slope* and *y-intercept*.
The *two-point form* requires knowing *two points* on the line.

It is important to memorize each of the forms of the equation of a line and be able to identify which form applies in a given situation.

## EXERCISES 7.4

1. Find the general form of the equation of the line passing through the point $(3, -1)$ with slope $-2$.
[*Hint:* First find the point-slope form and then transform it to the general form.]

2. Find the general form of the equation of the line passing through the point $(2, -4)$ parallel to a line with slope $1/3$. Also give the $x$-intercept and $y$-intercept.

3. Find the general form of the equation of the line passing through the point $(-1, -3)$ perpendicular to a line with slope $1/5$.

4. Find the general form of the equation of the line passing through the points $(7, -1)$ and $(5, 3)$.
[*Hint:* First find the two-point form and then transform it to the general form.]

5. Find the slope-intercept form of the equation of the line passing through the points $(-2, 3)$ and $(1, -3)$. What is the slope? The $y$-intercept?

**6.** Find the slope and $y$-intercept of the line with equation $5x - 2y + 4 = 0$.

**7.** Find the slope and $y$-intercept of the line with equation $5x + 4 = 0$.

**8.** Find the slope and $y$-intercept of the line with equation $-2y + 4 = 0$.

**9.** Determine whether the lines with equations $5x - 3y + 8 = 0$ and $3x + 5y - 7 = 0$ are parallel, perpendicular, or neither.

**10.** Determine whether the lines with equations $2x - y + 7 = 0$ and $-6x + 3y - 1 = 0$ are parallel, perpendicular, or neither.

**11.** Determine whether the lines with equations $2x + 1 = 0$ and $-3y - 4 = 0$ are parallel, perpendicular, or neither.

**12.** Find the general form of the equation of the line passing through $(-1, 4)$ parallel to the line with equation $-3x - y + 4 = 0$.

**13.** Find the general form of the equation of the line passing through $(-1, 4)$ perpendicular to the line with equation $-3x - y + 4 = 0$.

**14.** Find the general form of the equation of the line through $(-2, -4)$ with $m = -3$.

**15.** Find the general form of the equation of the line with $y$-intercept $(0, 2)$ and slope $-5$.

**16.** Find the general form of the equation of the line with $x$-intercept $(3, 0)$ and slope $1/2$.

**17.** Find the slope and $y$-intercept of the line with equation $6x + 2y - 10 = 0$.

**18.** Find the slope and $y$-intercept of the line with equation $3y + 9 = 0$.

**19.** Find the slope and $y$-intercept of the line with equation $x = 7$.

**20.** Consider the line which passes through the points (1, 6) and (5, −2).
(a) Find the slope of a line parallel to this line.

(b) Find the slope of a line perpendicular to this line.

**21.** Without trying to determine the actual value, indicate whether each line in the figure below has positive slope, negative slope, zero slope, or no slope.

(a) $l_1$ has ________ slope.

(b) $l_2$ has ________ slope.

(c) $l_3$ has ________ slope.

(d) $l_4$ has ________ slope.

(e) The $x$-axis has ________ slope.

(f) The $y$-axis has ________ slope.

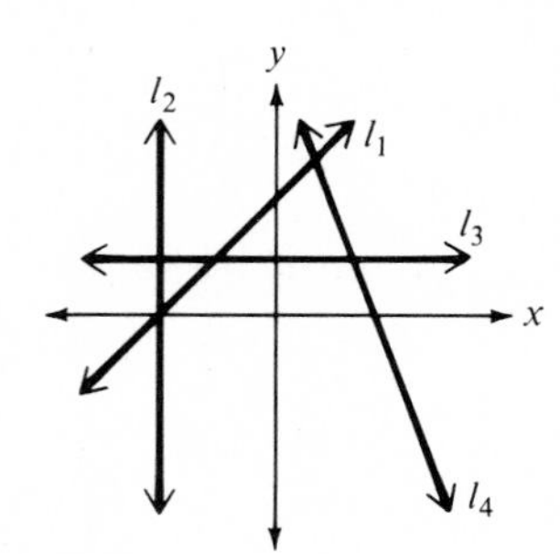


---

ANSWERS: **1.** $2x + y - 5 = 0$ **2.** $x - 3y - 14 = 0$, $x$-intercept (14, 0); $y$-intercept $\left(0, -\frac{14}{3}\right)$ **3.** $5x + y + 8 = 0$ **4.** $2x + y - 13 = 0$ **5.** $y = -2x - 1$; slope −2; $y$-intercept (0, −1) **6.** slope $\frac{5}{2}$; $y$-intercept (0, 2) **7.** no slope; no $y$-intercept **8.** slope 0; $y$-intercept (0, 2) **9.** perpendicular (slopes are $\frac{5}{3}$ and $-\frac{3}{5}$) **10.** parallel (both slopes are 2 and the $y$-intercepts are different) **11.** perpendicular (each is parallel to one of the axes) **12.** $3x + y - 1 = 0$ **13.** $x - 3y + 13 = 0$ **14.** $3x + y + 10 = 0$ **15.** $5x + y - 2 = 0$ **16.** $x - 2y - 3 = 0$ **17.** slope −3; $y$-intercept (0, 5) **18.** slope 0; $y$-intercept (0, −3) **19.** no slope; no $y$-intercept **20.** (a) −2 (b) $\frac{1}{2}$ **21.** (a) positive (b) no (c) zero (d) negative (e) zero (f) no

## 7.5 Distance And Midpoint Formulas

Two important formulas in algebra are the distance formula, used to compute the distance between two points, and the midpoint formula, used to compute the coordinates of a point midway between two points. The first of these formulas is a direct application of the **Pythagorean theorem** (the sum of the squares of the legs of a right triangle is equal to the square of the hypotenuse).

**DISTANCE FORMULA**

The distance between two points with coordinates $(x_1, y_1)$ and $(x_2, y_2)$ is given by

$$d = \sqrt{(x_1 - x_2)^2 + (y_1 - y_2)^2}.$$

If the points $P(x_1, y_1)$ and $Q(x_2, y_2)$ are both in the first quadrant as in Figure 7.22, the numbers $x_1 - x_2$ and $y_1 - y_2$ are the lengths of the legs of the triangle with vertices $P$, $Q$, and $R$. Applying the Pythagorean theorem, we obtain

$$d^2 = (x_1 - x_2)^2 + (y_1 - y_2)^2$$

so that

$$d = \sqrt{(x_1 - x_2)^2 + (y_1 - y_2)^2}$$

gives the length of the hypotenuse of the triangle—that is, the distance between the points $(x_1, y_1)$ and $(x_2, y_2)$. Since $(x_1 - x_2)$ and $(y_1 - y_2)$ are both squared, $(x_1 - x_2)^2 = (x_2 - x_1)^2$ and $(y_1 - y_2)^2 = (y_2 - y_1)^2$, it is immaterial which point we label $(x_1, y_1)$ and which point we label $(x_2, y_2)$. Although our particular points are in the first quadrant, the same results hold regardless of the location of the two points.

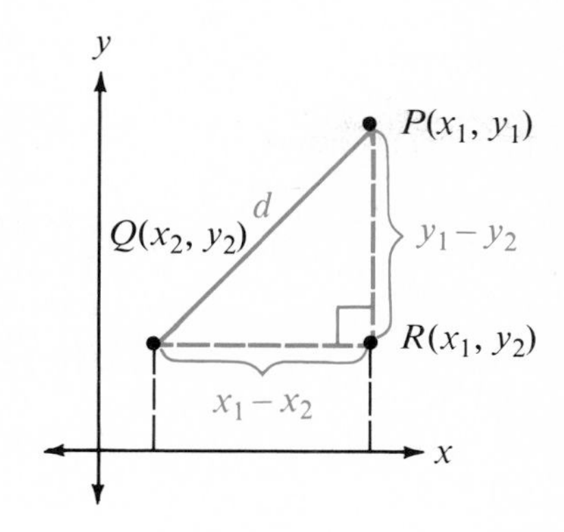


Figure 7.22

**EXAMPLE 1** Find the distance between (2, 5) and (6, 8).

Let $(x_1, y_1) = (2, 5)$ and $(x_2, y_2) = (6, 8)$ and substitute into the distance formula.

$$\begin{aligned} d &= \sqrt{(x_1 - x_2)^2 + (y_1 - y_2)^2} \\ &= \sqrt{(2-6)^2 + (5-8)^2} \\ &= \sqrt{(-4)^2 + (-3)^2} = \sqrt{16 + 9} = \sqrt{25} = 5 \end{aligned}$$

The distance between the given points is 5 units.

**EXAMPLE 2** Find the distance between (5, 5) and (2, 2).

Substituting into the distance formula,

$$\begin{aligned} d &= \sqrt{(5-2)^2 + (5-2)^2} \\ &= \sqrt{3^2 + 3^2} = \sqrt{2 \cdot 3^2} = 3\sqrt{2}. \end{aligned}$$

**EXAMPLE 3** Find the lengths of the sides of the triangle $ABC$ with vertices $A(3, 3)$, $B(9, 2)$ and $C(7, 5)$. Is the triangle a right triangle?

Let $a$, $b$, and $c$ be the sides as indicated in Figure 7.23 (on p. 346). Then apply the distance formula three times to find the lengths of $a$, $b$, and $c$.

$$\begin{aligned} a &= \sqrt{(9-7)^2 + (2-5)^2} = \sqrt{2^2 + (-3)^2} = \sqrt{4 + 9} = \sqrt{13} \\ b &= \sqrt{(7-3)^2 + (5-3)^2} = \sqrt{4^2 + 2^2} = \sqrt{16 + 4} = \sqrt{20} \\ &= \sqrt{4 \cdot 5} = 2\sqrt{5} \\ c &= \sqrt{(3-9)^2 + (3-2)^2} = \sqrt{(-6)^2 + 1^2} = \sqrt{36 + 1} = \sqrt{37} \end{aligned}$$

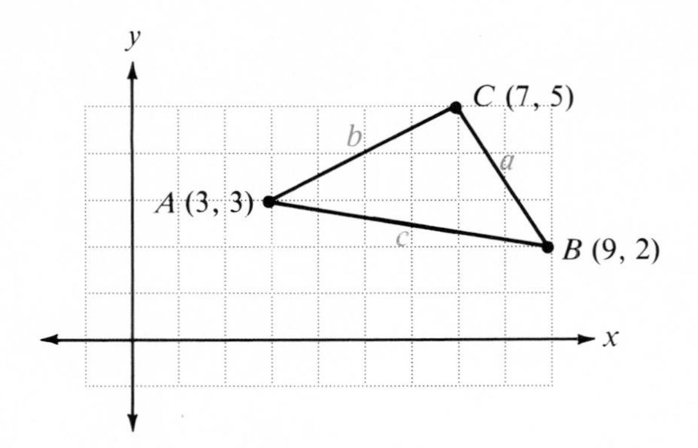


Figure 7.23

If the triangle is a right triangle, the sides must satisfy the conditions of the Pythagorean theorem. Since $c = \sqrt{37}$ is the longest side, $c$ is the hypotenuse (why?) and thus we would have to have $c^2 = a^2 + b^2$. But $c^2 = 37$, $a^2 = 13$ and $b^2 = 20$ so that

$$37 = c^2 \neq a^2 + b^2 = 13 + 20 = 33.$$

Hence the triangle cannot be a right triangle. Could we establish this fact another way with slopes? (Two of the slopes would have to be negative reciprocals of each other.)

The second formula, the midpoint formula, involves averaging the $x$-coordinates of two points and averaging the $y$-coordinates of the same two points. It is fairly easy to see in Figure 7.24 that the point midway between the two points (1, 2) and (5, 4) has coordinates (3, 3).

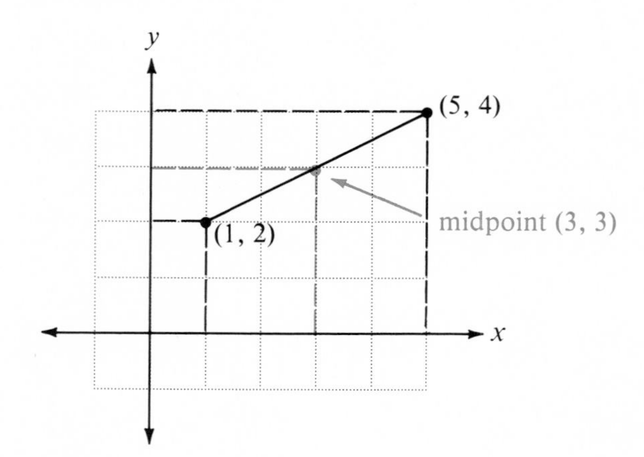


Figure 7.24

Using the distance formula, the distance between (1, 2) and (5, 4) is

$$d = \sqrt{(1-5)^2 + (2-4)^2} = \sqrt{(-4)^2 + (-2)^2} = \sqrt{16+4}$$
$$= \sqrt{20} = \sqrt{4 \cdot 5} = 2\sqrt{5}.$$

Also, the distance between (1, 2) and (3, 3) is

$$d_1 = \sqrt{(1-3)^2 + (2-3)^2} = \sqrt{(-2)^2 + (-1)^2} = \sqrt{4+1} = \sqrt{5}.$$

Thus $d_1 = \sqrt{5}$ is one half of $d = 2\sqrt{5}$, and (3, 3) *is* midway between (1, 2) and (5, 4). If we average the $x$-coordinates of these points we obtain

$$\frac{1+5}{2} = \frac{6}{2} = 3.$$

Similarly, the average of the $y$-coordinates of these points is

$$\frac{2+4}{2} = \frac{6}{2} = 3.$$

What we have in this example is a special case of the midpoint formula.

**MIDPOINT FORMULA**

The coordinates $(\bar{x}, \bar{y})$ of the midpoint of a line segment joining $(x_1, y_1)$ and $(x_2, y_2)$ are given by the **midpoint formula**

$$(\bar{x}, \bar{y}) = \left(\frac{x_1 + x_2}{2}, \frac{y_1 + y_2}{2}\right).$$

That is, the $x$-coordinate of the midpoint is the average of the $x$-coordinates of the points, and the $y$-coordinate of the midpoint is the average of the $y$-coordinates of the points.

CAUTION: In the distance formula we subtract $x$-coordinates and $y$-coordinates while in the midpoint formula we add $x$-coordinates and $y$-coordinates. Do not confuse these operations when working with these two formulas.

EXAMPLE 4 Find the midpoint between the points (−2, 1) and (6, −3).

Substitute into the midpoint formula.

$$(\bar{x}, \bar{y}) = \left(\frac{x_1 + x_2}{2}, \frac{y_1 + y_2}{2}\right) = \left(\frac{-2+6}{2}, \frac{1+(-3)}{2}\right)$$

$$= \left(\frac{4}{2}, \frac{-2}{2}\right) = (2, -1)$$

The midpoint is (2, −1).

EXAMPLE 5 Find the equation of the line perpendicular to the line that contains the points (−3, 2) and (3, −6), and passing through the point midway between these points.

The slope of the line joining (−3, 2) and (3, −6) is given by

$$m = \frac{-6-2}{3-(-3)} = \frac{-8}{6} = -\frac{4}{3}.$$

The slope of a line perpendicular to the line joining (−3, 2) and (3, −6) is given by

$$-\frac{1}{m} = -\frac{1}{-\frac{4}{3}} = \frac{3}{4}.$$

The point midway between (−3, 2) and (3, −6) has coordinates

$$(\bar{x}, \bar{y}) = \left(\frac{-3+3}{2}, \frac{2+(-6)}{2}\right) = \left(\frac{0}{2}, -\frac{4}{2}\right) = (0, -2).$$

Now, use the point-slope form of the equation of a line with (0, −2) as the point and 3/4 as the slope.

$$y - (-2) = \frac{3}{4}(x - 0)$$

$$y + 2 = \frac{3}{4}(x)$$

$$4(y + 2) = 3x$$
$$4y + 8 = 3x$$
$$0 = 3x - 4y - 8$$

The desired equation in general form is $3x - 4y - 8 = 0$.

The distance and midpoint formulas, as well as the forms of the equation of a line, must be memorized.

## EXERCISES 7.5

*Find the distance between the given points.*

**1.** (3, 6) and (−3, −2)

**2.** (3, 2) and (−2, −10)

**3.** (−2, 1) and (5, −3)

**4.** (2, 2) and (−2, −2)

*Find the midpoint of the line segment between the given points.*

**5.** (3, −4) and (1, 6)

**6.** (−2, −1) and (4, 1)

**7.** (5, −3) and (−2, −1)

**8.** (2, 2) and (−2, −2)

**9.** The point-slope form of the equation of a line passing through the point $(x_1, y_1)$ with slope $m$ is ____________________.

**10.** The slope-intercept form of the equation of a line with slope $m$ and $y$-intercept $(0, b)$ is ____________________.

**11.** The two-point form of the equation of a line passing through the points $(x_1, y_1)$ and $(x_2, y_2)$ is ____________________.

**12.** A line parallel to a line with slope $m$ has slope ____________________.

**13.** A line perpendicular to a line with slope $m$ has slope ____________________.

**14.** The distance $d$ between two points with coordinates $(x_1, y_1)$ and $(x_2, y_2)$ is given by $d =$ ____________________.

**15.** The coordinates $(\bar{x}, \bar{y})$ of the midpoint between two points with coordinates $(x_1, y_1)$ and $(x_2, y_2)$ are given by $(\bar{x}, \bar{y}) =$ ____________________.

**16.** Find the slope and $y$-intercept of the line with equation $6x + 3y - 9 = 0$.

**17.** Find the general form of the equation of the line passing through (−5, 2) with slope $m = -3$.

*Find the following for the two points* (1, 4) *and* (−7, −12).

**18.** The distance between the two points.

**19.** The midpoint of the line segment joining them.

**20.** The slope of the line joining them.

**21.** The general form of the equation of the line passing through the midpoint between them, perpendicular to the line joining them.

ANSWERS: **1.** 10 **2.** 13 **3.** $\sqrt{65}$ **4.** $4\sqrt{2}$ **5.** (2, 1) **6.** (1, 0) **7.** $\left(\frac{3}{2}, -2\right)$ **8.** (0, 0) **9.** $y - y_1 = m(x - x_1)$ **10.** $y = mx + b$ **11.** $y - y_1 = \frac{y_2 - y_1}{x_2 - x_1}(x - x_1)$ **12.** $m$ **13.** $-\frac{1}{m}$ **14.** $\sqrt{(x_1 - x_2)^2 + (y_1 - y_2)^2}$ **15.** $\left(\frac{x_1 + x_2}{2}, \frac{y_1 + y_2}{2}\right)$ **16.** slope $-2$; $y$-intercept (0, 3) **17.** $3x + y + 13 = 0$ **18.** $8\sqrt{5}$ **19.** $(-3, -4)$ **20.** 2 **21.** $x + 2y + 11 = 0$

## 7.6 Graphing Second-Degree Equations and Conics

We have seen that an equation of the form

$$ax + by + c = 0$$

where $a$, $b$, and $c$ are real numbers, $a$ and $b$ not both zero, has as its graph a straight line. Such equations are called **first-degree equations** (linear equations) in $x$ and $y$ since the variables appear only to the first power. A **second-degree equation in two variables** $x$ and $y$ can be written in the form

$$Ax^2 + Bxy + Cy^2 + Dx + Ey + F = 0$$

where at least one of $A$, $B$, or $C$ is not zero. (If all three are zero, the equation reduces to the first-degree equation $Dx + Ey + F = 0$.)

Some second-degree equations result in special curves called **conic sections,** which can be formed by intersecting a cone and a plane. The four most important conic sections are parabolas, circles, ellipses, and hyperbolas.

The first conic section we consider is a parabola. When a cone is cut by a plane as shown in Figure 7.25, the resulting curve is a **parabola.**

Recall that to graph an equation in two variables $x$ and $y$, we make a table of values and plot the corresponding points in a Cartesian coordinate system. When the points are connected with a line or curve, we obtain a reasonable approximation to the graph.

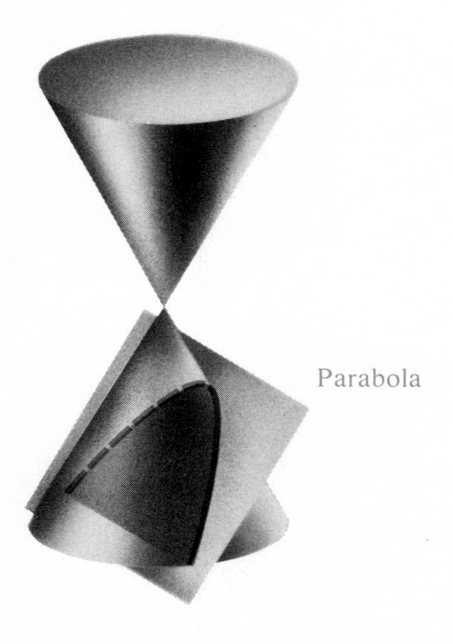


Figure 7.25

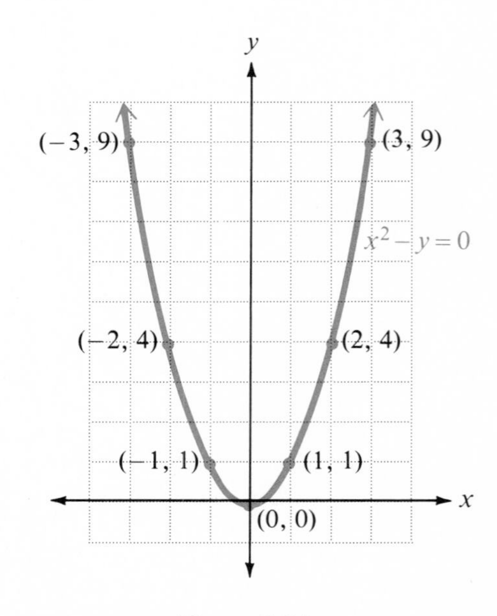


Figure 7.26

**EXAMPLE 1** Graph $x^2 - y = 0$.

It is often helpful to solve the equation for one of the variables. In this case, we solve for $y$.

$$y = x^2$$

To make the table of values,

| $x$ | 0 | 1 | −1 | 2 | −2 | 3 | −3 |
|---|---|---|---|---|---|---|---|
| $y$ | 0 | 1 | 1 | 4 | 4 | 9 | 9 |

substitute the value of $x$ from the top line of the table into the equation, compute the corresponding value of $y$, and list it in the table below the $x$ value. We plot the seven points

$$(0, 0), (1, 1), (-1, 1), (2, 4), (-2, 4), (3, 9), (-3, 9)$$

in a Cartesian coordinate system and connect them, as shown in Figure 7.26.

Any second-degree equation of the form

$$y = ax^2 + bx + c \quad (a \neq 0)$$

or

$$x = ay^2 + by + c \quad (a \neq 0)$$

has a parabola for its graph. It is helpful to have information such as this before drawing a graph.

**EXAMPLE 2** Graph $x = y^2 - 2y - 1$.

This time we substitute values of $y$ into the equation and simplify to obtain the corresponding values of $x$. The points in the table of values are plotted in Figure 7.27.

| $x$ | −1 | −2 | 2 | −1 | 7 | 2 | 7 |
|---|---|---|---|---|---|---|---|
| $y$ | 0 | 1 | −1 | 2 | −2 | 3 | 4 |

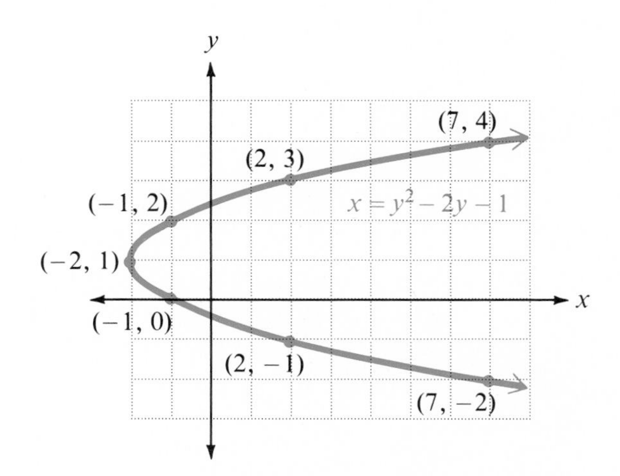


**Figure 7.27**

When a plane intersects a cone as shown in Figure 7.28, the resulting curve is a **circle.**

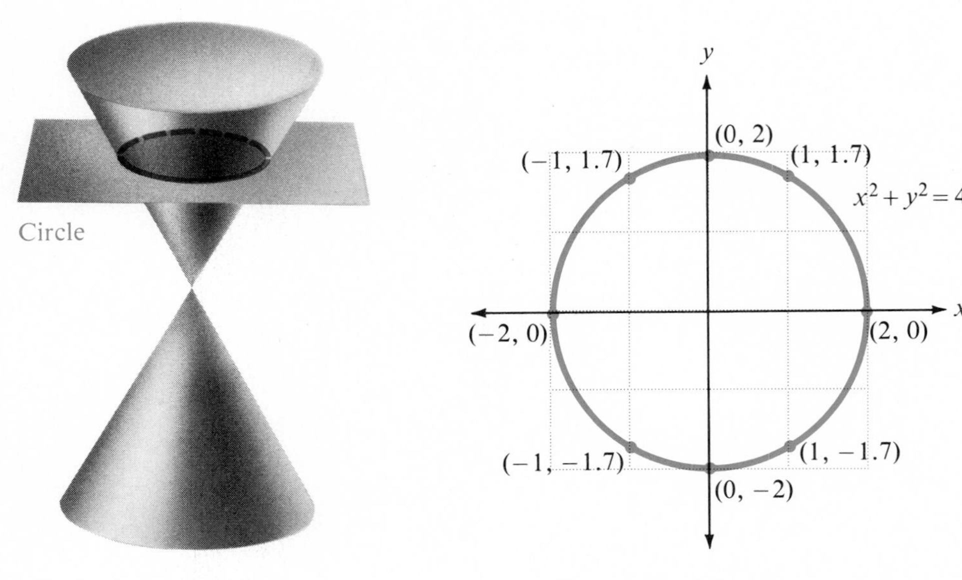


**Figure 7.28** **Figure 7.29**

EXAMPLE 3 Graph $x^2 + y^2 = 4$.

Solve the equation for $y$.

$$y^2 = 4 - x^2$$
$$y = \pm\sqrt{4 - x^2}$$

The points in the table of values are plotted in Figure 7.29 and connected by a smooth curve.

| $x$ | 0 | 1 | −1 | 2 | −2 |
|---|---|---|---|---|---|
| $y$ | ±2 | $\pm\sqrt{3} \approx \pm 1.7$ | $\pm\sqrt{3} \approx \pm 1.7$ | 0 | 0 |

A **circle** is defined to be the collection of all points in a plane which are a fixed distance $r$ from a given point $(h, k)$ in that plane. The given point is called the **center** of the circle and the fixed distance is called the length of the **radius** of the circle. In the preceding example, the center of the circle is the origin (0, 0) and the radius of the circle is $r = 2$. In general, an equation of the form

$$(x - h)^2 + (y - k)^2 = r^2$$

has as its graph a circle centered at $(h, k)$ with radius $r$. Notice that the equation

$$x^2 + y^2 = 4$$

can be written in this form.

$$(x - 0)^2 + (y - 0)^2 = 2^2$$

Knowledge of the graph in this case enables us to avoid extensive point plotting.

We simply locate the center, determine the length of the radius, and sketch the circle.

**EXAMPLE 4** Graph $(x - 1)^2 + (y + 2)^2 = 9$.

We rewrite the equation in the form $(x - h)^2 + (y - k)^2 = r^2$.

$$(x - 1)^2 + [y - (-2)]^2 = 3^2$$

Thus, the center is $(1, -2)$, and the radius has length 3. The graph is shown in Figure 7.30.

The third conic section we consider is an **ellipse,** formed by intersecting a cone and a plane as in Figure 7.31.

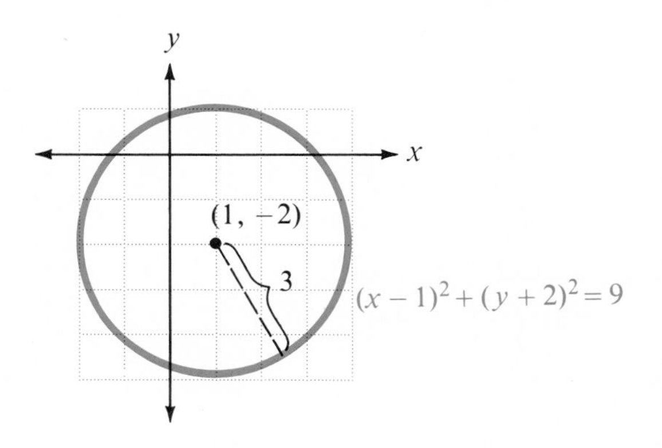


Figure 7.30

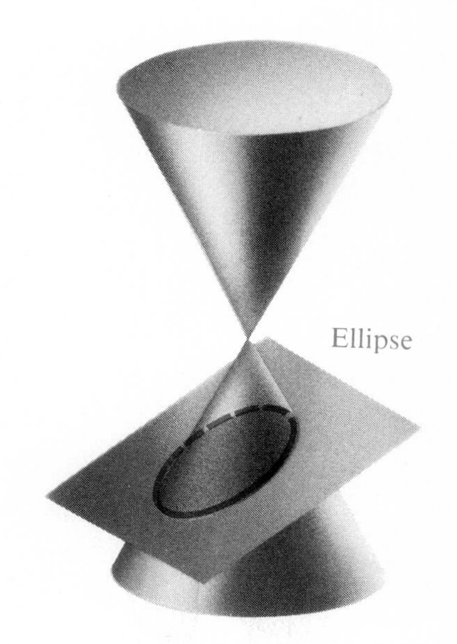


Figure 7.31

**EXAMPLE 5** Graph $\frac{x^2}{9} + \frac{y^2}{4} = 1$.

Solve for $y$.

$$\frac{y^2}{4} = 1 - \frac{x^2}{9} = \frac{9 - x^2}{9}$$

$$= \frac{1}{9}(9 - x^2)$$

$$y^2 = \frac{4}{9}(9 - x^2) \qquad \text{Multiply both sides by 4}$$

$$y = \pm\frac{2}{3}\sqrt{9 - x^2}$$

| $x$ | 0 | 1 | −1 | 3 | −3 |
|---|---|---|---|---|---|
| $y$ | $\pm 2$ | $\pm\frac{4}{3}\sqrt{2} \approx \pm 1.9$ | $\pm\frac{4}{3}\sqrt{2} \approx \pm 1.9$ | 0 | 0 |

The points in the table of values are plotted in Figure 7.32 and connected to form an ellipse.

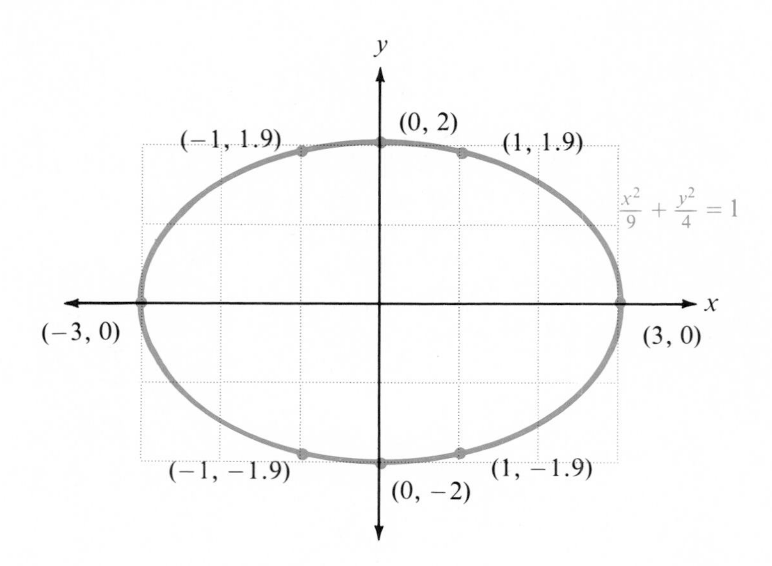


Figure 7.32

An ellipse like the one above is said to be **centered at the origin.** It has $x$-intercepts (3, 0) and (−3, 0) and $y$-intercepts (0, 2) and (0, −2). In general, an equation of the form

$$\frac{x^2}{a^2} + \frac{y^2}{b^2} = 1$$

has as its graph an ellipse centered at the origin with $x$-intercepts $(a, 0)$ and $(-a, 0)$, and $y$-intercepts $(0, b)$ and $(0, -b)$. With this knowledge, we can simply plot the four intercepts and connect them to form the desired graph.

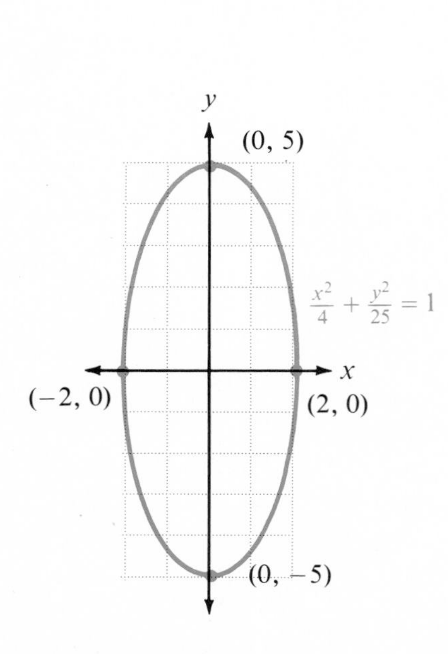


Figure 7.33

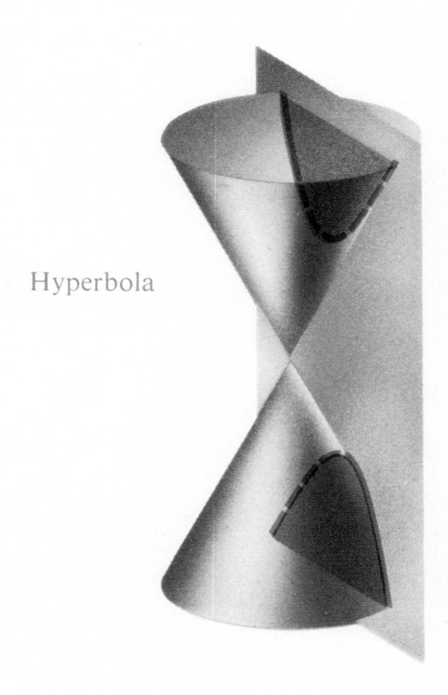


Figure 7.34

**EXAMPLE 6** Graph $\frac{x^2}{4}+\frac{y^2}{25}=1$.

Since $a^2 = 4$ and $b^2 = 25$, the intercepts are (2, 0), (−2, 0), (0, 5), and (0, −5). The graph is shown in Figure 7.33.

The final conic section to be discussed is a **hyperbola,** formed by intersecting a plane and a cone as in Figure 7.34.

**EXAMPLE 7** Graph $\frac{x^2}{9}-\frac{y^2}{4}=1$.

Solve the equation for $y$.

$$\frac{y^2}{4}=\frac{x^2}{9}-1=\frac{x^2-9}{9}$$

$$=\frac{1}{9}(x^2-9)$$

$$y^2=\frac{4}{9}(x^2-9) \qquad \text{Multiply both sides by 4}$$

$$y=\pm\frac{2}{3}\sqrt{x^2-9}$$

| $x$ | 3 | −3 | 4 | −4 | 5 | −5 |
|---|---|---|---|---|---|---|
| $y$ | 0 | 0 | $\pm\frac{2}{3}\sqrt{7}\approx\pm1.8$ | $\pm\frac{2}{3}\sqrt{7}\approx\pm1.8$ | $\pm\frac{8}{3}$ | $\pm\frac{8}{3}$ |

No values of $x$ between −3 and 3 are included in the table since for such numbers, $x^2 - 9$ is negative so that $\sqrt{x^2-9}$ is not a real number. The points in the table are plotted in Figure 7.35 and connected by a smooth curve.

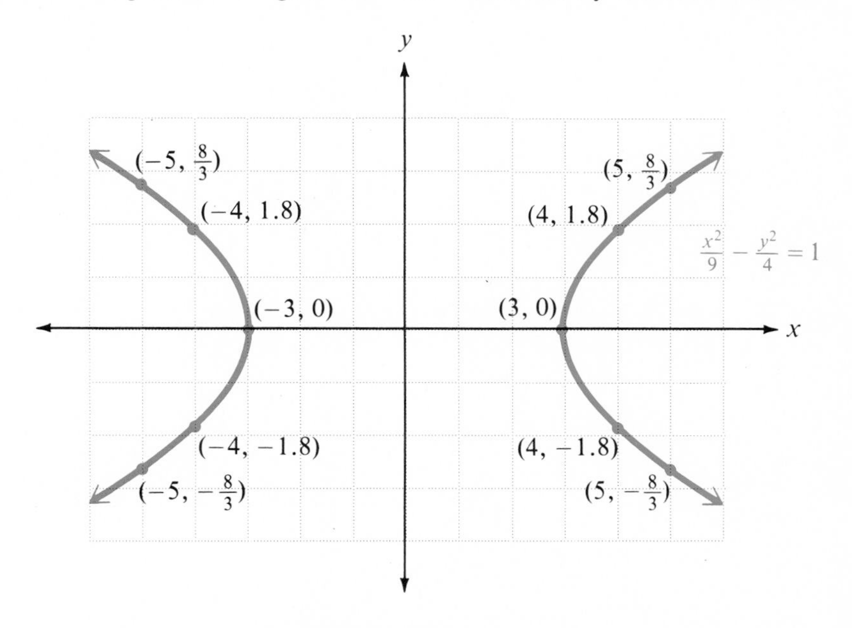


**Figure 7.35**

Associated with every hyperbola are two intersecting lines called the **asymptotes** of the hyperbola. An asymptote of a curve is a line which the curve approaches as the variable or variables increase or decrease in value. An easy way to find the asymptotes of a hyperbola is to use the rectangle whose sides are parallel to the axes and pass through the points $(a, 0)$, $(-a, 0)$, $(0, b)$, and $(0, -b)$. The rectangle for the hyperbola

$$\frac{x^2}{9} - \frac{y^2}{4} = 1 \quad (a = 3, b = 2)$$

is shown in Figure 7.36.

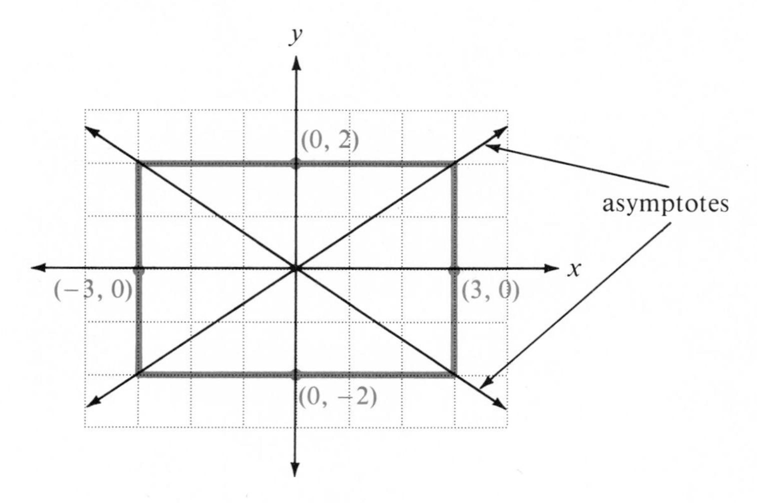


**Figure 7.36**

Extensions of the diagonals of the rectangle are the asymptotes of the hyperbola. If the hyperbola is plotted in a system containing the rectangle and the asymptotes, as in Figure 7.37, we see how the curve "gets closer and closer to" the asymptotes.

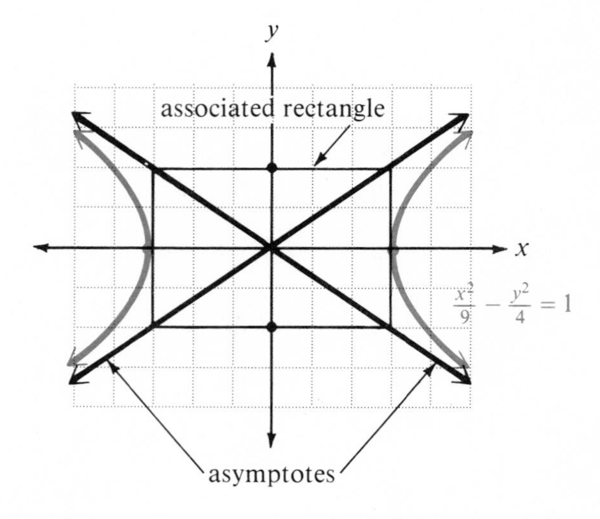


**Figure 7.37**

Remember that the asymptotes are not a part of the graph. They are simply used to construct the graph. The two basic forms of the equation of a hyperbola

**centered at the origin** (one whose asymptotes intersect at the origin) and their respective graphs are summarized in Figure 7.38.

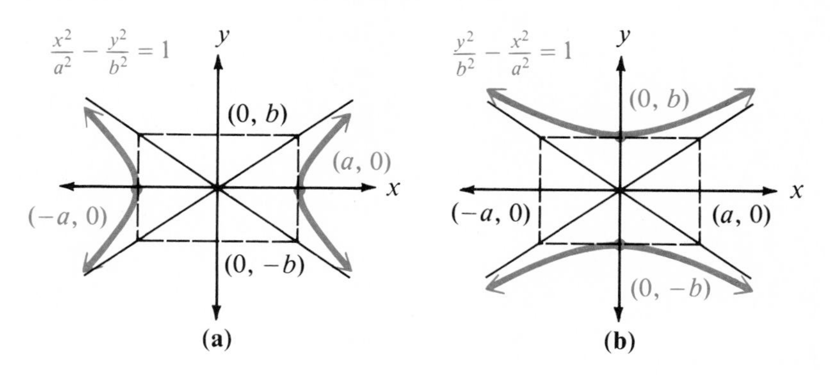


**Figure 7.38**

The hyperbola opens left and right when the coefficient of the $x^2$ term is positive, as in Figure 7.38(a), and it opens up and down when the coefficient of the $y^2$ term is positive, as in Figure 7.38(b).

## EXERCISES 7.6

*Graph each of the following.*

**1.** $x = y^2$

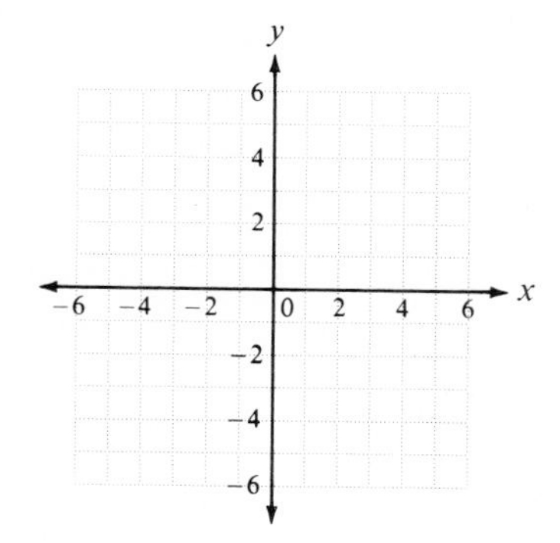

**2.** $y = -x^2$

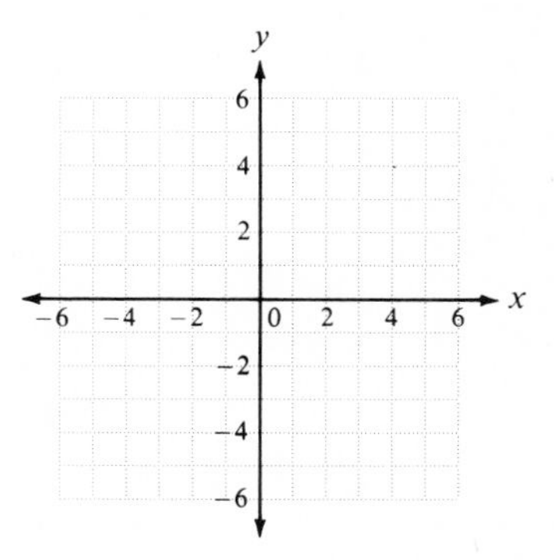

**3.** $x = -y^2$

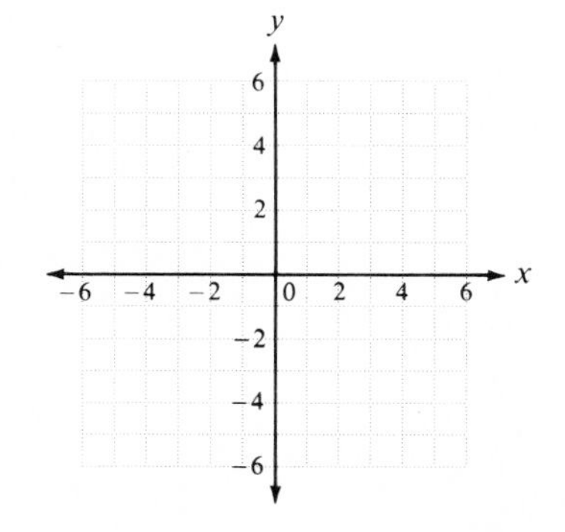

**4.** $y = x^2 - 2x + 3$

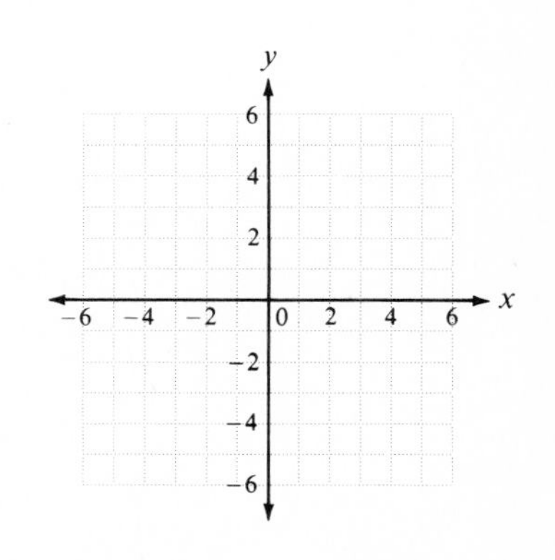

**5.** $x = y^2 - 2$

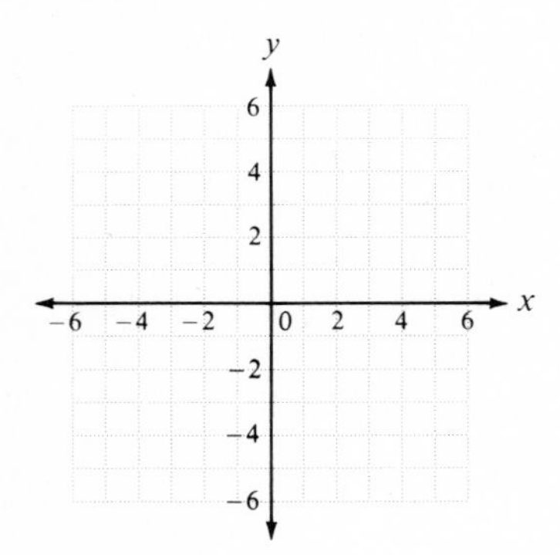

**6.** $x^2 + y^2 = 25$

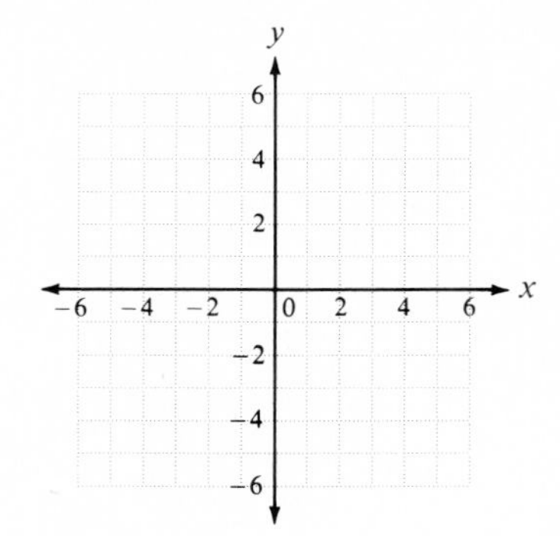

**7.** $(x - 2)^2 + (y + 3)^2 = 4$

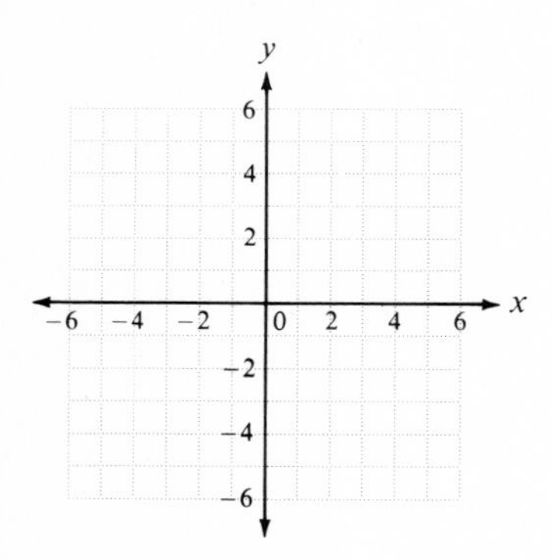

**8.** $\dfrac{x^2}{16} + \dfrac{y^2}{4} = 1$

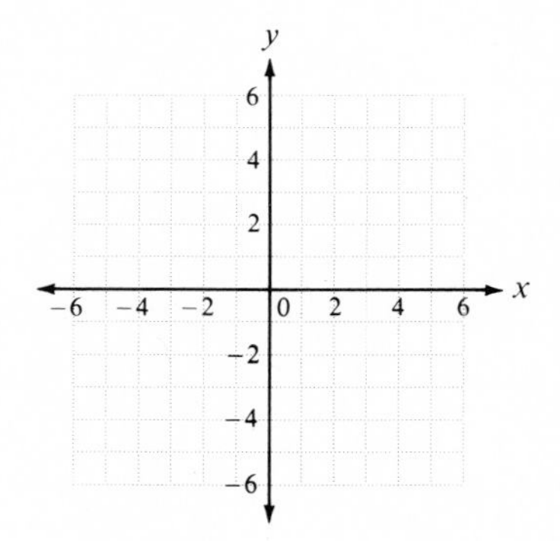

**9.** $\dfrac{x^2}{1} + \dfrac{y^2}{9} = 1$

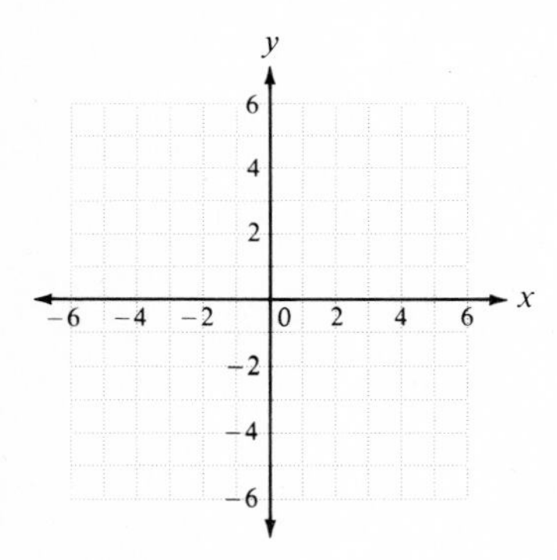

**10.** $\dfrac{x^2}{4} - \dfrac{y^2}{25} = 1$

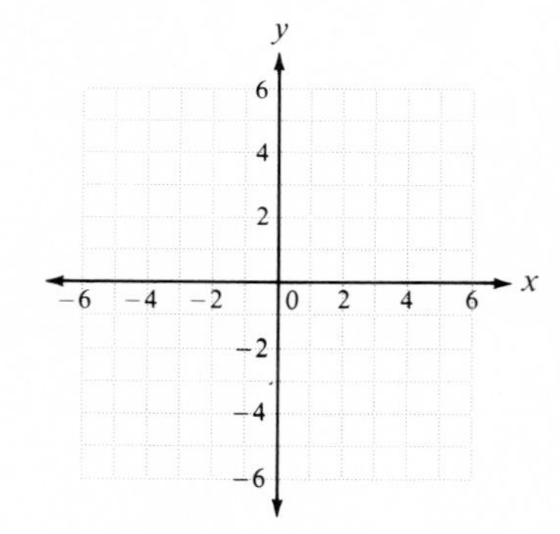

**11.** $\frac{y^2}{9} - \frac{x^2}{4} = 1$

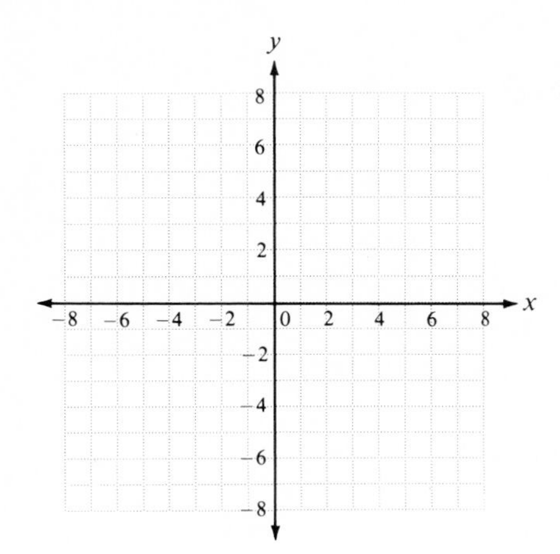

**12.** $xy = 8$

[*Hint:* Complete the table of values. The graph is a hyperbola with the coordinate axes as asymptotes.]

| $x$ | 1 | −1 | 2 | −2 | 3 | −3 | 4 | −4 | 8 | −8 |
|---|---|---|---|---|---|---|---|---|---|---|
| $y$ | | | | | | | | | | |

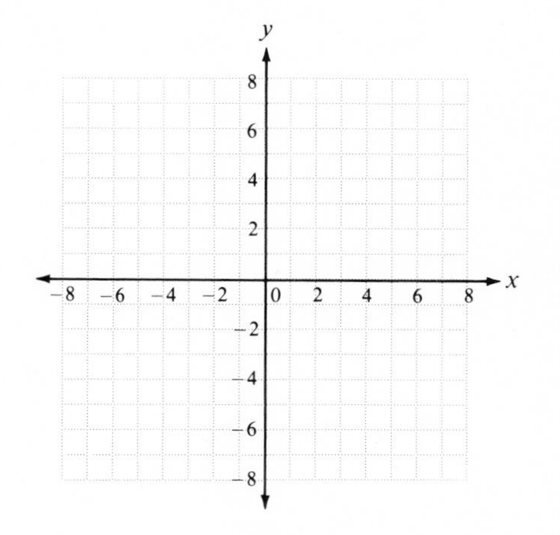

**13.** Find the slope and $y$-intercept of the line with equation $5x - 3y + 15 = 0$.

**14.** Find the general form of the equation of the line that is perpendicular to the line segment between $(3, -5)$ and $(7, -3)$ and passes through the midpoint of this line segment.

**15.** Find the distance between the points $(4, -7)$ and $(-2, -9)$.

**16.** Find the general form of the equation of the line passing through the points $(4, -7)$ and $(-2, -9)$.

**17.** Any line parallel to a line with slope $-4$ must have slope ________.

**18.** Any line perpendicular to a line with slope $-4$ must have slope ________.

---

ANSWERS:

**1.**

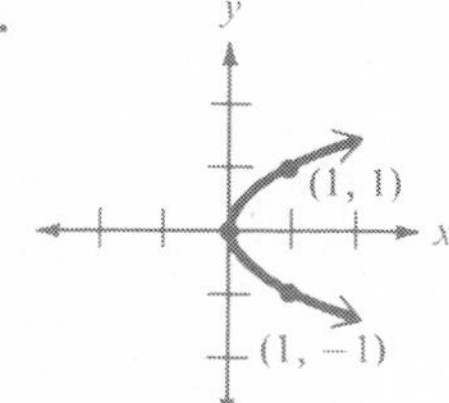


**2.**

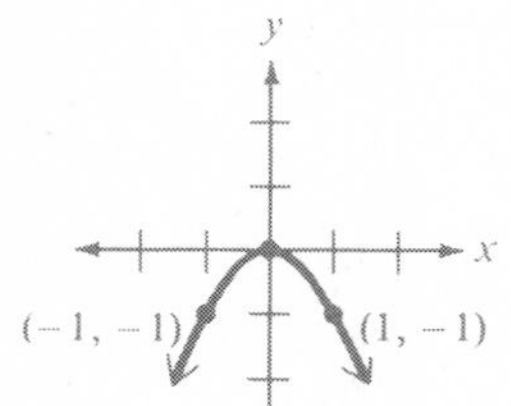


**3.**

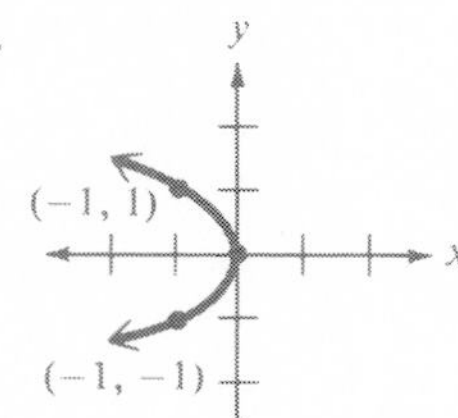


**4.**

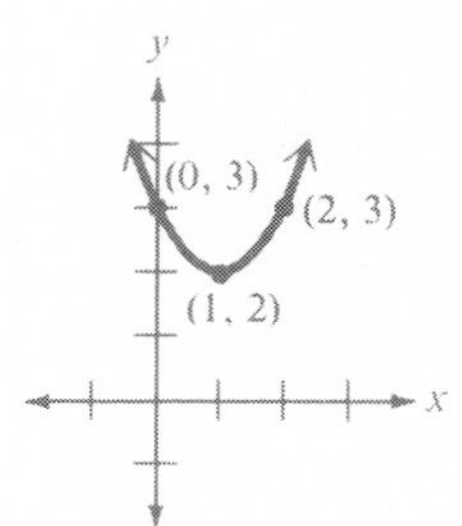


**5.**

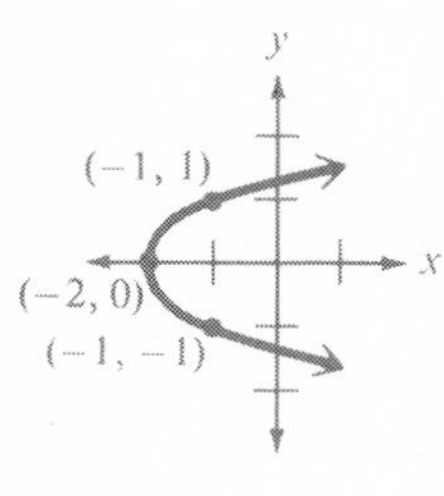


**6.**

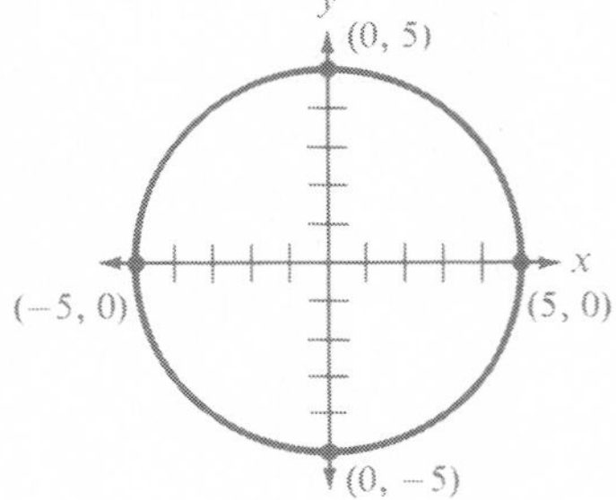


**7.**

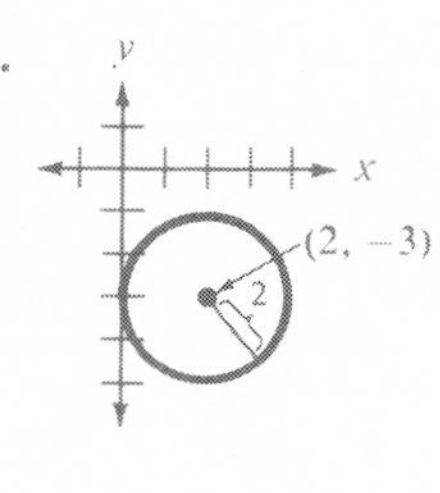


**8.**

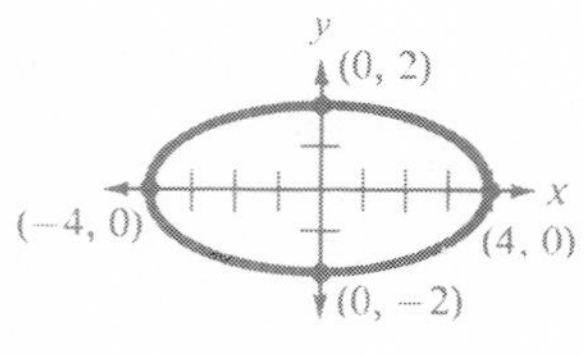


**9.**

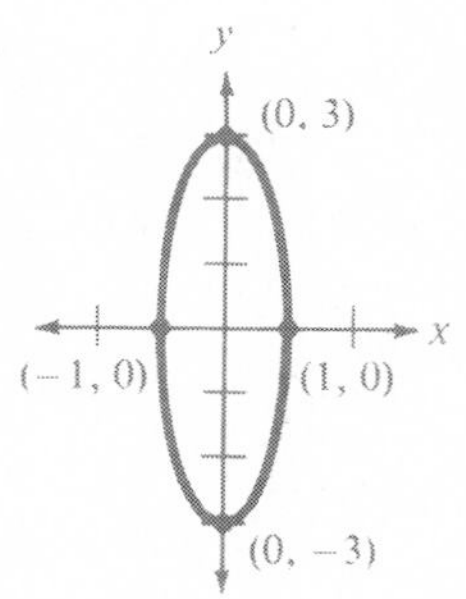


**10.**

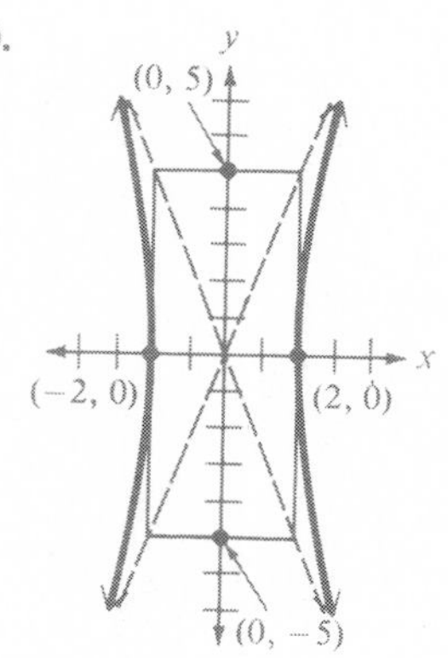


**11.**

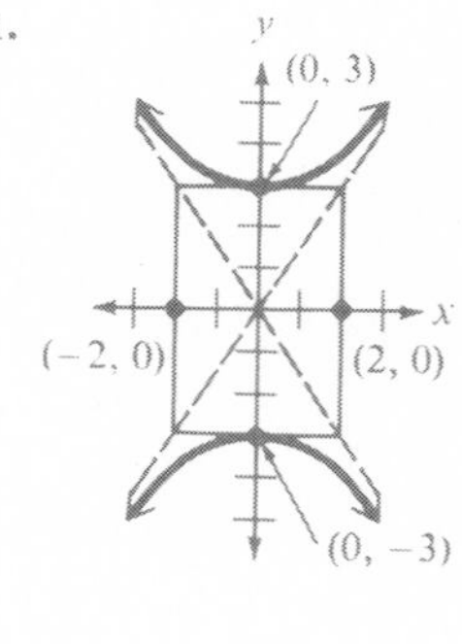


**12.**

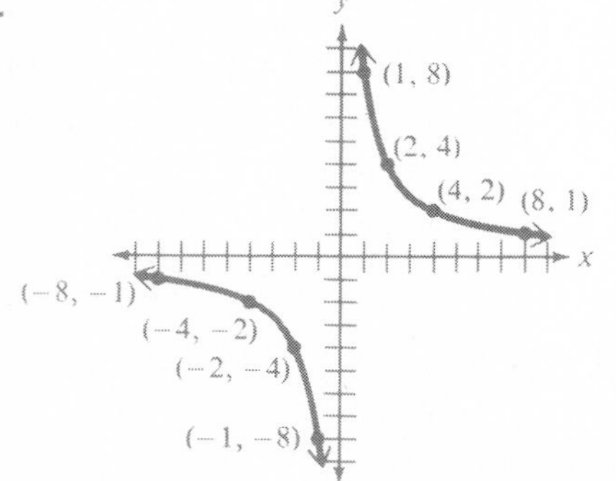


**13.** slope $\frac{5}{3}$; $y$-intercept $(0, 5)$

**14.** $2x + y - 6 = 0$

**15.** $2\sqrt{10}$

**16.** $x - 3y - 25 = 0$

**17.** $-4$

**18.** $\frac{1}{4}$

## 7.7 Functions and Graphs of Functions

A **function** is a correspondence or relationship between two sets of objects (often numbers). Consider the following example. To each of the starting five on a basketball team in a particular game, there corresponds a number, the number of points scored in that game. A table is often used to display this correspondence.

| *Players* | *Points scored* |
|---|---|
| Troy | 20 |
| Wayne | 19 |
| Mike | 12 |
| Dave | 15 |
| Rick | 8 |

For each player in the list on the left, there is one and only one (exactly one) number in the list on the right. It is this concept that designates a correspondence as a function. The set of elements in the first set (in our example, the players) is called the **domain** of the function. The second set (in our example, the points scored) is called the **range** of the function. We often use arrows to indicate the object in the range (the arrow points *to* it) which corresponds to the object in the domain (the arrow points *from* it).

| *Domain* | | *Range* |
|---|---|---|
| Troy | → | 20 |
| Wayne | → | 19 |
| Mike | → | 12 |
| Dave | → | 15 |
| Rick | → | 8 |

Suppose we keep the same domain (the starting five) and consider the function that establishes a correspondence between players and the number of rebounds in a particular game. This function could be indicated as follows.

| *Domain (Player)* | | *Range (Number of rebounds)* |
|---|---|---|
| Troy | → | 8 |
| Wayne | → | 8 |
| Mike | → | 11 |
| Dave | → | 7 |
| Rick | → | 2 |

Again, each player is paired with only one number. However, in this case the same number is paired with two different players (both Troy and Wayne had 8 rebounds). The definition of a function allows for this (which of course could happen in reality), but it does not allow a player to be identified with two different numbers in the range (which is impossible in reality). We usually do not list an object in the range more than once. Thus our function would take the following form.

| *Domain* | | *Range* |
|---|---|---|
| Troy | → | 8 |
| Wayne | → (to 8) | 11 |
| Mike | → (to 11) | 7 |
| Dave | → (to 7) | 2 |
| Rick | → (to 2) | |

We stress that an object in the range may have more than one arrow pointing *to* it, but an object in the domain can have only one arrow pointing *from* it. Before we consider other examples, we define a function more precisely.

A **function** is a correspondence or relationship between a first set of objects, called the **domain** of the function, and a second set of objects, the **range** of the function, such that every object in the domain corresponds to exactly one object in the range.

EXAMPLE 1

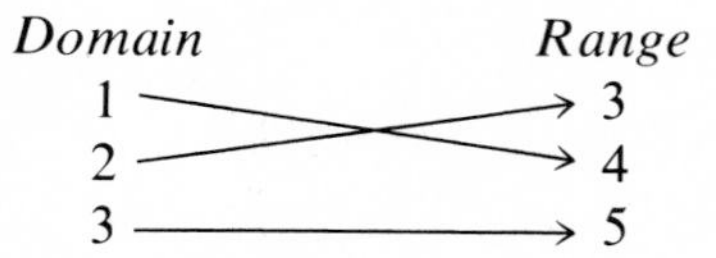


This *is* a function since each object in the domain corresponds to exactly one object in the range.

EXAMPLE 2

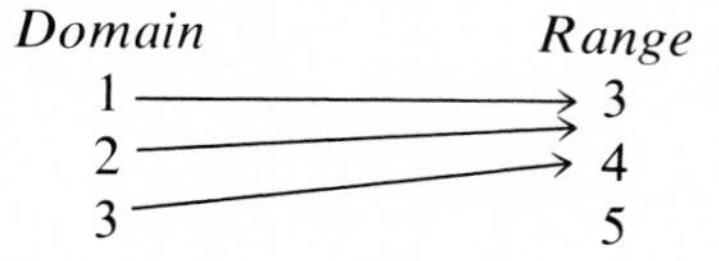


This *is* a function since 3 (in the range) can correspond to the two objects 1 and 2 (in the domain), and 5 need not correspond to any element in the domain.

EXAMPLE 3

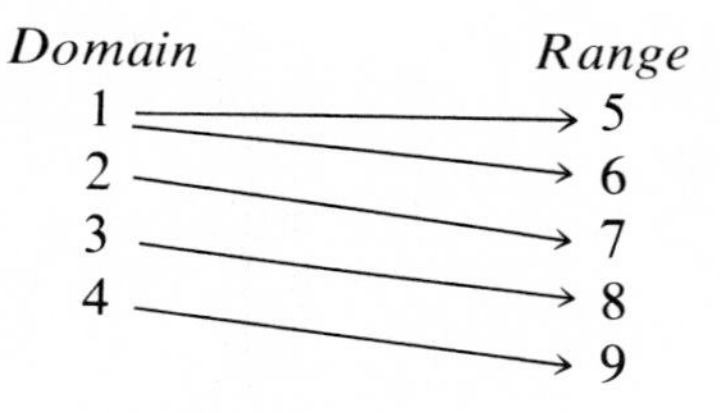


This *is not* a function since 1 (in the domain) corresponds to two objects, 5 and 6 (in the range).

EXAMPLE 4

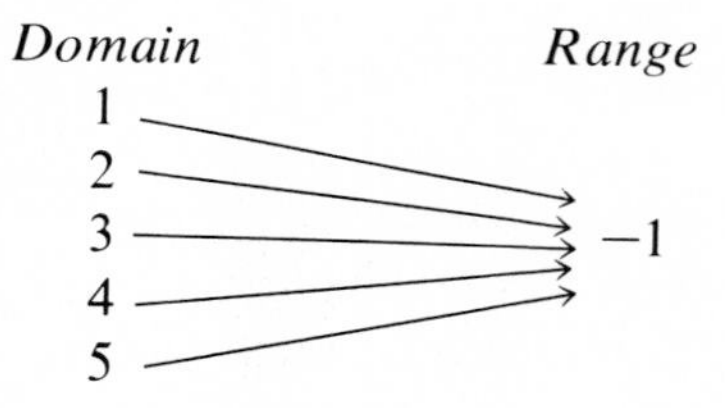


This *is* a function. (Why?)

EXAMPLE 5

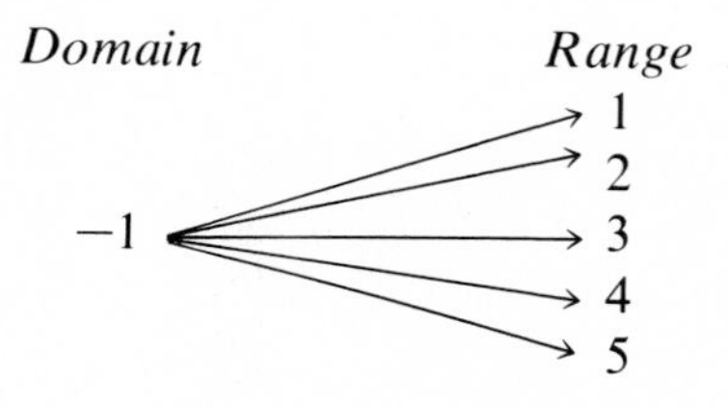


This *is not* a function. (Why?)

When the objects in the domain and range of a function are numbers, a collection of ordered pairs $(x, y)$ can be used to describe the function: if $x \longrightarrow y$, then the pair $(x, y)$ is included among the ordered pairs in the function. With this agreement, since ordered pairs of numbers can be plotted (graphed) in a Cartesian coordinate system, we have a way to **graph a function.**

**EXAMPLE 6** The following function is graphed in Figure 7.39.

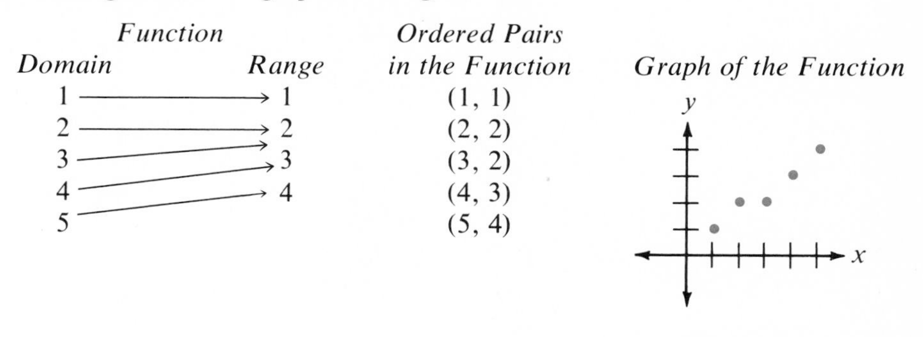


**Figure 7.39**

A correspondence which is not a function, for example

*Domain* 1, 2, 3 — *Range* 2, 3, 4 (1 → 2, 1 → 3, 2 → 4, 3 → 4)

Not a function since 1 is paired with both 2 and 3

may still be graphed. In this case we plot the points (1, 2), (1, 3), (2, 4), and (3, 4), as in Figure 7.40.

We can see from the figure that when graphing a correspondence which is not a function, two points on the graph (corresponding to (1, 2) and (1, 3) in our example) lie on the same vertical line $l$ parallel to the $y$-axis. On the other hand, any collection of points in a Cartesian coordinate system that has the property that no vertical line passes through two or more of the points can be considered the graph of a function.

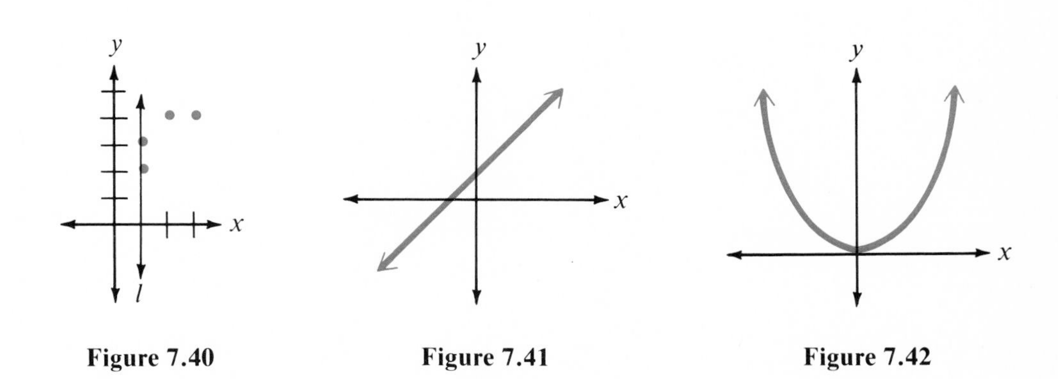


**Figure 7.40** **Figure 7.41** **Figure 7.42**

**EXAMPLE 7** The line in Figure 7.41 *is* the graph of a function since no vertical line can pass through more than one point on the graph.

**EXAMPLE 8** The curve in Figure 7.42 *is* the graph of a function since no vertical line can cross the graph more than once.

**EXAMPLE 9** The graph in Figure 7.43 *is not* the graph of a function since the vertical line $l$ crosses the graph twice. Note that two values of $y$ correspond to one value of $x$.

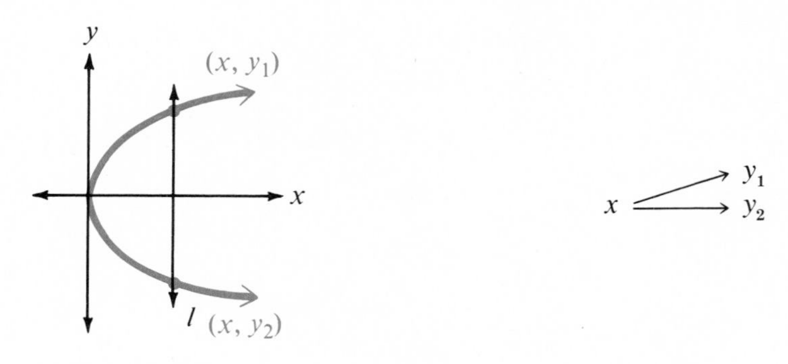


**Figure 7.43**

**EXAMPLE 10** The curve in Figure 7.44 *is not* the graph of a function, as is shown by the vertical line $l$. A given value of $x$ corresponds to three values of $y$.

Often, an equation is used to describe a function. In fact, nearly every linear equation

$$ax + by + c = 0$$

describes a function. The only linear equations that do not, have $b = 0$. For example, $2x - 3 = 0$ does not describe a function, as is clear from the graph of $2x - 3 = 0$ or $2x = 3$ or $x = 3/2$ in Figure 7.45. This cannot be the graph of a function since the graph itself is a vertical line that intersects the graph in two or more (in fact infinitely many) points. However, any other type of line in the plane will be the graph of a function.

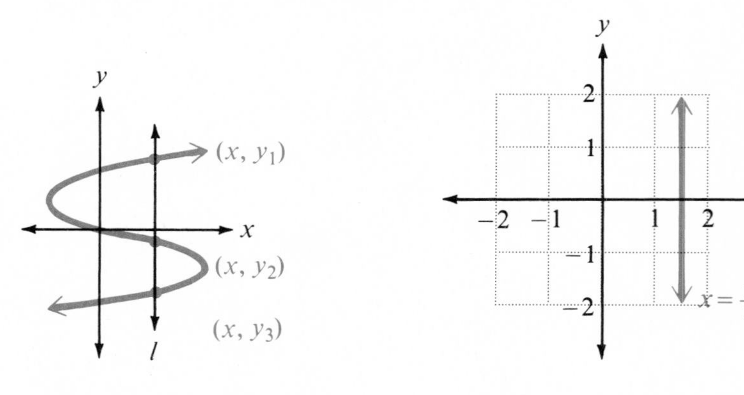


**Figure 7.44** **Figure 7.45**

**EXAMPLE 11** Graph the equation $y = x^2 + 1$ and determine if the equation describes a function.

We first construct a table of values.

| $x$ | 0 | 1 | −1 | 2 | −2 | 3 | −3 |
|---|---|---|---|---|---|---|---|
| $y$ | 1 | 2 | 2 | 5 | 5 | 10 | 10 |

(Remember that the value of $x$ in the top line is substituted into the equation and the corresponding $y$ value is then computed.) We plot the points corresponding to (0, 1), (1, 2), (−1, 2), (2, 5), (−2, 5), (3, 10), and (−3, 10) in a Cartesian coordinate system. Knowing that the graph must be a parabola ($x$ is squared and $y$ appears to the first power), we see that the graph in Figure 7.46 is the graph of a function.

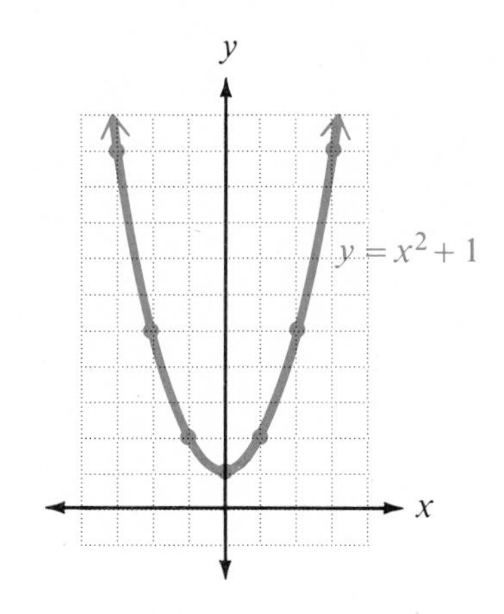


Figure 7.46

**EXAMPLE 12** Graph the equation $x = y^2 + 1$ and determine if the equation describes a function.

Constructing a table of values and knowing that the graph must be a parabola (why?) we obtain the graph in Figure 7.47.

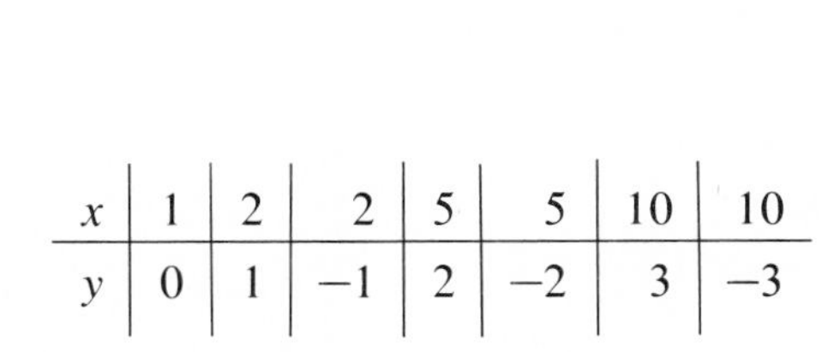

| $x$ | 1 | 2 | 2 | 5 | 5 | 10 | 10 |
|---|---|---|---|---|---|---|---|
| $y$ | 0 | 1 | −1 | 2 | −2 | 3 | −3 |

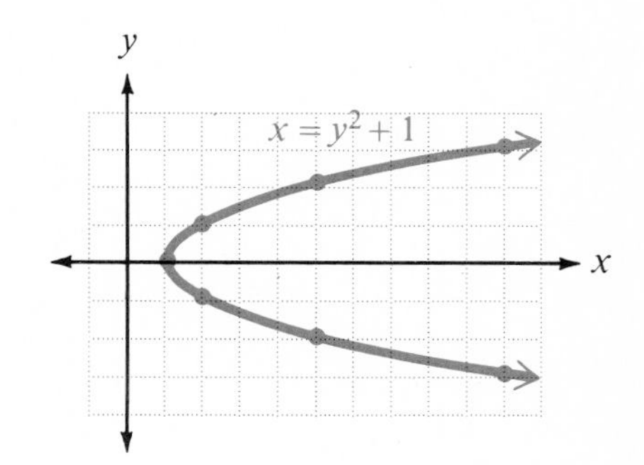


Figure 7.47

In this case the equation cannot be used to describe a function. (Why?)

## EXERCISES 7.7

1. A **(a)** ______________ is a correspondence between a first set called the **(b)** ______________ and a second set called the **(c)** ______________ with the property that every object in the first set corresponds to exactly **(d)** ______________ object in the second set.

**2.** If in a function $x$ corresponds to $y$ ($x \to y$), the ordered pair **(a)** ________ belongs to the function and the point in the plane that corresponds to this ordered pair is a point on the **(b)** ________ of the function.

*Which of the following are functions? (Explain.)*

**3.** 1 → 3; 2 → 4; 3 → 6; 4 → 5

**4.** 1 → 4; 2 → 4; 3 → 5; 4 → 5 (range column: 3, 4, 5, 6)

**5.** 1 → 3; 2 → 4; 3 → 5; 4 → 6

**6.** 1 → 3; 1 → 5; 2 → 6; 3 → 4; 4 → 4

**7.** 2 → −1; 2 → 0; 2 → 1 (domain column: 2, 4, 6, 7)

**8.** 2 → −1; 4 → −1; 6 → 0; 7 → 0 (range column: −1, 0, 1)

**9.** 2 → 0; 4 → −1; 6 → 1; 7 → 1

**10.** 2 → −1; 4 → 1; 6 → 1; 7 → 1 (range column: −1, 0, 1)

*Which of the following are graphs of functions? (Explain.)*

**11.**

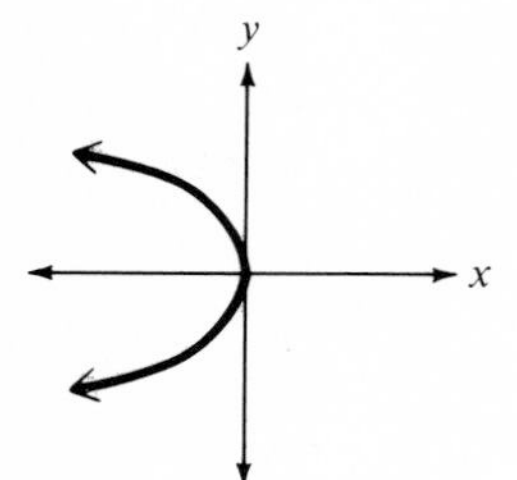


**12.**

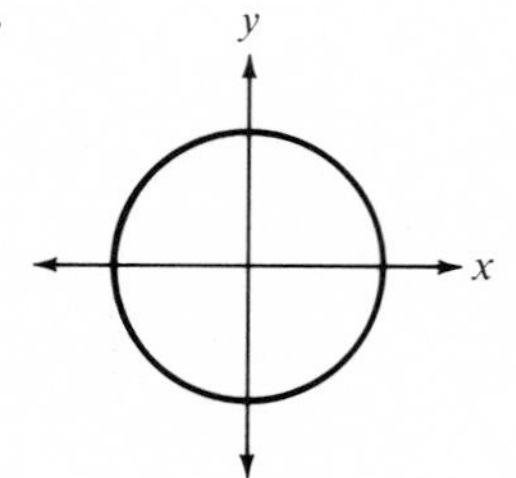


**13.**

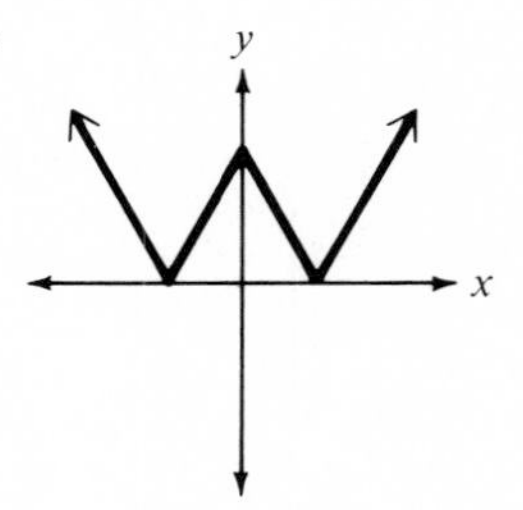


**14.**

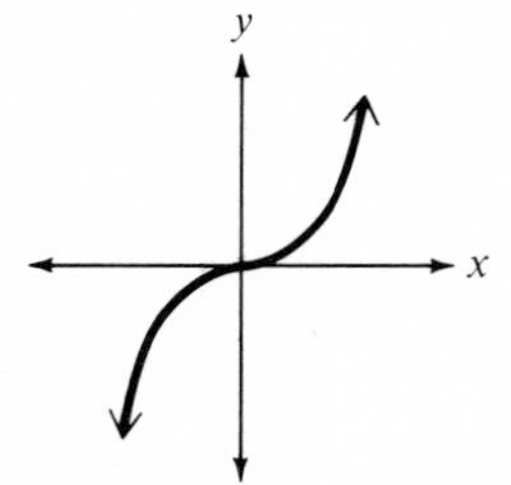


**15.**

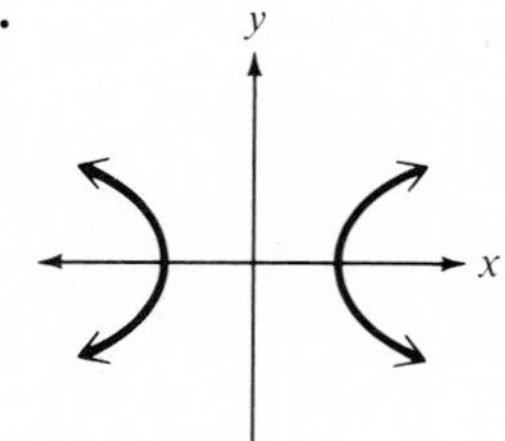


**16.**

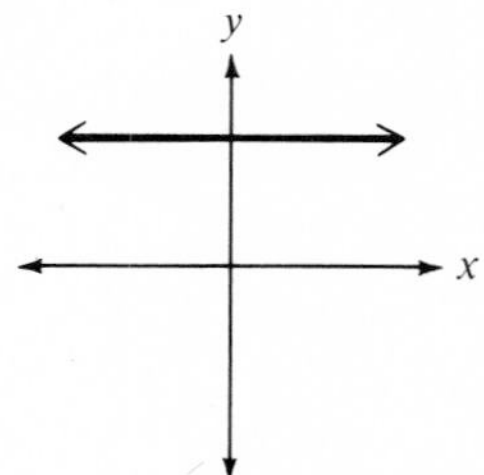


**17.**

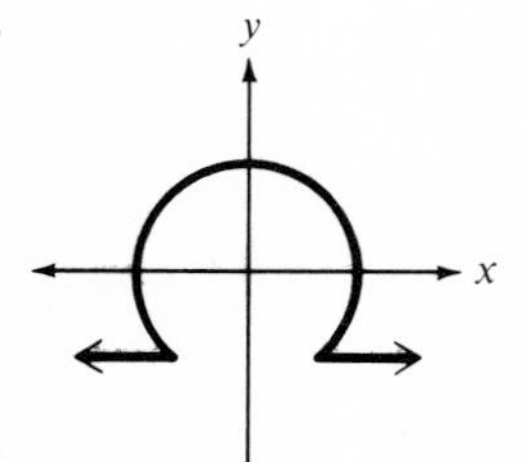


**18.**

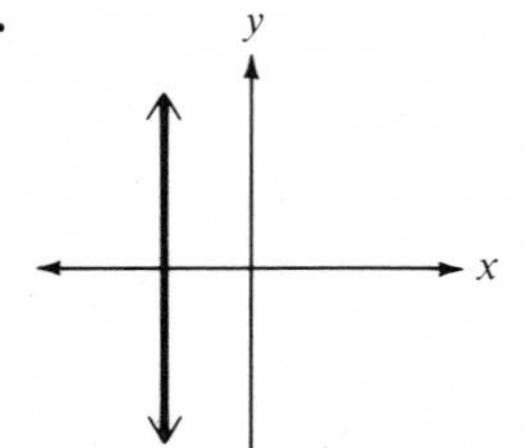


**19**

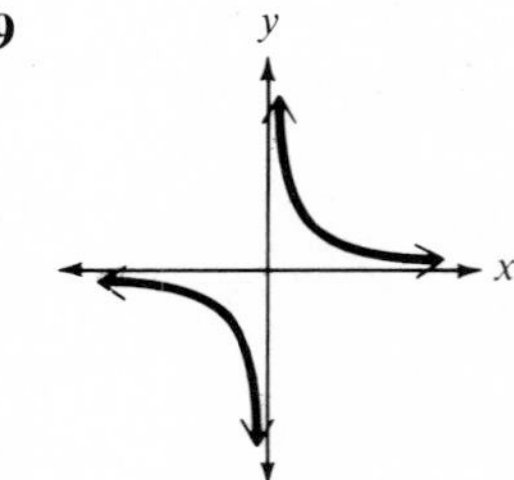

**20.** Graph the equation $3x + 5y - 15 = 0$ and determine if the equation describes a function.

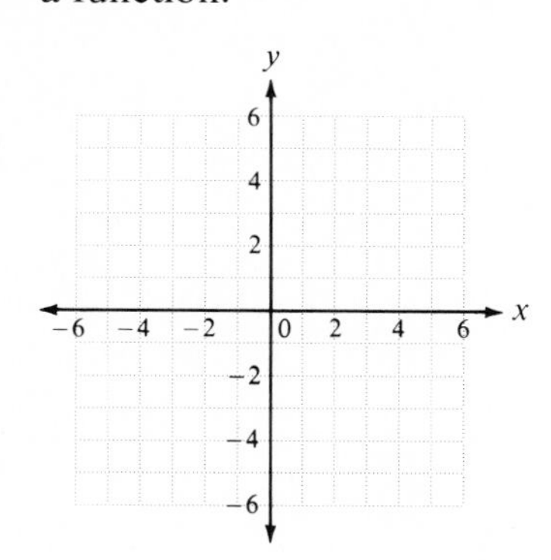

**21.** Graph the equation $3y - 12 = 0$ and determine if the equation describes a function.

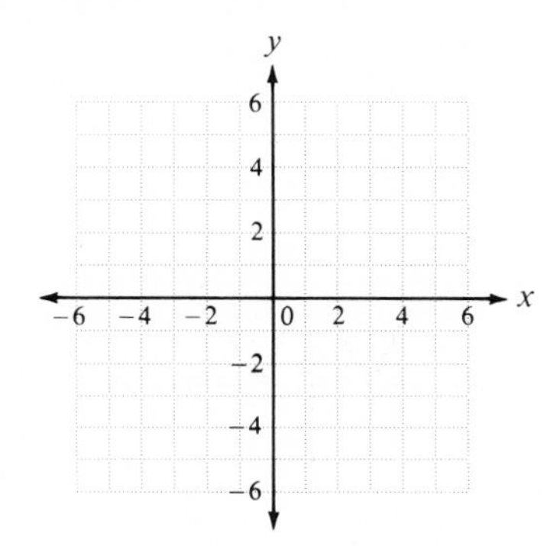

**22.** Graph the equation $y = x^3$ and determine if the equation describes a function.

| $x$ | 0 | 1 | −1 | 2 | −2 |
|---|---|---|---|---|---|
| $y$ | | | | | |

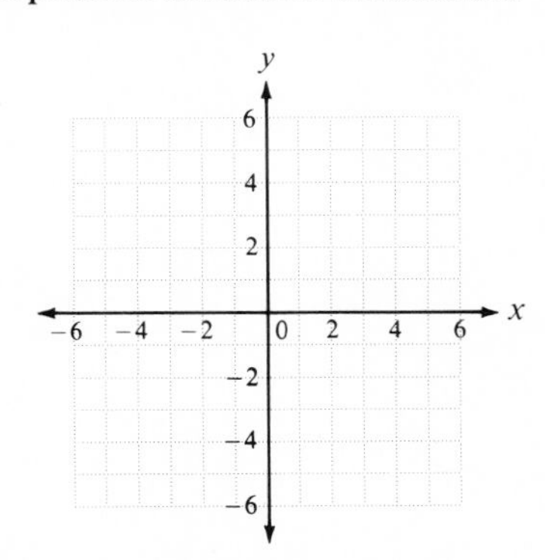

**23.** Graph the equation $y = -x^2$ and determine if the equation describes a function. [*Hint:* Remember that $-x^2$ *is not* $(-x)^2$]

| $x$ | 0 | 1 | −1 | 2 | −2 |
|---|---|---|---|---|---|
| $y$ | | | | | |

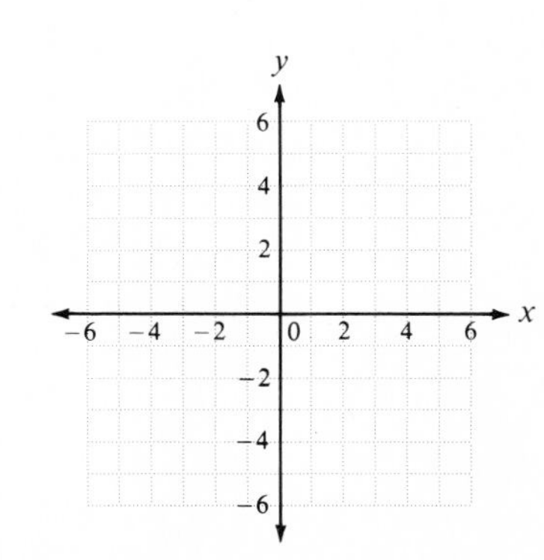

**24.** Graph the equation $x = -y^2$ and determine if the equation describes a function.

| $x$ | | | | | |
|---|---|---|---|---|---|
| $y$ | 0 | 1 | $-1$ | 2 | $-2$ |

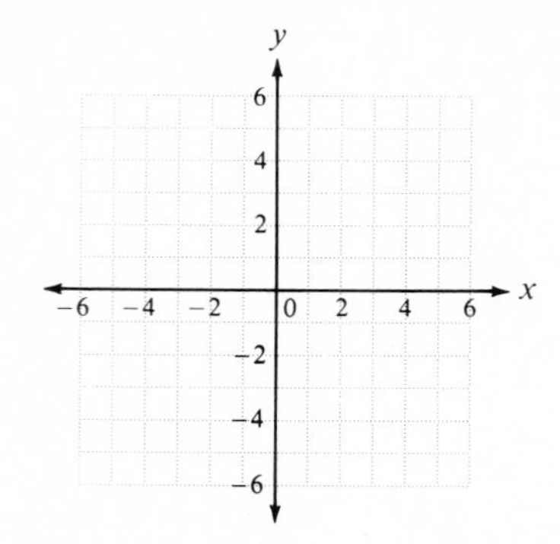

**25.** Graph the equation $x^2 + (y - 1)^2 = 9$ and determine if the equation describes a function.

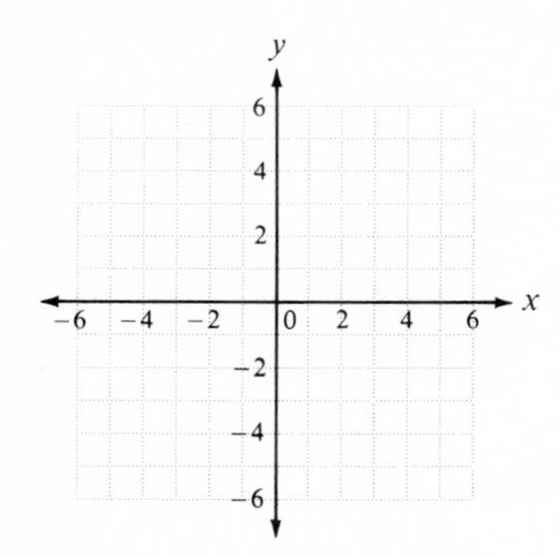

**26.** Graph the equation $\frac{x^2}{9} - \frac{y^2}{1} = 1$ and determine if the equation describes a function.

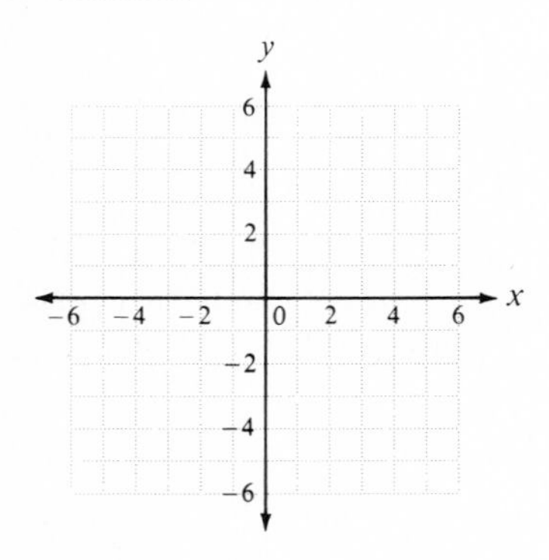

**27.** Graph the equation $\frac{x^2}{16} + \frac{y^2}{36} = 1$ and determine if the equation describes a function.

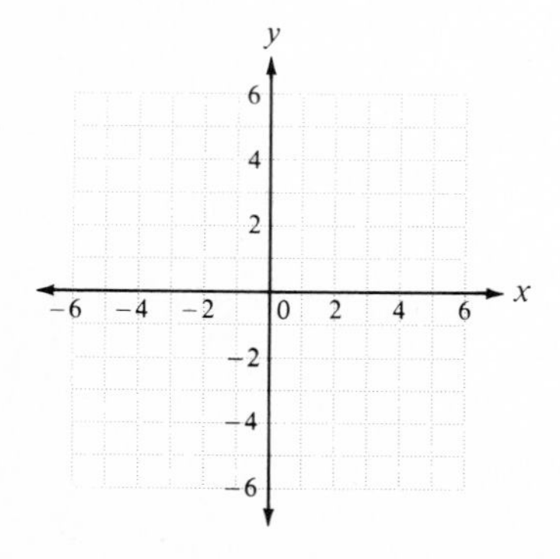

ANSWERS: **1.** (a) function (b) domain (c) range (d) one **2.** (a) $(x, y)$ (b) graph **3.** function **4.** function **5.** function **6.** not a function **7.** not a function **8.** function **9.** function **10.** function **11.** not the graph of a function **12.** not the graph of a function **13.** graph of a function **14.** graph of a function **15.** not the graph of a function **16.** graph of a function **17.** not the graph of a function **18.** not the graph of a function **19.** graph of a function

**20.** function

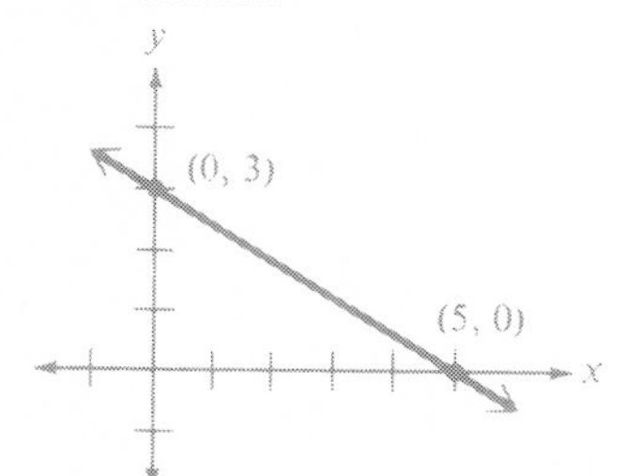


**21.** function

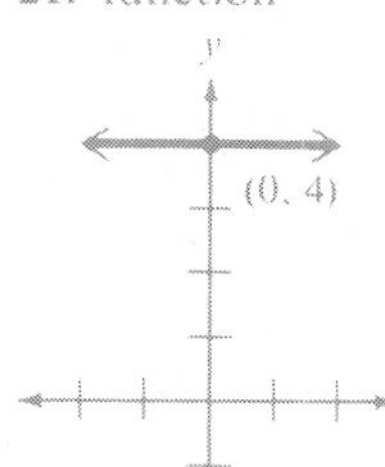


**22.** function

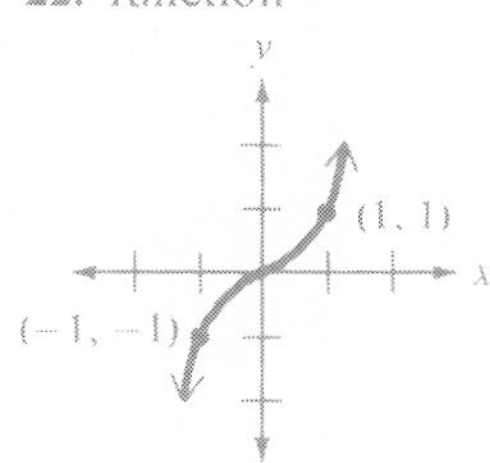


**23.** function

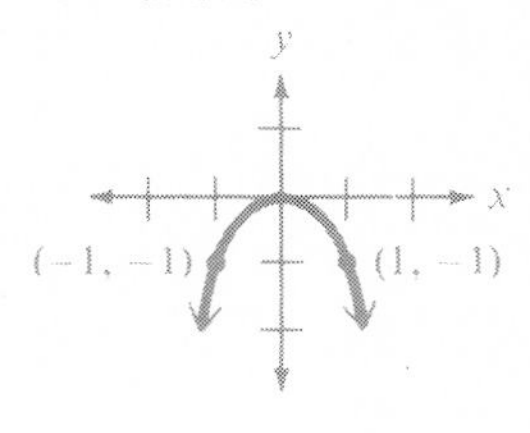


**24.** not a function

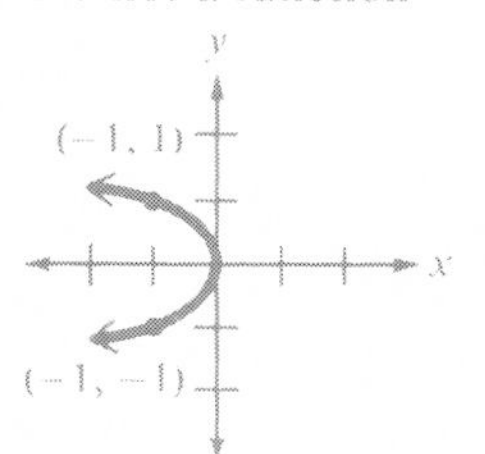


**25.** not a function

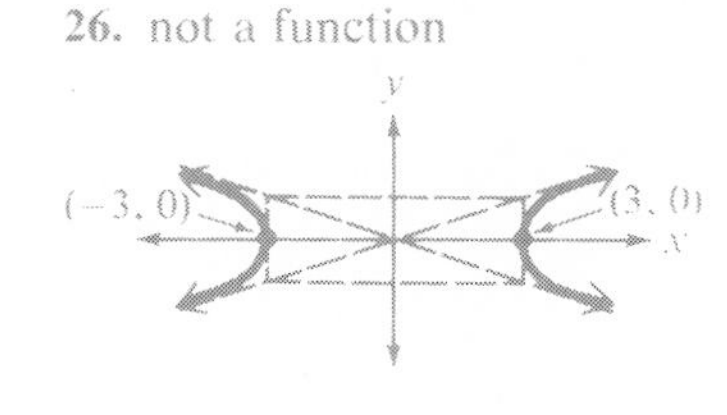


**26.** not a function



**27.** not a function

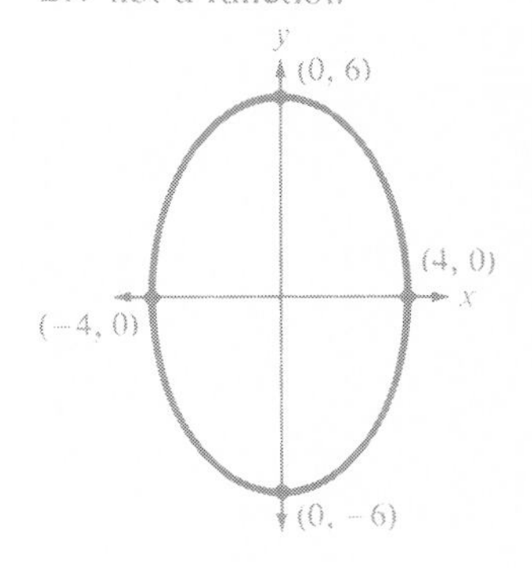


## 7.8 Functional Notation and Special Functions

Many times we use a letter to name a function ($f$ and $g$ are common choices). Suppose that a function $f$ is described by the equation

$$y = x^2 + 1.$$

We are often interested in determining the number $y$ which corresponds to a given number $x$. For example, we may ask the question:

"What number $y$ corresponds to $x = 2$ relative to the function $f$?"

This question can be shortened considerably if we use functional notation. We let

$$f(x),$$

read "$f$ of $x$," denote

"the number which corresponds to the number $x$ relative to the function $f$"

or

"the value of the function $f$ at $x$."

Using this notation the above question becomes:

"What is $f(2)$?"

In general, the $y$ value under the function $f$ is $f(x)$, that is $y = f(x)$. Hence in our example,

$$f(x) = x^2 + 1 \quad (y = x^2 + 1)$$

In order to **evaluate** the function at a particular value of $x$, we substitute the value into the formula

$$f(x) = x^2 + 1.$$

It is sometimes helpful to use parentheses in the formula, as in

$$f(\quad) = (\quad)^2 + 1.$$

To compute $f(2)$, for example, we place a 2 inside both sets of parentheses and perform the necessary simplification.

$$f(2) = (2)^2 + 1 = 4 + 1 = 5$$
$$f(-3) = (-3)^2 + 1 = 9 + 1 = 10$$
$$f(0) = (0)^2 + 1 = 0 + 1 = 1$$

Again, remember that $f(2)$ is the number that corresponds to 2 under $f$. That is, $2 \to f(2)$, or $(2, f(2))$ belongs to the function.

**EXAMPLE 1** Consider the function $g$ defined by

$$g(x) = 2x + 3 \quad [g(\quad) = 2(\quad) + 3].$$

**(a)** $g(0) = 2(0) + 3 = 0 + 3 = 3$ $\qquad 0 \to 3$; $(0, 3)$ belongs to $g$

**(b)** $g(4) = 2(4) + 3 = 8 + 3 = 11$ $\qquad 4 \to 11$; $(4, 11)$ belongs to $g$

**(c)** $g(-2) = 2(-2) + 3 = -4 + 3 = -1$ $\qquad -2 \to -1$; $(-2, -1)$ belongs to $g$

**(d)** $g(a) = 2(a) + 3 = 2a + 3$ $\qquad a \to 2a + 3$; $(a, 2a + 3)$ belongs to $g$

**(e)** $g(b + 1) = 2(b + 1) + 3 = 2b + 2 + 3 = 2b + 5$

Several functions, many of which we have already considered, are worthy of special mention. Any function that can be described by an equation such as $y = c$ ($c$ a constant), or

$$f(x) = c \quad (y = f(x))$$

is called a **constant function.** No matter what value (in the domain) $x$ assumes, the corresponding $y$ value, $f(x)$, is the same constant $c$. Since the graph of every equation of the form $y = c$ is a horizontal line, every constant function has as its graph a straight line parallel to the $x$-axis.

**EXAMPLE 2** The constant function $f$ defined by

$$f(x) = 3 \quad [f(\quad) = 3]$$

has the following values (why?) and is graphed in Figure 7.48.

$f(0) = 3$
$f(1) = 3$
$f(2) = 3$
$f(-1) = 3$
$f(a + 1) = 3$

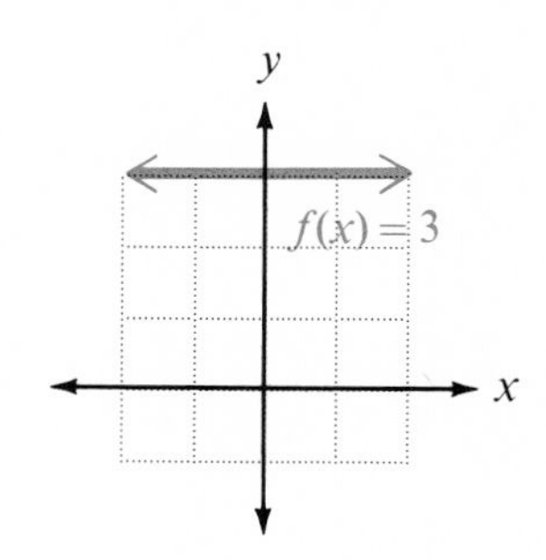


**Figure 7.48**

Any function that can be described by an equation of the form

$$f(x) = mx + b \qquad m \text{ and } b \text{ constants}$$

is called a **linear function.** Recall that every equation of the form $y = mx + b$ has as its graph a straight line (hence the term *linear* function). Also, if $m = 0$, the linear function $f(x) = mx + b$ becomes $f(x) = b$, a constant function that still has as its graph a straight line. Thus, a constant function is actually a special type of linear function.

**EXAMPLE 3** The linear function $f$ defined by

$$f(x) = 2x + 1 \quad [f(\quad) = 2(\quad) + 1]$$

has the following values.

$f(0) = 2(0) + 1 = 1$
$f(1) = 2(1) + 1 = 3$
$f(2) = 2(2) + 1 = 5$
$f(-1) = 2(-1) + 1 = -1$
$f(a + 1) = 2(a + 1) + 1 = 2a + 2 + 1 = 2a + 3$

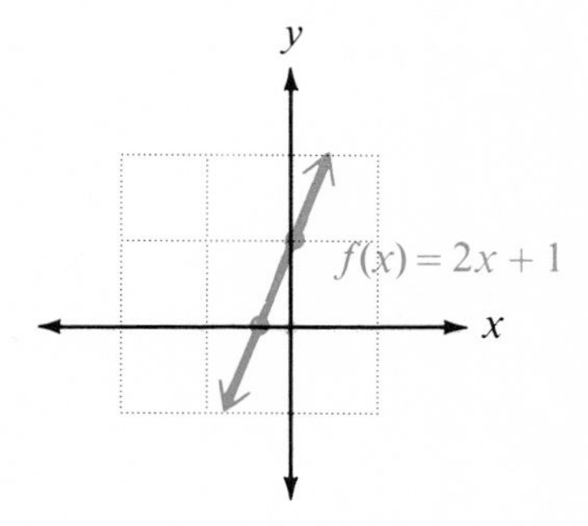


**Figure 7.49**

To graph the linear function $f(x) = 2x + 1$, we graph $y = 2x + 1$ as in Figure 7.49 by finding the intercepts $(0, 1)$ and $\left(-\frac{1}{2}, 0\right)$.

Any function that can be described by an equation of the form

$$f(x) = ax^2 + bx + c \quad (a, b, c \text{ constants}, a \neq 0)$$

is called a **quadratic function.** If $a$ were zero, we would lose the squared term and have a linear equation.

**EXAMPLE 4** The quadratic function $f$ defined by

$$f(x) = 2x^2 - x + 3 \qquad [f(\ \ ) = 2(\ \ )^2 - (\ \ ) + 3]$$

has the following values.

$$f(0) = 2(0)^2 - (0) + 3 = 2 \cdot 0 - 0 + 3 = 0 - 0 + 3 = 3$$
$$f(1) = 2(1)^2 - (1) + 3 = 2 \cdot 1 - 1 + 3 = 2 - 1 + 3 = 4$$
$$f(-1) = 2(-1)^2 - (-1) + 3 = 2 \cdot 1 + 1 + 3 = 2 + 1 + 3 = 6$$
$$f(3) = 2(3)^2 - (3) + 3 = 2 \cdot 9 - 3 + 3 = 18 - 3 + 3 = 18$$
$$f(-3) = 2(-3)^2 - (-3) + 3 = 2 \cdot 9 + 3 + 3 = 18 + 3 + 3 = 24$$
$$f(a) = 2(a)^2 - (a) + 3 = 2a^2 - a + 3$$
$$f(a + 1) = 2(a + 1)^2 - (a + 1) + 3 = 2(a^2 + 2a + 1) - (a + 1) + 3$$
$$= 2a^2 + 4a + 2 - a - 1 + 3 = 2a^2 + 3a + 4$$

The graph of a quadratic function is a parabola which either opens up, as in Figure 7.50, or down as in Figure 7.51, and never left nor right (such parabolas are not the graphs of functions). The graphs of the above two quadratic functions should be memorized. In this section we graph only quadratic functions with $a = 1$ or $-1$ and $b = 0$: functions of the forms

$$f(x) = x^2 + c \qquad \text{or} \qquad f(x) = -x^2 + c.$$

The general quadratic function is considered in the next section.

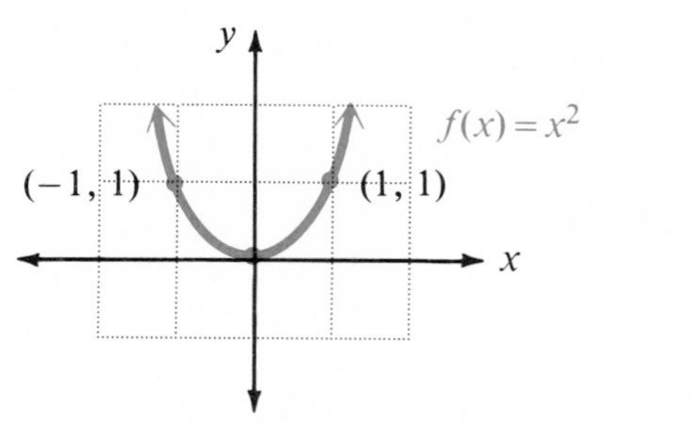


**Figure 7.50**

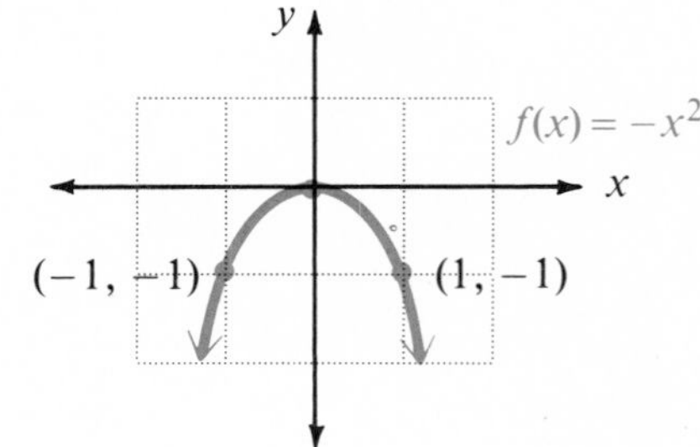


**Figure 7.51**

Study the graphs in Figure 7.52 carefully. Notice that the graph of $f(x) = x^2 + 2$ can be obtained by "sliding" the graph of $f(x) = x^2$ up 2 (+2) units. The graph of $f(x) = x^2 - 1$ can be obtained by "sliding" the graph of $f(x) = x^2$ down 1 (−1) unit. Similarly, the graph of $f(x) = -x^2 + 3$ can be obtained by "sliding" the graph of $f(x) = -x^2$ up 3 (+3) units, and the graph of $f(x) = -x^2 - 2$ can be obtained by "sliding" the graph of $f(x) = -x^2$ down 2 (−2) units. By using this information, we can quickly sketch the graphs of a multitude of quadratic functions by simply "moving" or "sliding" a memorized graph.

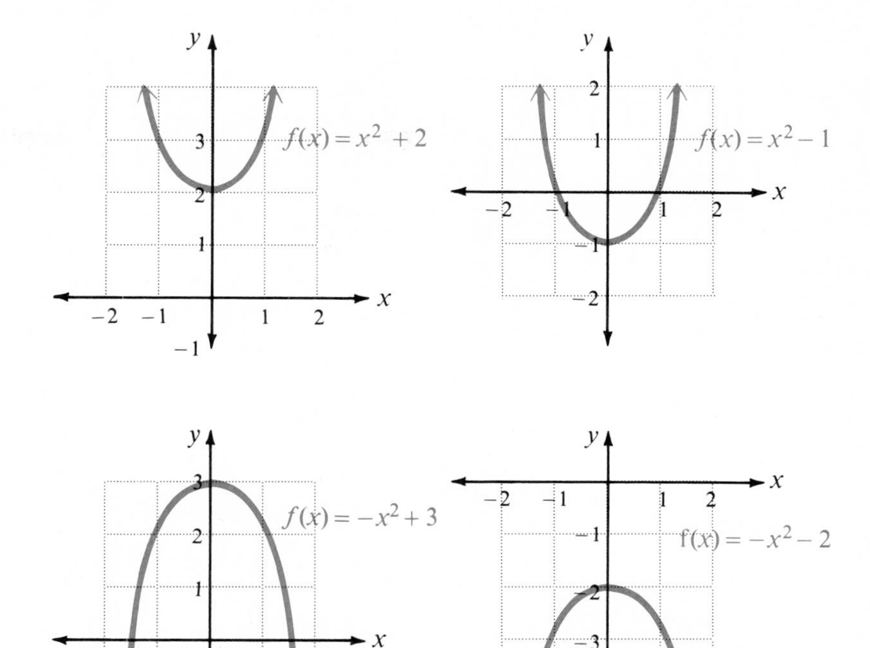


**Figure 7.52**

**EXAMPLE 5** Graph the quadratic function $f(x) = x^2 - 3$.

We begin with the graph of $f(x) = x^2$ and "slide" it down 3 units (notice the minus 3).

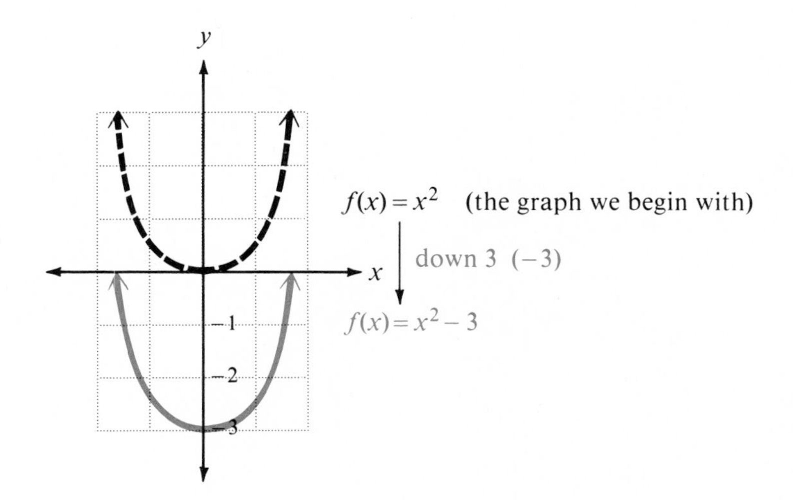


**Figure 7.53**

**EXAMPLE 6** Graph the quadratic function $f(x) = -x^2 + 2$.

In Figure 7.54 on p. 374, we begin with the graph of $f(x) = -x^2$ and "slide" it up 2 units (notice the plus 2).

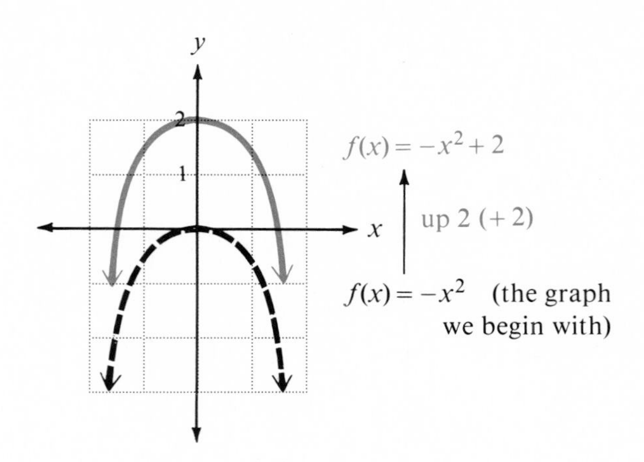


**Figure 7.54**

In summary, the graph of a constant function

$$f(x) = c$$

is a horizontal line, the graph of a linear function

$$f(x) = mx + b$$

is a straight line, and the graph of a quadratic function

$$f(x) = ax^2 + bx + c$$

is a parabola opening up or down. The only parabolas that we consider in this section are found by "sliding" up or down the graph of $f(x) = x^2$ or $f(x) = -x^2$.

## EXERCISES 7.8

*Evaluate each of the following.*

**1.** Let $f(x) = 2x + 5$.
- **(a)** $f(0) =$
- **(b)** $f(1) =$
- **(c)** $f(3) =$
- **(d)** $f(-2) =$
- **(e)** $f(a) =$
- **(f)** $f(b - 1) =$

**2.** Let $g(x) = x^2 - 3$.
- **(a)** $g(0) =$
- **(b)** $g(-1) =$
- **(c)** $g(3) =$
- **(d)** $g(-2) =$
- **(e)** $g(b) =$
- **(f)** $g(a + 1) =$

**3.** Let $f(x) = 2x^3 - 1$.
- **(a)** $f(0) =$
- **(b)** $f(-1) =$
- **(c)** $f(2) =$
- **(d)** $f(-3) =$
- **(e)** $f(a) =$
- **(f)** $f(b + 1) =$

**4.** Let $g(x) = -7$.
- **(a)** $g(0) =$
- **(b)** $g(-1) =$
- **(c)** $g(2) =$
- **(d)** $g(-3) =$
- **(e)** $g(a) =$
- **(f)** $g(b - 1) =$

**5.** Let $f(x) = -3x + 1$.
**(a)** $f(0) =$
**(b)** $f(-1) =$
**(c)** $f(2) =$
**(d)** $f(-3) =$
**(e)** $f(b) =$
**(f)** $f(a - 1) =$

**6.** Let $g(x) = 4x^2 - 3x + 1$.
**(a)** $g(0) =$
**(b)** $g(-1) =$
**(c)** $g(2) =$
**(d)** $g(-3) =$
**(e)** $g(a) =$
**(f)** $g(b + 1) =$

*State whether the given is a constant, linear, or quadratic function, and sketch its graph.*

**7.** $f(x) = -2x + 1$

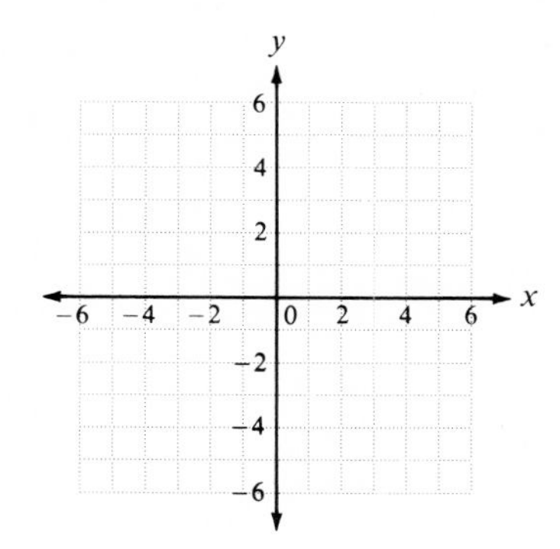


**8.** $f(x) = -x^2$

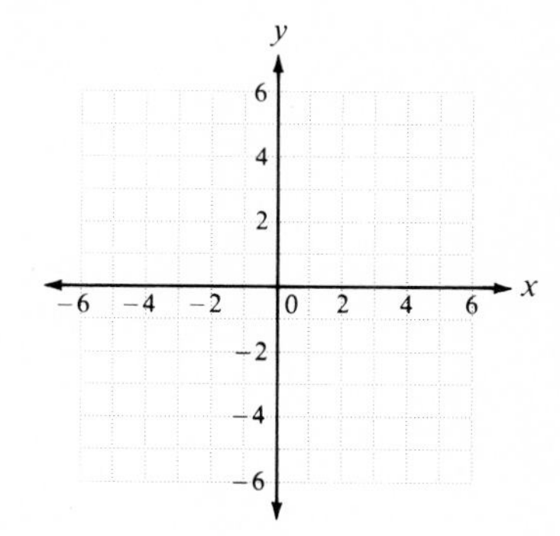


**9.** $f(x) = -2$

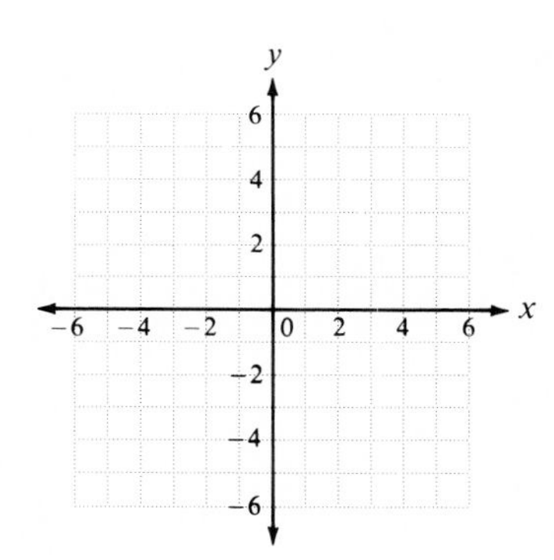


**10.** $f(x) = x^2$

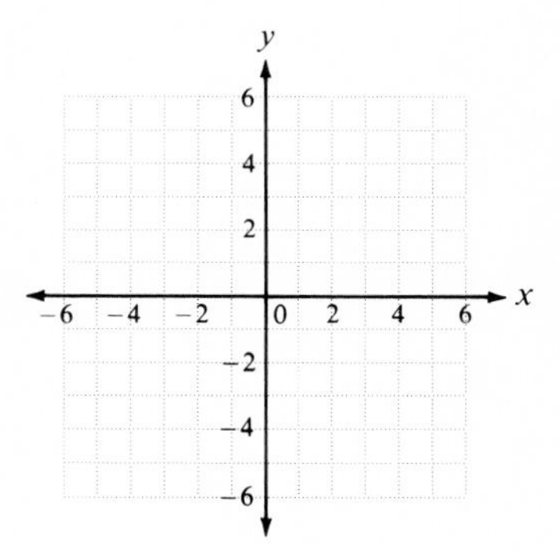


**11.** $f(x) = -x^2 + 1$

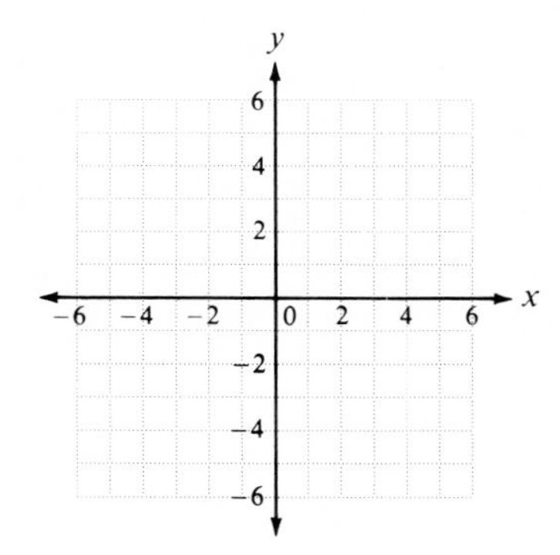


**12.** $f(x) = x^2 + 3$

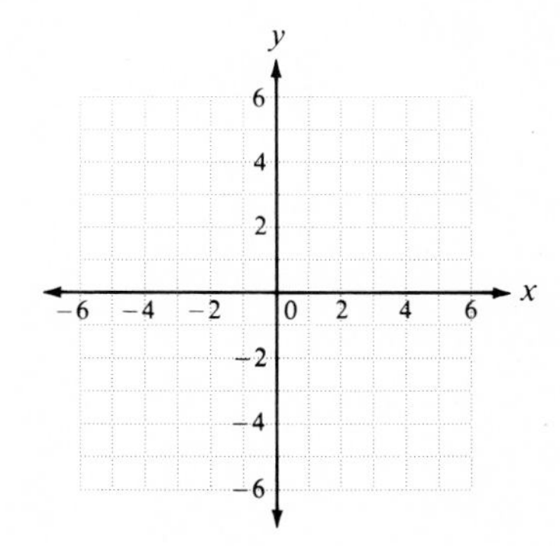


*Which of the following are functions? (Explain.)*

**13.**

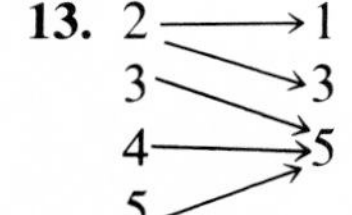


**14.** 2 → 1
3 → 3
4 → 7
5 → 7
(5 not mapped to)

**15.** 2 → 1
2 → 3
2 → 5
2 → 7
3
4
5

*Which of the following are graphs of functions? (Explain.)*

**16.**

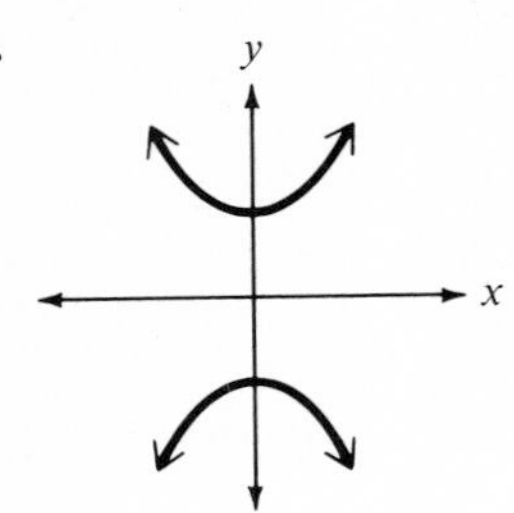


**17.**

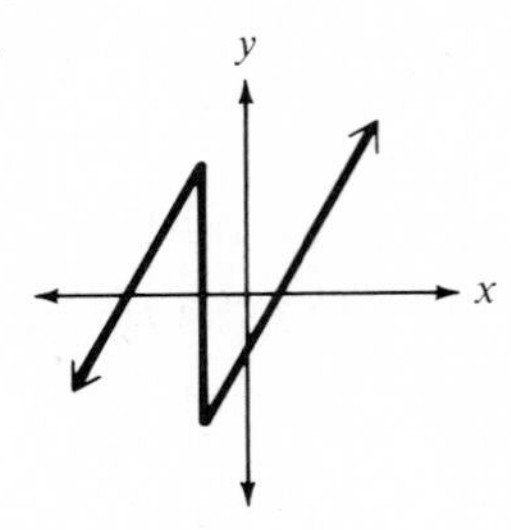


**18.**

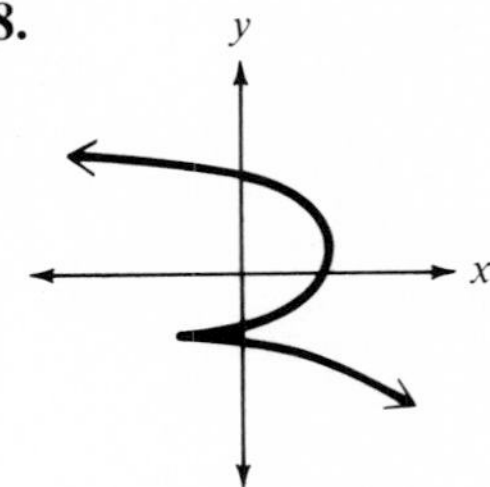


**19.** Find the midpoint of the line segment joining the points $(-8, 2)$ and $(3, -2)$.

**20.** Find the distance between the points $(3, -5)$ and $(2, -6)$.

---

**ANSWERS:** **1.** (a) 5 (b) 7 (c) 11 (d) 1 (e) $2a + 5$ (f) $2b + 3$ **2.** (a) $-3$ (b) $-2$ (c) 6 (d) 1 (e) $b^2 - 3$ (f) $a^2 + 2a - 2$ **3.** (a) $-1$ (b) $-3$ (c) 15 (d) $-55$ (e) $2a^3 - 1$ (f) $2(b + 1)^3 - 1$ **4.** (a) $-7$ (b) $-7$ (c) $-7$ (d) $-7$ (e) $-7$ (f) $-7$ **5.** (a) 1 (b) 4 (c) $-5$ (d) 10 (e) $-3b + 1$ (f) $-3a + 4$ **6.** (a) 1 (b) 8 (c) 11 (d) 46 (e) $4a^2 - 3a + 1$ (f) $4b^2 + 5b + 2$

**7.** linear function

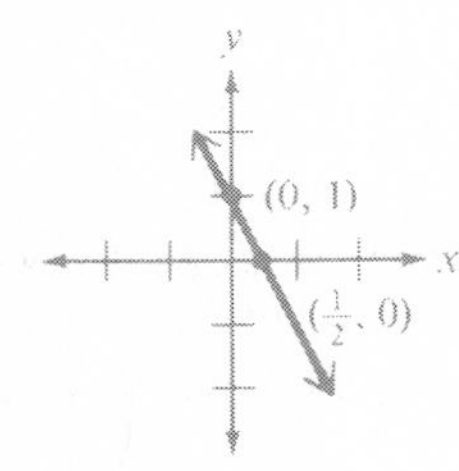


**8.** quadratic function

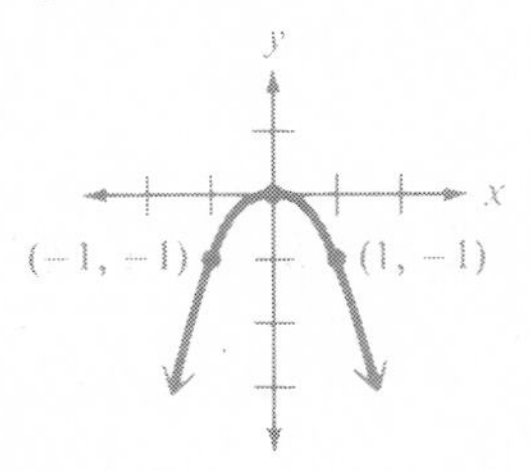


**9.** constant function

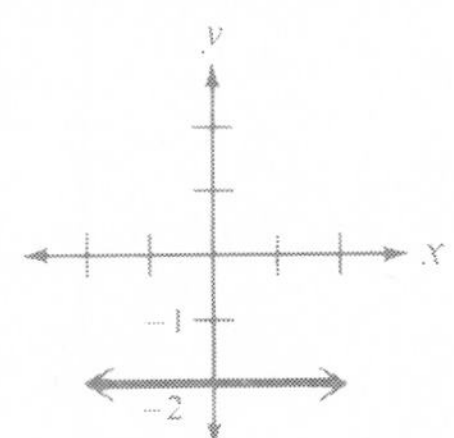


**10.** quadratic function

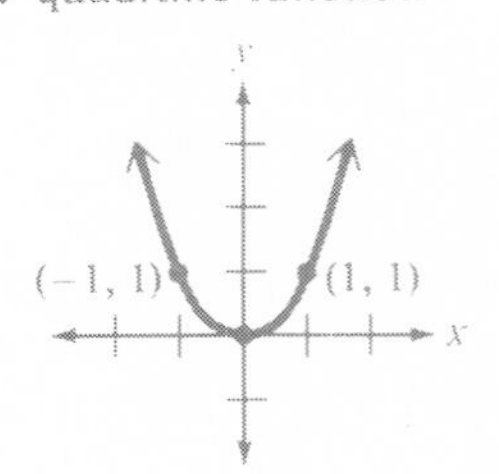


**11.** quadratic function

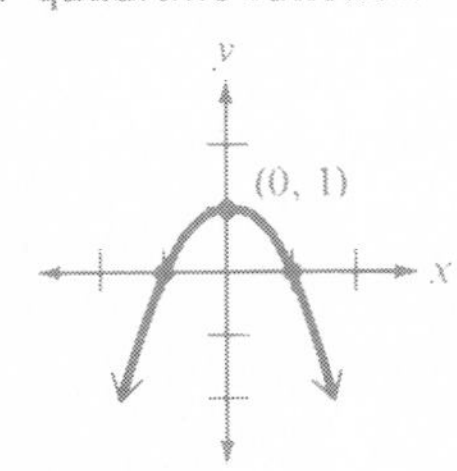


**12.** quadratic function

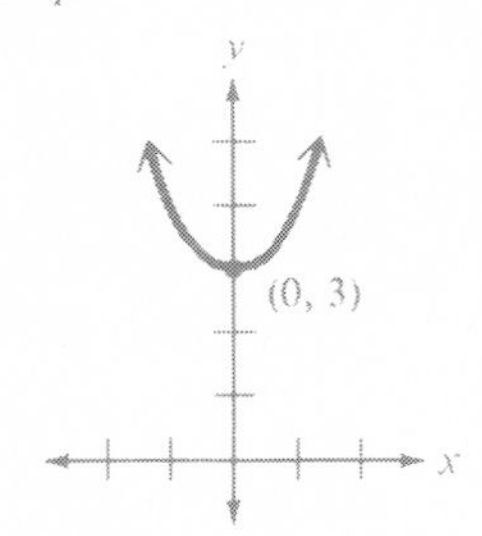


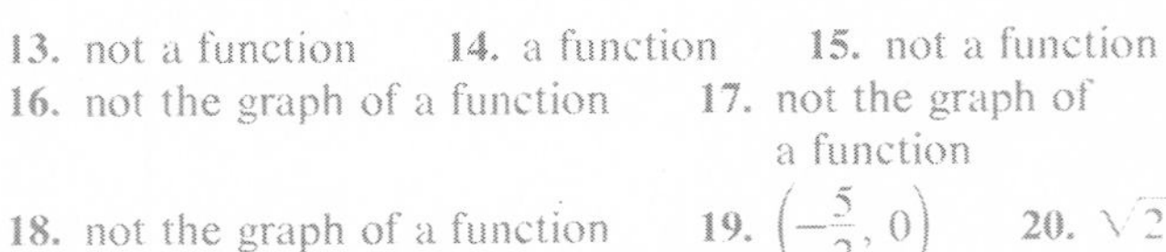
**13.** not a function **14.** a function **15.** not a function **16.** not the graph of a function **17.** not the graph of a function **18.** not the graph of a function **19.** $\left(-\frac{5}{2}, 0\right)$ **20.** $\sqrt{2}$

## 7.9 Quadratic Functions

In the previous section we saw that any function of the form

$$f(x) = ax^2 + bx + c \qquad (a \neq 0)$$

is a **quadratic function** and its graph is a **parabola.** We also graphed quadratic functions such as $f(x) = x^2$ and $g(x) = x^2 + 2$. In this section we graph general quadratic functions and develop several special techniques which will aid in the graphing process. First, however, we consider the following example in which we determine the graph by plotting several points.

**EXAMPLE 1** Graph $f(x) = x^2 - 2x - 3$.

| $x$ | 0 | 1 | −1 | 2 | −2 | 3 | 4 |
|---|---|---|---|---|---|---|---|
| $f(x)$ | −3 | −4 | 0 | −3 | 5 | 0 | 5 |

Note that the graph in Figure 7.55 crosses the $x$-axis ($y = 0$) at −1 and 3. The points (−1, 0) and (3, 0) are the $x$-intercepts. The graph has a minimum or low point at (1, −4), called the **vertex.** Also, for each point on the graph there is a corresponding point on the other side of the line $x = 1$. This line is called the **line of symmetry.**

In the above example it was easy to determine the values of $x$ for which the function is zero since they occurred at points that were plotted. However, if these values of $x$ had been irrational, it is unlikely that we would have discovered them by completing a table of values. In such cases, we can use the quadratic formula developed in Chapter 5 to determine them.

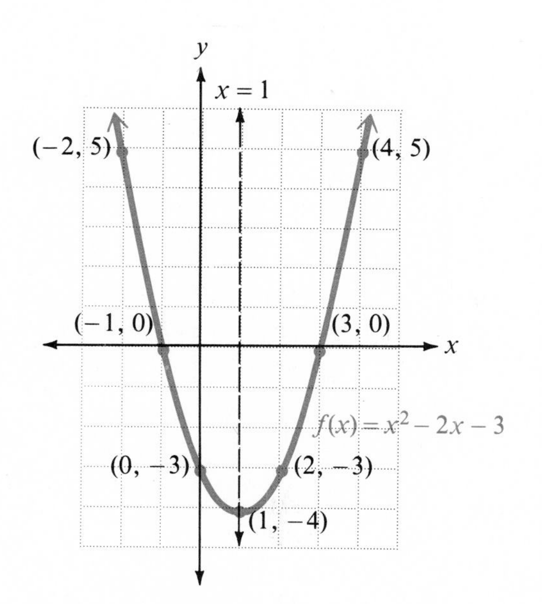


Figure 7.55

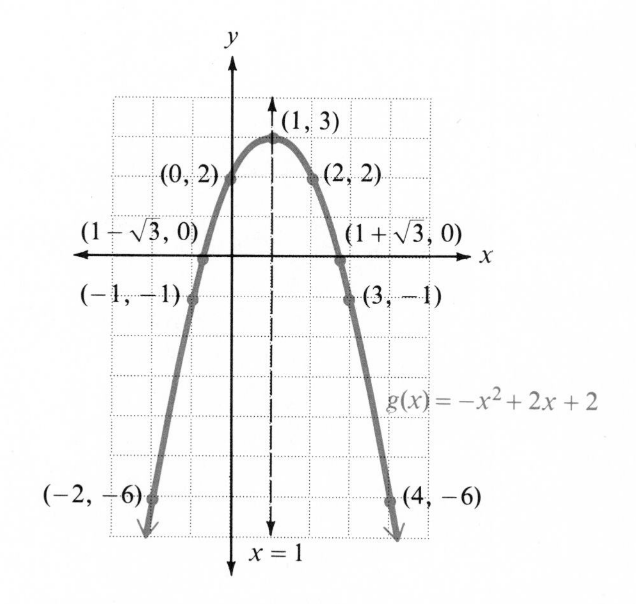


Figure 7.56

**EXAMPLE 2** Graph $g(x) = -x^2 + 2x + 2$ and determine the values of $x$ for which the function is zero.

| $x$ | 0 | 1 | −1 | 2 | −2 | 3 | 4 |
|---|---|---|---|---|---|---|---|
| $g(x)$ | 2 | 3 | −1 | 2 | −6 | −1 | −6 |

To find the values of $x$ for which the function $g$ is zero, we get $g(x)$ equal to zero and solve.

$$-x^2 + 2x + 2 = 0$$
$$x^2 - 2x - 2 = 0$$
$$x = \frac{-b \pm \sqrt{b^2 - 4ac}}{2a}$$
$$= \frac{-(-2) \pm \sqrt{(-2)^2 - 4(1)(-2)}}{2}$$
$$= \frac{2 \pm \sqrt{4 + 8}}{2} = \frac{2 \pm 2\sqrt{3}}{2}$$
$$= \frac{2(1 \pm \sqrt{3})}{2} = 1 \pm \sqrt{3} \approx 2.73 \text{ and } -0.73$$

Thus, the $x$-intercepts are $(1 + \sqrt{3}, 0)$ and $(1 - \sqrt{3}, 0)$. See Figure 7.56 on p. 377.

In the preceding examples, once the $x$-intercepts were found, we plotted several other points to determine the graph of the function. However, in many instances, it is helpful to know the **maximum (max)** or **minimum (min)** of the function (the largest or smallest value of $f(x)$). If the vertex (where the max or min occurs) has an integer for its $x$-coordinate, we may discover this value when plotting points. However, if it does not have an integer-valued $x$-coordinate, the process of completing the square is often used. (This process was explained in Chapter 5 to develop the quadratic formula.)

EXAMPLE 3 Graph $f(x) = x^2 + 5x + 4$, find the $x$-intercepts of the function, and find the vertex.

| $x$ | 0 | 1 | −1 | −2 | −3 | −4 | −5 | −6 |
|---|---|---|---|---|---|---|---|---|
| $f(x)$ | 4 | 10 | 0 | −2 | −2 | 0 | 4 | 10 |

The $x$-intercepts are $(-1, 0)$ and $(-4, 0)$, but this time the vertex is not at an integer value of $x$. We complete the square by adding and subtracting the square of one-half the coefficient of $x$.

$$f(x) = x^2 + 5x \qquad\qquad + 4 \qquad \text{Rewrite and leave space}$$

$$= x^2 + 5x + \frac{25}{4} - \frac{25}{4} + 4 \qquad \left(\frac{5}{2}\right)^2 = \frac{25}{4}, \text{ added and subtracted}$$

$$= \left(x + \frac{5}{2}\right)^2 - \frac{25}{4} + \frac{16}{4} \qquad x^2 + 5x + \frac{25}{4} \text{ is the perfect square } \left(x + \frac{5}{2}\right)^2$$

$$= \left(x + \frac{5}{2}\right)^2 - \frac{9}{4} \qquad -\frac{25}{4} + \frac{16}{4} = -\frac{9}{4}$$

When $f(x)$ is written in this form, we can see that the minimum value will occur when

$$\left(x + \frac{5}{2}\right)^2 = 0 \qquad \text{The least value } \left(x + \frac{5}{2}\right)^2 \text{ can have is zero}$$

$$x + \frac{5}{2} = 0$$

$$x = -\frac{5}{2}.$$

Thus, the minimum value of $f(x)$ occurs at $x = -\frac{5}{2}$ and is

$$f\left(-\frac{5}{2}\right) = \left(-\frac{5}{2} + \frac{5}{2}\right)^2 - \frac{9}{4}$$
$$= 0^2 - \frac{9}{4}$$
$$= -\frac{9}{4}.$$

Thus, the vertex is $(-\frac{5}{2}, -\frac{9}{4})$. Also, since the line of symmetry is a vertical line passing through the vertex, the equation of this line is $x = -\frac{5}{2}$ (See Figure 7.57).

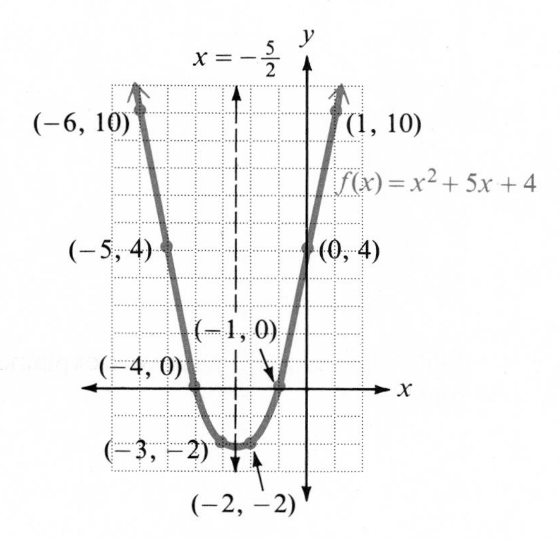


Figure 7.57

**The graph of a parabola**

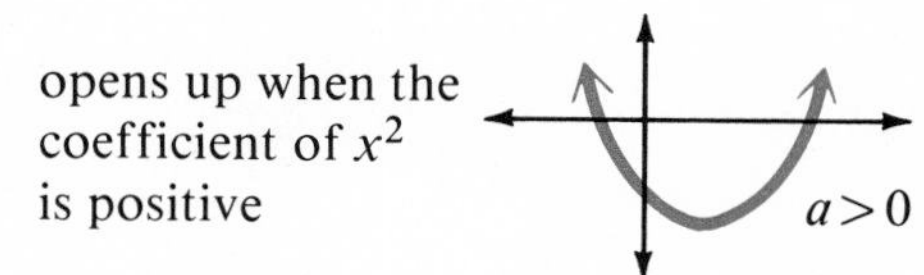


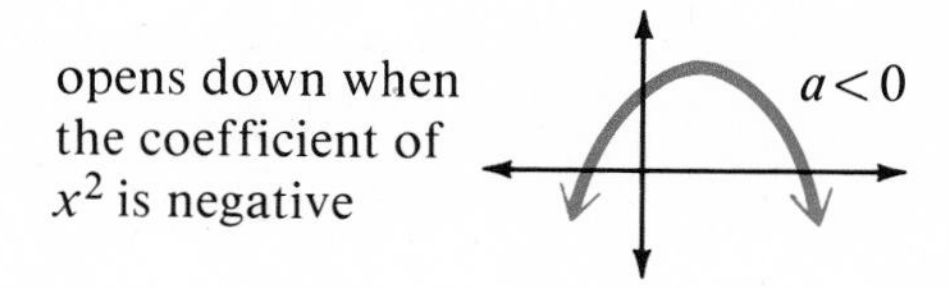


Figure 7.58

The results of the preceding three examples are summarized in Figure 7.58. If the graph opens up, there is a minimum but no maximum, and when it opens down, there is a maximum but no minimum.

Formulas for determining the vertex and the line of symmetry for a parabola can be derived by completing the square on the general quadratic function

$$f(x) = ax^2 + bx + c.$$

When this is done, we obtain

$$f(x) = a\left(x + \frac{b}{2a}\right)^2 + \frac{4ac - b^2}{4a}.$$

Thus, the vertex is given by

$$V\left(-\frac{b}{2a}, \frac{4ac - b^2}{4a}\right)$$

and the line of symmetry by

$$x = -\frac{b}{2a}.$$

In Example 3, $a = 1$, $b = 5$, and $c = 4$. Substituting in the formulas, we have

$$V\left(-\frac{b}{2a}, \frac{4ac - b^2}{4a}\right) = V\left(-\frac{5}{2(1)}, \frac{4(1)(4) - 5^2}{4(1)}\right) = V\left(-\frac{5}{2}, -\frac{9}{4}\right)$$

for the vertex, and the line of symmetry is

$$x = -\frac{b}{2a} = -\frac{5}{2},$$

the same results as before.

In order to find the vertex and line of symmetry of a given quadratic function, we may either complete the square or substitute the values of $a$, $b$, and $c$ into the formulas.

**EXAMPLE 4** Use the $x$-intercepts, the vertex, the line of symmetry, and a few other points to graph $g(x) = -2x^2 + 12x - 10$.

Since the coefficient of $x^2$ is negative ($a = -2$), the parabola opens downward. First we determine the intercepts.

$$\begin{aligned} -2x^2 + 12x - 10 &= 0 && g(x) = 0 \text{ gives intercepts} \\ -2(x^2 - 6x + 5) &= 0 && \text{Factor out } -2\text{, change all signs} \\ x^2 - 6x + 5 &= 0 && \text{Divide out } -2 \\ (x - 1)(x - 5) &= 0 \end{aligned}$$

$$x - 1 = 0 \quad \text{or} \quad x - 5 = 0$$
$$x = 1 \qquad\qquad x = 5$$

The $x$-intercepts are $(1, 0)$ and $(5, 0)$.

Now we determine the vertex and line of symmetry by completing the square. Recall that the method of completing the square requires that the coefficient of $x^2$ be 1.

$$\begin{aligned} g(x) &= -2x^2 + 12x - 10 \\ &= -2(x^2 - 6x) - 10 && \text{Factor coefficient of } x^2 \text{ out of first two terms leaving a coefficient of 1} \\ &= -2(x^2 - 6x + 9 - 9) - 10 && \left(\frac{6}{2}\right)^2 = 3^2 = 9 \\ &= -2(x^2 - 6x + 9) + (-2)(-9) - 10 && \text{Remove } -9 \text{ from parentheses and multiply by } -2 \\ &= -2(x - 3)^2 + 18 - 10 \\ &= -2(x - 3)^2 + 8 \end{aligned}$$

Hence, the maximum of the function occurs when

$$\begin{aligned} -2(x - 3)^2 &= 0 \\ x - 3 &= 0 \\ x &= 3 \end{aligned}$$

and the maximum is 8. Thus, the vertex is $(3, 8)$, and the line of symmetry is $x = 3$. Using the formulas developed above with $a = -2$, $b = 12$, and $c = -10$,

$$\begin{aligned} V\left(-\frac{b}{2a}, \frac{4ac - b^2}{4a}\right) &= V\left(-\frac{12}{2(-2)}, \frac{4(-2)(-10) - (12)^2}{4(-2)}\right) \\ &= V(3, 8) \end{aligned}$$

and
$$x = -\frac{b}{2a} = -\frac{12}{2(-2)} = 3.$$

Thus, if you know the formulas, you can obtain the vertex and line of symmetry more quickly.

We now know the intercepts (1, 0) and (5, 0), the vertex, (3, 8), and with two more points, one on each side of the line of symmetry, $x = 3$, we can plot the graph in Figure 7.59.

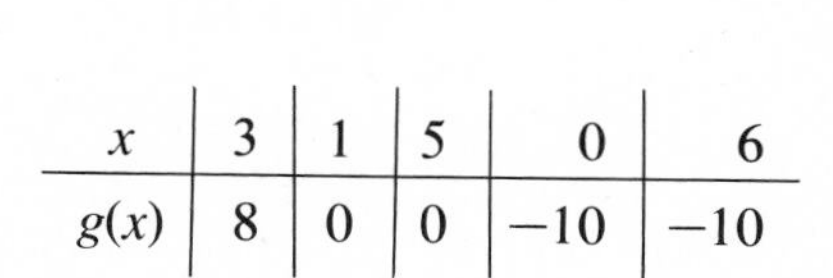

| $x$ | 3 | 1 | 5 | 0 | 6 |
|---|---|---|---|---|---|
| $g(x)$ | 8 | 0 | 0 | −10 | −10 |

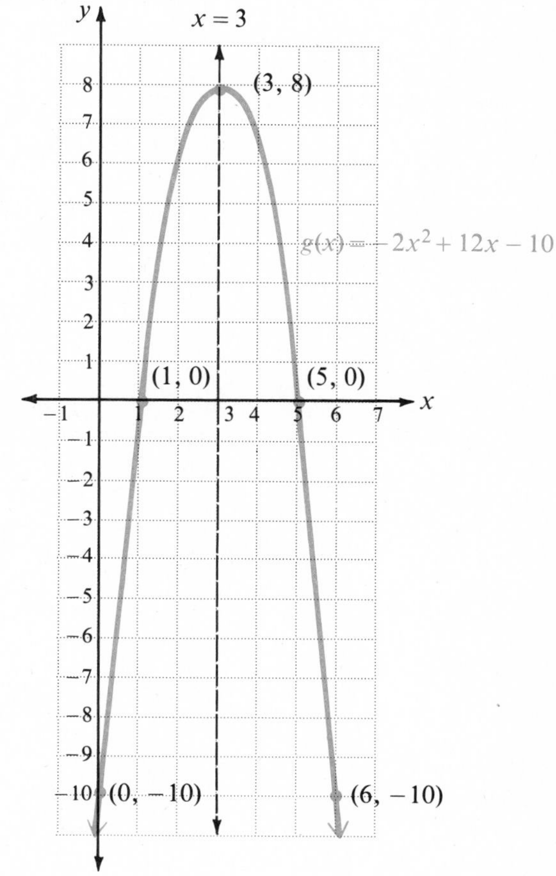


**Figure 7.59**

**TO GRAPH THE QUADRATIC FUNCTION $f(x) = ax^2 + bx + c$**

1. Determine whether the graph opens upward ($a > 0$) or downward ($a < 0$).
2. Determine the $x$-intercepts (if such exist) by solving the quadratic equation
$$ax^2 + bx + c = 0.$$
3. Determine the vertex and line of symmetry by completing the square or by using the formulas
$$V\left(-\frac{b}{2a}, \frac{4ac - b^2}{4a}\right) \quad \text{and} \quad x = -\frac{b}{2a}.$$
4. Plot the vertex, the $x$-intercepts, and two other points on each side of the line of symmetry.

If the curve does not cross the $x$-axis, we must determine other points on the graph to replace the missing intercepts as in the following example.

**EXAMPLE 5** Graph $f(x) = 3x^2 + 6x + 4$.

The graph opens upward since $a = 3 > 0$. First we try to find the $x$-intercepts by solving $3x^2 + 6x + 4 = 0$. The equation does not factor so we use the quadratic formula with $a = 3$, $b = 6$, and $c = 4$.

$$\begin{aligned} x &= \frac{-b \pm \sqrt{b^2 - 4ac}}{2a} \\ &= \frac{-6 \pm \sqrt{(6)^2 - 4(3)(4)}}{2(3)} \\ &= \frac{-6 \pm \sqrt{36 - 48}}{6} \\ &= \frac{-6 \pm \sqrt{-12}}{6} \end{aligned}$$

Since $\sqrt{-12}$ is not a real number there are no $x$-intercepts. Remember that our coordinate axes are only used to plot real numbers (never complex ones such as $\sqrt{-12}$).

Next we determine the vertex and line of symmetry by using the formulas.

$$\begin{aligned} V\left(-\frac{b}{2a}, \frac{4ac - b^2}{4a}\right) &= V\left(-\frac{6}{2(3)}, \frac{4(3)(4) - (6)^2}{4(3)}\right) \\ &= V(-1, 1) \end{aligned}$$

and

$$x = -\frac{b}{2a} = -\frac{6}{2(3)} = -1$$

Finally, we determine the additional points in the table of values, and graph the equation shown in Figure 7.60.

| $x$ | $-1$ | $0$ | $-2$ | $1$ | $-3$ |
|---|---|---|---|---|---|
| $f(x)$ | $1$ | $4$ | $4$ | $13$ | $13$ |

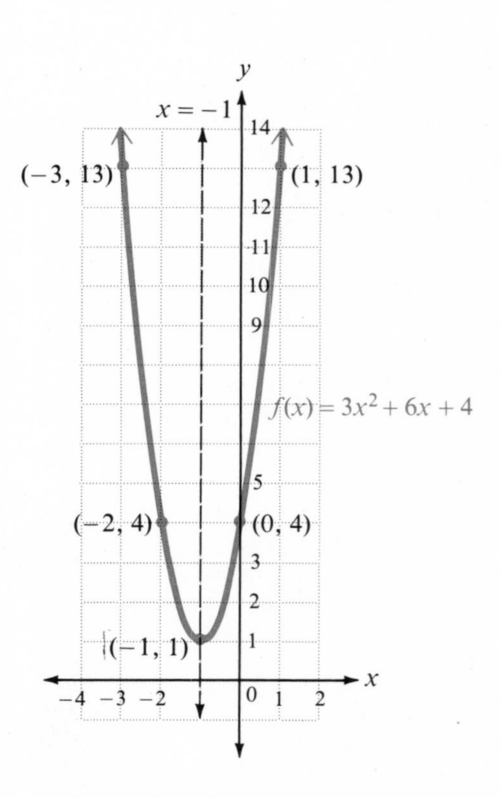


Figure 7.60

## EXERCISES 7.9

*Without graphing, find the x-intercepts (if they exist) and the vertex of each function.*

**1.** $f(x) = x^2 - 4x + 3$

**2.** $g(x) = x^2 + 4x + 3$

**3.** $f(x) = x^2 + 2x + 2$

**4.** $g(x) = x^2 + 8x$

**5.** $f(x) = -x^2 + 5x - 6$

**6.** $g(x) = 3x^2 - 6x + 3$

**7.** $f(x) = -2x^2 + 4x - 3$

**8.** $g(x) = 3x^2 + 5x + 1$

*Find the x-intercepts (if they exist), the vertex, and graph of the function.*

**9.** $f(x) = x^2 + 2x - 3$

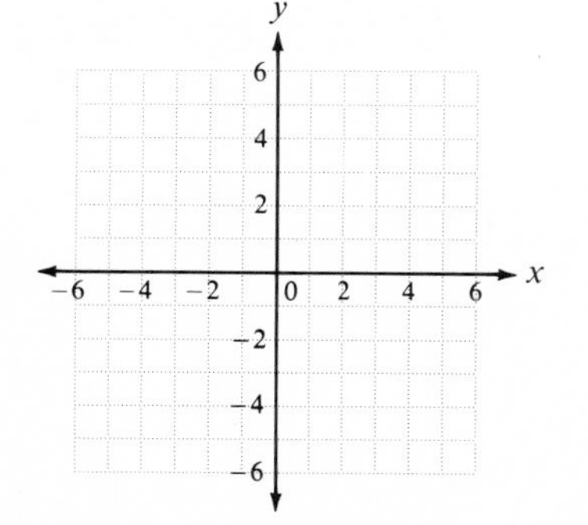


**10.** $g(x) = -x^2 - 2x + 3$

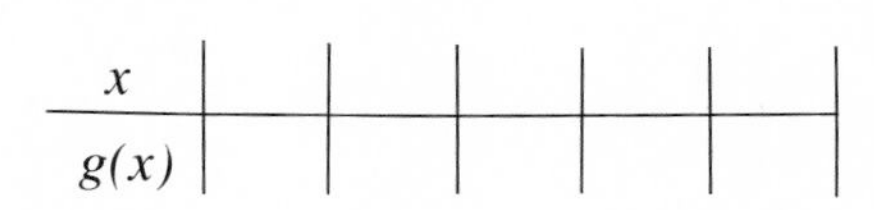


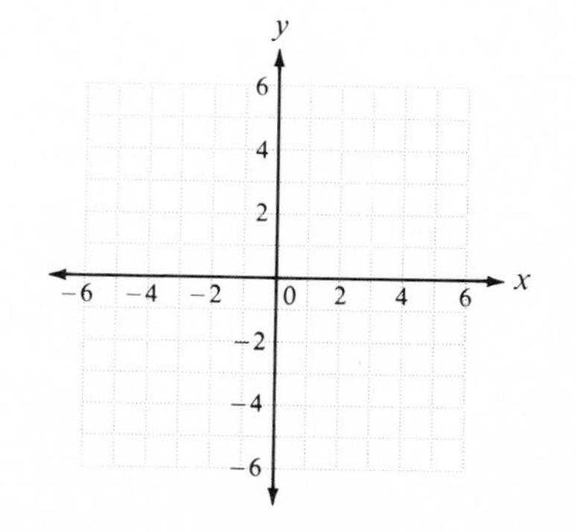

**11.** $f(x) = x^2 - 4x$

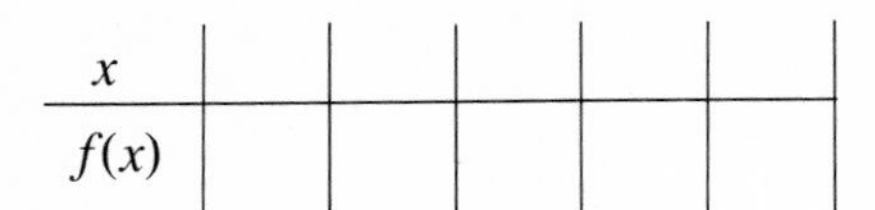

| $x$ | | | | | |
|---|---|---|---|---|---|
| $f(x)$ | | | | | |

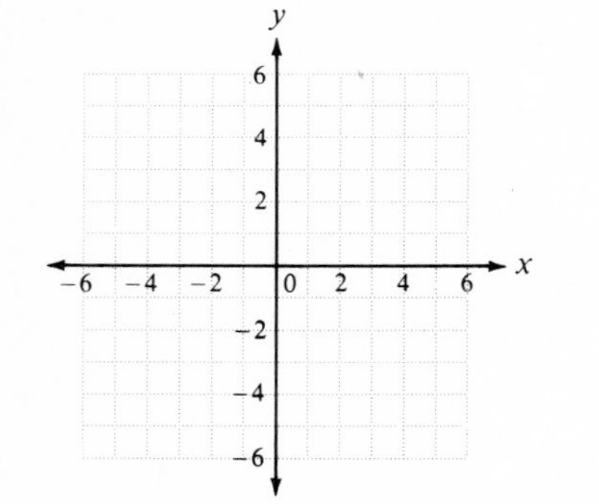


**12.** $g(x) = -x^2 + 4x$

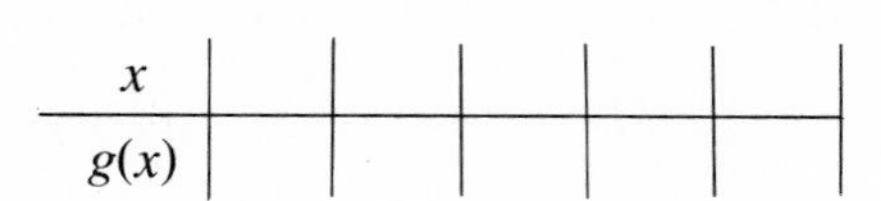

| $x$ | | | | | |
|---|---|---|---|---|---|
| $g(x)$ | | | | | |

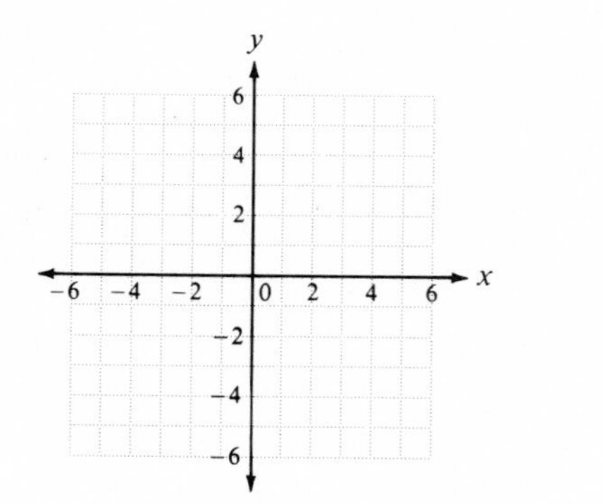


**13.** $f(x) = -4x^2 - 8x - 4$

| $x$ | | | | | |
|---|---|---|---|---|---|
| $f(x)$ | | | | | |

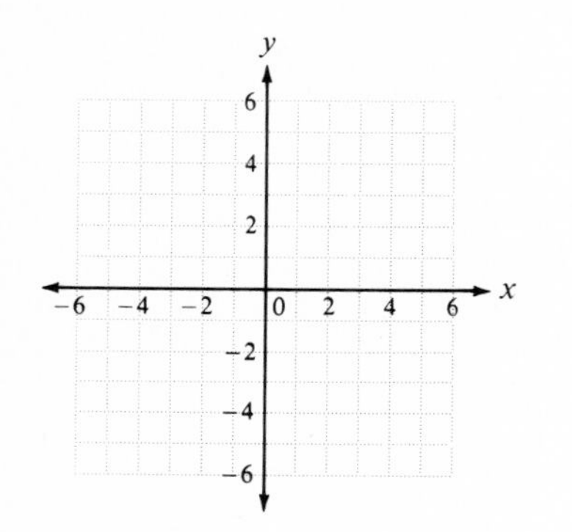


**14.** $g(x) = 2x^2 - 8x + 5$

| $x$ | | | | | |
|---|---|---|---|---|---|
| $g(x)$ | | | | | |

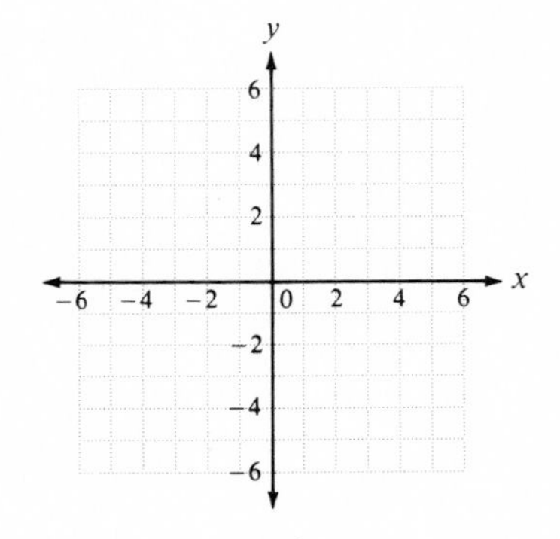


**15.** $f(x) = 2x^2 - 8x + 11$

| $x$ | | | | | |
|---|---|---|---|---|---|
| $f(x)$ | | | | | |

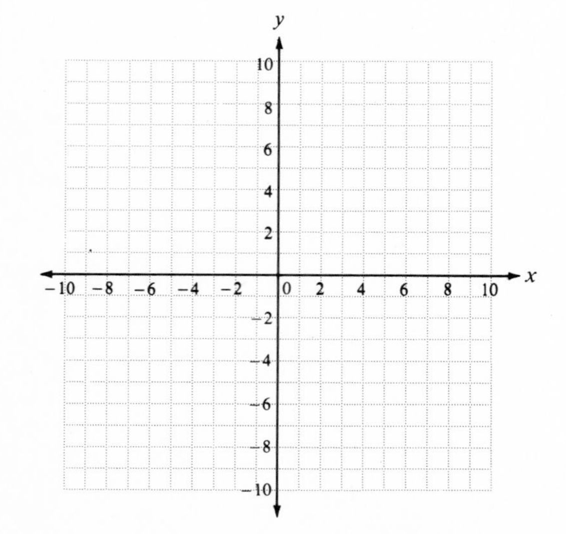


**16.** $g(x) = -2x^2 + 5x + 7$

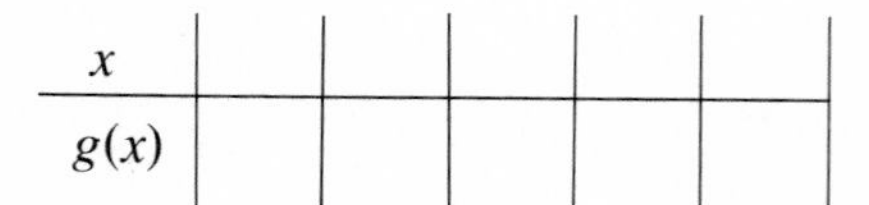

| $x$ | | | | | |
|---|---|---|---|---|---|
| $g(x)$ | | | | | |

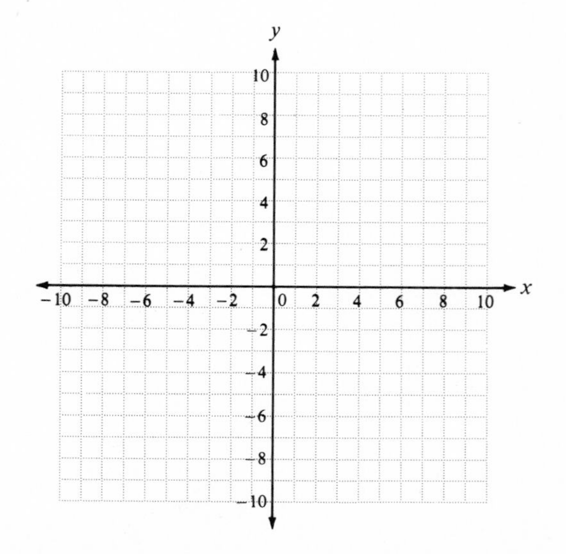

ANSWERS: **1.** (1, 0) and (3, 0); (2, −1) **2.** (−1, 0) and (−3, 0); (−2, −1) **3.** no $x$-intercepts; (−1, 1) **4.** (0, 0) and (−8, 0); (−4, −16) **5.** (2, 0) and (3, 0); $\left(\frac{5}{2}, \frac{1}{4}\right)$ **6.** (1, 0) is the only $x$-intercept; (1, 0) **7.** no $x$-intercepts; (1, −1) **8.** $\left(\frac{-5+\sqrt{13}}{6}, 0\right)$ and $\left(\frac{-5-\sqrt{13}}{6}, 0\right)$; $\left(-\frac{5}{6}, -\frac{13}{12}\right)$

**9.**

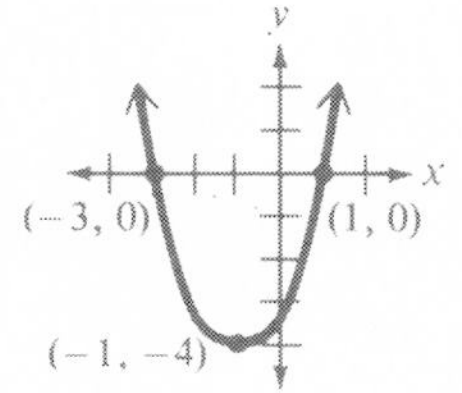


**10.**

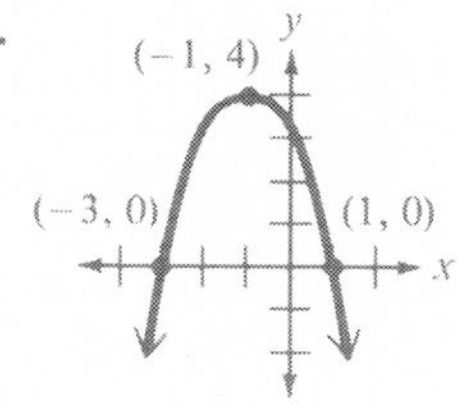


**11.**

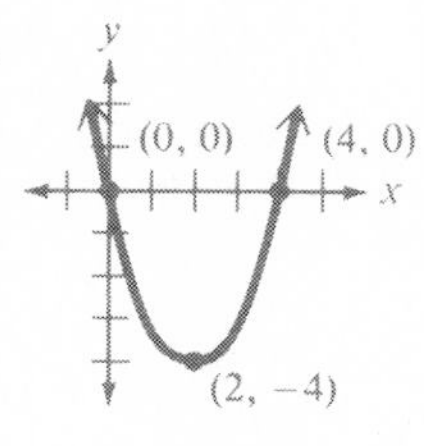


**12.**

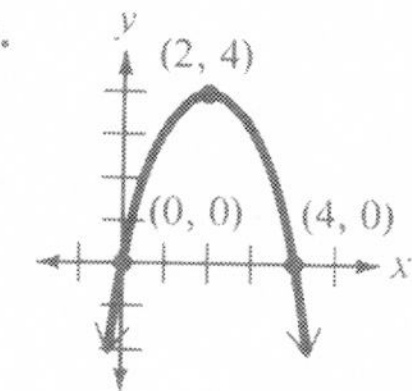


**13.**

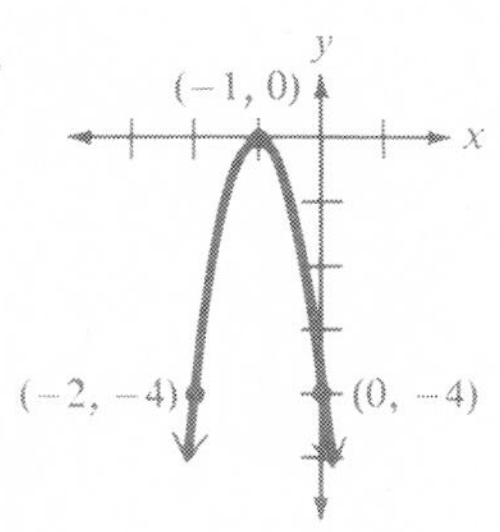


**14.**

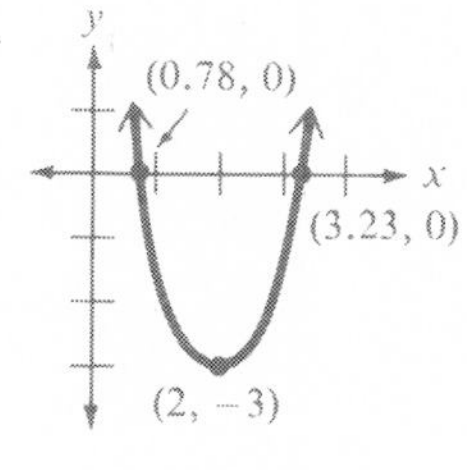


**15.**

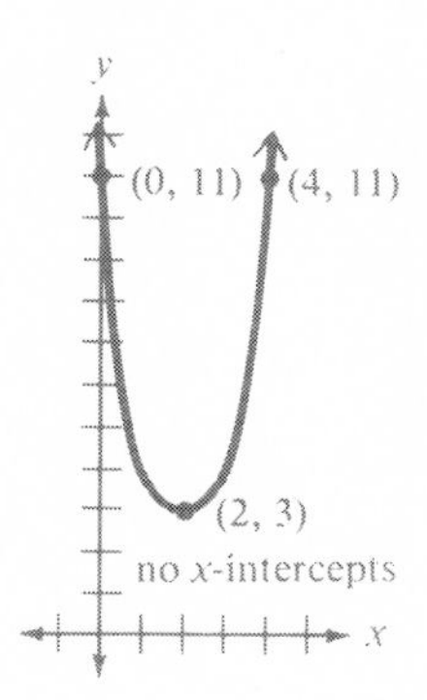


**16.**

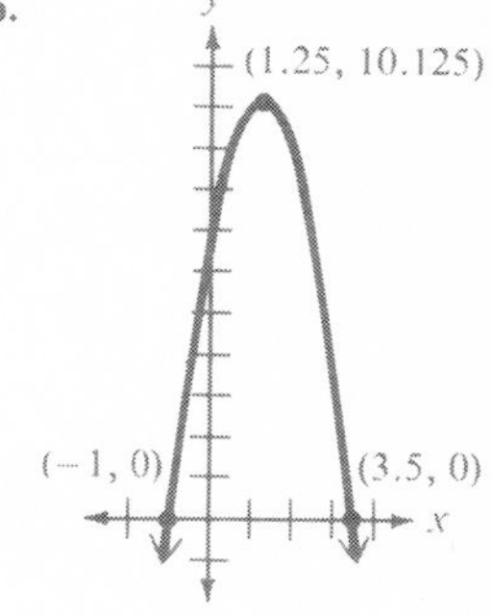


## Chapter 7 Summary

### Words and Phrases for Review

[7.1] number line
plot
origin
Cartesian coordinate system
horizontal axis ($x$-axis)
vertical axis ($y$-axis)
ordered pair
$x$-coordinate (abscissa)
$y$-coordinate (ordinate)
quadrant

[7.2] graph
linear (first-degree) equation
$x$-intercept
$y$-intercept

[7.3] slope
collinear
parallel
perpendicular

[7.6] second-degree equation
conic section
parabola
circle
ellipse
hyperbola
asymptote

[7.7] function
domain
range

[7.9] vertex of a parabola
line of symmetry

### Brief Reminders

[7.1] To graph an equation in a Cartesian coordinate system, first construct a table of values.

[7.2] The general form of the equation of a line is $ax + by + c = 0$. If $a = 0$, the line is parallel to the $x$-axis with $y$-intercept $(0, -c/b)$. If $b = 0$, the line is parallel to the $y$-axis with $x$-intercept $(-c/a, 0)$.

[7.3] **1.** The slope of a line passing through points with coordinates $(x_1, y_1)$ and $(x_2, y_2)$ is $m = (y_2 - y_1)/(x_2 - x_1)$.

**2.** If $m_1$ and $m_2$ are the slopes of two lines and if $m_1 = m_2$, then the lines are parallel; if $m_1 = -1/m_2$, then the lines are perpendicular.

[7.4] **1.** Point-slope form: $y - y_1 = m(x - x_1)$.

**2.** Two-point form: $y - y_1 = \dfrac{y_2 - y_1}{x_2 - x_1}(x - x_1)$.

**3.** Slope-intercept form: $y = mx + b$.

[7.5] **1.** Distance between points $(x_1, y_1)$ and $(x_2, y_2)$:
$$d = \sqrt{(x_1 - x_2)^2 + (y_1 - y_2)^2}.$$

**2.** Midpoint of the line segment joining $(x_1, y_1)$ and $(x_2, y_2)$:

$$(\bar{x}, \bar{y}) = \left(\frac{x_1 + x_2}{2}, \frac{y_1 + y_2}{2}\right).$$

**3.** Remember to *subtract* coordinates in the distance formula and to *add* them in the midpoint formula.

[7.6] **1.** General second-degree equation:
$Ax^2 + Bxy + Cy^2 + Dx + Ey + F = 0$.

**2.** Parabola: $y = ax^2 + bx + c$ or $x = ay^2 + by + c$

**3.** Circle: $(x - h)^2 + (y - k)^2 = r^2$

**4.** Ellipse: $\dfrac{x^2}{a^2} + \dfrac{y^2}{b^2} = 1$

**5.** Hyperbola: $\dfrac{x^2}{a^2} - \dfrac{y^2}{b^2} = 1$, $\dfrac{y^2}{b^2} - \dfrac{x^2}{a^2} = 1$, $xy = c$

[7.7] A function is a correspondence between a domain and range in such a way that each object in the domain corresponds to *exactly one* object in the range.

[7.8] **1.** Constant function: $f(x) = c$.

**2.** Linear function: $f(x) = mx + b$

**3.** Quadratic function: $f(x) = ax^2 + bx + c \quad (a \neq 0)$

[7.9] The graph of a quadratic function $f(x) = ax^2 + bx + c \ (a \neq 0)$ is a parabola.

**1.** The parabola opens upward if $a > 0$ and downward if $a < 0$.

**2.** The vertex can be found by completing the square or by substitution in $(-b/2a, (4ac - b^2)/4a)$.

**3.** The line of symmetry of the graph is the line $x = -b/2a$.

**4.** The $x$-intercepts (if such exist) are found by solving the quadratic equation $ax^2 + bx + c = 0$.

## CHAPTER 7 REVIEW EXERCISES

[7.1] **1.** Graphs of equations and functions are plotted in a ______________ ______________.

**2.** The ordered pair $(-2, 3)$ has $x$-coordinate **(a)** ________, $y$-coordinate **(b)** ________, and the point identified with the pair is located in quadrant **(c)** ________.

**3.** The perpendicular lines in a coordinate system are called ________.

**4.** The set of points in a Cartesian coordinate system which corresponds to the solutions of the equation is called the ______________ of the equation.

[7.2] **5.** Every linear equation of the form $x = a$ ($a$ a constant) has as its graph a straight line parallel to the ______________.

**6.** Every linear equation of the form $y = b$ ($b$ a constant) has as its graph a straight line parallel to the ______________.

**7.** The general form of a linear equation is ______________.

**8.** Another name for a first-degree equation is a ______________.

**9.** The points at which the graph of a linear equation cross the axes are called the ______________.

**10.** If $(0, 3)$ is a point on the graph of a linear equation, the point is called the ______________.

**11.** If $(2, 0)$ is a point on the graph of a linear equation, the point is called the ______________.

**12.** The line with equation $x = 0$ is the ______________.

**13.** The line with equation $y = 0$ is the ______________.

[7.3] **14.** The slope of the line joining $(x_1, y_1)$ and $(x_2, y_2)$ is given by $m =$ ______.

**15.** A line that "slopes" from upper left to lower right has slope a ______ ______________ number.

**16.** A line that "slopes" from lower left to upper right has slope a ______ ______________ number.

**17.** A vertical line has ________ slope.

**18.** A horizontal line has ________ slope.

**19.** If a line has slope $m$, every parallel line has slope **(a)** ________ and every perpendicular line has slope **(b)** ________.

[7.4] **20.** The point-slope form of the equation of a line having slope $m$ and passing through the point $(x_1, y_1)$ is ______________.

**21.** The slope-intercept form of the equation of a line with slope $m$ and $y$-intercept $(0, b)$ is ______________.

22. If a linear equation is solved for $y$, the resulting equation is the **(a)** __________________ form of the equation of the line and the coefficient of $x$ is the **(b)** __________________ of the line while the constant term is the $y$-coordinate of the **(c)** __________________ of the line.

23. The two-point form of the equation of a line passing through $(x_1, y_1)$ and $(x_2, y_2)$ is __________________.

[7.5] 24. The distance between two points with coordinates $(x_1, y_1)$ and $(x_2, y_2)$ is $d =$ __________________.

25. The coordinates of the midpoint of the line segment joining $(x_1, y_1)$ and $(x_2, y_2)$ are $(\bar{x}, \bar{y}) =$ __________________.

[7.6] 26. The graph of a second-degree equation that is a curve formed by intersecting a plane with a cone is called a(n) __________________.

27. An equation of the form $y = ax^2 + bx + c$ has a(n) __________________ for its graph.

28. An equation of the form $\frac{x^2}{a^2} + \frac{y^2}{b^2} = 1$ has a(n) __________________ for its graph.

29. An equation of the form $(x - h)^2 + (y - k)^2 = 1$ has a(n) ____________ for its graph.

30. An equation of the form $\frac{x^2}{a^2} - \frac{y^2}{b^2} = 1$ has a(n) __________________ for its graph.

[7.7] 31. A **(a)** __________________ is a correspondence between a first set called the **(b)** __________________ and a second set called the **(c)** __________________ with the property that every object in the first set corresponds to exactly **(d)** __________________ object in the second set.

32. A graph will not be the graph of a function if a vertical line can be drawn passing through __________________ point on the graph.

[7.8] 33. A function described by an equation of the form $f(x) = mx + b$ is called a __________________ function.

34. A function described by an equation of the form $f(x) = c$ is called a __________________ function.

35. The graph of a linear function is always a __________________.

36. The graph of $f(x) = x^2 + 7$ can be sketched by "sliding" the graph of $f(x) = x^2$ __________________ 7 units.

**37.** The graph of $f(x) = -x^2 - 5$ can be sketched by "sliding" the graph of $f(x) = -x^2$ ______________ 5 units.

[7.9] **38.** A function described by an equation of the form $f(x) = ax^2 + bx + c$ is called a ______________ function.

**39.** The graph of a quadratic function is always a ______________.

[7.1] **40.** Give the coordinates of the points $A$, $B$, $C$, and $D$. In which quadrant is each point located?

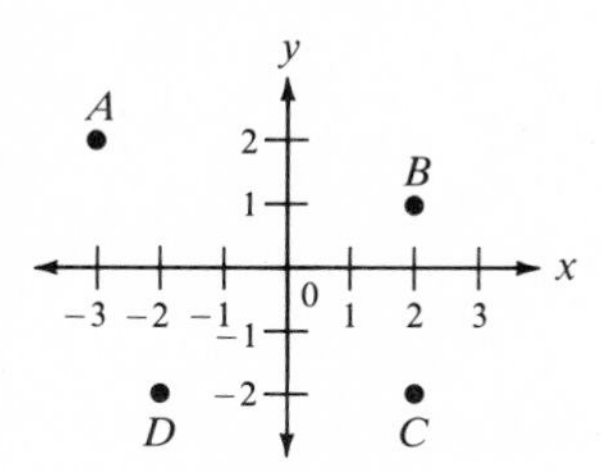


[7.2] *Determine the intercepts, slope, and graph of the following equations.*

**41.** $x + 2y - 4 = 0$

$x$-intercept =
$y$-intercept =
slope =

**42.** $2x + 1 = 0$

$x$-intercept =
$y$-intercept =
slope =

**43.** $y - 3 = 0$

$x$-intercept =
$y$-intercept =
slope =

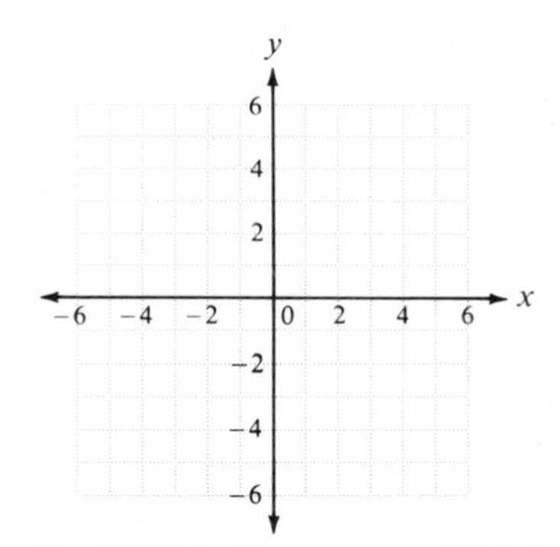

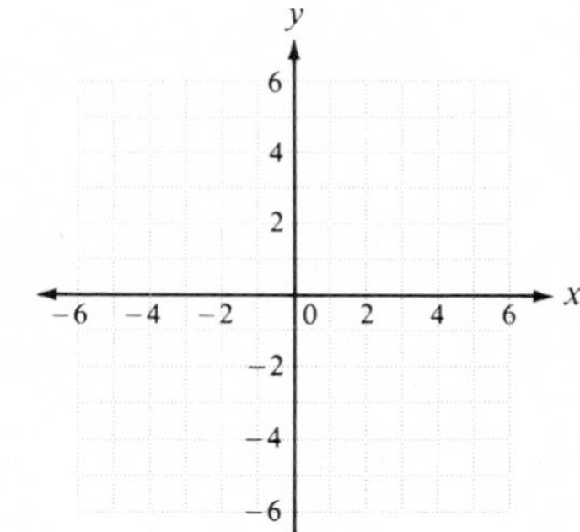

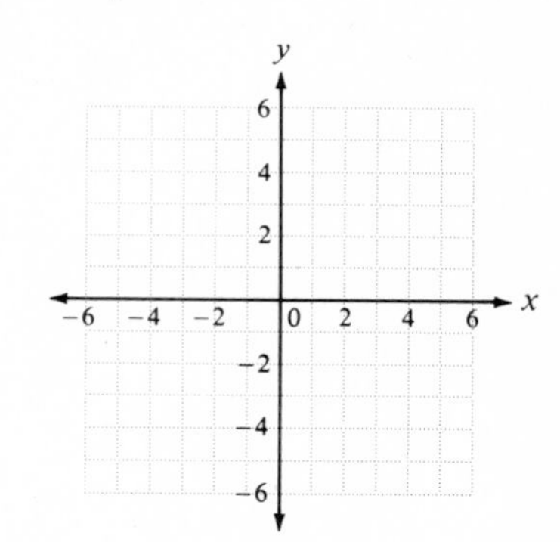

[7.3] *Are the given pairs of lines parallel, perpendicular, or neither?*

**44.** $3x - y + 2 = 0$
$x + 3y - 7 = 0$

**45.** $x + 2 = 0$
$y - 2 = 0$

**46.** $2x - 5y + 2 = 0$
$-4x + 10y - 2 = 0$

[7.4] **47.** Find the general form of the equation of the line with slope $-4$ and passing through $(-1, 3)$.

**48.** Find the slope-intercept form of the equation of the line with equation $4x + 2y - 10 = 0$. What is the slope? The $y$-intercept?

**49.** Find the general form of the equation of the line passing through the points $(-2, 5)$ and $(6, 9)$.

**50.** Find the general form of the equation of the line which passes through the point $(-3, 1)$ and is perpendicular to the line $2x + y - 3 = 0$.

**51.** What are the intercepts and the slope of the line with equation $3y - 15 = 0$?

[7.5] **52.** What is the distance between the points $(4, -1)$ and $(-2, -1)$?

**53.** What is the midpoint of the line segment joining $(1, -4)$ and $(3, 2)$?

[7.6] *Graph the following equations.*

**54.** $(x + 3)^2 + (y - 1)^2 = 9$

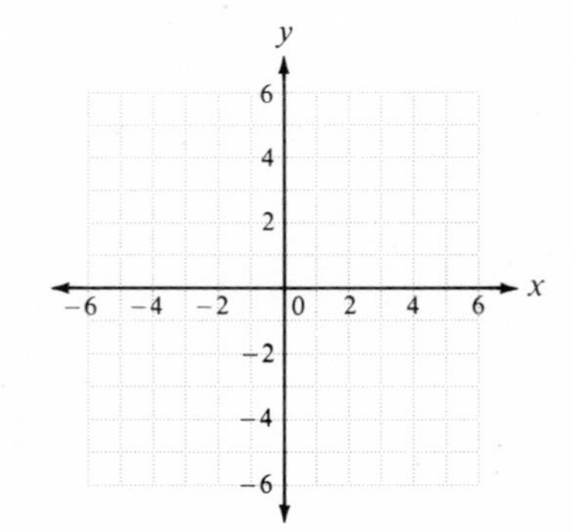


**55.** $\dfrac{x^2}{25} + \dfrac{y^2}{4} = 1$

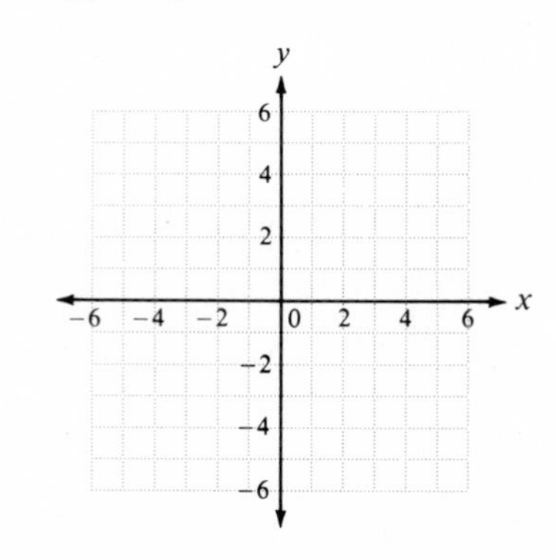

**56.** $\frac{x^2}{4} - \frac{y^2}{1} = 1$

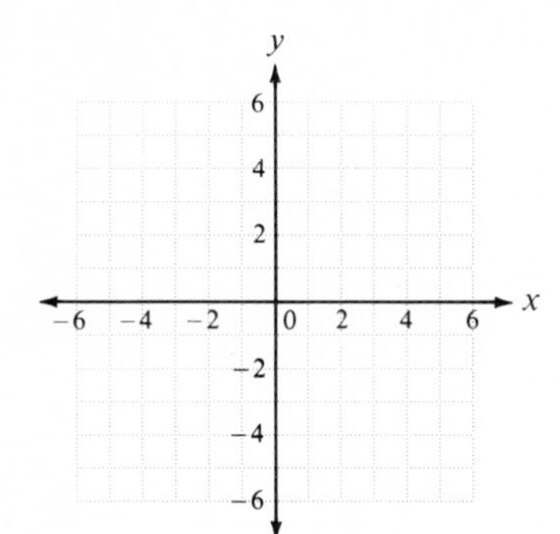

**57.** $\frac{y^2}{1} - \frac{x^2}{4} = 1$

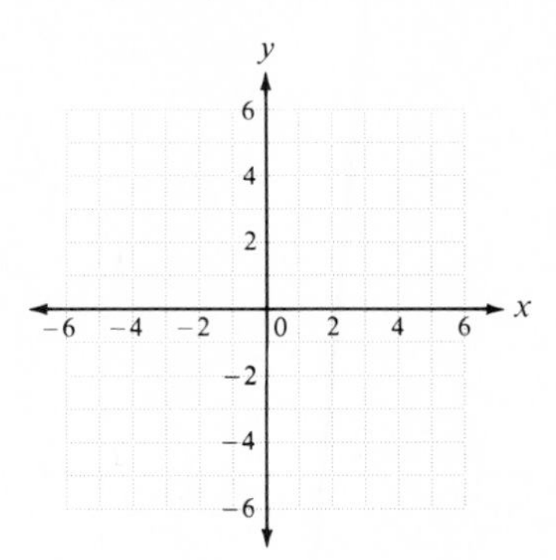

[7.7] *Which of the following are functions?*

**58.**

| | |
|---|---|
| 1 | 3 |
| 2 | 4 |
| 3 | 5 |
| 4 | 6 |

**59.**

| | |
|---|---|
| 1 | 3 |
| 2 | 4 |
| 3 | 5 |
| 4 | 6 |

**60.**

| | |
|---|---|
| 1 | 3 |
| 2 | 4 |
| 3 | 5 |
| 4 | 6 |

*Which of the following are graphs of functions?*

**61.**

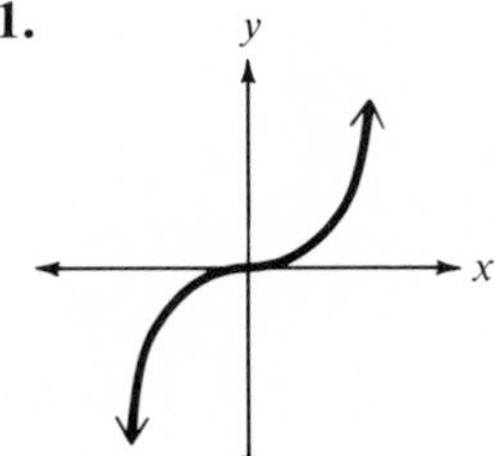

**62.**

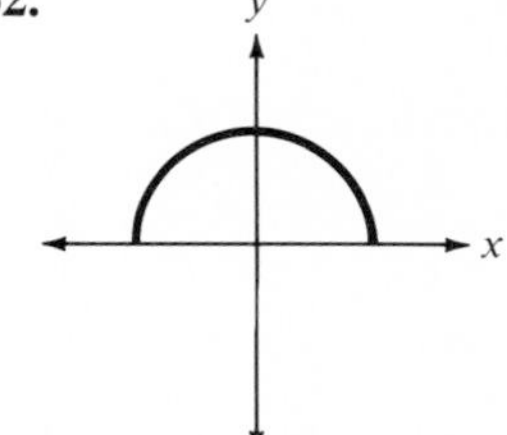

**63.**

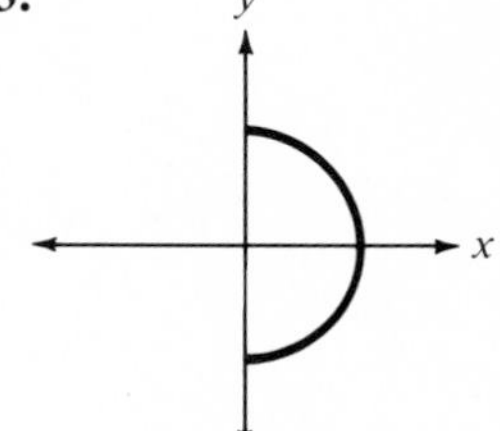

[7.8] *For the given function:* **(a)** Find $f(0)$ **(b)** Find $f(1)$ **(c)** Find $f(-2)$ **(d)** Find $f(b)$ **(e)** Find $f(a - 1)$ **(f)** Is this a constant, linear, or quadratic function? **(g)** Give the graph of the function.

**64.** $f(x) = -2$

**(a)** $f(0) =$
**(b)** $f(1) =$
**(c)** $f(-2) =$
**(d)** $f(b) =$
**(e)** $f(a - 1) =$
**(f)**

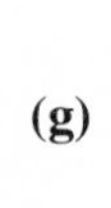

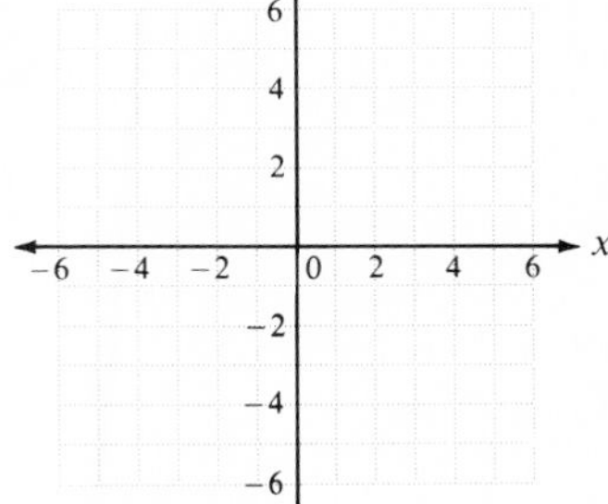

**65.** $f(x) = -x^2 + 3$

**(a)** $f(0) =$
**(b)** $f(1) =$
**(c)** $f(-2) =$
**(d)** $f(b) =$
**(e)** $f(a - 1) =$
**(f)**

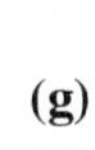

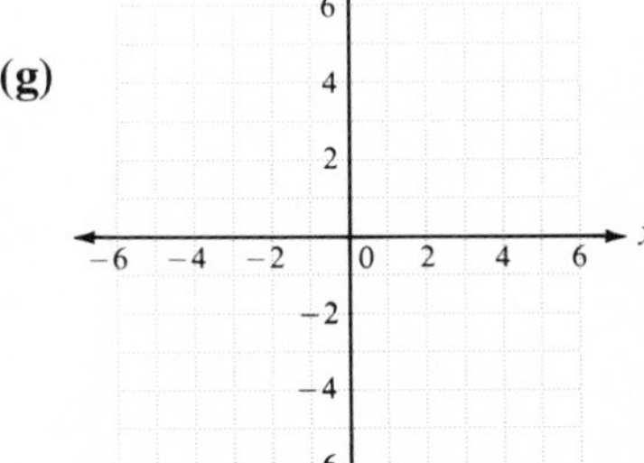

**66.** $f(x) = 4x - 2$

**(a)** $f(0) =$
**(b)** $f(1) =$
**(c)** $f(-2) =$
**(d)** $f(b) =$
**(e)** $f(a - 1) =$
**(f)**

**(g)**

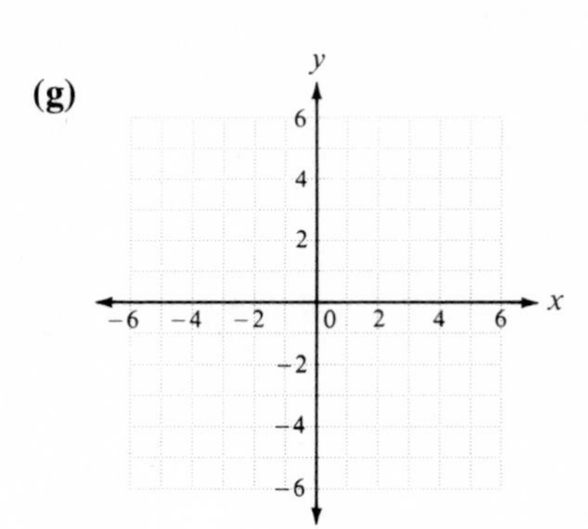


[7.9] *Find the x-intercepts (if they exist), the vertex, and graph of the function.*

**67.** $f(x) = x^2 - 2x - 3$

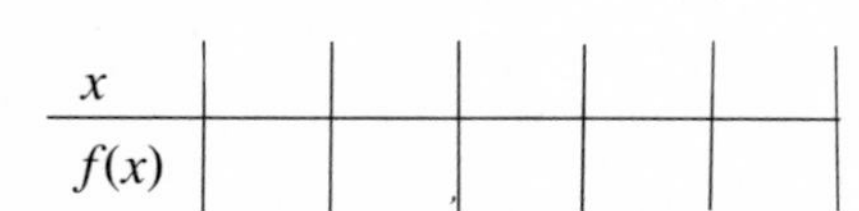

| $x$ | | | | | | |
|---|---|---|---|---|---|---|
| $f(x)$ | | | | | | |

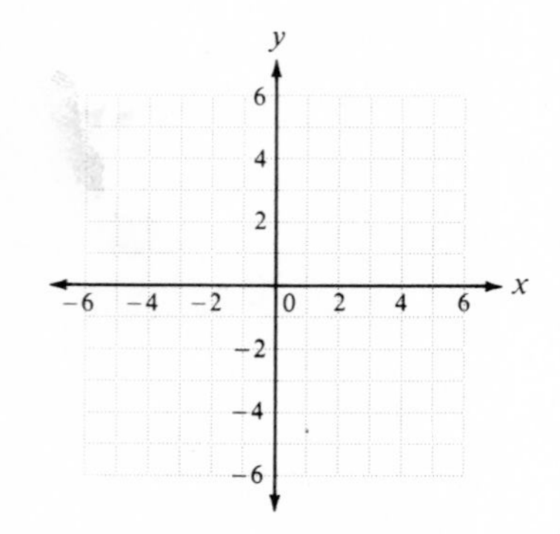


**68.** $g(x) = -x^2 + 4x - 5$

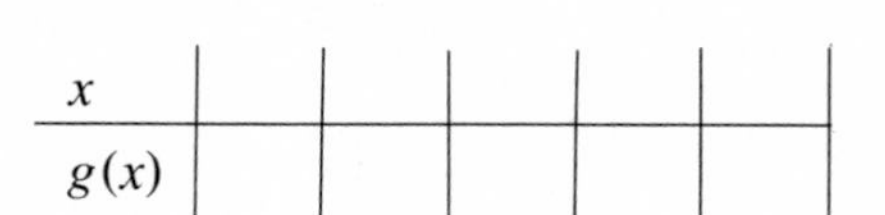

| $x$ | | | | | | |
|---|---|---|---|---|---|---|
| $g(x)$ | | | | | | |

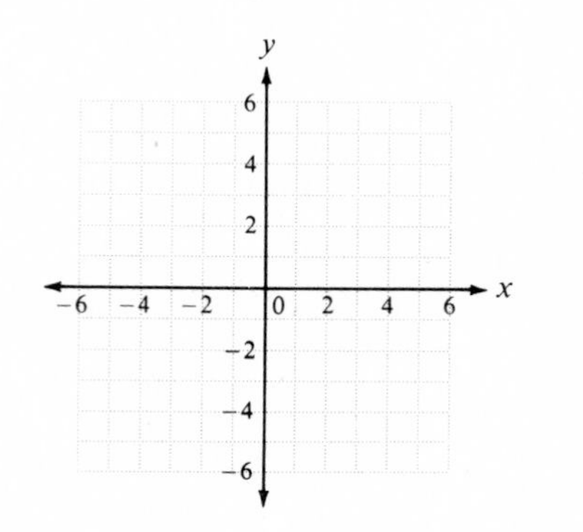


---

**ANSWERS:** **1.** Cartesian coordinate system **2.** **(a)** −2 **(b)** 3 **(c)** II **3.** axes **4.** graph **5.** $y$-axis **6.** $x$-axis **7.** $ax + by + c = 0$ **8.** linear equation **9.** intercepts **10.** $y$-intercept **11.** $x$-intercept **12.** $y$-axis **13.** $x$-axis **14.** $m = \frac{y_2 - y_1}{x_2 - x_1}$ **15.** negative **16.** positive **17.** no **18.** zero **19.** **(a)** $m$ **(b)** $-\frac{1}{m}$ **20.** $y - y_1 = m(x - x_1)$ **21.** $y = mx + b$ **22.** **(a)** slope-intercept **(b)** slope **(c)** $y$-intercept **23.** $y - y_1 = \frac{y_2 - y_1}{x_2 - x_1}(x - x_1)$ **24.** $\sqrt{(x_1 - x_2)^2 + (y_1 - y_2)^2}$ **25.** $\left(\frac{x_1 + x_2}{2}, \frac{y_1 + y_2}{2}\right)$ **26.** conic section **27.** parabola **28.** ellipse **29.** circle **30.** hyperbola **31.** **(a)** function **(b)** domain **(c)** range **(d)** one **32.** more than one **33.** linear **34.** constant **35.** straight line **36.** up **37.** down **38.** quadratic **39.** parabola **40.** $A(-3, 2)$ in II; $B(2, 1)$ in I; $C(2, -2)$ in IV; $D(-2, -2)$ in III

**41.** $x$-intercept (4, 0); $y$-intercept (0, 2); slope $-\frac{1}{2}$

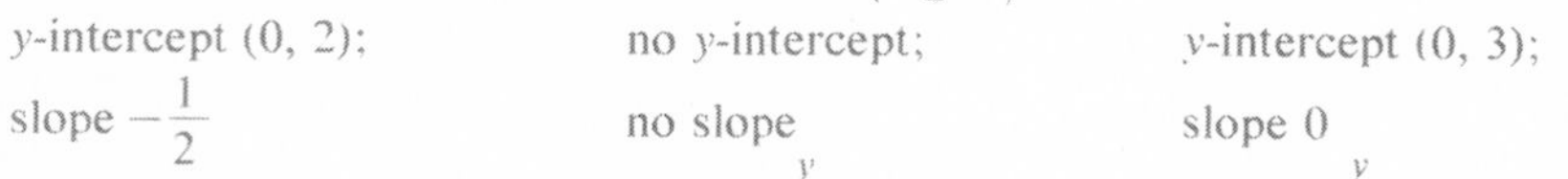

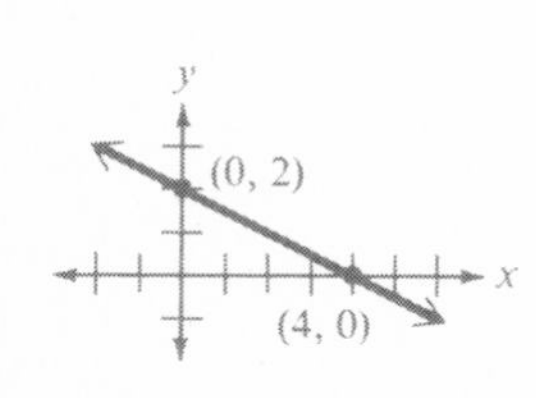


**42.** $x$-intercept $\left(-\frac{1}{2}, 0\right)$; no $y$-intercept; no slope

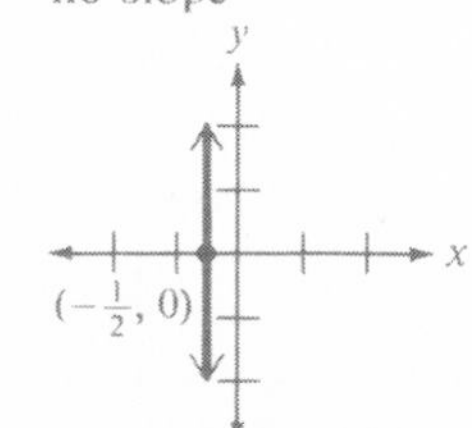


**43.** no $x$-intercept; $y$-intercept (0, 3); slope 0

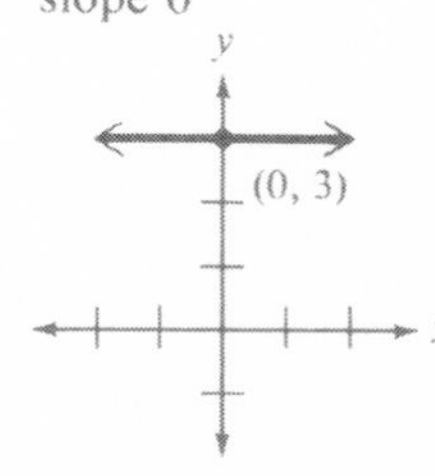


**44.** perpendicular $\left(\text{slopes are 3 and } -\frac{1}{3}\right)$ **45.** perpendicular (each is parallel to an axis) **46.** parallel $\left(\text{both have slope } \frac{2}{5}\right)$ **47.** $4x + y + 1 = 0$ **48.** $y = -2x + 5$; slope −2; $y$-intercept (0, 5) **49.** $x - 2y + 12 = 0$ **50.** $x - 2y + 5 = 0$

**51.** no $x$-intercept; $y$-intercept (0, 5); slope 0 **52.** 6 **53.** (2, −1)

**54.**

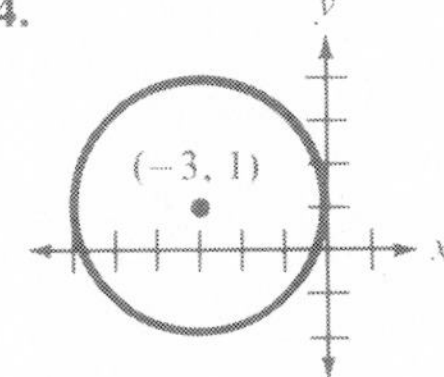


**55.**

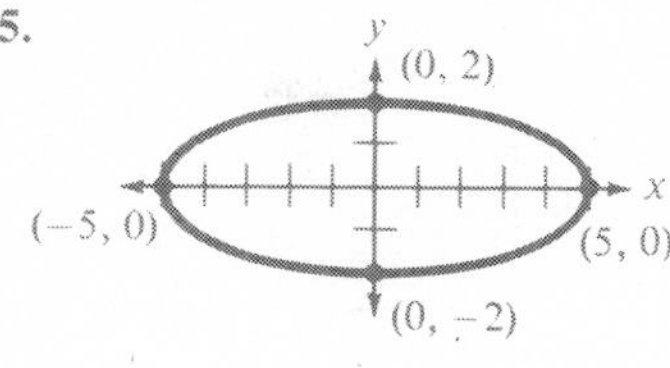


**56.**

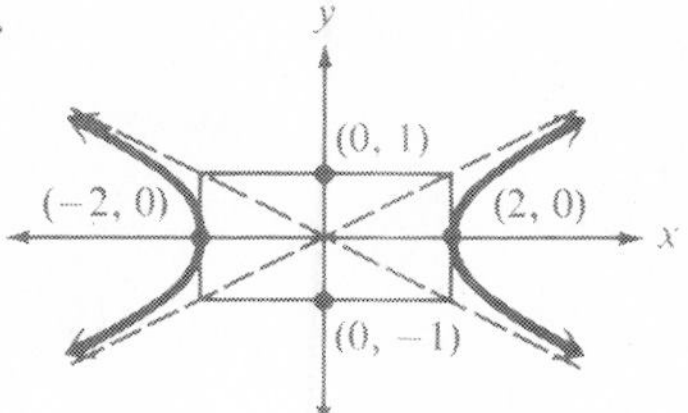


**57.**

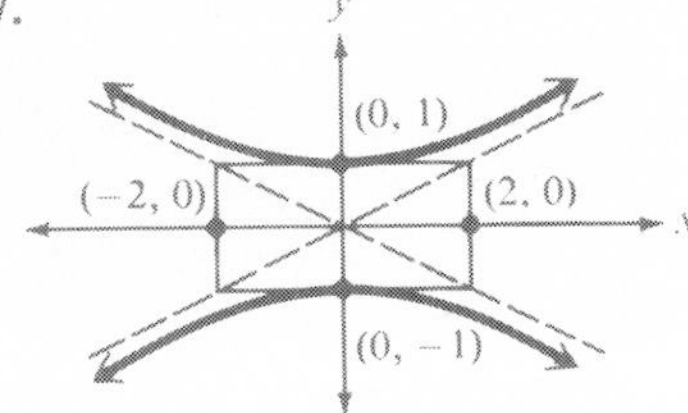


**58.** not a function **59.** function **60.** not a function
**61.** function **62.** function **63.** not a function

**64.** (a) $f(0) = -2$
(b) $f(1) = -2$
(c) $f(-2) = -2$
(d) $f(b) = -2$
(e) $f(a - 1) = -2$
(f) constant
(g)

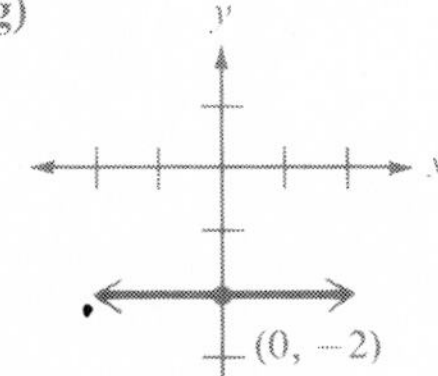


**65.** (a) $f(0) = 3$
(b) $f(1) = 2$
(c) $f(-2) = -1$
(d) $f(b) = -b^2 + 3$
(e) $f(a - 1) = -a^2 + 2a + 2$
(f) quadratic
(g)

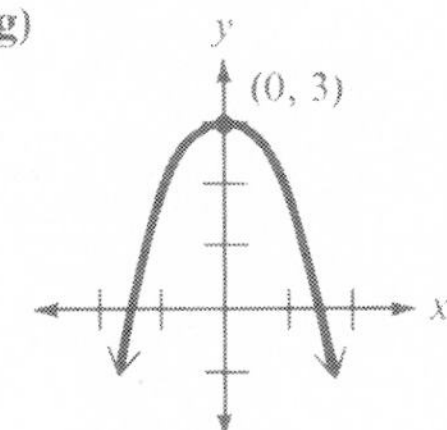


**66.** (a) $f(0) = -2$
(b) $f(1) = 2$
(c) $f(-2) = -10$
(d) $f(b) = 4b - 2$
(e) $f(a - 1) = 4a - 6$
(f) linear
(g)

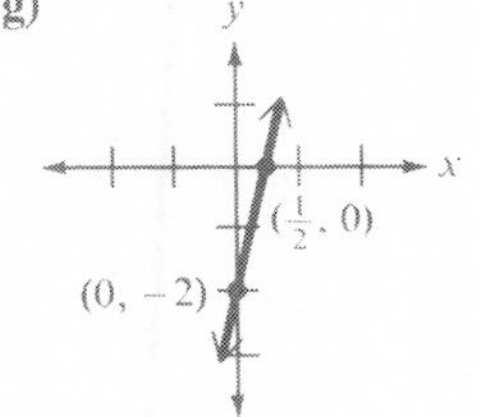


**67.** $x$-intercepts (3, 0) and (−1, 0); vertex (1, −4)

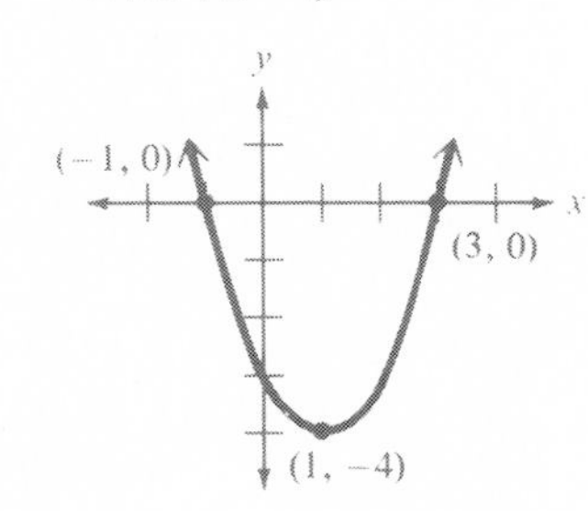


**68.** no $x$-intercepts; vertex (2, −1)

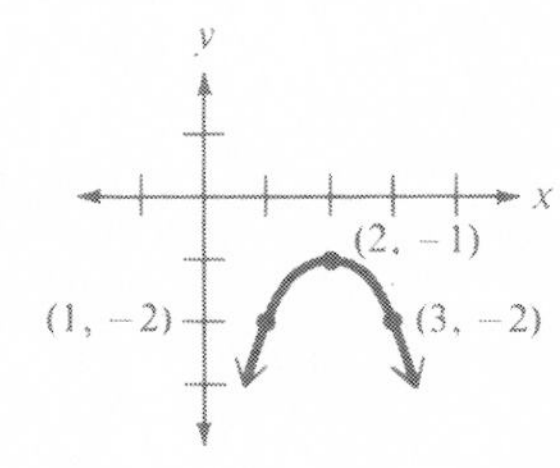

# 8

# Systems of Equations and Inequalities

## 8.1 Parallel, Coinciding, and Intersecting Lines

When two linear equations are graphed in the same Cartesian coordinate system, one of three possibilities occurs:

1. The lines intersect in exactly one point.
2. The lines are parallel (do not intersect at any point).
3. The lines coincide (the two equations represent the same line).

As we consider a **system** (pair) of two linear equations, it is helpful for us to be able to determine which of the above cases applies. Consider the following three systems of linear equations and their corresponding graphs in Figure 8.1.

| | | |
|---|---|---|
| (A) $2x - y + 4 = 0$ | (C) $2x - y + 4 = 0$ | (E) $2x - y + 4 = 0$ |
| (B) $6x - 3y + 12 = 0$ | (D) $2x - y - 2 = 0$ | (F) $x + y - 1 = 0$ |

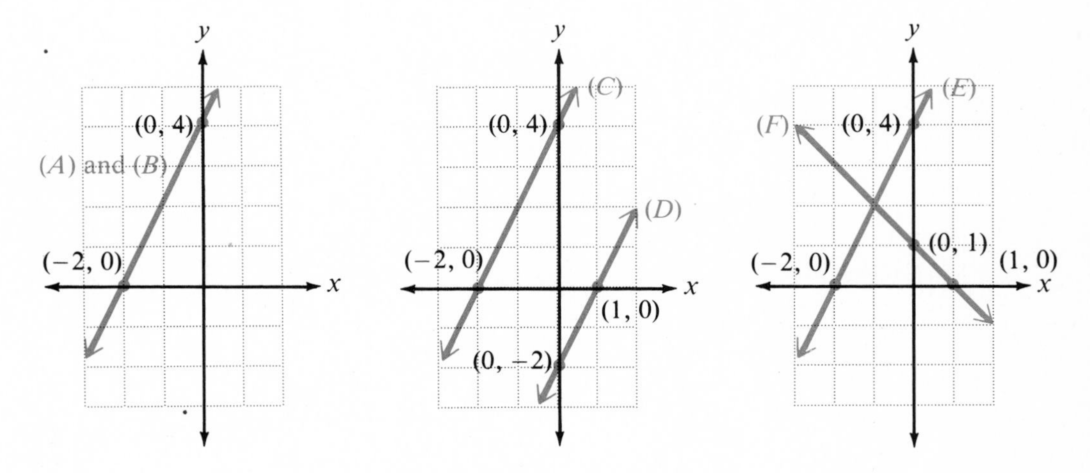


**Figure 8.1**

In the first system, lines (A) and (B) coincide; in the second system, lines (C) and (D) are parallel; and in the third system, lines (E) and (F) intersect in one point. Let us write each equation in slope-intercept form (solve each equation for $y$).

| | | |
|---|---|---|
| (A) $y = 2x + 4$ | (C) $y = 2x + 4$ | (E) $y = 2x + 4$ |
| (B) $y = 2x + 4$ | (D) $y = 2x - 2$ | (F) $y = -x + 1$ |

In slope-intercept form, the equations can be compared more easily. Recall that the coefficient of $x$ is the slope of the line, and that parallel lines must have the same slope.

**TO DETERMINE THE NATURE OF THE GRAPHS OF A SYSTEM OF EQUATIONS**

1. Write each equation in slope-intercept form and identify the slope and $y$-intercept of each.
2. If the slopes are equal and the $y$-intercepts are equal, the graphs (lines) coincide (the equations are equivalent).
3. If the slopes are equal and the $y$-intercepts are unequal, the graphs (lines) are parallel.
4. If the slopes are unequal, the graphs (lines) intersect in exactly one point.

The rule is not applicable when one (or both) of the equations in a system is of the form $ax = c$, since such equations cannot be solved for $y$. However, recalling that the graph of an equation of this form is always a line parallel to the $y$-axis, parallel or coinciding lines result only if both equations in the system are of this type.

**EXAMPLE 1** Determine the nature of the graphs of the equations in the following system.

$$\begin{aligned} 3x + y - 4 &= 0 \\ x - 2y + 1 &= 0 \end{aligned}$$

Solve each for $y$ (write in slope-intercept form).

$$\begin{aligned} 3x + y - 4 &= 0 \\ y &= -3x + 4 \end{aligned} \qquad \begin{aligned} x - 2y + 1 &= 0 \\ -2y &= -x - 1 \\ y &= \frac{1}{2}x + \frac{1}{2} \end{aligned}$$

Since the slopes are $-3$ and $\frac{1}{2}$, the graphs (lines) intersect in one point.

**EXAMPLE 2** Determine the nature of the graphs of the equations in the following system.

$$\begin{aligned} 4x - 3y + 2 &= 0 \\ -8x + 6y - 4 &= 0 \end{aligned}$$

Write each in slope-intercept form.

$$\begin{aligned} 4x - 3y + 2 &= 0 \\ -3y &= -4x - 2 \\ y &= \frac{4}{3}x + \frac{2}{3} \end{aligned} \qquad \begin{aligned} -8x + 6y - 4 &= 0 \\ 6y &= 8x + 4 \\ y &= \frac{8}{6}x + \frac{4}{6} \\ y &= \frac{4}{3}x + \frac{2}{3} \end{aligned}$$

Reduce fractions

Since the slopes are both $\frac{4}{3}$ and the $y$-intercepts are both $\left(0, \frac{2}{3}\right)$, the graphs (lines) coincide.

EXAMPLE 3 Determine the nature of the graphs of the equations in the following system.

$$\begin{aligned} x - 2y + 1 &= 0 \\ -3x + 6y - 1 &= 0 \end{aligned}$$

Write each in slope-intercept form.

$$\begin{aligned} x - 2y + 1 &= 0 \\ -2y &= -x - 1 \\ y &= \frac{1}{2}x + \frac{1}{2} \end{aligned}$$

$$\begin{aligned} -3x + 6y - 1 &= 0 \\ 6y &= 3x + 1 \\ y &= \frac{3}{6}x + \frac{1}{6} \\ y &= \frac{1}{2}x + \frac{1}{6} \quad \text{Reduce fraction} \end{aligned}$$

Since the slopes are both $\frac{1}{2}$ but the $y$-intercepts, $\left(0, \frac{1}{2}\right)$ and $\left(0, \frac{1}{6}\right)$, are not the same, the graphs (lines) are parallel.

EXAMPLE 4 Determine the nature of the graphs of the equations in the following system.

$$\begin{aligned} x + y - 3 &= 0 \\ 2x \quad\;\; - 6 &= 0 \end{aligned}$$

$2x - 6 = 0$ cannot be written in slope-intercept form, but we observe that $2x - 6 = 0$ becomes $x = 3$ (why?) which has as its graph a line parallel to the $y$-axis passing through $x$-intercept (3, 0). Since a $y$ term is present in $x + y - 3 = 0$, the graph of this equation cannot be a line parallel to the $y$-axis. Thus the two graphs must intersect in one point.

EXAMPLE 5 Determine the nature of the graphs of the equations in the following system.

$$\begin{aligned} 3y + 6 &= 0 \\ 2x - y + 5 &= 0 \end{aligned}$$

Write each in slope-intercept form.

$$\begin{aligned} 3y + 6 &= 0 \\ 3y &= -6 \\ y &= -2 \\ y &= 0 \cdot x - 2 \end{aligned}$$

$$\begin{aligned} 2x - y + 5 &= 0 \\ -y &= -2x - 5 \\ y &= 2x + 5 \end{aligned}$$

Since the slope of the first line is 0 (the line is parallel to the $x$-axis) and the slope of the second line is 2, the graphs (lines) intersect in one point.

## EXERCISES 8.1

1. When the slopes of the lines in a system of equations are equal but the $y$-intercepts are unequal, the graphs are ______________________ lines.

**2.** When the slopes of the lines in a system of equations are unequal, the graphs are ________________ lines.

**3.** When the slopes of the lines in a system of equations are equal and the $y$-intercepts are also equal, the graphs are ________________ lines.

**4.** If the $y$ term is missing in both equations in a system of equations, the graphs are **(a)** ________________ lines or are lines **(b)** ________________ to the $y$-axis.

**5.** If the $y$ term is missing in only one of the equations in a system of equations, the graphs are ________________ lines.

*Without graphing, determine the nature of the graphs of the equations in the given system.*

**6.** $-5x + 2y + 3 = 0$
$-5x - 2y + 3 = 0$

**7.** $-5x + 2y + 3 = 0$
$5x - 2y - 3 = 0$

**8.** $-5x + 2y + 3 = 0$
$10x - 4y + 3 = 0$

**9.** $x + y - 3 = 0$
$x - y + 3 = 0$

**10.** $x = y + 3$
$y = x + 3$

**11.** $3x - y + 4 = 0$
$y - 5 = 0$

**12.** $2x = 8$
$8y = 2$

**13.** $3x + 3y + 3 = 0$
$5x + 5y + 5 = 0$

**14.** $2x + y = 7$
$2x \quad = 3$

**15.** $8y + 7 = -3x$
$-6x + 14 = 16y$

ANSWERS: 1. parallel 2. intersecting 3. coinciding 4. (a) coinciding (b) parallel 5. intersecting 6. intersecting 7. coinciding 8. parallel 9. intersecting 10. parallel 11. intersecting 12. intersecting 13. coinciding 14. intersecting 15. parallel

## 8.2 Systems of Two Linear Equations in Two Variables

Recall that a linear equation (in the two variables $x$ and $y$) such as

$$x - y = 1$$

has infinitely many solutions, some of which are $(0, -1)$, $(1, 0)$, $(-1, -2)$, $(2, 1)$, and $(5, 4)$. The graph of this equation is a straight line, every point of which corresponds to a solution of the equation. Suppose that we had a second linear equation

$$2x + y = 5.$$

This equation too has infinitely many solutions, some of which are $(0, 5)$, $(1, 3)$, $(-1, 7)$, $(2, 1)$, and $(-3, 11)$.

A **system of two linear equations in two variables** is a pair of linear equations such as the following.

$$\begin{aligned} x - y &= 1 \\ 2x + y &= 5 \end{aligned}$$

We usually refer to such a pair as simply a **system of equations.** A **solution** to a system of equations is a pair of numbers which is a solution to both equations. For our particular example, since $(2, 1)$ is listed as a solution to both equations, $(2, 1)$ is a solution to the system of equations. If we write both equations in slope-intercept form,

$$\begin{aligned} x - y &= 1 \\ -y &= -x + 1 \\ y &= x - 1 \end{aligned} \qquad\qquad \begin{aligned} 2x + y &= 5 \\ y &= -2x + 5 \end{aligned}$$

we can see that the slopes are unequal so that the graphs of the equations are intersecting lines. Graphing both equations in the same Cartesian coordinate system in Figure 8.2, we see that the point of intersection corresponds to the solution $(2, 1)$ to the system.

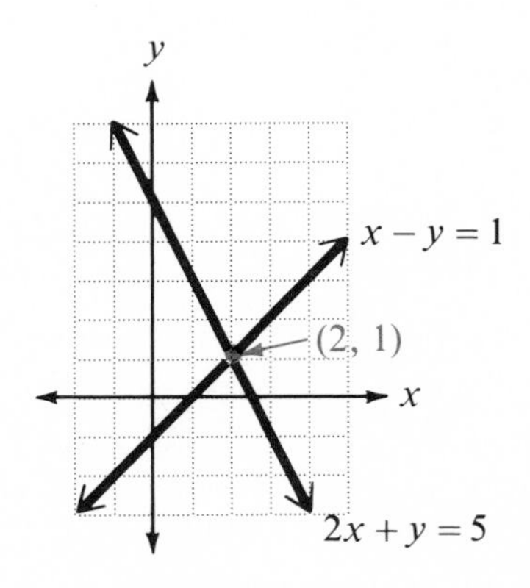


**Figure 8.2**

If a system of equations is such that the graphs of the lines are parallel, the system has no solution. If the graphs of the lines coincide, then any solution to either equation is also a solution to the other equation, and the system of equations has infinitely many solutions.

**NUMBER OF SOLUTIONS OF A SYSTEM OF EQUATIONS**

1. There is no solution if the graphs of the equations are parallel.
2. There is exactly one solution if the graphs of the equations intersect in one point.
3. There are infinitely many solutions if the graphs of the equations coincide.

**EXAMPLE 1** Without solving the system, determine the number of solutions.

$$\begin{aligned} 3x - 2y + 1 &= 0 \\ -4y + 6x &= -2 \end{aligned}$$

Solve each equation for $y$.

$$-2y = -3x - 1 \qquad\qquad -4y = -6x - 2$$

$$y = \frac{3}{2}x + \frac{1}{2} \qquad\qquad y = \frac{6}{4}x + \frac{2}{4} = \frac{3}{2}x + \frac{1}{2}$$

Since the lines coincide (why?), the system has infinitely many solutions.

**EXAMPLE 2** Without solving the system, determine the number of solutions.

$$\begin{aligned} 2x - 5y + 1 &= 0 \\ 15y - 6x &= 1 \end{aligned}$$

Solve each equation for $y$.

$$-5y = -2x - 1 \qquad\qquad 15y = 6x + 1$$

$$y = \frac{2}{5}x + \frac{1}{5} \qquad\qquad y = \frac{6}{15}x + \frac{1}{15} = \frac{2}{5}x + \frac{1}{15}$$

Since the lines are parallel (why?), the system has no solution.

**EXAMPLE 3** Without solving the system, determine the number of solutions.

$$\begin{aligned} 2x - 5 &= 0 \\ 2x + y &= 5 \end{aligned}$$

Since the first equation has as its graph a line with no slope (parallel to the $y$-axis), and the second equation has as its graph a line with slope $-2$ (why?), the lines intersect so that the system has exactly one solution.

**EXAMPLE 4** Determine if $(2, -1)$ is a solution to the following system.

$$\begin{aligned} 2x + y &= 3 \\ x - y &= 1 \end{aligned}$$

To do this we determine (by substitution) whether $(2, -1)$ is a solution to each equation.

$$\begin{aligned} 2(2) + (-1) &\stackrel{?}{=} 3 \\ 4 - 1 &\stackrel{?}{=} 3 \\ 3 &= 3 \end{aligned} \qquad\qquad \begin{aligned} (2) - (-1) &\stackrel{?}{=} 1 \\ 2 + 1 &\stackrel{?}{=} 1 \\ 3 &\neq 1 \end{aligned}$$

Although $(2, -1)$ is a solution to the first equation, it is not a solution to the system since it is not a solution to the second equation.

## EXERCISES 8.2

**1.** In a system of equations, if the lines coincide, the system has ____________________ solution(s).

**2.** In a system of equations, if the lines intersect, the system has ____________________ solution(s).

**3.** In a system of equations, if the lines are parallel, the system has ____________________ solution(s).

**4.** Is $(3, 0)$ a solution to the given systems?

**(a)** $\begin{aligned} 3x - 2y - 9 &= 0 \\ x + 8y - 3 &= 0 \end{aligned}$

**(b)** $\begin{aligned} x + y + 1 &= 4 \\ 2x - y + 2 &= 4 \end{aligned}$

**(c)** $\begin{aligned} x &= 3 \\ 2x - 4y - 6 &= 0 \end{aligned}$

**5.** Is $\left(\frac{1}{2}, -5\right)$ a solution to the given systems?

**(a)** $\begin{aligned} 2x + y + 4 &= 0 \\ 8x - 3y - 11 &= 0 \end{aligned}$

**(b)** $\begin{aligned} 2x - 1 &= 0 \\ y + 5 &= 0 \end{aligned}$

**(c)** $\begin{aligned} 3y + 15 &= 0 \\ 4x - y - 7 &= 0 \end{aligned}$

*Without solving, determine the number of solutions to the given system.*

**6.** $\begin{aligned} 5x - 3y + 7 &= 0 \\ -5x + 3y - 7 &= 0 \end{aligned}$

**7.** $\begin{aligned} 5x - 3y + 7 &= 0 \\ 3x - 5y - 7 &= 0 \end{aligned}$

**8.** $2x + 7 = 0$
$2x + 7y = 5$

**9.** $2x + 7 = 0$
$7x + 2 = 0$

**10.** $x + y + 1 = 0$
$y + x - 1 = 0$

**11.** $3x + 3y + 3 = 0$
$5x + 5y + 5 = 0$

**12.** $y - 6 = 0$
$x + 6 = 0$

**13.** $x + 2y + 3 = 0$
$3x + 2y + 1 = 0$

**14.** Given the system $x + y = 4$
$3x - y = 0$

**(a)** How many solutions does the system have? **(b)** In the same Cartesian coordinate system, carefully graph each equation. **(c)** By using the graphs, how would you estimate the solution? **(d)** What does the solution appear to be? **(e)** How would you verify your claim?

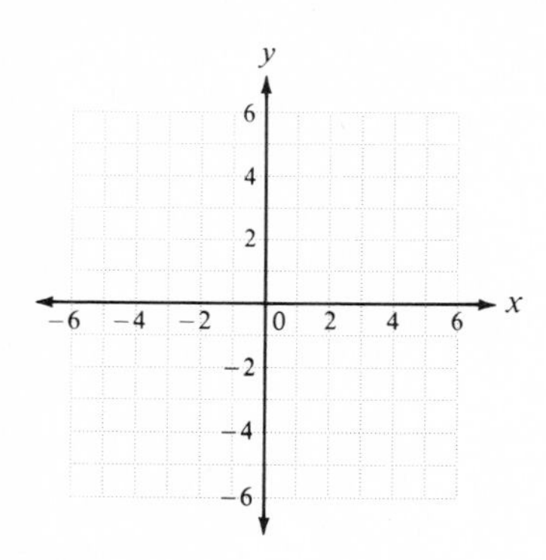


---

ANSWERS: **1.** infinitely many **2.** exactly one **3.** no **4. (a)** yes **(b)** no **(c)** yes (substitute (3, 0) into each equation) **5. (a)** no **(b)** yes **(c)** yes **6.** infinitely many **7.** exactly one **8.** exactly one **9.** none **10.** none **11.** infinitely many **12.** exactly one **13.** exactly one
**14. (a)** exactly one

**(b)**
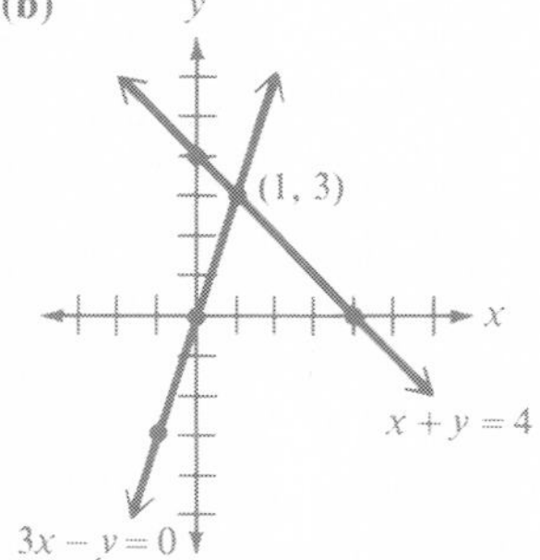


**(c)** determine the point of intersection
**(d)** appears to be (1, 3)
**(e)** substitute (1, 3) into both equations

$(1) + (3) = 4$ $\quad 3(1) - (3) = 0$
$4 = 4$ $\quad 3 - 3 = 0$
$0 = 0$

(1, 3) is the solution.

## 8.3 Solving Systems: The Substitution Method

Although we can now determine the number of solutions to a system of equations, we have not yet considered a good method for finding a particular solution. One method (not a particularly good one) was illustrated in Exercise 14 in the preceding section. However, solving a system by graphing takes too much time and determining a solution depends on the accuracy of our graphing techniques. We now turn our attention to solving a system algebraically by using the **method of substitution.** The system

$$\begin{aligned} x + y &= 4 \\ 3x - y &= 0 \end{aligned}$$

has exactly one solution. (Why?) Solve the first equation for $x$

$$x = -y + 4$$

and substitute this value of $x$ into the second equation

$$3(-y + 4) - y = 0.$$

The result is an equation in the single variable $y$ which can be solved for $y$.

$$\begin{aligned} -3y + 12 - y &= 0 \\ -4y &= -12 \\ y &= 3 \end{aligned}$$

Substitute this value of $y$ into either of the first equations, say the second

$$3x - (3) = 0$$

and solve for $x$.

$$\begin{aligned} 3x &= 3 \\ x &= 1 \end{aligned}$$

We obtain the solution (1, 3) to the system. It is always wise to check by substituting (1, 3) into each of the original equations (Do this.). Compare your work here with the results of Exercise 14 in the preceding section.

**TO SOLVE A SYSTEM OF EQUATIONS USING THE SUBSTITUTION METHOD**

1. Solve one of the equations for one of the variables.
2. Substitute that value of the variable in the *remaining* equation.
3. Solve this new equation and substitute the numerical solution into either of the two *original* equations to find the numerical value of the second variable.
4. Check your solution in both original equations.

EXAMPLE 1 Solve the following system by the substitution method.

$$2y - x - 2 = 0$$
$$2x + y - 6 = 0$$

We can solve either equation for either variable, but we can avoid fractions by solving the second equation for $y$,

$$y = -2x + 6,$$

and substituting $-2x + 6$ for $y$ in the first equation.

$$\begin{aligned} 2(-2x + 6) - x - 2 &= 0 \\ -4x + 12 - x - 2 &= 0 \\ -5x + 10 &= 0 \\ -5x &= -10 \\ x &= 2 \end{aligned}$$

Now substitute 2 for $x$ in the second equation.

$$\begin{aligned} 2(2) + y - 6 &= 0 \\ 4 + y - 6 &= 0 \\ y - 2 &= 0 \\ y &= 2 \end{aligned}$$

The solution is (2, 2). (Check in both equations.)

Instead of substituting 2 for $x$ in the second equation, we could substitute 2 for $x$ in $y = -2x + 6$. This would give $y = 2$ in one less step.

EXAMPLE 2 Solve the following system by the substitution method.

$$2x - y = 1$$
$$-4x + 2y = -2$$

Solve the first equation for $y$,

$$y = 2x - 1,$$

and substitute this expression into the second equation.

$$\begin{aligned} -4x + 2(2x - 1) &= -2 \\ -4x + 4x - 2 &= -2 \\ -2 &= -2 \end{aligned}$$

When we obtain an identity such as this, we should return to the original system and examine the nature of the solutions by solving each equation for $y$.

$$\begin{aligned} 2x - y &= 1 \\ -y &= -2x + 1 \\ y &= 2x - 1 \end{aligned} \qquad\qquad \begin{aligned} -4x + 2y &= -2 \\ 2y &= 4x - 2 \\ y &= 2x - 1 \end{aligned}$$

The slopes and $y$-intercepts are equal, so the lines coincide and there are infinitely many solutions to the system. In fact, any pair of numbers

$$(x, 2x - 1) \qquad (\text{since } y = 2x - 1)$$

($x$ any real number) is a solution to the system. Several representative solutions are $(0, -1)$, $(1, 1)$, $(2, 3)$, and $(-1, -3)$.

**EXAMPLE 3** Solve the system by the substitution method.

$$2x + 4y = 1$$
$$x + 2y = 8$$

Solve the second equation for $x$,

$$x = -2y + 8,$$

and substitute into the first equation.

$$2(-2y + 8) + 4y = 1$$
$$-4y + 16 + 4y = 1$$
$$16 = 1$$

When we obtain a contradiction such as this, we should return to the original system and examine the nature of the solutions by solving each equation for $y$.

$$2x + 4y = 1$$
$$4y = -2x + 1$$
$$y = -\frac{2}{4}x + \frac{1}{4}$$
$$y = -\frac{1}{2}x + \frac{1}{4}$$

$$x + 2y = 8$$
$$2y = -x + 8$$
$$y = -\frac{1}{2}x + 4$$

The slopes are equal and the $y$-intercepts are unequal, so the lines are parallel and there are no solutions to the system.

The above examples lead us to the following:

**WHEN USING THE SUBSTITUTION METHOD**

1. If an identity results, the lines coincide and there are infinitely many solutions to the system.
2. If a contradiction results, the lines are parallel and there is no solution to the system.

## EXERCISES 8.3

1. If the two lines in a system coincide, how many solutions does the system have?

2. If the two lines in a system are parallel, how many solutions does the system have?

3. If the two lines in a system intersect, how many solutions does the system have?

4. Suppose in the process of solving a system of equations using the substitution method we obtain $3 = 0$. What does this tell us about the system?

**5.** Suppose in the process of solving a system of equations using the substitution method we obtain $2x = 2x$. What does this tell us about the system?

*Solve using the substitution method.*

**6.** $x + 2y = -1$
$2x - 3y = 12$

**7.** $4x - y = -3$
$2x - 3y = 1$

**8.** $2x + 5y = 10$
$x - 3y = 16$

**9.** $2x + 4y = -12$
$3x + 5y = -14$

**10.** $2x - 3y = 2$
$-4x + 6y = 2$

**11.** $x - 4 = 0$
$x + y = 1$
[*Hint:* $x$ is always 4 so substitute to obtain $y$.]

**12.** $2y - 6 = 0$
$2x + y = -1$

**13.** $5x - 5y = 10$
$x - y = 2$

**14.** $3y - 9 = 0$
$4 - 2x = 0$

**15.** $3x + 5y = -2$
$5x + 3y = 2$

**16.** Which variable in which equation is easiest to solve for in Exercise 15? (The next section will offer us a method for solving a system which is a better method for solving systems such as in Exercise 15.)

---

ANSWERS: **1.** infinitely many **2.** none **3.** exactly one **4.** It has no solution; the lines are parallel. **5.** It has infinitely many solutions; the lines coincide. **6.** $(3, -2)$ **7.** $(-1, -1)$ **8.** $(10, -2)$ **9.** $(2, -4)$ **10.** no solution **11.** $(4, -3)$ **12.** $(-2, 3)$ **13.** infinitely many solutions of the form $(x, x-2)$ for $x$ any real number **14.** $(2, 3)$ **15.** $(1, -1)$ **16.** No one variable is better to solve for than any other; each choice results in fractions.

## 8.4 Solving Systems: The Addition-Subtraction Method

An alternate method of solving systems of equations that is perhaps superior to substitution for systems such as

$$\begin{aligned} 3x + 5y &= -2 \\ 5x + 3y &= 2 \end{aligned}$$

is known as the **addition-subtraction method.** Solving the above system (Exercise 15 in the previous section) using the substitution method involves fractions regardless of our choice of variable and choice of equation. The addition-subtraction method enables us to avoid computations with fractions. Recall that if the same expression is added to or subtracted from both sides of an equation the results are equal.

Before we solve the above system, we consider a somewhat simpler system.

$$\begin{aligned} x + y &= 4 \\ 2x - y &= 5 \end{aligned}$$

Observe that $2x - y$ and 5 are both names or expressions for the same number. If we add 5 to both sides of the first equation by adding $2x - y$ to the left member and 5 to the right member (add equals to equals), the results are equal.

$$\begin{aligned} (x + y) + (2x - y) &= 4 + 5 \\ 3x &= 9 \end{aligned}$$

Generally, we add vertically and avoid the intermediate steps.

$$\begin{array}{r} x + y = 4 \\ \underline{2x - y = 5} \\ 3x \qquad = 9 \\ x = 3 \end{array}$$

By adding in this case, we were able to "eliminate" the $y$ term and obtain an equation in the one variable $x$. Once we know that $x$ is 3, we substitute this value into one of the two original equations (just as in the substitution method) to find the value of $y$. Substituting in the first equation,

$$\begin{aligned}(3) + y &= 4\\ y &= 1.\end{aligned}$$

Thus the solution is (3, 1). (Check this in both of the equations in the given system.)

At times, addition or subtraction alone will not eliminate a variable and yield an equation in only one variable. When this happens, we may have to multiply one or both equations by a number that makes the coefficients of one of the variables negatives of each other. Then by adding we eliminate a variable and proceed as before. An example will make this clear.

**EXAMPLE 1** Solve by the addition-subtraction method.

$$\begin{aligned}5x - 2y &= -13\\ 2x + y &= 11\end{aligned}$$

By adding or subtracting immediately we do not eliminate a variable. However, if we multiply both sides of the second equation by 2 (using the multiplication rule), the resulting equation is

$$4x + 2y = 22.$$

Add this equation to the first.

$$\begin{aligned}5x - 2y &= -13\\ 4x + 2y &= 22\\ \hline 9x \quad &= 9\\ x &= 1\end{aligned}$$

Substitute 1 for $x$ in the second equation.

$$\begin{aligned}2(1) + y &= 11\\ 2 + y &= 11\\ y &= 9\end{aligned}$$

The solution is (1, 9). (Check by substitution.)

At times the multiplication rule must be applied to both equations before adding or subtracting. Let us solve our original system.

$$\begin{aligned}3x + 5y &= -2\\ 5x + 3y &= 2\end{aligned}$$

Multiply the first equation by $-5$ and the second equation by 3 (making the coefficients of the $x$ terms $-15$ and 15, respectively).

$$\begin{aligned}-15x - 25y &= 10\\ 15x + 9y &= 6\end{aligned}$$

Add to eliminate $x$.

$$\begin{aligned}-16y &= 16\\ y &= -1\end{aligned}$$

Substitute $-1$ for $y$ in the first equation.

$$\begin{aligned} 3x + 5(-1) &= -2 \\ 3x - 5 &= -2 \\ 3x &= 3 \\ x &= 1 \end{aligned}$$

The solution is $(1, -1)$. (Check by substitution.)

Compare this example with your work in solving the same system using the substitution method. Which method is easier and affords less chance for making an error?

**TO SOLVE A SYSTEM OF EQUATIONS USING THE ADDITION-SUBTRACTION METHOD**

1. Apply the multiplication rule to one or both equations (if necessary) to transform them so that addition or subtraction will eliminate a variable.
2. Solve the resulting single variable equation and substitute this value into one of the original equations and solve.
3. Check your answer by substitution in both original equations.

**EXAMPLE 2** Solve by the addition-subtraction method.

$$\begin{aligned} 7x - 2y &= -11 \\ 4x + 3y &= 2 \end{aligned}$$

Multiply the first equation by 3 and the second by 2.

$$\begin{aligned} 21x - 6y &= -33 \\ 8x + 6y &= 4 \end{aligned}$$

Add to eliminate $y$.

$$\begin{aligned} 29x \qquad &= -29 \\ x &= -1 \end{aligned}$$

Substitute $-1$ for $x$ in the second equation.

$$\begin{aligned} 4(-1) + 3y &= 2 \\ -4 + 3y &= 2 \\ 3y &= 6 \\ y &= 2 \end{aligned}$$

The solution is $(-1, 2)$. (Check by substitution.)

**EXAMPLE 3** Solve by the addition-subtraction method.

$$\begin{aligned} 3x - 2y &= 5 \\ -3x + 2y &= -5 \end{aligned}$$

Notice that we can eliminate the variable $x$ by addition.

$$\begin{array}{r} 3x - 2y = \phantom{-}5 \\ -3x + 2y = -5 \\ \hline 0 = \phantom{-}0 \end{array}$$

However, we have actually eliminated both variables and obtained an identity. Just as with the method of substitution, this indicates that there are infinitely many solutions. (Verify that the lines do coincide.) If we solve the first equation for $y$, we obtain

$$-2y = -3x + 5$$
$$y = \frac{3}{2}x - \frac{5}{2}.$$

Thus the solutions to the system are of the form $\left(x, \frac{3}{2}x - \frac{5}{2}\right)$ where $x$ is any real number.

**EXAMPLE 4** Solve by the addition-subtraction method.

$$\begin{aligned} 2x - \phantom{2}y &= \phantom{-}3 \\ -4x + 2y &= -1 \end{aligned}$$

Multiplying the first equation by 2 to obtain

$$4x - 2y = 6$$

and adding the result to the second equation we obtain

$$0 = 5.$$

Again both variables have been eliminated, but this time a contradiction results. As before, when a contradiction is obtained, the system has no solution. (Verify that the lines are parallel.)

## EXERCISES 8.4

1. Suppose that in the process of solving a system of equations using the addition-subtraction rule we obtain $0 = 2$. What does this tell us about the system?

2. Suppose that in the process of solving a system of equations using the addition-subtraction rule we obtain $0 = 0$. What does this tell us about the system?

*Solve using the addition-subtraction method.*

3. $3x + 2y = 4$
   $3x - 3y = 9$

4. $2x - 5y = 1$
   $4x + 2y = 14$

**5.** $2x - 7y = 24$
$3x + 2y = -14$

**6.** $5x - 7y = 3$
$-10x + 14y = -1$

**7.** $4x - 7y = 11$
$-3x + 5y = -8$

**8.** $3x + 3y = 12$
$7x + 7y = 28$

*Solve using the substitution method* or *the addition-subtraction method (whichever seems appropriate).*

**9.** $x + 2y = -9$
$3x + 4y = -17$

**10.** $x + 2 = 0$
$3x + 2y = 10$

**11.** $5x - 11y = -6$
$11x - 5y = 6$

**12.** $3y - 15 = 0$
$3x - 15y = 0$

**13.** $3x + y = -1$
$-4x + 2y = -7$

**14.** $7x - 7y = 7$
$-13x + 13y = -13$

**ANSWERS:** **1.** It has no solution; the lines are parallel. **2.** It has infinitely many solutions; the lines coincide. **3.** $(2, -1)$ **4.** $(3, 1)$ **5.** $(-2, -4)$ **6.** no solution **7.** $(1, -1)$ **8.** infinitely many solutions of the form $(x, 4 - x)$, $x$ any real number **9.** $(1, -5)$ **10.** $(-2, 8)$ **11.** $(1, 1)$ **12.** $(25, 5)$ **13.** $\left(\frac{1}{2}, -\frac{5}{2}\right)$ **14.** infinitely many solutions of the form $(x, x - 1)$, $x$ any real number.

## 8.5 Word Problems Resulting in Systems of Two Equations

In previous chapters we worked several kinds of number, age, geometry, and motion problems. Some of these (as well as other problems) can be simplified by using two variables (instead of one) and translating them into a system of equations. Several examples illustrate this idea.

EXAMPLE 1 The sum of two numbers is 11, and twice one number minus three times the second is equal to 2. Find the numbers.

Let $x =$ one number
$y =$ the other number

Then $2x - 3y = 2$ Twice one minus three times the other is 2
$x + y = 11$ Their sum is 11

In any word problem, we attempt to solve the system by the simplest method. In this case, we solve the second equation for $x$,

$$x = 11 - y,$$

and substitute into the first.

$$\begin{aligned} 2(11 - y) - 3y &= 2 \\ 22 - 2y - 3y &= 2 \\ -5y &= -20 \\ y &= 4 \end{aligned}$$

Substitute 4 for $y$ in the second equation.

$$\begin{aligned} x + 4 &= 11 \\ x &= 7 \end{aligned}$$

The numbers are 7 and 4. (Check.)

EXAMPLE 2 The sum of Jan's and Sue's ages is 15. In 3 years, Jan will be twice as old as Sue. What are their present ages?

Let $x =$ Jan's age
$y =$ Sue's age
$x + 3 =$ Jan's age in 3 years
$y + 3 =$ Sue's age in 3 years

Since in 3 years Jan will be twice as old as Sue, $x + 3$ must be twice $y + 3$.

$$x + 3 = 2(y + 3)$$
$$x + 3 = 2y + 6$$
$$x - 2y = 3$$

Also, $x + y = 15$ since the sum of their ages is 15. Thus, we must solve the following system.

$$x - 2y = 3$$
$$x + y = 15$$

Solve the first equation for $x$,

$$x = 2y + 3,$$

and substitute into the second.

$$(2y + 3) + y = 15$$
$$2y + 3 + y = 15$$
$$3y = 12$$
$$y = 4$$

Substitute 4 for $y$ in the second equation.

$$x + 4 = 15$$
$$x = 11$$

Jan is 11 and Sue is 4. (Check.)

EXAMPLE 3 Two angles are **complementary** (that is, their sum is 90°). If one is 10° more than seven times the other, find the angles.

Let $x =$ one angle
$y =$ the second angle

Then

$$x + y = 90 \quad \text{Why?}$$

If $x$ is 10° more than seven times $y$, we would add 10° to $7y$ to obtain $x$. That is

$$x = 7y + 10.$$

The system

$$x + y = 90$$
$$x - 7y = 10$$

can be solved by substituting $7y + 10$ for $x$ in $x + y = 90$.

$$(7y + 10) + y = 90$$
$$8y + 10 = 90$$
$$8y = 80$$
$$y = 10$$
$$x = 7(10) + 10 = 70 + 10 = 80$$

The angles are 10° and 80°. (Check.)

**EXAMPLE 4** By going 20 mph for one period of time and then 30 mph for another, a boater traveled from Glen Canyon Dam to the end of Lake Powell, a distance of 180 miles. If she had gone 21 mph throughout the same period of time, she would only have reached Hite Marina, a distance of 147 miles. How many hours did she travel at each speed?

Let $x$ = the number of hours that she traveled at 20 mph
$y$ = the number of hours that she traveled at 30 mph

Since

$$(\text{distance}) = (\text{rate}) \cdot (\text{time})$$

it follows that

$$\begin{aligned} \text{distance traveled at 20 mph} &= 20x \\ \text{distance traveled at 30 mph} &= 30y. \end{aligned}$$

The total distance traveled is the sum of the two distances traveled at the two rates, so the first equation is

$$20x + 30y = 180.$$

She would have traveled 147 miles at a rate of 21 mph for the total time $x + y$. Thus the second equation is

$$21(x + y) = 147 \qquad \text{or} \qquad 21x + 21y = 147.$$

It is usually wise to simplify equations before solving. Dividing the first equation by 10 and the second by 7, we obtain the following system.

$$\begin{aligned} 2x + 3y &= 18 \\ 3x + 3y &= 21 \end{aligned}$$

Subtract the first from the second.

$$x = 3$$

Then

$$\begin{aligned} 2(3) + 3y &= 18 \\ 3y &= 12 \\ y &= 4. \end{aligned}$$

She traveled 3 hours at 20 mph and 4 hours at 30 mph.

An important kind of word problem that results in a system of equations is known as a **combination problem.** Generally, two quantities are combined, and the two equations we seek are formed by considering two different combinations. Example 4 is a combination problem. Values of quantities are often important. The **total value** of something which is composed of a number of units each of the same value can be computed by

$$\textbf{(total value)} = \textbf{(value per unit)} \cdot \textbf{(number of units).}$$

For example, the total value of 8 lbs of chicken worth \$1.10 per pound is

$$\begin{aligned} (\text{total value}) &= (\text{value per pound}) \cdot (\text{number of pounds}) \\ &= (\$1.10) \cdot (8) \\ &= \$8.80. \end{aligned}$$

Similarly, the value of a collection of 12 quarters is

$$\begin{aligned}(\text{total value}) &= (\text{value per quarter}) \cdot (\text{number of quarters}) \\ &= (25¢) \cdot (12) \\ &= 300¢ = \$3.00.\end{aligned}$$

The next example illustrates how calculation of values is used to solve a combination problem.

**EXAMPLE 5** Burford bought 5 shirts (of the same value) and 4 pair of socks (of the same value) for \$87. He returned to the same store a week later and purchased (at the same prices) 2 more shirts and 6 pair of socks for \$48. Find the price of each.

Let $x =$ price of one shirt
$y =$ price of one pair of socks
$5x =$ value of 5 shirts [(value) = (price per shirt) · (number of shirts)]
$4y =$ value of 4 pair of socks

Thus

$$5x + 4y = 87.$$

Similarly,

$$2x + 6y = 48.$$

Divide the second equation by 2 to simplify.

$$x + 3y = 24$$

Solve for $x$,

$$x = 24 - 3y,$$

and substitute into the first equation.

$$\begin{aligned}5(24 - 3y) + 4y &= 87 \\ 120 - 15y + 4y &= 87 \\ -11y &= -33 \\ y &= 3 \\ x &= 24 - 3y \\ &= 24 - 3(3) = 24 - 9 = 15\end{aligned}$$

Each shirt cost \$15 and each pair of socks cost \$3.

A special type of combination problem is called a **mixture problem.** Generally, two quantities are mixed together, and the two equations we seek might be called a *quantity equation* and a *value equation*.

**EXAMPLE 6** A man wishes to mix candy selling for \$1.50 per pound with nuts selling for \$1.00 per pound to obtain a party mix to be sold for \$1.20 per pound. How many pounds of each must be used to obtain 50 pounds of the mixture?

Let $x =$ the number of pounds of candy
$y =$ the number of pounds of nuts

The first equation we obtain is the quantity equation.

(number lbs of candy) + (number lbs of nuts) = (number lbs of mixture)

This translates to

$$x + y = 50.$$

Next we obtain the value equation.

(total value of candy) + (total value of nuts) = (total value of mixture)

That is, the value of a mixture is equal to the sum of the values of its parts. To avoid decimals, we convert all monetary units to cents. Then

$$\begin{aligned} 150 \cdot x &= \text{total value of candy} = (150¢ \text{ per lb}) \cdot (\text{number of lb}) \\ 100 \cdot y &= \text{total value of nuts} = (100¢ \text{ per lb}) \cdot (\text{number of lb}) \\ 120 \cdot 50 &= \text{total value of mixture} = (120¢ \text{ per lb}) \cdot (\text{number of lb}) \end{aligned}$$

Thus, the value equation is

$$150x + 100y = 120 \cdot 50$$

or

$$3x + 2y = 120. \qquad \text{Dividing through by 50}$$

The system we must solve is

$$\begin{aligned} x + y &= 50 \\ 3x + 2y &= 120. \end{aligned}$$

Solve the first equation for $x$,

$$x = 50 - y,$$

and substitute into the second.

$$\begin{aligned} 3(50 - y) + 2y &= 120 \\ 150 - 3y + 2y &= 120 \\ -y &= -30 \\ y &= 30 \\ x &= 50 - y = 50 - 30 = 20 \end{aligned}$$

The man must use 20 lbs of candy and 30 lbs of nuts to obtain the desired mixture of 50 lbs.

**EXAMPLE 7** Becky has 30 coins consisting of dimes and quarters. If the total value is $4.20, find the number of dimes and the number of quarters in the collection.

Let $x$ = the number of dimes in the collection
$y$ = the number of quarters in the collection

The quantity equation

(number of dimes) + (number of quarters) = (number of coins in collection)

translates to

$$x + y = 30.$$

The value equation

$$(\text{value of dimes}) + (\text{value of quarters}) = (\text{value of collection})$$

becomes

$$10x + 25y = 420. \qquad \text{Convert to cents}$$

Solve the first equation for $x$, $x = 30 - y$, and substitute.

$$\begin{aligned} 10(30 - y) + 25y &= 420 \\ 300 - 10y + 25y &= 420 \\ 15y &= 120 \\ y &= 8 \\ x &= 30 - y = 30 - 8 = 22 \end{aligned}$$

Becky has 22 dimes and 8 quarters.

**EXAMPLE 8** A chemist has two solutions each containing a certain percentage of acid. If solution $A$ is 5% acid and solution $B$ is 15% acid, how much of each should be mixed to obtain 20 L (liters) of a solution which is 12% acid?

Let $x$ = number of liters of the 5% acid solution
$y$ = number of liters of the 15% acid solution

The quantity equation is

$$x + y = 20. \qquad (\text{Why?})$$

The value equation this time equates the amount of acid in the solutions.

$$\begin{aligned} .05x &= \text{the amount of acid in } x \text{ L of the 5\% solution} \\ .15y &= \text{the amount of acid in } y \text{ L of the 15\% solution} \\ (.12)(20) &= \text{the amount of acid in 20 L of the 12\% mixture} \end{aligned}$$

Thus we have

$$.05x + .15y = .12 \cdot 20$$

or

$$5x + 15y = 240. \qquad \text{Clear all decimals}$$

Solve the first equation for $x$, $x = 20 - y$, and substitute.

$$\begin{aligned} 5(20 - y) + 15y &= 240 \\ 100 - 5y + 15y &= 240 \\ 10y &= 140 \\ y &= 14 \\ x &= 20 - y = 20 - 14 = 6 \end{aligned}$$

The chemist should mix 6 L of the 5% solution with 14 L of the 15% solution.

## EXERCISES 8.5

*Solve.*

**1.** The sum of two numbers is 50 and their difference is 16. Find the numbers.

**2.** Pete is 7 years younger than Jim. Three years from now the sum of their ages will be 33. How old is each now?

**3.** Two angles are supplementary (their sum is 180°), and one is 18° more than five times the other. Find the angles.

**4.** By traveling 30 mph for one period of time and then 40 mph for another, Shirley traveled 230 miles. Had she gone 10 mph faster throughout, she would have traveled 300 miles. How many hours did she travel at each rate?

**5.** Four books (of the same kind) and 6 pens (of the same kind) cost $9.00. Three books and 9 pens also cost $9.00. Find the cost of one book and of one pen.

**6.** A candy mix sells for \$1.10 per pound. If the mix is composed of two kinds of candy, one worth 90¢ per lb and the other worth \$1.50 per pound, how many pounds of each would be in a 30-pound mixture?

**7.** A collection of dimes and nickels is worth \$3.75. If there are 55 coins in the collection, how many of each are there?

**8.** If there were 450 people at a play, the total receipts were \$600 and the admission price was \$2.00 for adults and 75¢ for children, how many adults and how many children were in attendance?

**9.** Mr. Smith is 43 years old. If the sum of Mr. Smith's two daughters' ages is 23, and if three times the age of the younger plus the age of the older daughter is equal to the age of Mr. Smith, how old is each?

**10.** A 40-ft rope is cut in two pieces. One piece is 1 ft more than twice the other. How long is each piece?

**11.** In a right triangle, one acute angle is 6° less than three times the other. Find each acute angle.

**12.** A collection of quarters and nickels is worth \$3.50. If there are 30 coins in the collection, how many of each are there?

**13.** A woman wishes to mix two blends of coffee, one selling for \$1.80 per pound and the other for \$2.40 per pound, to obtain a 20-pound mixture selling for \$2.10 per pound. How many pounds of each must she use?

**14.** A chemist has one solution which is 50% acid and a second which is 25% acid. How much of each should he use to make 10 L of a 40%-acid solution?

**15.** The perimeter of a rectangular field is 192 yards. If the length is 8 more than the width, find the dimensions.

ANSWERS: 1. 17, 33 2. Pete is 10, Jim is 17 3. 27°, 153° 4. 5 hr at 30 mph, 2 hr at 40 mph 5. book is \$1.50, pen is 50¢ 6. 20 lbs of 90¢ candy, 10 lbs of \$1.50 candy 7. 20 dimes, 35 nickels 8. 210 adults, 240 children 9. 10 yr, 13 yr 10. 13 ft, 27 ft 11. 24°, 66° 12. 10 quarters, 20 nickels 13. 10 lbs of each 14. 4 L of 25% solution, 6 L of 50% solution 15. 52 yd, 44 yd

## 8.6 Systems of Three Linear Equations in Three Variables

An equation of the form

$$ax + by + cz = d$$

when $a$, $b$, $c$, and $d$ are constant real numbers and $x$, $y$, and $z$ are variables, is called a **linear equation in three variables.** The term *linear equation* in this context is a bit misleading (perhaps *first-degree equation* is better) since the graph of a linear equation in three variables is actually a plane in space. However, our work with linear equations in three variables will be purely algebraic, patterned after similar situations involving linear equations in two variables.

A **solution** to a linear equation such as

$$2x + y - 3z = 3$$

is an **ordered triple** of numbers. For example, $(1, -2, -1)$ is a solution to the above equation since if $x$, $y$, and $z$ are replaced with $1, -2$, and $-1$, respectively, we obtain a true equation.

$$\begin{aligned} 2(1) + (-2) - 3(-1) &\stackrel{?}{=} 3 \\ 2 - 2 + 3 &\stackrel{?}{=} 3 \\ 3 &= 3 \end{aligned}$$

(The *order* in the ordered triple will always refer to the $x$ value first, the $y$ value second, and the $z$ value third.)

EXAMPLE 1 Are $(2, 1, 1)$ and $(-3, -1, 4)$ solutions to $x + 3y - 2z = 3$?

Substitute 2 for $x$, 1 for $y$, and 1 for $z$.

$$\begin{aligned} (2) + 3(1) - 2(1) &\stackrel{?}{=} 3 \\ 2 + 3 - 2 &\stackrel{?}{=} 3 \\ 3 &= 3 \end{aligned}$$

$(2, 1, 1)$ is a solution to the equation. Substitute $-3$ for $x$, $-1$ for $y$ and 4 for $z$.

$$\begin{aligned} (-3) + 3(-1) - 2(4) &\stackrel{?}{=} 3 \\ -3 - 3 - 8 &\stackrel{?}{=} 3 \\ -14 &\neq 3 \end{aligned}$$

$(-3, -1, 4)$ is not a solution to the equation.

A **system of three linear equations in three variables** is a trio of linear equations such as the following.

$$\begin{aligned} x + y + z &= 2 \\ -2x - y + z &= 1 \\ x - 2y - z &= 1 \end{aligned}$$

When there can be no confusion, we refer to the above system simply as a **system of equations.** A **solution** to such a system is an ordered triple of numbers which is a solution to all three equations. It is easy to verify (by substitution) that $(1, -1, 2)$ is a solution to each of the above equations, and hence is a solution to the system.

There are three possibilities for the solutions of a system of three equations in three variables: exactly one solution, no solution, or infinitely many solutions. We will only be concerned with solutions to systems which have exactly one or no solutions.

We could solve a system by using a method of substitution or addition-subtraction. Alternatively, a combination method of elimination and reduction to a system of two equations in two variables is perhaps the easiest method. The next rule summarizes the basic steps of this technique.

**TO SOLVE A SYSTEM OF THREE EQUATIONS IN THREE VARIABLES**

1. Select any two of the three equations and by using the addition-subtraction rule eliminate one of the variables.
2. Use the equation in the original system which was not used in the first step together with either of the other two equations and eliminate the *same* variable as in the first step.
3. Solve the resulting pair of equations in two variables.
4. Substitute the values obtained in Step 3 into any of the three original equations to obtain the value of the third variable.
5. Check the solution (ordered triple) in all three of the original equations.
6. If, at any step, a contradiction is obtained, the system has no solution. If, at any step, an identity is obtained, the system has infinitely many solutions.

We shall illustrate the rule as we solve the system of three equations in three variables given above. Notice how we use letters to identify the various equations and indicate the steps in the solution process.

EXAMPLE 2 Solve the following system.

$$\begin{aligned} &\text{(A)} & x + y + z &= 2 \\ &\text{(B)} & -2x - y + z &= 1 \\ &\text{(C)} & x - 2y - z &= 1 \end{aligned}$$

If we select equations (A) and (B) and subtract (B) from (A), we eliminate $z$.

$$\begin{array}{lrl} \text{(A)} & x + \phantom{2}y + z = 2 & \\ \text{(B)} & -2x - \phantom{2}y + z = 1 & \\ \hline \text{(D)} & 3x + 2y \phantom{+ z} = 1 & \text{(A)} - \text{(B)} = \text{(D)} \end{array}$$

We now use equation (C) with (A) and add to eliminate $z$ again.

$$\begin{array}{lrl} \text{(A)} & x + \phantom{2}y + z = 2 & \\ \text{(C)} & x - 2y - z = 1 & \\ \hline \text{(E)} & 2x - \phantom{2}y \phantom{- z} = 3 & \text{(A)} + \text{(C)} = \text{(E)} \end{array}$$

Thus we obtain the following system of two equations in $x$ and $y$.

$$\begin{array}{lr} \text{(D)} & 3x + 2y = 1 \\ \text{(E)} & 2x - \phantom{2}y = 3 \end{array}$$

If we multiply (E) by 2 and add to (D) $y$ is eliminated.

$$\begin{array}{rrl} \text{(D)} & 3x + 2y = 1 & \\ 2\text{(E)} & 4x - 2y = 6 & \\ \hline \text{(F)} & 7x \phantom{+ 2y} = 7 & \text{(D)} + 2\text{(E)} = \text{(F)} \\ & x = 1 & \end{array}$$

Substitute 1 for $x$ in (E).

$$\begin{aligned} 2(1) - y &= 3 \\ -y &= 1 \\ y &= -1 \end{aligned}$$

Substitute 1 for $x$ and $-1$ for $y$ in (A).

$$\begin{aligned} 1 + (-1) + z &= 2 \\ z &= 2 \end{aligned}$$

The solution is $(1, -1, 2)$.

Check:

$$\begin{array}{lr} \text{(A)} & 1 + (-1) + 2 \stackrel{?}{=} 2 \\ & 2 = 2 \\ \text{(B)} & -2(1) - (-1) + 2 \stackrel{?}{=} 1 \\ & -2 + 1 + 2 \stackrel{?}{=} 1 \\ & 1 = 1 \\ \text{(C)} & 1 - 2(-1) - 2 \stackrel{?}{=} 1 \\ & 1 + 2 - 2 \stackrel{?}{=} 1 \\ & 1 = 1 \end{array}$$

$(1, -1, 2)$ checks in all three equations.

**EXAMPLE 3** Solve the following system.

$$\begin{array}{lr} \text{(A)} & 3x + 2y + 3z = 3 \\ \text{(B)} & 4x - 5y + 7z = 1 \\ \text{(C)} & 2x + 3y - 2z = 6 \end{array}$$

We can eliminate $x$ from (A) and (B) by adding 4 times (A) to $-3$ times (B).

$$\begin{array}{rrl} 4\text{(A)} & 12x + \phantom{1}8y + 12z = 12 & \\ -3\text{(B)} & -12x + 15y - 21z = -3 & \\ \hline \text{(D)} & 23y - \phantom{1}9z = \phantom{-}9 & 4\text{(A)} - 3\text{(B)} = \text{(D)} \end{array}$$

We can eliminate $x$ from (B) and (C) by adding (B) to $-2$ times (C).

$$\begin{array}{lrl} \text{(B)} & 4x - 5y + 7z = & 1 \\ -2\text{(C)} & -4x - 6y + 4z = & -12 \\ \hline \text{(E)} & -11y + 11z = & -11 \end{array} \qquad \text{(B)} - 2\text{(C)} = \text{(E)}$$

Simplify (E) by multiplying through by $-\frac{1}{11}$ (or dividing by $-11$).

$$\text{(F)} \quad y - z = 1 \qquad \text{(F)} - \frac{1}{11}\text{(E)}$$

Thus we must solve the following system.

$$\begin{array}{ll} \text{(D)} & 23y - 9z = 9 \\ \text{(F)} & y - z = 1 \end{array}$$

Solve (F) for $y$, obtaining $y = 1 + z$, and substitute into (D).

$$\begin{aligned} 23(1 + z) - 9z &= 9 \\ 23 + 23z - 9z &= 9 \\ 14z &= -14 \\ z &= -1 \end{aligned}$$

Substitute $-1$ for $z$ in (F).

$$\begin{aligned} y - (-1) &= 1 \\ y + 1 &= 1 \\ y &= 0 \end{aligned}$$

Substitute 0 for $y$ and $-1$ for $z$ in (A).

$$\begin{aligned} 3x + 2(0) + 3(-1) &= 3 \\ 3x - 3 &= 3 \\ 3x &= 6 \\ x &= 2 \end{aligned}$$

The solution is $(2, 0, -1)$. (Check in all three equations.)

**EXAMPLE 4** Solve the system.

$$\begin{array}{ll} \text{(A)} & 3x - y + 2z = 4 \\ \text{(B)} & -6x + 2y - 4z = 1 \\ \text{(C)} & 5x - 3y + 8z = 0 \end{array}$$

If we try to eliminate $y$ by adding 2 times (A) to (B)

$$\begin{array}{ll} 2\text{(A)} & 6x - 2y + 4z = 8 \\ \text{(B)} & -6x + 2y - 4z = 1 \\ \hline & 0 = 9 \end{array}$$

we obtain a contradiction, so we know that the system has no solution.

**EXAMPLE 5** Solve the system.

$$\begin{array}{ll} \text{(A)} & x + y + z = 6 \\ \text{(B)} & x - z = -2 \\ \text{(C)} & y + 3z = 11 \end{array}$$

When a variable is missing from one or more equations, we can shorten our work a bit. Since $x$ does not appear in (C), if we subtract (B) from (A) to eliminate $x$, the resulting equation can be paired with (C) immediately.

$$\begin{array}{lrl} \text{(A)} & x + y + z = & 6 \\ \text{(B)} & x \quad - z = & -2 \\ \hline \text{(D)} & y + 2z = & 8 \end{array} \qquad \text{(A)} - \text{(B)} = \text{(D)}$$

Thus, we must solve the following system.

$$\begin{array}{lrl} \text{(C)} & y + 3z = & 11 \\ \text{(D)} & y + 2z = & 8 \end{array}$$

Subtracting (D) from (C), we eliminate $y$.

$$\text{(E)} \quad z = 3 \qquad \text{(C)} - \text{(D)} = \text{(E)}$$

Substitute 3 for $z$ in (C).

$$\begin{aligned} y + 3(3) &= 11 \\ y + 9 &= 11 \\ y &= 2 \end{aligned}$$

Substitute 3 for $z$ in (B).

$$\begin{aligned} x - 3 &= -2 \\ x &= 1 \end{aligned}$$

The solution is (1, 2, 3). (Check in all three equations.)

## EXERCISES 8.6

*Solve the following systems using the procedures illustrated in the examples.*

**1.** $\begin{aligned} x + y + z &= 2 \\ -x + y - 2z &= -5 \\ 2x - y + z &= 1 \end{aligned}$

**2.** $\begin{aligned} x + 2y + 3z &= 6 \\ -2x + y - z &= -2 \\ -x + 3y - 2z &= 0 \end{aligned}$

**3.** $\begin{aligned} x - 2y - z &= 2 \\ 2x - y + z &= 7 \\ 3x + 2y + z &= 2 \end{aligned}$

**4.** $\begin{aligned} x + y + z &= 8 \\ x - y - z &= 0 \\ x + 2y + z &= 9 \end{aligned}$

**5.**
$$\begin{aligned} 3x + y - z &= 4 \\ y - 2z &= 5 \\ 2x + z &= -1 \end{aligned}$$

**6.**
$$\begin{aligned} 2x + 3y &= 13 \\ 5x - 2z &= -2 \\ 4y - z &= 6 \end{aligned}$$

**7.**
$$\begin{aligned} 4x - y + 3z &= 3 \\ -8x + 2y - 6z &= -7 \\ x + 5y - z &= 2 \end{aligned}$$

**8.**
$$\begin{aligned} 2x - 3y + 4z &= 3 \\ 5x + 4y - 2z &= 7 \\ -3x + 2y - 5z &= -6 \end{aligned}$$

**9.** Murphy is 6 years older than Jim. In 5 years, the sum of their ages will be 90. How old is each?

**10.** By traveling 50 mph for one period of time and then 55 mph for another, Pam traveled 310 miles. If she had gone 5 mph faster throughout, she would have traveled 340 miles. How many hours did she travel at each rate?

**11.** A collection of nickels and dimes is worth \$3.10. If there are 47 coins in the collection, how many of each are there?

**12.** If 2 lbs of nuts and 3 lbs of candy cost a total of \$5.40, and 3 lbs of nuts and 4 lbs of candy cost a total of \$7.50, find the cost of 1 lb of nuts and of 1 lb of candy.

**13.** A chemist has one solution which is 10% acid and a second which is 5% acid. How much of each should be used to make 20 L of an 8%-acid solution?

**14.** A boat travels 48 miles downstream in 2 hours and returns the 48 miles upstream in 3 hours. Find the speed of the boat and the speed of the stream. [*Hint:* If $x =$ speed of boat and $y =$ speed of stream, $x + y =$ speed going downstream and $x - y =$ speed going upstream. Use $d = r \cdot t$ with $d = 48$, twice.]

---

ANSWERS: **1.** $(-1, 0, 3)$ **2.** $(1, 1, 1)$ **3.** $(1, -2, 3)$ **4.** $(4, 1, 3)$ **5.** $(0, 3, -1)$ **6.** $(2, 3, 6)$ **7.** no solution **8.** $(1, 1, 1)$ **9.** Murphy is 43, Jim is 37 **10.** 4 hr at 50 mph and 2 hr at 55 mph **11.** 32 nickels, 15 dimes **12.** nuts are 90¢ per lb, candy is $1.20 per lb **13.** 8 L of 5% solution and 12 L of 10% solution **14.** boat speed is 20 mph, stream speed is 4 mph

## 8.7 Word Problems Resulting in Systems of Three Equations

A word problem can often be translated into a system of three equations in three variables. We then solve the resulting system of equations. Several examples will illustrate this notion.

**EXAMPLE 1** The sum of three numbers is 9. The first, minus the second, plus twice the third is 7. Twice the first, plus twice the second, minus the third is 6. Find the numbers.

Let $x =$ the first number
$y =$ the second number
$z =$ the third number

The first sentence translates to

$$x + y + z = 9.$$

The second sentence translates to

$$x - y + 2z = 7.$$

The third sentence translates to

$$2x + 2y - z = 6.$$

Thus, we must solve the following system.

$$\begin{aligned} &(A) \quad x + y + z = 9 \\ &(B) \quad x - y + 2z = 7 \\ &(C) \quad 2x + 2y - z = 6 \end{aligned}$$

Adding (A) and (B) we eliminate $y$.

$$\begin{array}{lrcl} (A) & x + y + z &=& 9 \\ (B) & x - y + 2z &=& 7 \\ \hline (D) & 2x \qquad + 3z &=& 16 \end{array} \qquad (A) + (B) = (D)$$

If we add (C) to $-2$ times (A) we also eliminate $y$.

$$\begin{array}{lrcl} -2(A) & -2x - 2y - 2z &=& -18 \\ (C) & 2x + 2y - z &=& 6 \\ \hline (E) & -3z &=& -12 \end{array} \qquad -2(A) + (C) = (E)$$

Actually, we not only eliminated $y$ but also $x$. This is extremely helpful since now all we must do is solve (E) for $z$ and begin substituting back in the other equations.

$$\begin{aligned} (E) \quad -3z &= -12 \\ z &= 4 \end{aligned}$$

Substitute 4 for $z$ in (D).

$$\begin{aligned} 2x + 3(4) &= 16 \\ 2x &= 4 \\ x &= 2 \end{aligned}$$

Substitute 2 for $x$ and 4 for $z$ in (A).

$$\begin{aligned} 2 + y + 4 &= 9 \\ y + 6 &= 9 \\ y &= 3 \end{aligned}$$

The numbers are 2, 3, and 4. (Check.)

**EXAMPLE 2** Pete has a collection of nickels, dimes, and quarters with total value \$11.50. There are seven times as many dimes as quarters, and the total number of coins is 120. How many of each are there?

Let $x =$ the number of nickels in the collection
$y =$ the number of dimes in the collection
$z =$ the number of quarters in the collection

The system of equations we must solve is the following.

| | | |
|---|---|---|
| (A) | $5x + 10y + 25z = 1150$ | The total value is \$11.50, converted to cents |
| (B) | $x + y + z = 120$ | The total number of coins is 120 |
| (C) | $7z = y$ | Multiply the number of quarters by seven to obtain the number of dimes |

If we substitute $7z$ for $y$ in (A) and (B), the resulting two equations will be in $x$ and $z$ only.

$$5x + 10(7z) + 25z = 1150$$
$$5x + 70z + 25z = 1150$$
$$5x + 95z = 1150$$
$$\text{(D)} \quad x + 19z = 230 \quad \text{Divide by 5}$$

and

$$x + (7z) + z = 120$$
$$\text{(E)} \quad x + 8z = 120$$

Thus, we must solve the following system.

$$\text{(D)} \quad x + 19z = 230$$
$$\text{(E)} \quad x + \phantom{1}8z = 120$$

Subtract (E) from (D).

$$11z = 110$$
$$z = 10$$

Substitute 10 for $z$ in (E).

$$x + 8(10) = 120$$
$$x + 80 = 120$$
$$x = \phantom{1}40$$

Substitute 10 for $z$ in (C).

$$7(10) = y$$
$$70 = y$$

There are 40 nickels, 70 dimes, and 10 quarters in the collection. (Check.)

**EXAMPLE 3** In a triangle, the largest angle is 70° more than the smallest angle, and the remaining angle is 10° more than three times the smallest angle. Find the measure of each angle.

Let $x$ = the measure of the smallest angle
$y$ = the measure of the middle angle
$z$ = the measure of the largest angle

Since $z$ is 70° more than $x$, we have

$$x + 70 = z$$

or

$$\text{(A)} \quad x - z = -70.$$

Since $y$ is 10° more than 3 times $x$,

$$3x + 10 = y$$

or

$$\text{(B)} \quad 3x - y = -10.$$

Finally, the sum of the measures of the angles of a triangle is 180°.

$$\text{(C)} \quad x + y + z = 180$$

Hence we must solve the following system.

$$\begin{aligned} &(A) \quad x \quad\quad - z = -70 \\ &(B) \quad 3x - y \quad\quad = -10 \\ &(C) \quad x + y + z = 180 \end{aligned}$$

$y$ is missing in (A), and if we add (B) and (C), the result is also an equation with $y$ missing.

$$(D) \quad 4x + z = 170 \qquad (B) + (C) = (D)$$

Hence we must solve

$$\begin{aligned} &(A) \quad x - z = -70 \\ &(D) \quad 4x + z = 170. \end{aligned}$$

Add (A) and (D).

$$\begin{aligned} 5x &= 100 \\ x &= 20 \end{aligned}$$

Substitute 20 for $x$ in (A).

$$\begin{aligned} 20 - z &= -70 \\ -z &= -90 \\ z &= 90 \end{aligned}$$

Substitute 20 for $x$ in (B).

$$\begin{aligned} 3(20) - y &= -10 \\ 60 - y &= -10 \\ -y &= -70 \\ y &= 70 \end{aligned}$$

The angles have measures 20°, 70°, and 90°. (Check.)

**EXAMPLE 4** Joe has $10,000 split into three separate investments. Part of the money is invested in certificates at 7%, part in bonds at 8% and the rest invested in his brother Bob's private business. If the business earns 6%, the total earnings from all the investments will amount to $660. However, if the business slumps, he loses 6% on his investment, and the total earnings from the three will amount to only $60. How much is invested in each category?

Let $x$ = the number of dollars invested in certificates
$y$ = the number of dollars invested in bonds
$z$ = the number of dollars invested in Bob's business

Then we must solve

$$\begin{aligned} x + y + z &= 10000 && \text{Total invested is \$10,000} \\ .07x + .08y + .06z &= 660 && \text{Amount earned if business earns 6\%} \\ .07x + .08y - .06z &= 60 && \text{Amount earned if business loses; note the minus } .06z \end{aligned}$$

It is best to clear all decimals and then solve.

$$\begin{aligned} &(A) \quad x + y + z = 10000 \\ &(B) \quad 7x + 8y + 6z = 66000 && \text{Multiply by 100} \\ &(C) \quad 7x + 8y - 6z = 6000 && \text{Multiply by 100} \end{aligned}$$

Adding (B) and (C) eliminates $z$.

$$\text{(D)}\quad 14x + 16y = 72000 \qquad \text{(B)} + \text{(C)} = \text{(D)}$$

Adding (B) to $-6$ times (A) will also eliminate $z$.

$$\text{(E)}\quad x + 2y = 6000 \qquad \text{(B)} - 6\text{(A)} = \text{(E)}$$

Solve (E) for $x$, $x = 6000 - 2y$, and substitute into (D).

$$\begin{aligned} 14(6000 - 2y) + 16y &= 72000 \\ 84000 - 28y + 16y &= 72000 \\ -12y &= -12000 \\ y &= 1000 \end{aligned}$$

Substitute 1000 for $y$ in (E).

$$\begin{aligned} x + 2(1000) &= 6000 \\ x + 2000 &= 6000 \\ x &= 4000 \end{aligned}$$

Substitute 4000 for $x$ and 1000 for $y$ in (A).

$$\begin{aligned} 4000 + 1000 + z &= 10000 \\ 5000 + z &= 10000 \\ z &= 5000 \end{aligned}$$

Joe has invested \$4000 in certificates, \$1000 in bonds, and \$5000 in his brother's business. (Check.)

## EXERCISES 8.7

Solve.

1. The sum of three numbers is 6. The first, plus twice the second, minus the third is $-3$. Three times the first, minus the second, plus twice the third is 18. Find the three numbers.

2. A collection of 90 coins consisting of nickels, dimes, and quarters has a value of \$11.50. If the number of quarters is twice the number of nickels, find the number of each type of coin.

**3.** The smallest angle of a triangle is one-third the middle-sized angle, and the largest angle is 5° more than the middle-sized angle. Find the measure of each.

**4.** Anita has $10,000 split into three separate investments. Part is invested in a mutual fund which earns 8%, part is in time certificates which earn 7%, and the rest is invested in a business. If the business does well, it will earn 10% and her total earnings will amount to $840. If the business loses 2% her total earnings will amount to only $360. How much is invested in each category?

**5.** The average of a student's three scores is 78. If the first is 10 points less than the second and the third is 4 points more than the second, find all three scores. [*Hint:* Could the first sentence translate to $\frac{x+y+z}{3} = 78$?]

**6.** Find three numbers such that the sum of the first and second is −1, the sum of the second and third is −4, and the sum of the first and third is 7.

**7.** An airplane flies 900 miles in 2 hours with the wind and 1050 miles in 3 hours against the wind. Find the speed of the plane in still air and the speed of the wind.

**8.** Solve.

$$\begin{aligned} x + y \quad &= 1 \\ -y + z &= -1 \\ x \quad + z &= 2 \end{aligned}$$

---

ANSWERS: **1.** $(3, -1, 4)$ **2.** 10 nickels, 60 dimes, 20 quarters **3.** 25°, 75°, 80° **4.** \$2000 in a mutual fund, \$4000 in certificates, \$4000 in a business **5.** 70, 80, 84 **6.** $(5, -6, 2)$ **7.** airplane speed is 400 mph, wind speed is 50 mph **8.** no solution

## 8.8 Linear Inequalities in Two Variables

In the second chapter we solved inequalities in one variable, such as

$$2x + 1 < 3 \qquad \text{and} \qquad x - 2 \geq 2(x - 4),$$

and also graphed the solutions to such inequalities on a number line. For example, if we solve

$$\begin{aligned} 2x + 1 &< 3 \\ 2x &< 2 \\ x &< 1 \end{aligned}$$

the solution is $x < 1$ and is graphed in Figure 8.3.

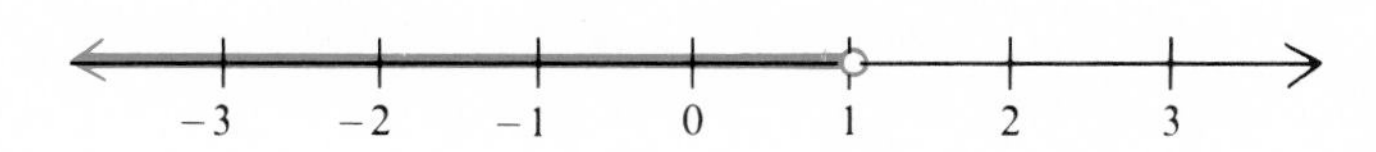


**Figure 8.3**

We now consider inequalities in two variables in which the variables are raised only to the first power. Such inequalities are called **linear inequalities in two variables.** For example,

$$2x + y < -1$$

is a linear inequality in the two variables $x$ and $y$. A **solution** to such an inequality is an ordered pair of numbers which when substituted for $x$ and $y$ yields a true statement. Thus $(-1, 0)$ is a solution to $2x + y < -1$ since by replacing $x$ with $-1$ and $y$ with 0 we obtain

$$\begin{aligned} 2(-1) + 0 &< -1 \\ -2 &< -1 \quad \text{which is true.} \end{aligned}$$

On the other hand, $(4, -3)$ is not a solution since

$$\begin{aligned} 2(4) + (-3) &< -1 \\ 8 - 3 &< -1 \\ 5 &< -1 \quad \text{is false.} \end{aligned}$$

**EXAMPLE 1** Determine whether $(-1, -3)$, $(1, 1)$ and $(-2, 0)$, are solutions to $3x - 2y \geq 1$.

$$\begin{aligned} 3(-1) - 2(-3) &\geq 1 \\ -3 + 6 &\geq 1 \\ 3 &\geq 1 \quad \text{True} \end{aligned}$$

$(-1, -3)$ is a solution.

$$\begin{aligned} 3(1) - 2(1) &\geq 1 \\ 3 - 2 &\geq 1 \\ 1 &\geq 1 \quad \text{True} \end{aligned}$$

$(1, 1)$ is a solution.

$$\begin{aligned} 3(-2) - 2(0) &\geq 1 \\ -6 - 0 &\geq 1 \\ -6 &\geq 1 \quad \text{False} \end{aligned}$$

$(-2, 0)$ is not a solution.

The set of all solutions to a linear inequality in two variables can be displayed in a Cartesian coordinate system. If possible, solve the inequality for $y$. For example,

$$\begin{aligned} -4x - 2y &\leq 6 \\ -2y &\leq 4x + 6 && \text{Add } 4x \text{ to both sides} \\ \left(-\frac{1}{2}\right) \cdot (-2y) &\geq \left(-\frac{1}{2}\right)(4x + 6) && \text{Multiply by } -\frac{1}{2} \text{ and reverse inequality} \\ y &\geq -2x - 3. \end{aligned}$$

If we replace the inequality symbol with the equal sign, the resulting equation is

$$y = -2x - 3.$$

The graph of this equation, in Figure 8.4, is a line with slope $-2$ and $y$-intercept $(0, -3)$. Also, the $x$-intercept is $(-3/2, 0)$. (Why?)

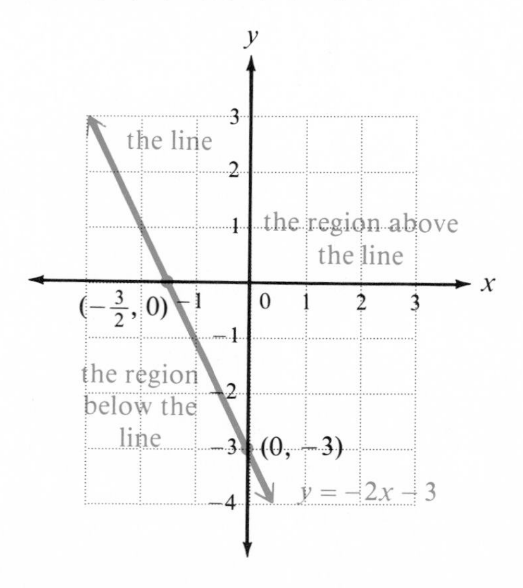


**Figure 8.4**

Every straight line with slope separates the plane into three distinct regions: the line itself, the area "above" the line, and the area "below" the line. The graph of every inequality as well as the equation itself can be described by a combination of these three regions.

| *Relation* | *Graph* |
|---|---|
| $y = -2x - 3$ | all points *on* the line itself |
| $y < -2x - 3$ | all points *below* the line |
| $y \leq -2x - 3$ | all points *on or below* the line |
| $y > -2x - 3$ | all points *above* the line |
| $y \geq -2x - 3$ | all points *on or above* the line |

We agree to "dash" the line when it *is not* part of the graph and leave the line solid when it *is* part of the graph. To indicate the area either above or below the line, we shade the region. The four graphs in Figure 8.5 show the various possibilities.

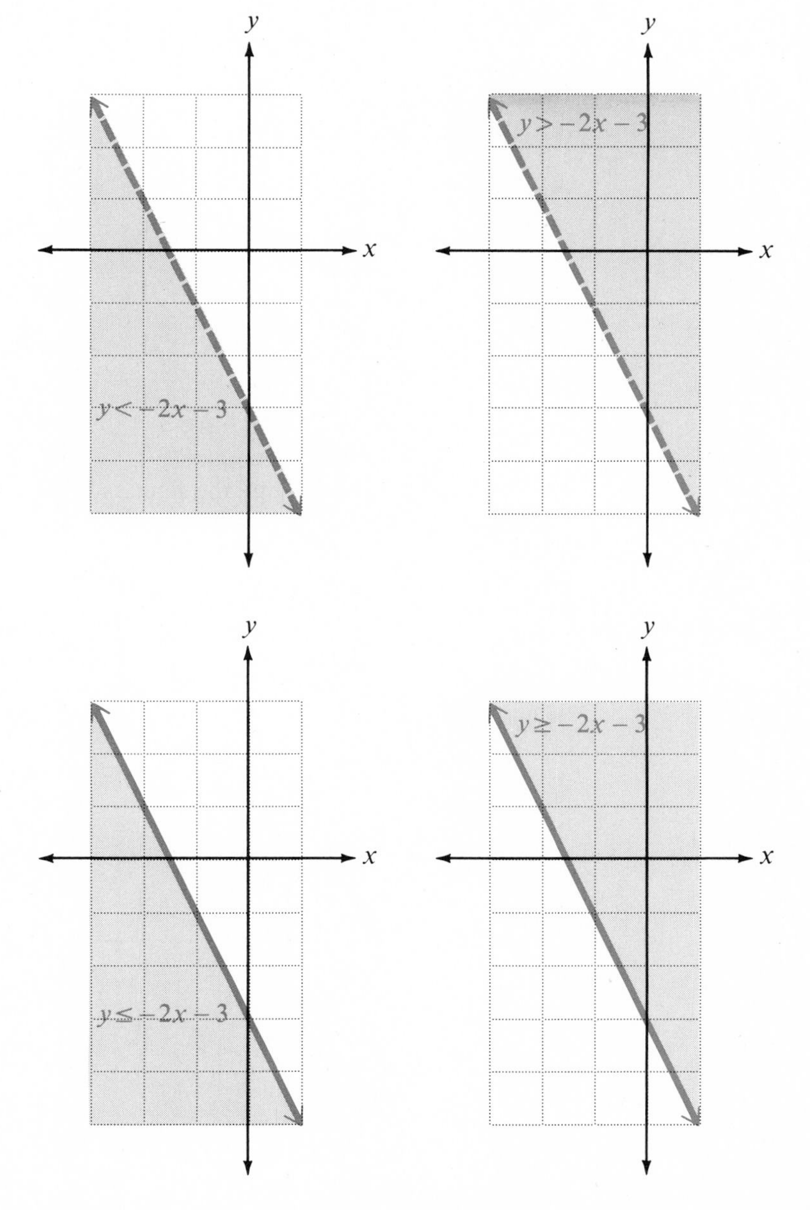


Figure 8.5

Every linear inequality in which the $y$-term is present (that is, for which the associated linear equation has as its graph a line with slope) can be graphed by following the procedures illustrated above and made precise in the next rule.

**TO GRAPH A LINEAR INEQUALITY IN TWO VARIABLES**

1. Solve the inequality for $y$ (if possible), putting $y$ on the left, and replace the inequality sign with an equal sign.
2. Graph the resulting equation using a solid line if the inequality is either $\leq$ or $\geq$.
3. Graph the resulting equation using a dashed line if the inequality is either $<$ or $>$.
4. Shade the region above the line if, when solved for $y$, the inequality is either $>$ or $\geq$.
5. Shade the region below the line if, when solved for $y$, the inequality is either $<$ or $\leq$.

**EXAMPLE 2** Graph $x - 3y < 6$.

Solve the inequality for $y$.

$$x - 3y < 6$$
$$-3y < -x + 6$$
$$y > \frac{1}{3}x - 2 \qquad \text{Reverse the inequality}$$

Replacing the $>$ with $=$ we have $y = \frac{1}{3}x - 2$. The intercepts are $(0, -2)$ and $(6, 0)$. (Why?) Since the inequality is $>$, we graph the line using a dashed line and shade the region *above* the line, as in Figure 8.6. We can check our work by testing a point in the shaded region. For example, if we substitute $(0, 0)$ into $x - 3y < 6$, we obtain

$$0 - 3(0) < 6$$
$$0 < 6$$

which is true.

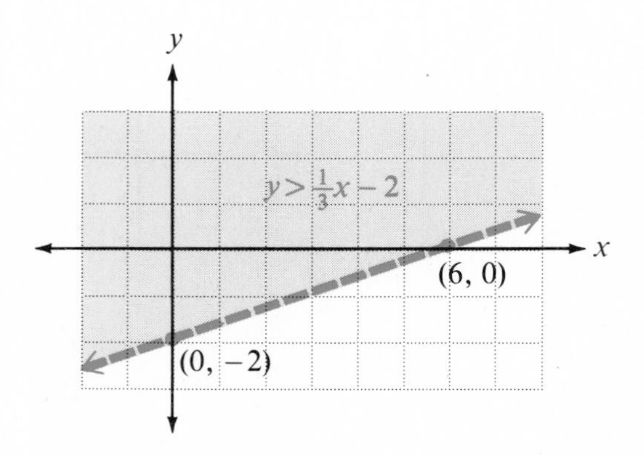


**Figure 8.6**

Linear inequalities in two variables for which the $y$ term is missing, such as

$$2x - 4 \geq 0$$

are graphed similarly. We first solve the inequality for $x$.

$$2x - 4 \geq 0$$
$$2x \geq 4$$
$$x \geq 2$$

Replace $\geq$ with $=$ and graph $x = 2$, which is a line parallel to the $y$-axis passing through $x$-intercept (2, 0). Since the inequality is $\geq$, we draw a solid line and shade the region to the right (if it were $\leq$ we would shade the region to the left). The graph of $2x - 4 \geq 0$ ($x \geq 2$) is shown in Figure 8.7. The graph of $2x - 4 < 0$ ($x < 2$) is shown in Figure 8.8.

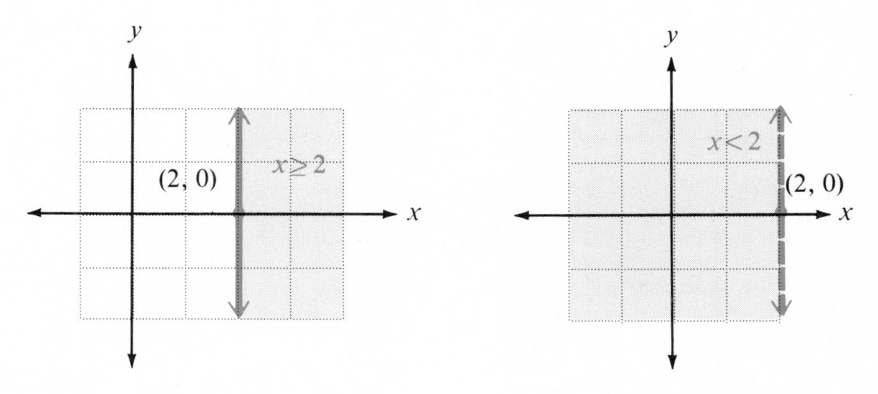


**Figure 8.7** **Figure 8.8**

EXAMPLE 3 Graph $3y + 3 < 0$.

Since the $y$ term is present, we solve for $y$.

$$3y + 3 < 0$$
$$3y < -3$$
$$y < -1$$

Since the inequality is $<$, we graph $y = -1$ using a dashed line and shade the region below the line. (See Figure 8.9.)

The graph of $3y + 3 \geq 0$ ($y \geq -1$) is shown in Figure 8.10.

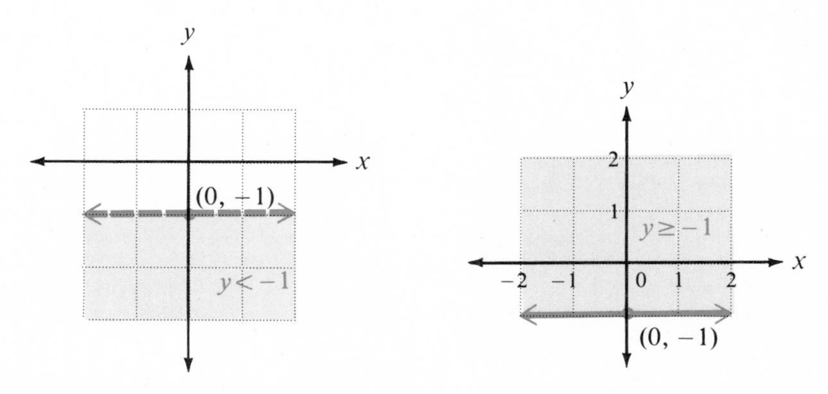


**Figure 8.9** **Figure 8.10**

## EXERCISES 8.8

*Graph the linear inequalities in two variables.*

**1.** $x + y > 2$

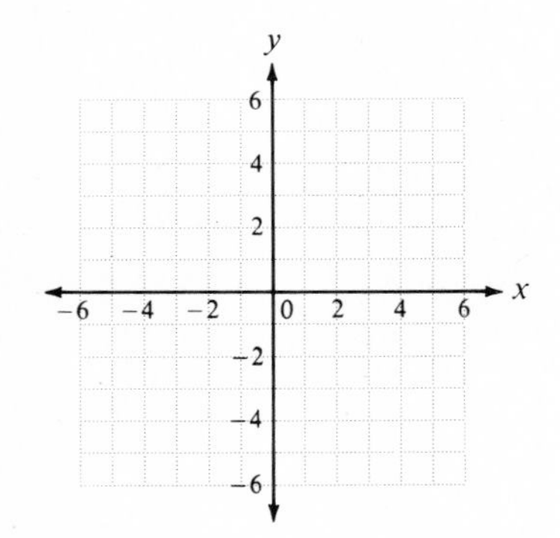

**2.** $x + y \leq 2$

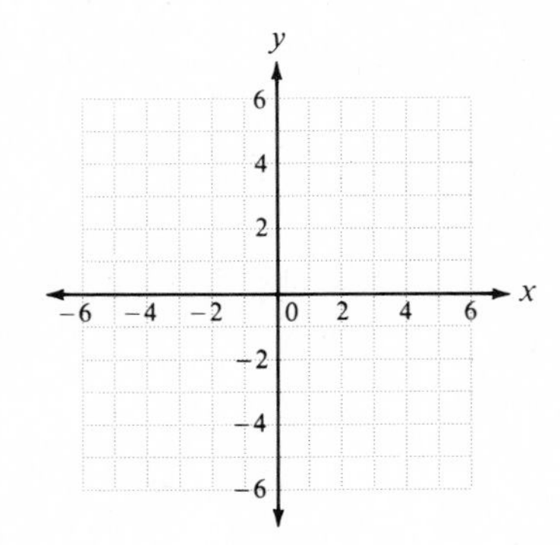

**3.** $x - y \geq -1$

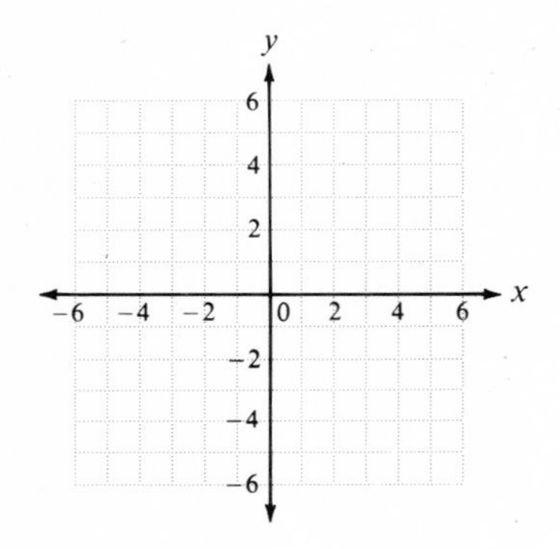

**4.** $x - y < -1$

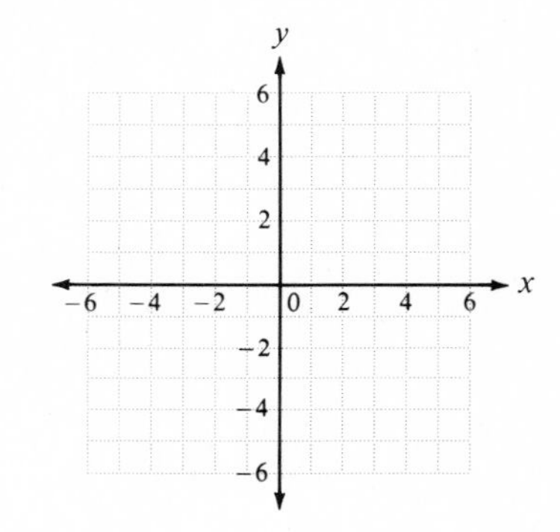

**5.** $3x + 4y < 12$

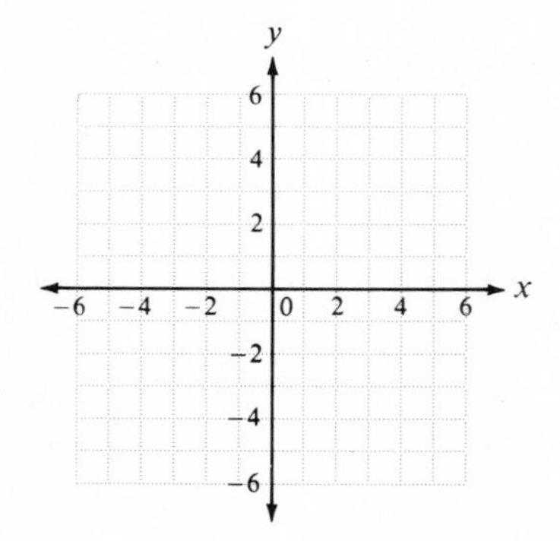

**6.** $3x + 4y \geq 12$

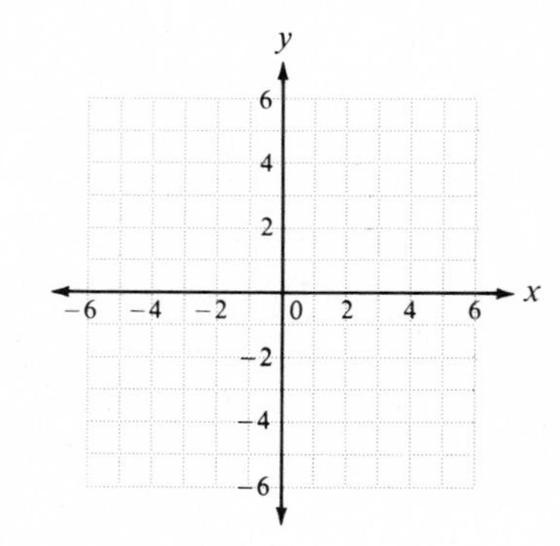

**7.** $4x + 8 > 0$

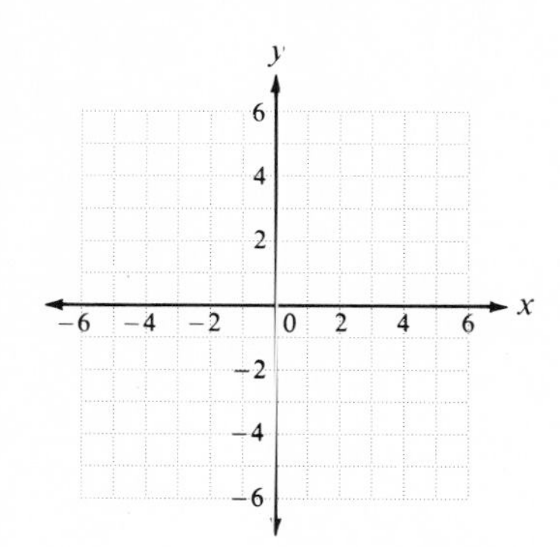

**8.** $4x + 8 \leq 0$

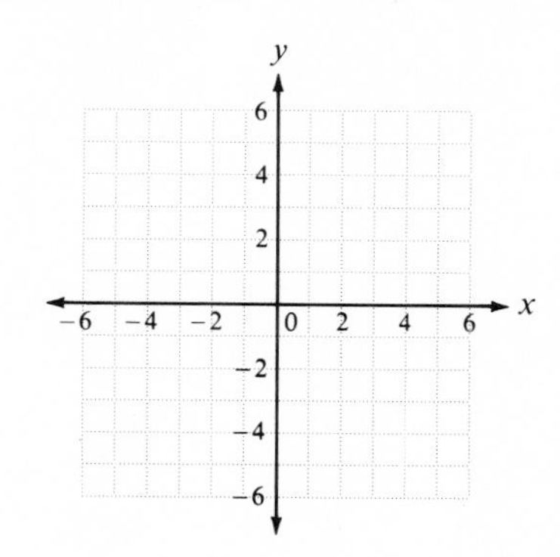

**9.** $2y - 6 \leq 0$

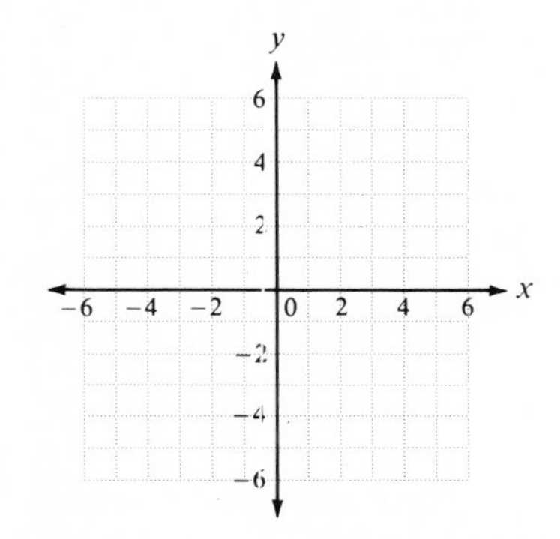

**10.** $2y - 6 > 0$

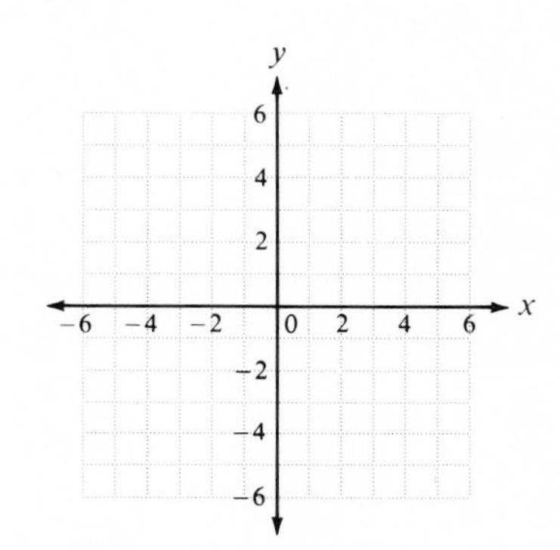

**11.** $2x + y < 0$

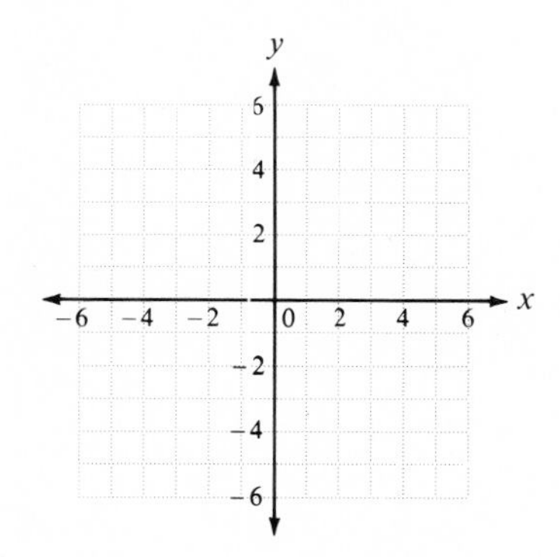

**12.** $2x + y \geq 0$

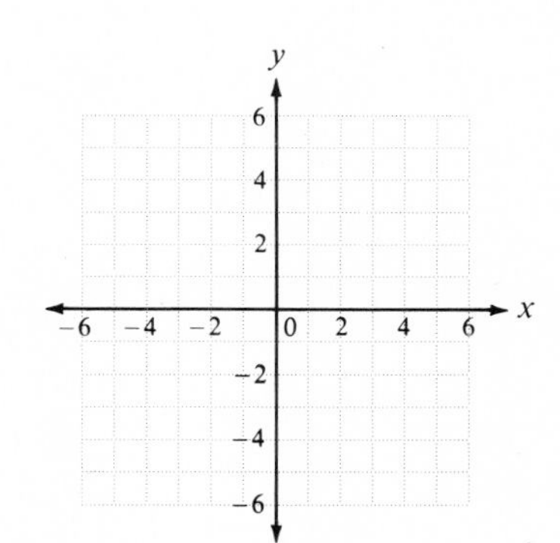

**13.** The largest angle of a triangle is twice the smallest, and the remaining angle is 20° less than the largest. Find the measure of each angle.

**14.** Find three numbers such that the sum of the first and second is twice the third, the first plus the third is 2, and the second minus the third is $-4$.

**15.** Chuck had \$5000 split into two investments. The part invested in bonds at 8% together with the other part invested in certificates at 7% earned \$370 one year. How much was invested in each category?

---

ANSWERS:

**1.**

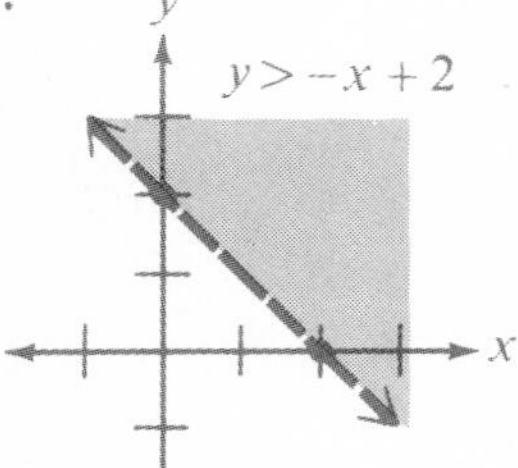


**2.**

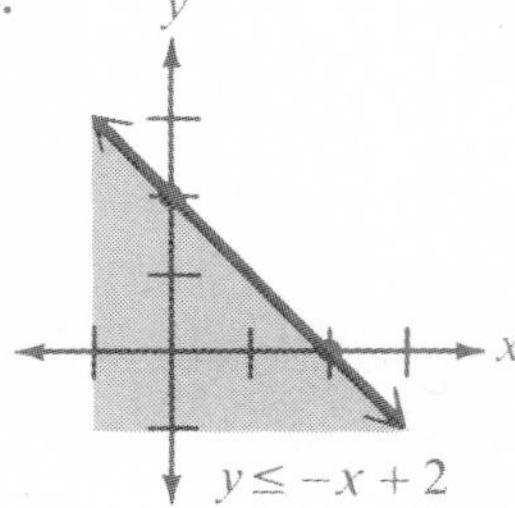


**3.**

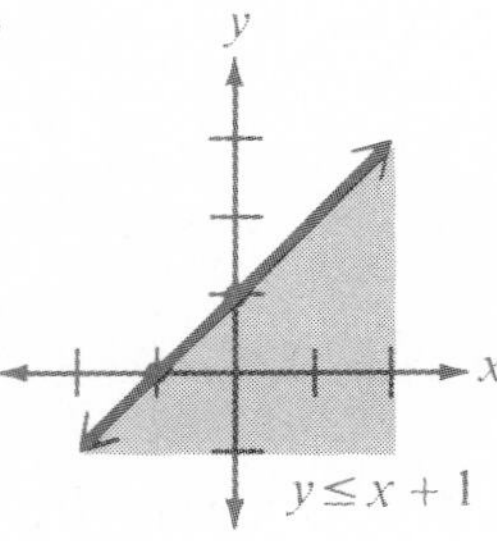


**4.**

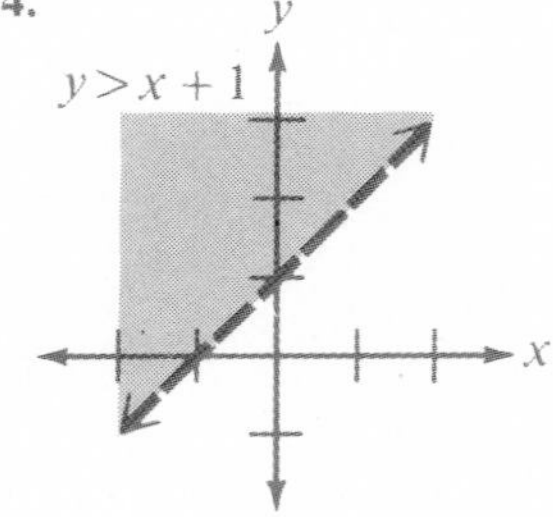


**5.**

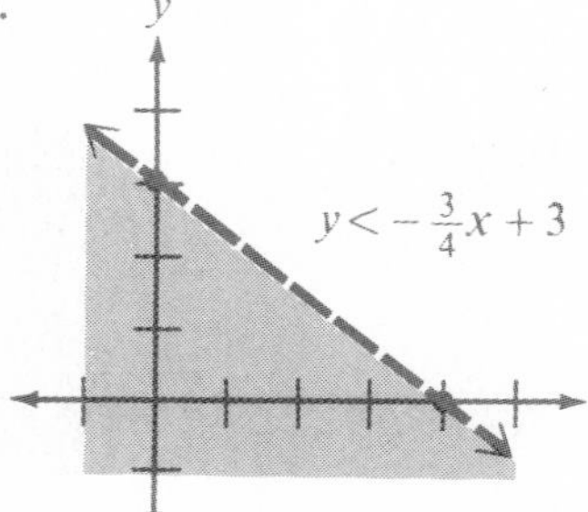


**6.**

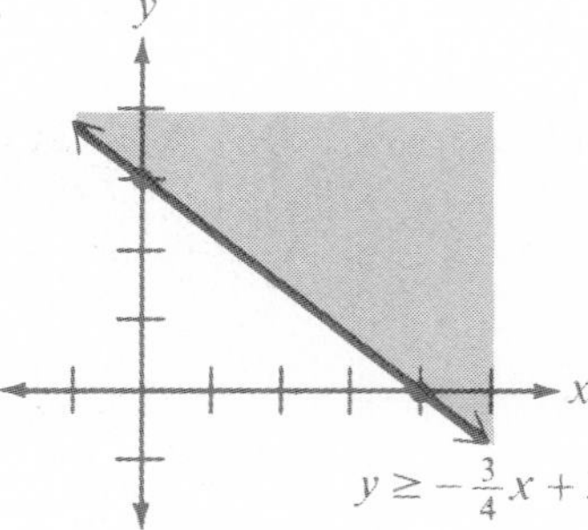


**7.**

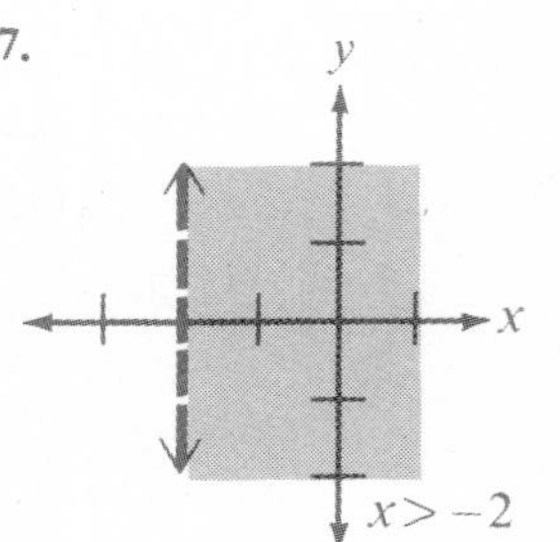


**8.**

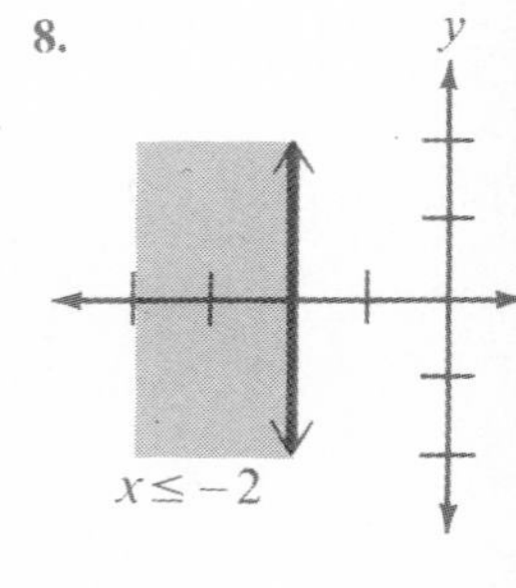


**9.**

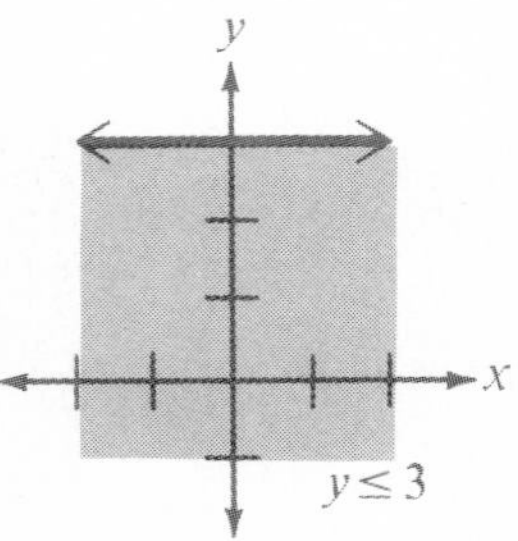


**10.**

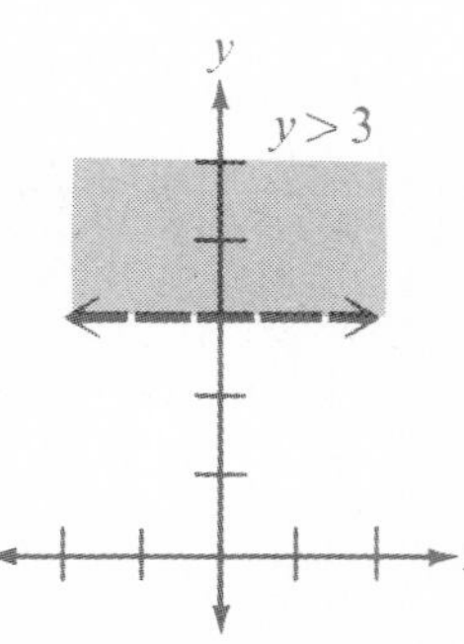


**11.**

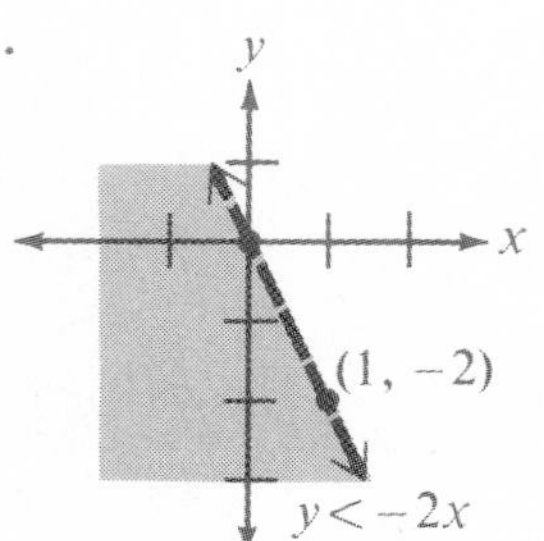


**12.**

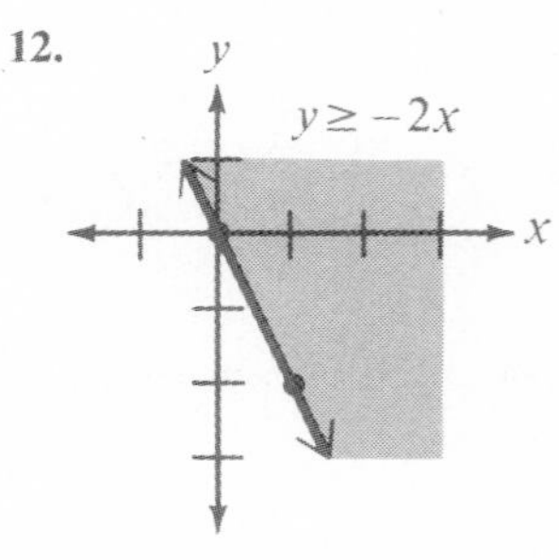


**13.** 40°, 60°, 80° **14.** $(3, -5, -1)$ **15.** \$2000 in bonds, \$3000 in certificates

## 8.9 Systems of Linear Inequalities in Two Variables

When we graph a linear inequality in two variables, the points on the graph correspond to ordered pairs of numbers which are solutions to the inequality. A **system of two linear inequalities in two variables** is simply a pair of linear inequalities in two variables. A **solution** to such a system is an ordered pair of numbers which is a solution to both inequalities in the system. Graphing a system of inequalities is outlined in the following rule.

**TO GRAPH A SYSTEM OF TWO LINEAR INEQUALITIES IN TWO VARIABLES**

1. Graph the first inequality and shade the graph using vertical lines.
2. Graph the second inequality and shade the graph using horizontal lines.
3. In addition to points on either or both lines (if any exist), the graph consists of those points in the cross-hatched region since these points correspond to solutions to both inequalities.

**EXAMPLE 1** Graph the following system of inequalities.

$$3x + 4y < 12$$
$$y - x \geq 1$$

Solve each inequality for $y$.

$$3x + 4y < 12 \qquad\qquad y - x \geq 1$$
$$4y < -3x + 12 \qquad\qquad y \geq x + 1$$
$$y < -\frac{3}{4}x + 3$$

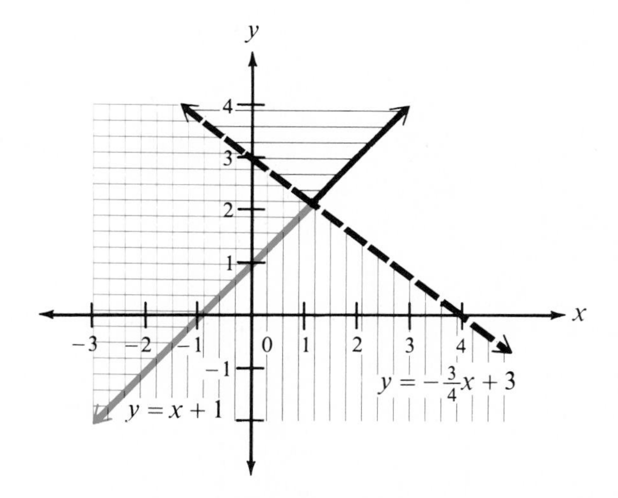


Figure 8.11

$y < -\frac{3}{4}x + 3$ is shaded with vertical lines, $y \geq x + 1$ is shaded with horizontal lines, and the graph of the system is cross-hatched.

**EXAMPLE 2** Graph $3x - y > 6$
$3x - 3 \geq 0$.

$$\begin{aligned} 3x - y &> 6 \\ -y &> -3x + 6 \\ y &< 3x - 6 \end{aligned} \qquad \begin{aligned} 3x - 3 &\geq 0 \\ 3x &\geq 3 \\ x &\geq 1 \end{aligned}$$

In Figure 8.12, $y < 3x - 6$ is shaded vertically, $x \geq 1$ is shaded horizontally, and the graph of the system is cross-hatched.

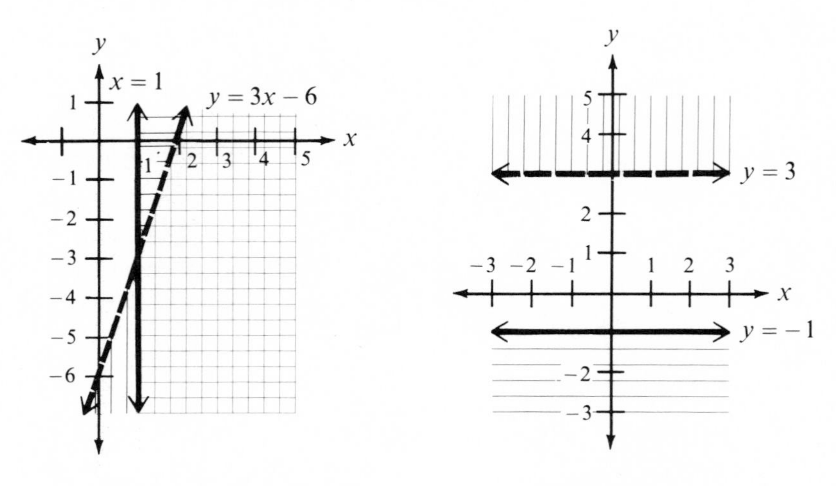


Figure 8.12

Figure 8.13

**EXAMPLE 3** Graph the system.

$$\begin{aligned} 2y - 6 &> 0 \\ y + 1 &\leq 0 \end{aligned}$$

$$\begin{aligned} 2y - 6 &> 0 \\ 2y &> 6 \\ y &> 3 \end{aligned} \qquad \begin{aligned} y + 1 &\leq 0 \\ y &\leq -1 \end{aligned}$$

$y > 3$ is shaded vertically and $y \leq -1$ is shaded horizontally in Figure 8.13. Since there is no overlap between these regions (that is, there is no region which is cross-hatched), there is no solution to this system.

**EXAMPLE 4** Graph the following system.

$$\begin{aligned} x &\geq 0 \\ y &\geq 0 \end{aligned}$$

The graph of the inequality $x \geq 0$ consists of the points to the right of the $y$-axis together with the points on the $y$-axis, while the graph of $y \geq 0$ includes the points on and above the $x$-axis. The region where these two graphs overlap is the graph of the system; it includes all the points in the first quadrant and the points on the positive axes. This region is cross-hatched in Figure 8.14.

The points in the second, third, and fourth quadrants may be described in a similar fashion. The systems which describe these points are shown in Figure 8.15 with their graphs.

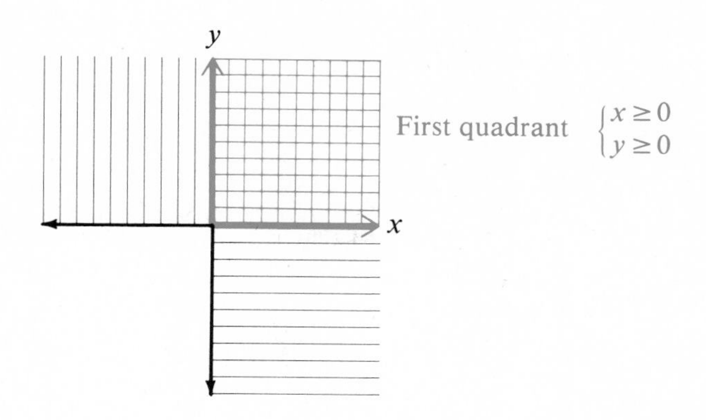


**Figure 8.14**

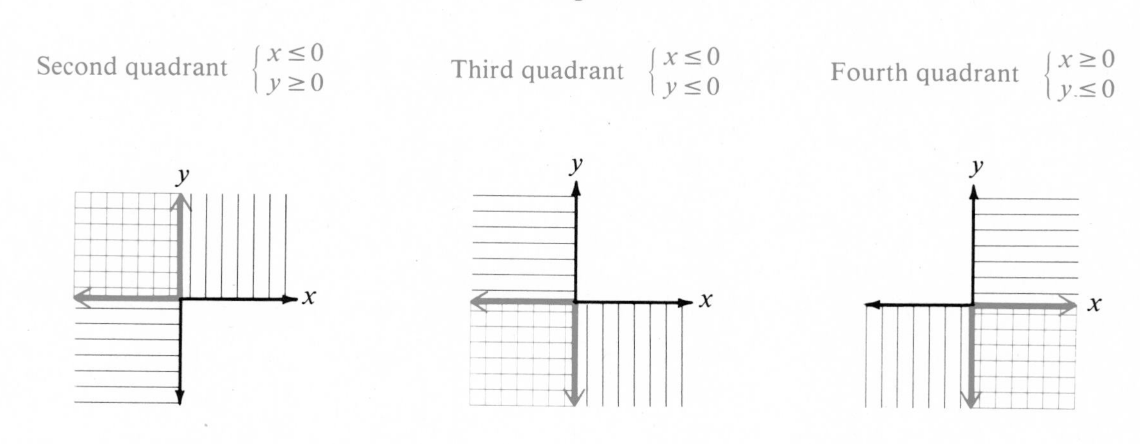


**Figure 8.15**

Systems involving three or more inequalities are often of interest. To solve such systems, either we need a more complex system of shading or else a better form of recognition must be introduced. If a system includes any of the four pairs of inequalities above, we must determine the portion of the plane which is in the particular quadrant and the region described by the remaining inequalities in the system.

EXAMPLE 5 Graph the following system.

$$\begin{aligned} x &\geq 0 \\ y &\geq 0 \\ y + \frac{1}{2}x &< 1 \end{aligned}$$

The first two inequalities describe the points in the first quadrant and on the positive axes. Thus, we are interested in points which also satisfy $y + \frac{1}{2}x < 1$ or

$$y < -\frac{1}{2}x + 1.$$

The intercepts of $y = -\frac{1}{2}x + 1$ are (0, 1) and (2, 0). The solution to the system is the set of all points below the line $y = -\frac{1}{2}x + 1$ which are also in the first quadrant, as shown in the shaded triangular region in Figure 8.16.

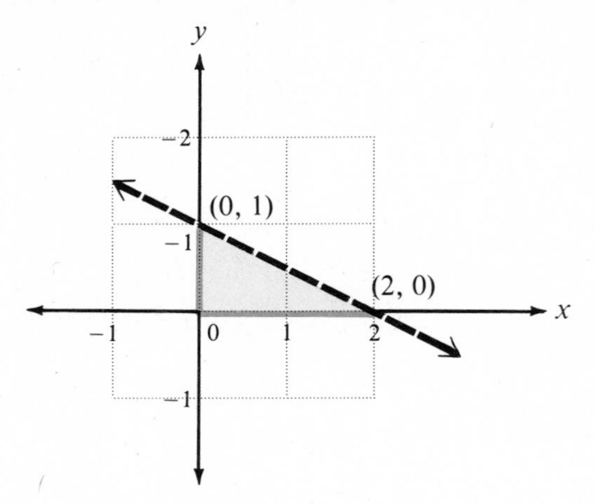


**Figure 8.16**

EXAMPLE 6 Graph the following system.

$$\begin{aligned} x &\geq 0 \\ y &\leq 0 \\ y &> 2x - 4 \\ 2x &\leq 3 \end{aligned}$$

Since the first two inequalities describe the points in the fourth quadrant, we begin by graphing the lines $y = 2x - 4$ and $2x = 3$. The point of intersection of the two lines, $(3/2, -1)$, can be found by substituting $x = 3/2$ into $y = 2x - 4$ and solving for $y$. The region which satisfies the two inequalities

$$y > 2x - 4 \qquad \text{and} \qquad 2x \leq 3$$

is cross-hatched in Figure 8.17(a). The solution to the original system is that portion of the plane in the fourth quadrant (described by $x \geq 0$ and $y \leq 0$) which also satisfies the two inequalities above, as shown in Figure 8.17(b).

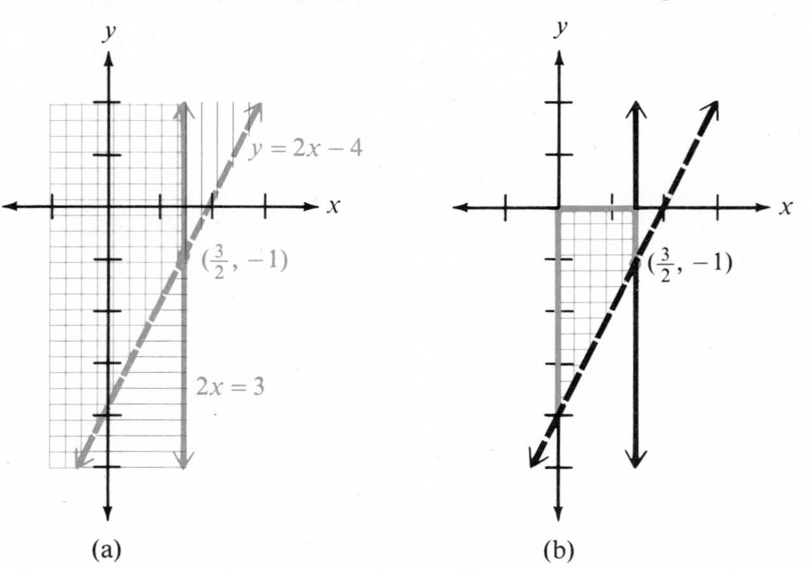


**Figure 8.17**

The system in the preceding example could also have been written as follows.

$$0 \leq x \leq \frac{3}{2}$$
$$y \leq 0$$
$$y > 2x - 4$$

The first inequality describes points in the strip between (and on) the $y$-axis and the line $x = \frac{3}{2}$. A system often has one of the variables restricted between two such values.

Regions defined by systems such as those in the last two examples play an important role in a problem-solving method called **linear programming.** If you are preparing for a career in business, you will study this method.

## EXERCISES 8.9

*Graph the following systems of inequalities.*

**1.** $3x - 2y \geq 6$
$x + 2y > -2$

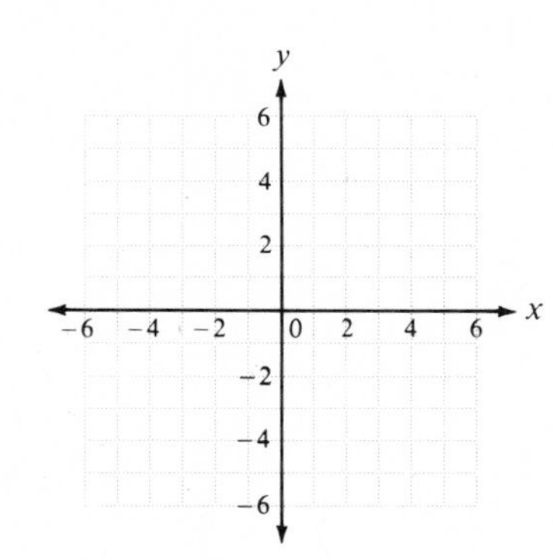


**2.** $3x - 2y < 6$
$x + 2y \geq -2$

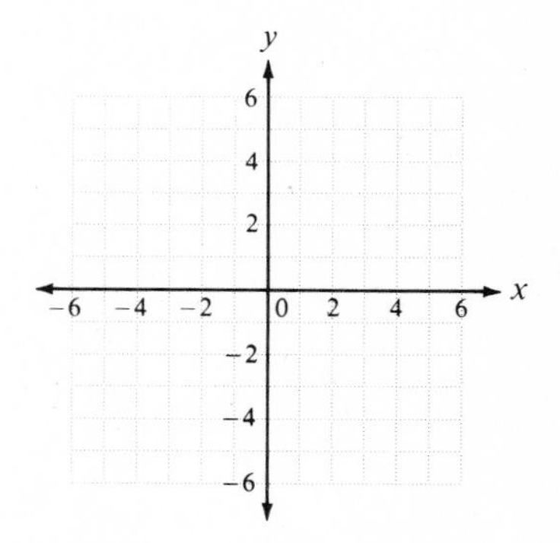


**3.** $3x - 2y \leq 6$
$x + 2y < -2$

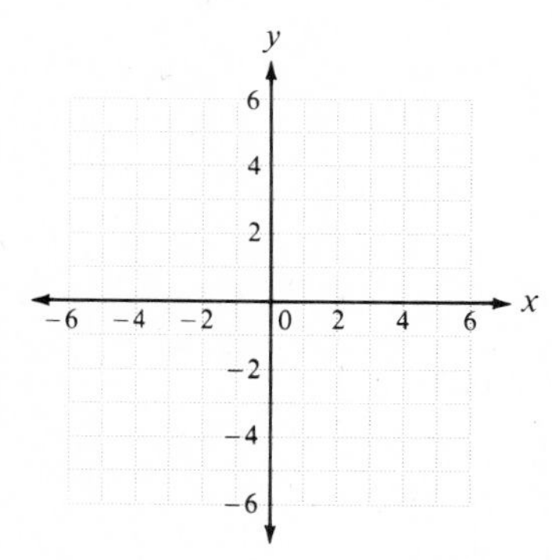


**4.** $x - y < 1$
$2x - 2y \geq -4$

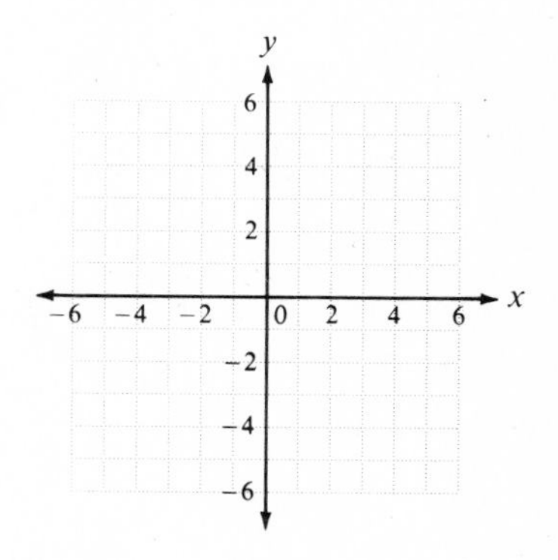


**5.** $x + 1 \leq 0$
$3y + 6 > 0$

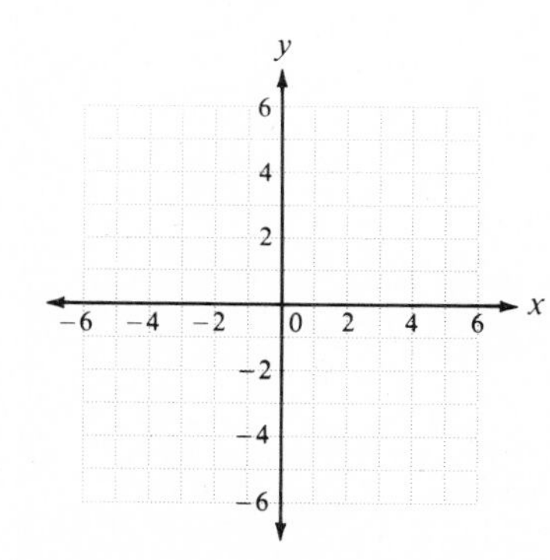


**6.** $x - y > 1$
$2x - 2y \leq -4$

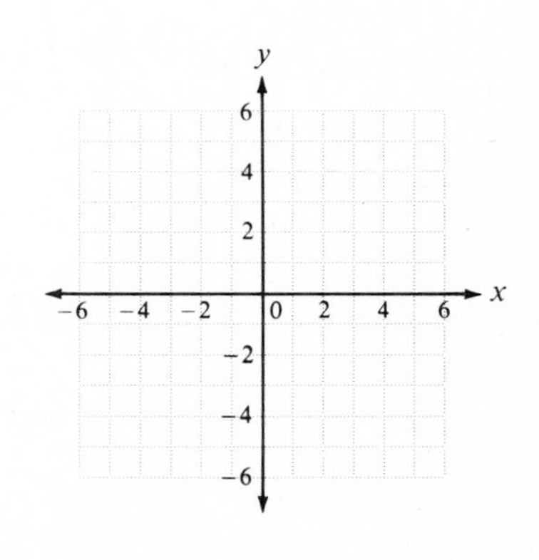

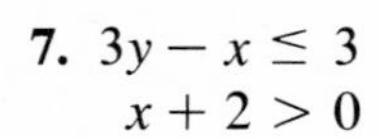

**7.** $3y - x \le 3$
$x + 2 > 0$

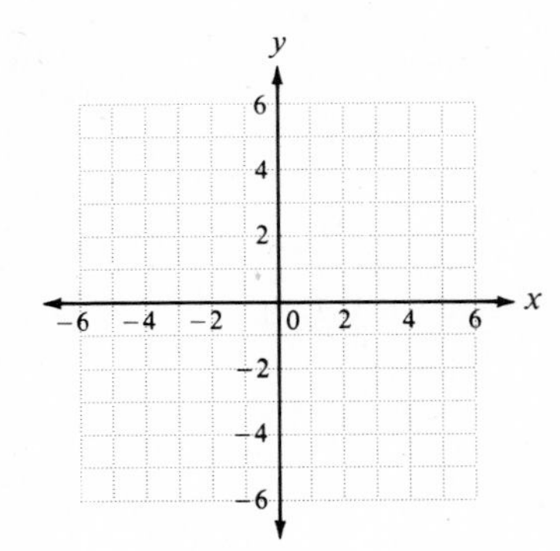


**8.** $y \ge x$
$y \ge -x$

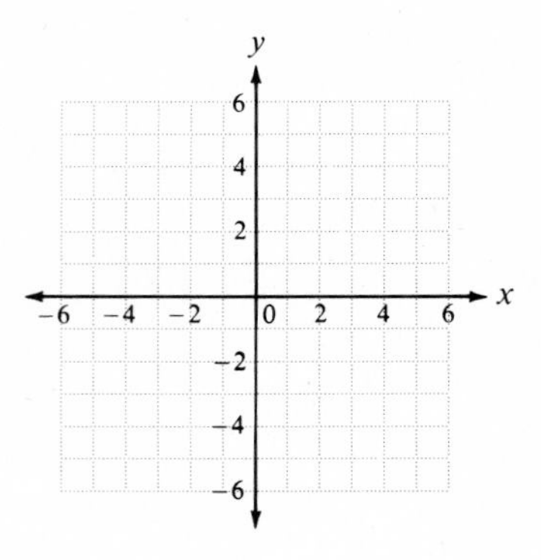


**9.** $x - 2 > 0$
$2 - x > 0$

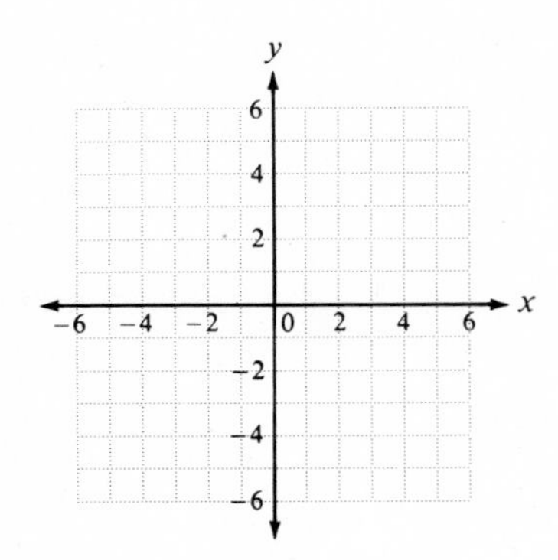


**10.** $x \ge 0$
$y \ge 0$
$x + y < 1$

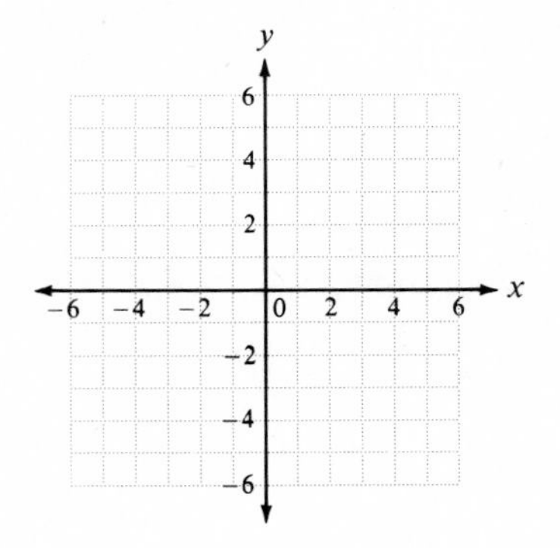


**11.** $0 \le x \le 3$
$y \ge 0$
$2y - x \le 4$

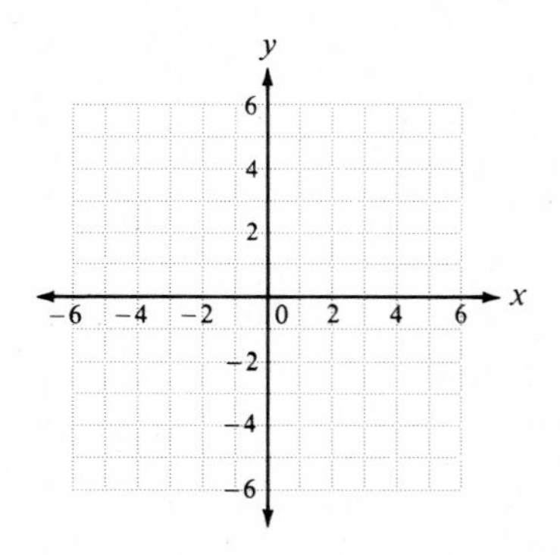


**12.** $0 \le x \le 1$
$y < 2 - x$
$y \ge -1$

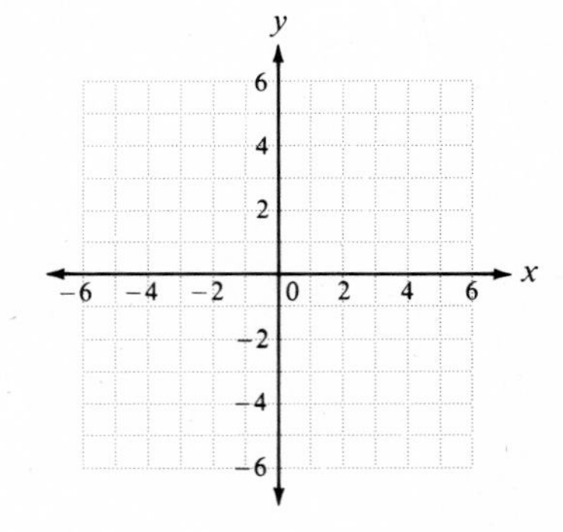


---

ANSWERS:

1.

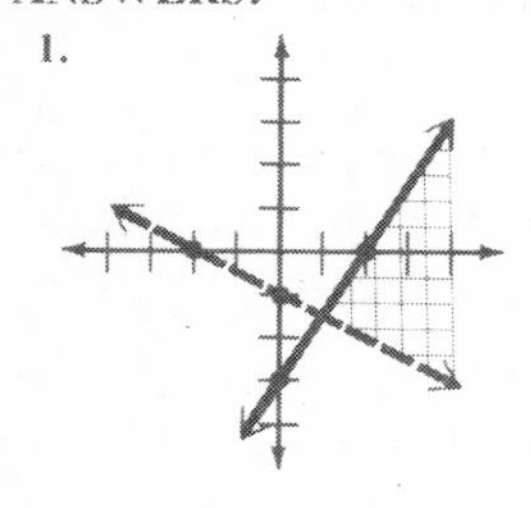

2.

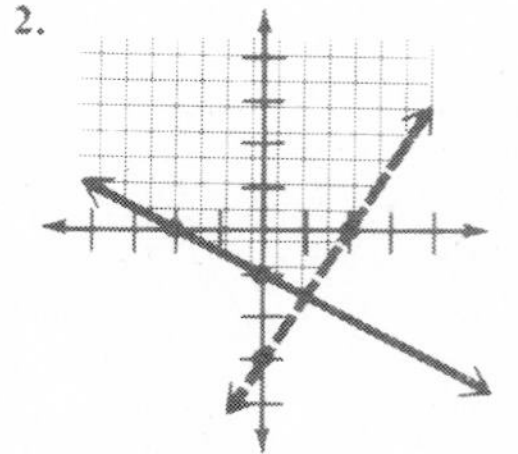

3.

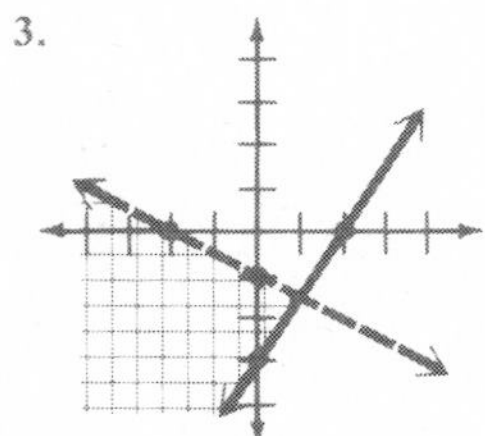

4.

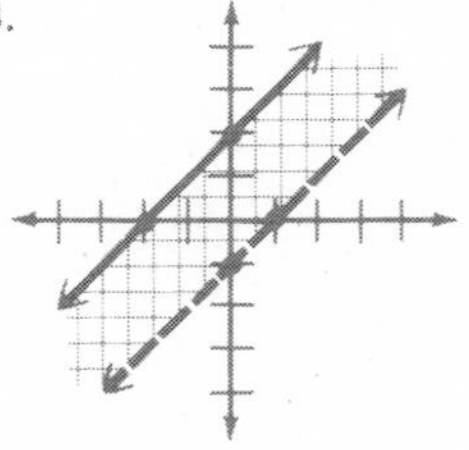

5.

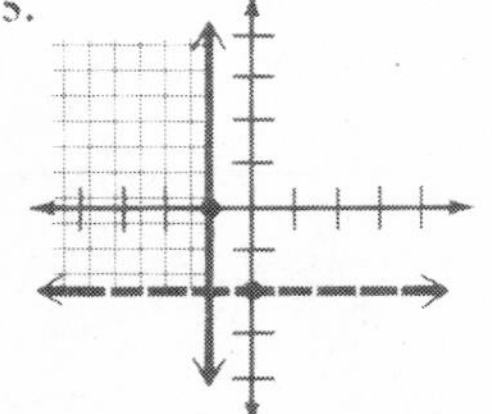

6.

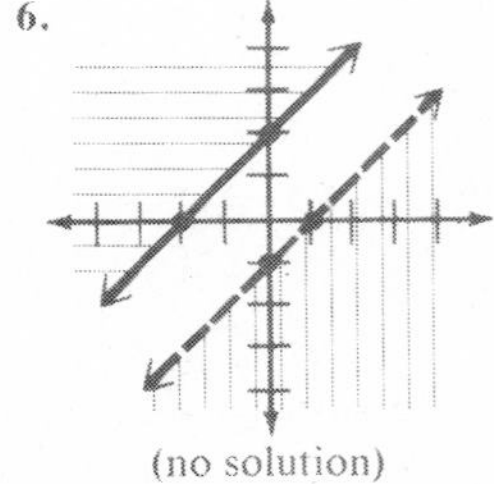

(no solution)

7.

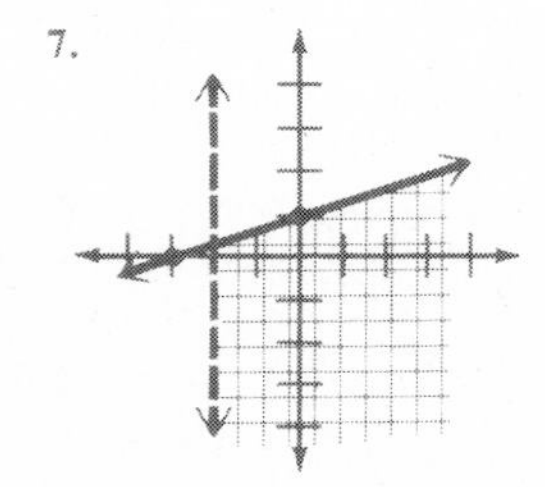

8.

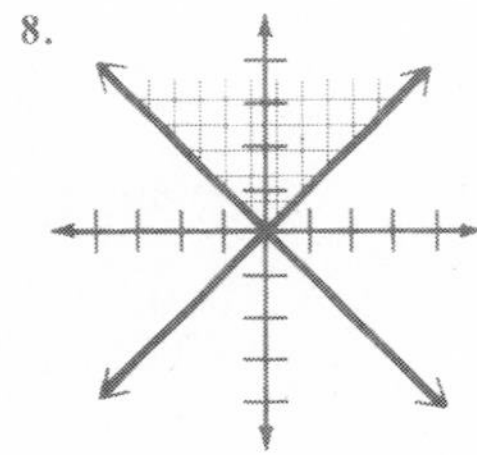

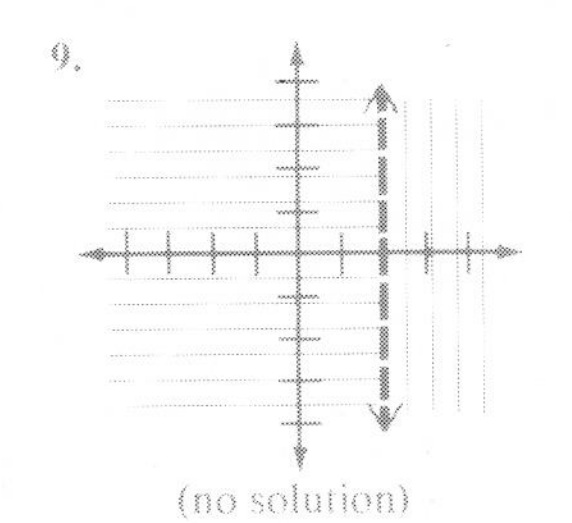


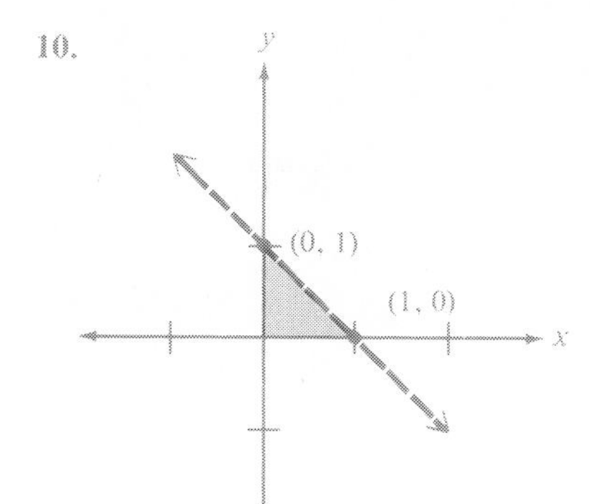


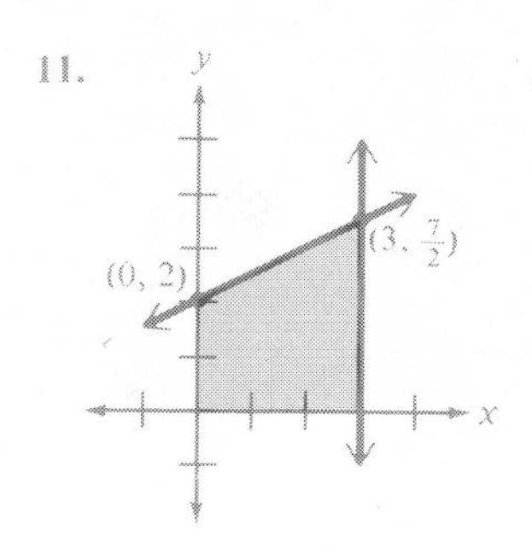


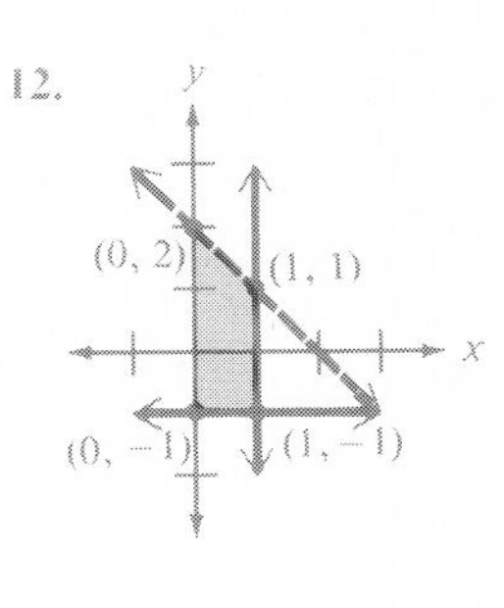


## Chapter 8 Summary

### Words and Phrases for Review

[8.1] system of equations
coinciding lines
parallel lines
intersecting lines

[8.2] solution (ordered pair)

[8.6] solution (ordered triple)

[8.8] linear inequality

[8.9] system of inequalities
linear programming

### Brief Reminders

[8.1–8.2] A system of two linear equations has: no solution if the graphs of the lines are parallel, exactly one solution if the graphs of the lines intersect, infinitely many solutions if the graphs of the lines coincide.

[8.3–8.4] **1.** To solve a system of equations, use either the substitution method or the addition-subtraction method.

**2.** If an identity (such as $0 = 0$) results when solving a system, the system has infinitely many solutions.

**3.** If a contradiction (such as $5 = 0$) results when solving a system, the system has no solution.

[8.6] An equation of the form $ax + by + cz = d$ is a linear equation in three variables. Any solution to a system of three linear equations in three variables is an ordered triple of numbers which is a solution of each equation.

[8.8] To graph a linear inequality in two variables in which the $y$-term is present: solve for $y$, putting $y$ on the left, and replace the inequality symbol with an equal sign; graph the equation using a solid line if the inequality is either $\leq$ or $\geq$ and a dashed line if it is either $<$ or $>$; shade the region above the line if the inequality is either $>$ or $\geq$ and below the line if it is either $<$ or $\leq$.

[8.9] To graph a system of two linear inequalities in two variables: graph the first and shade using vertical lines; graph the second and shade using horizontal lines; in addition to possible points on either or both lines, the graph consists of points in the cross-hatched region.

## CHAPTER 8 REVIEW EXERCISES

[8.1] **1.** When the slopes of the lines in a system of equations are unequal, the graphs are ______________ lines.

**2.** When the slopes of the lines in a system of equations are equal but the $y$-intercepts are unequal, the graphs are ______________ lines.

**3.** When the slopes of the lines in a system of equations are equal and the $y$-intercepts are also equal, the graphs are ______________ lines.

**4.** If the $y$-term is missing in both equations in a system, either the graphs coincide or else both are ______________ to the $y$-axis.

**5.** If the $y$ term is missing in only one of the equations in a system, the graphs are ______________ lines.

[8.2] **6.** In a system of two equations in two variables, if the lines intersect, the system has ______________ solution(s).

**7.** In a system of two equations in two variables, if the lines are parallel, the system has ________ solution(s).

**8.** In a system of two equations in two variables, if the lines coincide, the system has ______________ solution(s).

[8.3] **9.** In the process of solving a system of two equations in two variables, suppose we obtain (a) $3 = 0$ or (b) $0 = 0$. What does each tell us about the system?

[8.8] **10.** When graphing a linear inequality such as $2x + 3y \geq 5$ the first step is to solve for ________.

**11.** In the graph of $y > 3x - 1$, would the line with equation $y = 3x - 1$ be (a) solid or (b) dashed?

**12.** In the graph of $y \geq 3x + 2$, the region (a) above or (b) below the line with equation $y = 3x + 2$ would be shaded.

[8.9] **13.** If in a system of two linear inequalities we shade vertically the region corresponding to the first inequality and shade horizontally the region corresponding to the second inequality, the desired graph of the system appears ______________.

[8.2] *Without solving, determine the number of solutions to the given system and justify your answer.*

**14.** $2x - y = 7$
$x - 2y = 7$

**15.** $3x + 4y = 12$
$-6x - 8y = -12$

**16.** $2x + 2y = 2$
$5x + 5y = 5$

**17.** Determine if $(1, -2)$ is a solution to the system.

$$\begin{aligned} 2x + 3y &= -4 \\ -3x + y &= 5 \end{aligned}$$

[8.3–8.4] *Solve.*

**18.** $3x - y = 2$
$-6x + 2y = 2$

**19.** $3x - y = 6$
$2x + 5y = -13$

**20.** $5x + 3y = -1$
$2x - 4y = -16$

**21.** $x + 7y = -1$
$-2x - 14y = 2$

[8.5] **22.** The sum of two numbers is 13 and their difference is 3. Find the numbers.

**23.** A man bought 2 shirts of the same value and 5 pair of pants of the same value for $94. Later he returned and bought 3 shirts and 1 pair of pants for $50. How much does one shirt cost and how much does one pair of pants cost?

**24.** A chemist has one solution which is 15% salt and another which is 20% salt. How many gallons of each should she mix together to obtain 50 gallons of a solution which is 18% salt?

**25.** Gail has a collection of 80 coins made up of nickels and dimes. If the value of the collection is $6.75, how many of each are there?

[8.6] *Solve.*

**26.**
$$\begin{aligned} x + y + z &= 0 \\ 3x - y + z &= -4 \\ 2x + y - 3z &= -10 \end{aligned}$$

**27.**
$$\begin{aligned} x + 2y \qquad &= 3 \\ y - 2z &= -5 \\ 2x \qquad + z &= 5 \end{aligned}$$

**28.** Determine if (−1, 2, 3) is a solution to the system.

$$\begin{aligned} x + y + z &= 4 \\ 3x - y + 2z &= 1 \\ -2x - y + z &= 3 \end{aligned}$$

[8.7] **29.** The average of a student's three scores is 80. The sum of the first and second scores is 150, and the third score is 5 more than the second. Find all three scores.

**30.** Bill has \$10,000 split into three separate investments. Part is invested in stocks which earn 8%, part is in a savings account which earns 7%, and the rest is invested in a business. If the business does well, it will earn 10% and his total earnings on the three investments will amount to \$810. If the business does poorly, he will lose 4% and his total earnings will amount to \$390. How much is invested in each category?

[8.8] *Graph the following in a Cartesian coordinate system.*

**31.** $x - 2y \leq -2$

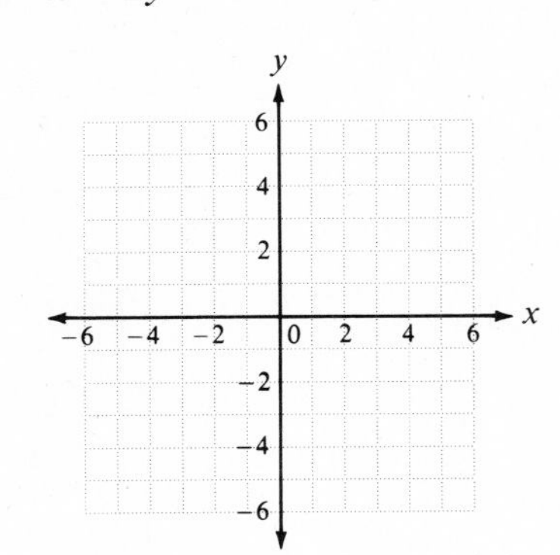


**32.** $2y + 4 > 0$

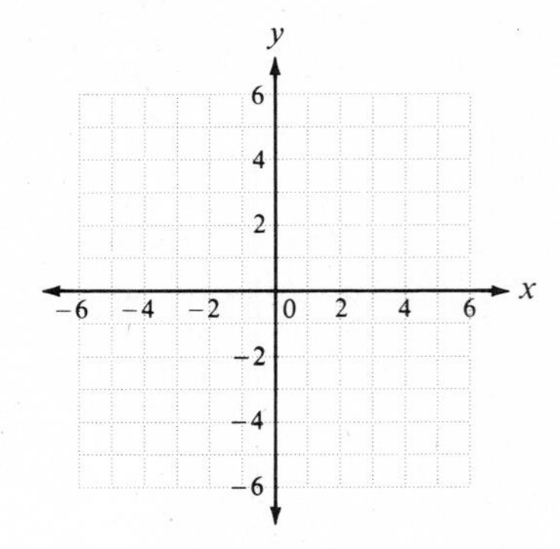


**33.** $3 - 3x \geq 0$

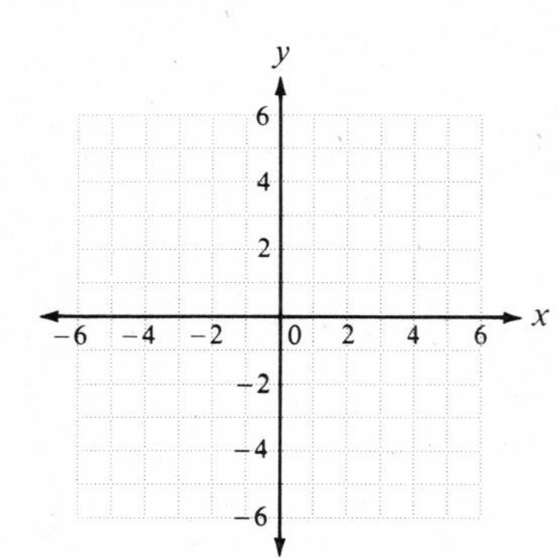


**34.** $3x + 2y < 6$

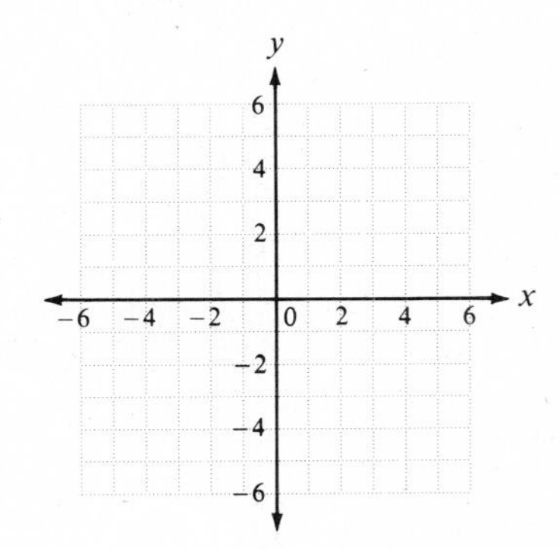


[8.9] *Graph the following in a Cartesian coordinate system.*

**35.** $3x + 2y < 6$
$2y + 4 > 0$

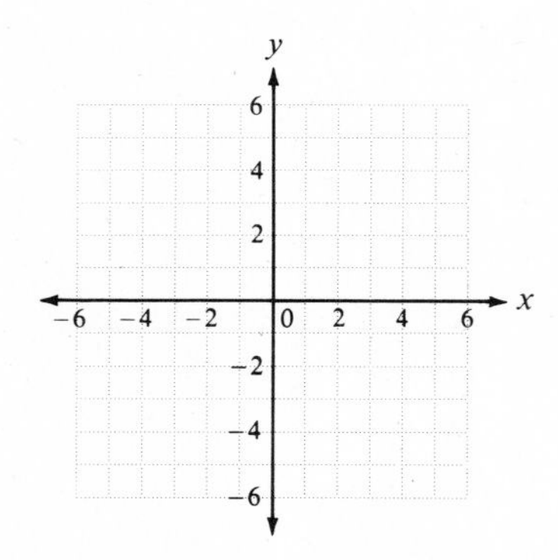


**36.** $y \geq x + 1$
$y < x - 1$

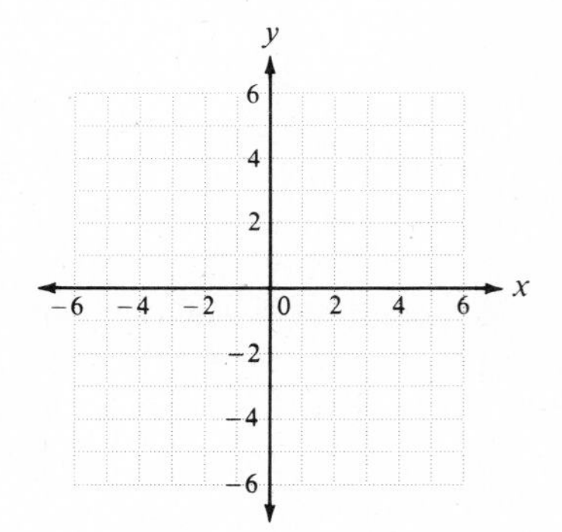


**37.** $1 \leq x \leq 2$
$y \geq 0$
$x - y \geq -1$

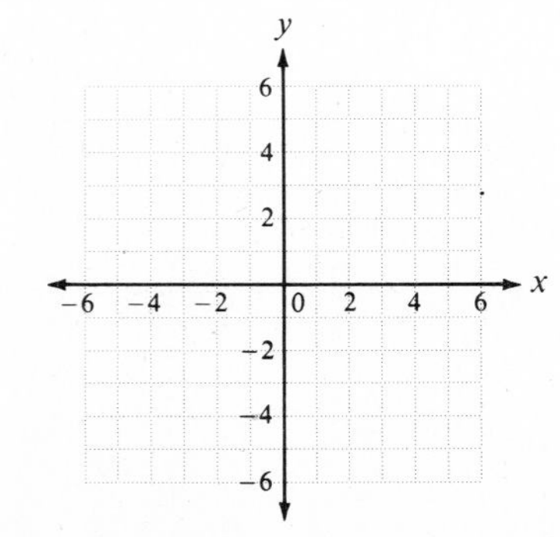

**ANSWERS:** **1.** intersecting **2.** parallel **3.** coinciding **4.** parallel **5.** intersecting **6.** exactly one **7.** no **8.** infinitely many **9.** **(a)** no solution (lines are parallel) **(b)** infinitely many solutions (lines coincide) **10.** $y$ **11.** dashed **12.** above **13.** cross-hatched **14.** exactly one (lines are intersecting) **15.** no solution (lines are parallel) **16.** infinitely many (lines coincide) **17.** no **18.** no solution **19.** $(1, -3)$ **20.** $(-2, 3)$ **21.** infinitely many solutions of the form $\left(x, -\frac{1}{7}x - \frac{1}{7}\right)$, $x$ any number **22.** 8, 5 **23.** shirt is \$12, pants are \$14 **24.** 20 gal of 15%, 30 gal of 20% **25.** 55 dimes and 25 nickels **26.** $(-2, 0, 2)$ **27.** $(1, 1, 3)$ **28.** yes **29.** 65, 85, 90 **30.** \$2000 in stocks, \$5000 in savings, \$3000 in business

**31.**

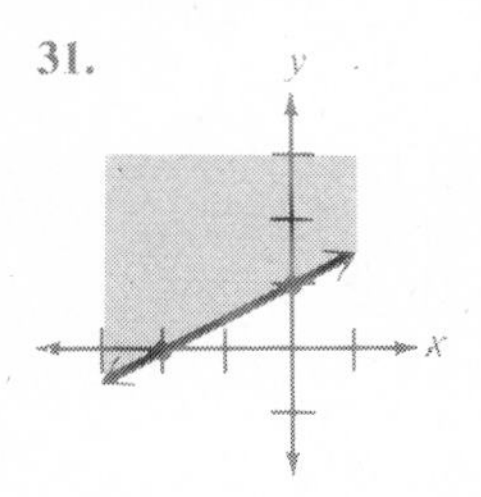


**32.**

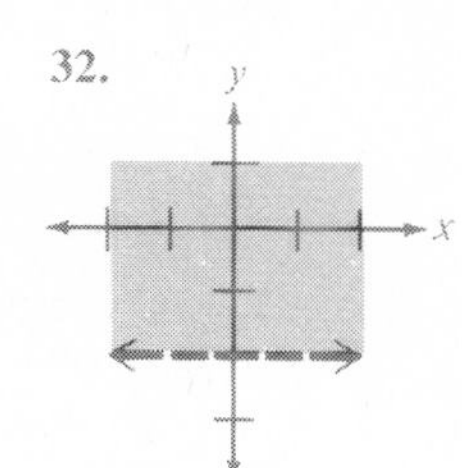


**33.**

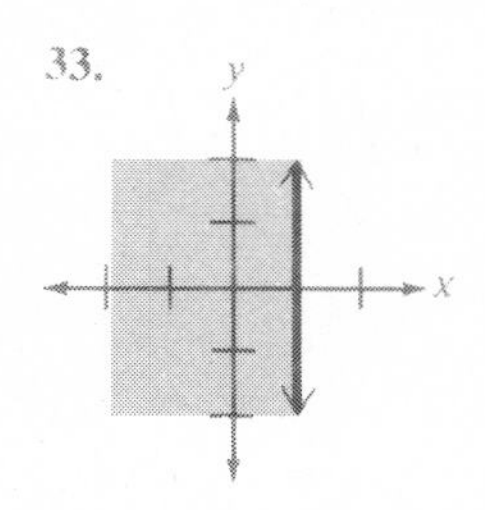


**34.**

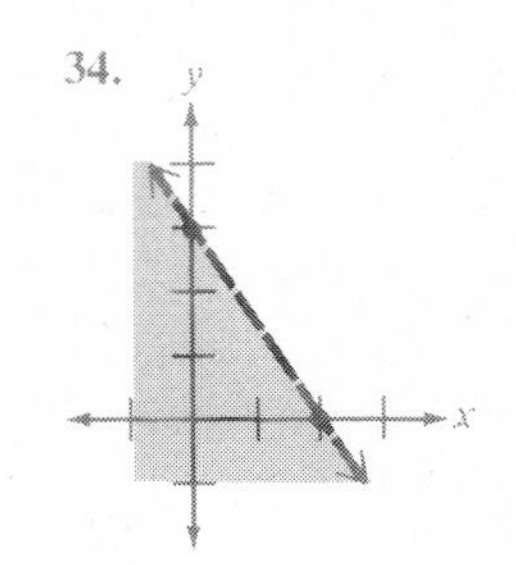


**35.**

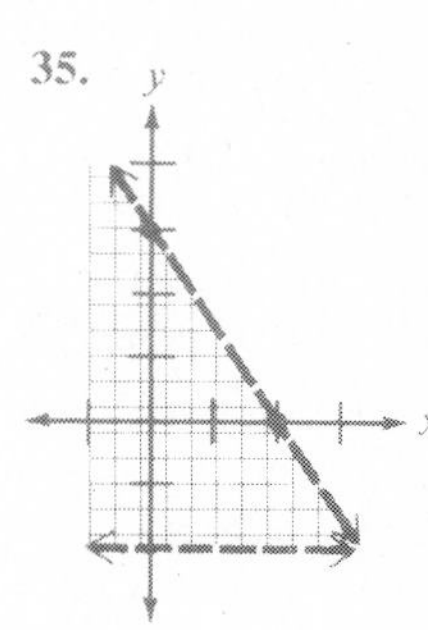


**36.**

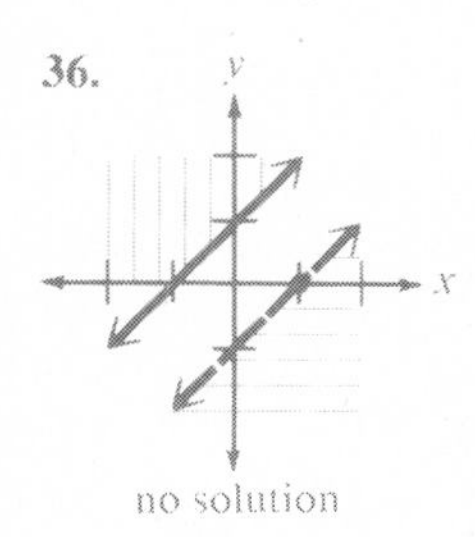


**37.**

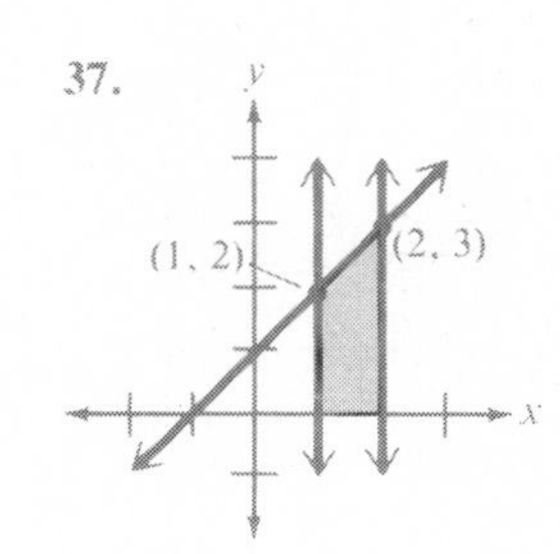

# 9

# Logarithms and Exponentials

## 9.1 Basic Definitions

Historically, logarithms were developed to help carry out complicated numerical calculations. With the advent of computers and hand calculators, however, computation using logarithms is no longer of much interest. Nevertheless, logarithmic and exponential equations and functions are still very important in mathematics today.

In previous chapters we solved equations such as

$$3x + 2 = 7$$

and

$$2x^2 - 5x - 3 = 0.$$

In these equations, the variable does not appear in an exponent. We now consider equations such as

$$3^x = 9$$

in which the variable does appear in an exponent. The equation-solving rules we have already learned are of little help in equations of this type. At present, we must discover the solution by inspection. In this case, the problem is not too difficult, and we would probably recognize that when 3 is raised to the second power, the result is 9. Thus, the solution to the equation is

$$x = 2. \qquad (3^2 = 9)$$

Alternatively, we might give the solution verbally.

$x$ is the exponent on 3 which gives the number 9

Traditionally, the word *logarithm* has been used instead of *exponent* so that the solution might be written

$x$ is the logarithm on 3 which gives the number 9.

Since 3 is called the *base* in the expression $3^x$, the solution can also be written in the form

$x$ is the logarithm of 9 using 3 as the base

or

$x$ is the logarithm to the base 3 of 9.

This final statement is usually symbolized by

$$x = \log_3 9.$$

Basically, we have shown that

$$3^x = 9 \qquad \text{and} \qquad x = \log_3 9$$

are two forms of the same equation, with $3^x = 9$ called the **exponential form** and $x = \log_3 9$ called the **logarithmic form.** That is, they are equivalent equations, with the logarithmic form having the desirable quality of being "solved" for the variable $x$. In this case, one additional equivalent equation is

$$x = 2 \qquad (2 = \log_3 9)$$

which is even more desirable than

$$x = \log_3 9.$$

For now, however, we are not always able to make this one additional simplification.

EXAMPLE 1 Solve $5^x = 10$.

Since $x$ is the logarithm (exponent) to the base 5 of 10, we have

$$x = \log_5 10.$$

In this case, we cannot find a simpler expression for $x$.

Our discussion is summarized in the following definition.

**DEFINITION** Let $a$, $x$, and $y$ be numbers, $a > 0$, $a \neq 1$, which satisfy the exponential equation

$$a^x = y.$$

Then $x$ is the **logarithm to the base $a$ of $y$,** and the equivalent logarithmic equation is

$$x = \log_a y.$$

EXAMPLE 2 Solve $2^x = 32$.

Since $2^5 = 32$, $x = 5$. Alternatively, since $x$ is the logarithm (exponent) to the base 2 of 32, we have

$$x = \log_2 32 \qquad (5 = \log_2 32)$$

**EXAMPLE 3** Write $7^x = c$ in logarithmic form.

Since $x$ is the logarithm (exponent) to the base 7 of $c$,

$$x = \log_7 c.$$

**EXAMPLE 4** Write $u = \log_3 8$ in exponential form.

Since $u$ is the logarithm (exponent) to the base 3 of 8,

$$3^u = 8.$$

When converting an equation from exponential to logarithmic form, or from logarithmic to exponential form, it may help to remember that in both cases the base is written below the level of the logarithm (exponent), as indicated in the diagram.

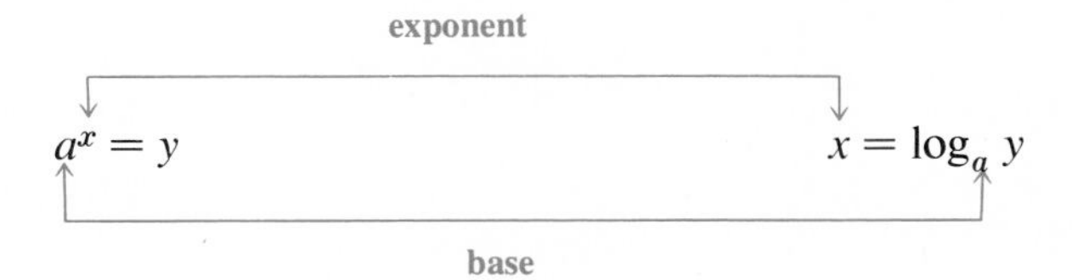


Since solutions to simple exponential equations can often be determined by direct inspection, solutions to some logarithmic equations are best found by converting to exponential form.

**EXAMPLE 5** Determine the numerical value of $x$ in $x = \log_4 16$.

We first convert $x = \log_4 16$ to exponential form.

$$4^x = 16$$

At this point, it might be clear that $x$ must be 2 since $4^2 = 16$. By writing 16 as a power of 4, $16 = 4^2$, we have

$$4^x = 4^2.$$

In this form, it is obvious that $x$ is 2 since the bases in both members of the equation are 4, forcing the exponents in both members to be equal.

**EXAMPLE 6** Determine the numerical value of $x$ in $\log_x \frac{1}{8} = -3$.

Convert to exponential form.

$$x^{-3} = \frac{1}{8}$$

$$x^{-3} = 2^{-3} \qquad \frac{1}{8} = \frac{1}{2^3} = 2^{-3}$$

It is now clear that $x$ is 2 since the exponents in both members of the equation are $-3$, forcing the bases in both members to be equal.

## EXERCISES 9.1

*Convert each equation to logarithmic form.*

**1.** $2^3 = 8$

**2.** $a^5 = 64$

**3.** $5^x = 100$

**4.** $10^3 = 1000$

**5.** $m = n^p$

**6.** $b^y = x$

**7.** $8^0 = 1$

**8.** $7^{1/2} = c$

**9.** $y^a = x$

**10.** $3^{-1} = \frac{1}{3}$

**11.** $10^{-2} = .01$

**12.** $u^{-7} = v$

*Convert each equation to exponential form.*

**13.** $\log_a 3 = x$

**14.** $\log_3 81 = 4$

**15.** $\log_{10} 100 = 2$

**16.** $\log_{10} (.1) = -1$

**17.** $\log_8 4 = \frac{2}{3}$

**18.** $\log_a a = 1$

**19.** $\log_a 1 = 0$

**20.** $u = \log_v 7$

**21.** $m = \log_p n$

**22.** $a = \log_y x$

**23.** $\log_8 32 = \frac{5}{3}$

**24.** $\log_b \left(\frac{1}{b}\right) = -1$

*Determine the numerical value of x in each of the following exponential equations.*

**25.** $2^x = 8$

**26.** $5^x = 25$

**27.** $3^x = 81$

**28.** $5^x = \frac{1}{5}$

**29.** $10^x = 1000$

**30.** $10^x = .0001$

**31.** $x^3 = 27$

**32.** $x^{-1} = \frac{1}{4}$

**33.** $x^{1/2} = 3$

*Determine the numerical value of x by first converting to the equivalent exponential equation.*

**34.** $x = \log_2 16$

**35.** $x = \log_{10} 1000$

**36.** $x = \log_8 2$

**37.** $\log_x 25 = 2$

**38.** $\log_x \frac{1}{4} = -1$

**39.** $\log_x 11 = 1$

**40.** $\log_3 x = 2$

**41.** $\log_4 x = \frac{1}{2}$

**42.** $\log_2 x = 5$

ANSWERS: **1.** $\log_2 8 = 3$ **2.** $\log_a 64 = 5$ **3.** $\log_5 100 = x$ **4.** $\log_{10} 1000 = 3$ **5.** $\log_n m = p$ **6.** $\log_b x = y$ **7.** $\log_8 1 = 0$ **8.** $\log_7 c = \frac{1}{2}$ **9.** $\log_y x = a$ **10.** $\log_3 \frac{1}{3} = -1$ **11.** $\log_{10} (.01) = -2$ **12.** $\log_u v = -7$ **13.** $a^x = 3$ **14.** $3^4 = 81$ **15.** $10^2 = 100$ **16.** $10^{-1} = .1$ **17.** $8^{2/3} = 4$ **18.** $a^1 = a$ **19.** $a^0 = 1$ **20.** $v^u = 7$ **21.** $p^m = n$ **22.** $y^a = x$ **23.** $8^{5/3} = 32$ **24.** $b^{-1} = \frac{1}{b}$ **25.** $x = 3$ **26.** $x = 2$ **27.** $x = 4$ **28.** $x = -1$ **29.** $x = 3$ **30.** $x = -4$ **31.** $x = 3$ **32.** $x = 4$ **33.** $x = 9$ **34.** $x = 4$ **35.** $x = 3$ **36.** $x = \frac{1}{3}$ **37.** $x = 5$ **38.** $x = 4$ **39.** $x = 11$ **40.** $x = 9$ **41.** $x = 2$ **42.** $x = 32$

## 9.2 Properties of Logarithms

Since logarithms are exponents, the basic rules of exponents can be used to develop several useful properties of logarithms.

Suppose that $a$, $x$, and $y$ are numbers, and $a$ is a suitable base for a logarithm ($a > 0$, $a \neq 1$). The product rule for exponents is

$$a^x \cdot a^y = a^{x+y}.$$

Set

$$u = a^x \quad \text{and} \quad v = a^y$$

and write each in logarithmic form.

$$x = \log_a u \quad \text{and} \quad y = \log_a v$$

Then

$$u \cdot v = a^x \cdot a^y = a^{x+y}.$$

Write $u \cdot v = a^{x+y}$ in logarithmic form.

$$\log_a (u \cdot v) = x + y$$

Substitute $\log_a u$ for $x$ and $\log_a v$ for $y$.

$$\log_a (u \cdot v) = \log_a u + \log_a v$$

This gives us the **product rule** for logarithms.

**PRODUCT RULE**

For any base $a$ ($a > 0$, $a \neq 1$) and positive numbers $u$ and $v$,

$$\log_a (u \cdot v) = \log_a u + \log_a v.$$

(The log of a product is the sum of the logs.)

EXAMPLE 1 Express $\log_2 (4 \cdot 8)$ as a sum of logarithms.

$$\log_2 (4 \cdot 8) = \log_2 4 + \log_2 8 \qquad \text{Product rule}$$

In this case, we can verify the equation directly.

$$\begin{aligned} &\log_2 4 \cdot 8 = \log_2 32 = 5 && 2^5 = 32 \\ &\log_2 4 = 2 && 2^2 = 4 \\ &\log_2 8 = 3 && 2^3 = 8 \end{aligned}$$

Hence

$$\log_2 4 \cdot 8 = 5 = 2 + 3 = \log_2 4 + \log_2 8.$$

EXAMPLE 2 Express $\log_a 3 + \log_a w$ as a single logarithm.

$$\log_a 3 + \log_a w = \log_a (3 \cdot w) \qquad \text{Product rule in reverse}$$

CAUTION: Do not conclude that the sum of two logs is the log of a sum.

$$\log_a u + \log_a v \neq \log_a (u + v)$$

For example,

$$\log_a 6 = \log_a 2 \cdot 3 = \log_a 2 + \log_a 3 \neq \log_a (2 + 3) = \log_a 5.$$

If $a$ is a base for a logarithm, and $x$ and $y$ are numbers, the quotient rule for exponents is

$$\frac{a^x}{a^y} = a^{x-y}.$$

As before, we set

$$u = a^x \qquad \text{and} \qquad v = a^y$$

and express these in logarithmic form.

$$x = \log_a u \qquad \text{and} \qquad y = \log_a v$$

Then

$$\frac{u}{v} = \frac{a^x}{a^y} = a^{x-y}.$$

Writing $\frac{u}{v} = a^{x-y}$ in logarithmic form, we obtain

$$\log_a \frac{u}{v} = x - y.$$

Substituting $\log_a u$ for $x$ and $\log_a v$ for $y$,

$$\log_a \frac{u}{v} = \log_a u - \log_a v.$$

This is the **quotient rule** for logarithms.

**QUOTIENT RULE**

For any base $a$ ($a > 0$, $a \neq 1$) and positive numbers $u$ and $v$,

$$\log_a \frac{u}{v} = \log_a u - \log_a v.$$

(The log of a quotient is the difference of the logs.)

**EXAMPLE 3** Express $\log_2 \frac{32}{8}$ as a difference of logarithms.

$$\log_2 \frac{32}{8} = \log_2 32 - \log_2 8 \qquad \text{Quotient rule}$$

In this case, we can verify the equation directly.

$$\log_2 \frac{32}{8} = \log_2 4 = 2 \qquad 2^2 = 4$$

$$\log_2 32 = 5 \qquad 2^5 = 32$$

$$\log_2 8 = 3 \qquad 2^3 = 8$$

Hence

$$\log_2 \frac{32}{8} = 2 = 5 - 3 = \log_2 32 - \log_2 8.$$

**EXAMPLE 4** Express $\log_a 3 - \log_a w$ as a single logarithm.

$$\log_a 3 - \log_a w = \log_a \frac{3}{w} \qquad \text{Quotient rule in reverse}$$

**CAUTION:** Do not conclude that the difference of two logs is the log of the difference.

$$\log_a u - \log_a v \quad \neq \quad \log_a (u - v)$$

For example,

$$\log_a 3 = \log_a \frac{6}{2} = \log_a 6 - \log_a 2 \quad \neq \quad \log_a (6 - 2) = \log_a 4.$$

Assume that

$$u = a^x$$

and raise both sides of this equation to the $c$th power.

$$u^c = (a^x)^c$$

By the power rule for exponents,

$$u^c = (a^x)^c = a^{x \cdot c}.$$

Converting $u^c = a^{x \cdot c}$ to logarithmic form, we obtain

$$\log_a u^c = x \cdot c.$$

But since $u = a^x$, $x = \log_a u$ so that

$$\log_a u^c = x \cdot c = (\log_a u) \cdot c = c \cdot \log_a u.$$

Thus, we have the **power rule** for logarithms.

**POWER RULE**

For any base $a$ ($a > 0$, $a \neq 1$), any positive number $u$ and any number $c$,

$$\log_a u^c = c \cdot \log_a u.$$

(The log of a number to a power is the power times the log of the number.)

**EXAMPLE 5** Express $\log_2 4^3$ as a product.

$$\log_2 4^3 = 3 \cdot \log_2 4 \qquad \text{Power rule}$$

We can verify the above equation directly.

$$\log_2 4^3 = \log_2 64 = 6 \qquad 2^6 = 64$$
$$3 \cdot \log_2 4 = 3 \cdot 2 = 6 \qquad 2^2 = 4$$

**EXAMPLE 6** Express $\log_2 \sqrt[3]{64}$ as a product.

First we write $\sqrt[3]{64}$ as $64^{1/3}$. Then

$$\log_2 \sqrt[3]{64} = \log_2 64^{1/3} = \frac{1}{3} \cdot \log_2 64. \qquad \text{Power rule}$$

The truth of this equation is clear since

$$\log_2 \sqrt[3]{64} = \log_2 4 = 2 \qquad \sqrt[3]{64} = 4$$

and

$$\frac{1}{3} \cdot \log_2 64 = \frac{1}{3} \cdot 6 = 2. \qquad 2^6 = 64$$

Logarithms of expressions which involve more complicated products, quotients, or powers can be simplified by using a combination of the rules.

**EXAMPLE 7** Express $\log_a \frac{x\sqrt{y}}{z^3}$ in terms of logarithms of $x$, $y$, and $z$.

$$\log_a \frac{x\sqrt{y}}{z^3} = \log_a \frac{x \cdot y^{1/2}}{z^3} \qquad \text{Convert radical to fractional exponent}$$
$$= \log_a (x \cdot y^{1/2}) - \log_a z^3 \qquad \text{Quotient rule}$$
$$= \log_a x + \log_a y^{1/2} - \log_a z^3 \qquad \text{Product rule}$$
$$= \log_a x + \frac{1}{2} \cdot \log_a y - 3 \cdot \log_a z \qquad \text{Power rule}$$

A combination of logarithms can often be simplified to a single logarithm by using the various rules in reverse.

**EXAMPLE 8** Express $\frac{1}{3}\log_a x - 5 \cdot \log_a y + \log_a z$ as a single logarithm.

$$\begin{aligned}
\frac{1}{3}\log_a x - 5 \cdot \log_a y + \log_a z &= \log_a x^{1/3} - \log_a y^5 + \log_a z && \text{Always use power rule first} \\
&= \log_a \frac{x^{1/3}}{y^5} + \log_a z && \text{Quotient rule} \\
&= \log_a \left(\frac{x^{1/3}}{y^5} \cdot z\right) && \text{Product rule} \\
&= \log_a \frac{z\sqrt[3]{x}}{y^5} && \text{Simplify}
\end{aligned}$$

**EXAMPLE 9** Given that $\log_a 2 = 0.3010$ and $\log_a 3 = 0.4771$, find the following logarithms.

**(a)** $\log_a 6 = \log_a 2 \cdot 3 = \log_a 2 + \log_a 3 = 0.3010 + 0.4771 = 0.7781$

**(b)** $\log_a 9 = \log_a 3^2 = 2 \cdot \log_a 3 = 2(0.4771) = 0.9542$

**(c)** $\log_a \frac{3}{2} = \log_a 3 - \log_a 2 = 0.4771 - 0.3010 = 0.1761$

## EXERCISES 9.2

*Express as a sum of logarithms.*

**1.** $\log_3 (9 \cdot 80)$ **2.** $\log_5 (25 \cdot u)$ **3.** $\log_c (a \cdot b)$

*Express as a difference of logarithms.*

**4.** $\log_4 \frac{7}{30}$ **5.** $\log_2 \frac{12}{v}$ **6.** $\log_c \frac{a}{b}$

*Express as a product.*

**7.** $\log_2 x^5$ **8.** $\log_3 \sqrt{y}$ **9.** $\log_a b^{-3}$

*Express as a single logarithm.*

**10.** $\log_4 3 + \log_4 8$ **11.** $\log_a 5 + \log_a b$ **12.** $\log_b x + \log_b y$

**13.** $\log_2 5 - \log_2 9$ **14.** $\log_b a - \log_b 8$ **15.** $\log_c x - \log_c y$

*Express in terms of logarithms of x, y, and z.*

**16.** $\log_a \frac{x^3y}{z}$ **17.** $\log_a xy^2\sqrt{z}$

**18.** $\log_a \sqrt{\dfrac{x}{yz}}$

**19.** $\log_a \sqrt[3]{\dfrac{xy^2}{z}}$

*Express as a single logarithm.*

**20.** $2 \cdot \log_a x - \log_a y$

**21.** $3 \cdot \log_a y + 2 \cdot \log_a x$

**22.** $\log_a x - 3 \cdot \log_a y$

**23.** $\dfrac{1}{2} \log_a y - \log_a x + 2 \cdot \log_a z$

**24.** $3 \cdot \log_a x - \log_a y - \log_a z$

**25.** $\dfrac{1}{2} (\log_a x - \log_a y + 3 \cdot \log_a z)$

*Given that $\log_a 3 = 0.4771$ and $\log_a 5 = 0.6990$, find the following logarithms.*

**26.** $\log_a 15$

**27.** $\log_a \dfrac{5}{3}$

**28.** $\log_a 5^2$

**29.** $\log_a 125$

**30.** $\log_a 75$

**31.** $\log_a \dfrac{25}{3}$

*Determine whether the following are true or false.*

**32.** $\log_a u \cdot v = \log_a u + \log_a v$

**33.** $\log_a \dfrac{u}{v} = \log_a u - \log_a v$

**34.** $\dfrac{\log_a u}{\log_a v} = \log_a u - \log_a v$

**35.** $\log_a a = 1$

**36.** $\log_a 1 = 0$

**37.** $\log_a u^c = (\log_a c)(\log_a u)$

**38.** $\log_a (u + v) = \log_a u + \log_a v$

**39.** $\log_a u \cdot v = (\log_a u)(\log_a v)$

**40.** Convert $x^a = y$ to logarithmic form.

**41.** Convert $\log_b m = p$ to exponential form.

*Determine the numerical value of x in the following.*

**42.** $2^x = 128$

**43.** $x = \log_3 81$

**44.** $\log_x \frac{1}{8} = -3$

**45.** $\log_7 x = 2$

---

ANSWERS: **1.** $\log_3 9 + \log_3 80$ **2.** $\log_5 25 + \log_5 u$ **3.** $\log_c a + \log_c b$ **4.** $\log_4 7 - \log_4 30$ **5.** $\log_2 12 - \log_2 v$ **6.** $\log_c a - \log_c b$ **7.** $5 \cdot \log_2 x$ **8.** $\frac{1}{2} \log_3 y$ **9.** $-3 \cdot \log_a b$ **10.** $\log_4 (3 \cdot 8)$ **11.** $\log_a 5b$ **12.** $\log_b xy$ **13.** $\log_2 \frac{5}{9}$ **14.** $\log_b \frac{a}{8}$ **15.** $\log_c \frac{x}{y}$ **16.** $3 \cdot \log_a x + \log_a y - \log_a z$ **17.** $\log_a x + 2 \cdot \log_a y + \frac{1}{2} \log_a z$ **18.** $\frac{1}{2} (\log_a x - \log_a y - \log_a z)$ **19.** $\frac{1}{3} (\log_a x + 2 \cdot \log_a y - \log_a z)$ **20.** $\log_a \frac{x^2}{y}$ **21.** $\log_a y^3x^2$ **22.** $\log_a \frac{x}{y^3}$ **23.** $\log_a \frac{z^2 \sqrt{y}}{x}$ **24.** $\log_a \frac{x^3}{yz}$ **25.** $\log_a \sqrt{\frac{xz^3}{y}}$ **26.** 1.1761 **27.** 0.2219 **28.** 1.3980 **29.** 2.0970 **30.** 1.8751 **31.** 0.9209 **32.** true **33.** true **34.** false **35.** true (convert to exponential form) **36.** true (convert to exponential form) **37.** false **38.** false **39.** false **40.** $a = \log_x y$ **41.** $b^p = m$ **42.** $x = 7$ **43.** $x = 4$ **44.** $x = 2$ **45.** $x = 49$

## 9.3 Common Logarithms

Since $1^y = x$ can be true only if $x$ is 1 ($1^y = 1$ for every $y$), $y = \log_1 x$ has very little importance. Thus, when a base for logarithms is considered, 1 must be omitted. However, every positive number except 1 can be used. Since our number system is based on 10, logarithms to the base 10 were commonly used for computational purposes. As a result, base 10 logarithms are called **common logarithms.** For convenience, we omit the base number 10 in common logarithmic expressions. For example, we write

$$\log 235$$

instead of

$$\log_{10} 235.$$

Before considering the basic concepts associated with common logarithms, we mention two important properties of logarithms in general. Since

$$a^0 = 1 \qquad \text{(for } a \text{ any number except 0)}$$

rewriting this equation in logarithmic form, we have

$$\log_a 1 = 0.$$

Likewise, since

$$a^1 = a \qquad \text{(for } a \text{ any number)}$$

converting to logarithmic form, we have

$$\log_a a = 1.$$

These two results are summarized in the next rule.

If $a$ is any base,

1. $\log_a a = 1 \qquad (a^1 = a)$
2. $\log_a 1 = 0 \qquad (a^0 = 1)$

**EXAMPLE 1** **(a)** $\log_5 5 = 1$ **(b)** $\log_5 1 = 0$ **(c)** $\log_{10} 10 = \log 10 = 1$

The fundamental property that makes common logarithms useful for computation is the fact that every positive number can be expressed as a power of 10. Consider the following list of integer powers of 10 and their logarithmic form.

| Power of 10 | Logarithmic form |
|---|---|
| $10^5 = 100{,}000$ | $\log_{10} 100{,}000 = \log 100{,}000 = 5$ |
| $10^4 = 10{,}000$ | $\log_{10} 10{,}000 = \log 10{,}000 = 4$ |
| $10^3 = 1000$ | $\log_{10} 1000 = \log 1000 = 3$ |
| $10^2 = 100$ | $\log_{10} 100 = \log 100 = 2$ |
| $10^1 = 10$ | $\log_{10} 10 = \log 10 = 1$ |
| $10^0 = 1$ | $\log_{10} 1 = \log 1 = 0$ |
| $10^{-1} = .1$ | $\log_{10} .1 = \log .1 = -1$ |
| $10^{-2} = .01$ | $\log_{10} .01 = \log .01 = -2$ |
| $10^{-3} = .001$ | $\log_{10} .001 = \log .001 = -3$ |
| $10^{-4} = .0001$ | $\log_{10} .0001 = \log .0001 = -4$ |
| $10^{-5} = .00001$ | $\log_{10} .00001 = \log .00001 = -5$ |

Several observations can be made from the list. The logarithm of 1 is 0, the logarithm of a number greater than 1 is positive, and the logarithm of a number less than 1 is negative. On the assumption that numbers between two integer powers of 10 have logarithms between the logarithms of the integer powers of 10 (that is, between the two integers), the log of a number between 1 and 10 must be a number between 0 and 1. For example,

$$10^0 = 1 < 5 < 10 = 10^1$$

so that

$$0 = \log 1 < \log 5 < \log 10 = 1.$$

Thus, the logarithm of 5 must be a positive decimal between 0 and 1. Similarly, the logarithm of a number between 10 and 100 must be a number between 1 and 2. That is, it must be 1 plus a positive decimal between 0 and 1.

Thus, we can accept the fact, based on these observations, that the common logarithm of any positive number consists of two parts, an integer and a positive decimal between 0 and 1. The integer part is called the **characteristic** of the logarithm, and the positive-decimal part is called the **mantissa** of the logarithm.

EXAMPLE 2 Give the characteristic and mantissa of the following logarithms.

**(a)** $\log 123 = 2.0899$

2 is the characteristic and .0899 is the mantissa.

**(b)** $\log 1000 = 3.0000$

3 is the characteristic and .0000 is the mantissa.

Determining the logarithm of any positive number, that is, determining the characteristic and the mantissa of the logarithm, is much simpler if the number is first expressed in scientific notation. Recall from Chapter 4 that a number is expressed in scientific notation if it is written as the product of a number between 1 and 10, and an integer power of 10. For example,

$$123 = 1.23 \times 10^2 \qquad \text{and} \qquad .123 = 1.23 \times 10^{-1}$$

are expressed in scientific notation. By the product rule for logarithms,

$$\begin{aligned} \log 123 = \log (1.23 \times 10^2) &= \log 1.23 + \log 10^2 \\ &= \log 1.23 + 2 \qquad \text{Log } 10^2 = 2 \end{aligned}$$

and

$$\begin{aligned} \log .123 = \log (1.23 \times 10^{-1}) &= \log 1.23 + \log 10^{-1} \\ &= \log 1.23 + (-1) \qquad \text{Log } 10^{-1} = -1 \\ &= \log 1.23 - 1. \end{aligned}$$

Thus, to find the logarithm of any number, we only need to be able to find logarithms of numbers between 1 and 10. As we saw, logarithms of such numbers are always positive decimals between 0 and 1. If we are given that

$$\log 1.23 = .0899$$

then

$$\begin{aligned} \log 123 = \log (1.23 \times 10^2) &= \log 1.23 + \log 10^2 \\ &= \log 1.23 + 2 = .0899 + 2 \\ &= 2.0899. \end{aligned}$$

In other words, .0899, the logarithm of 1.23, is the mantissa of log 123 while 2 is its characteristic. Table 1 in the back of this text lists the logarithms of numbers between 1 and 10. This table, together with the above observations, enable us to determine logarithms of all numbers (at least close approximations to these logarithms). The left-hand column in Table 1 shows numbers 1.0 through 9.9 while the numbers 0 through 9 head each of the other columns. To find log 1.23, we look down the left column to 1.2, read across that row to the column headed 3, and find the decimal .0899.

| $n$ | 0 | 1 | 2 | 3 | 4 | 5 | 6 | 7 | 8 | 9 |
|---|---|---|---|---|---|---|---|---|---|---|
| 1.0 | .0000 | .0043 | .0086 | .0128 | .0170 | .0212 | .0253 | .0294 | .0334 | .0374 |
| 1.1 | .0414 | .0453 | .0492 | .0531 | .0569 | .0607 | .0645 | .0682 | .0719 | .0755 |
| 1.2 | .0792 | .0828 | .0864 | .0899 | .0934 | .0969 | .1004 | .1038 | .1072 | .1106 |
| 1.3 | .1139 | .1173 | .1206 | .1239 | .1271 | .1303 | .1335 | .1367 | .1399 | .1430 |
| 1.4 | .1461 | .1492 | .1523 | .1553 | .1584 | .1614 | .1644 | .1673 | .1703 | .1732 |

**TO FIND THE COMMON LOGARITHM OF A GIVEN NUMBER IN A TABLE**

1. Write the number in scientific notation.
2. Use Table 1 to find the logarithm of the number between 1 and 10 (the mantissa of the logarithm of the original number).
3. The logarithm of the original number is the characteristic (the exponent on 10) plus the mantissa.

**EXAMPLE 3** Find log 45,600.

$$\begin{aligned} \log 45600 &= \log(4.56 \times 10^4) && \text{Scientific notation} \\ &= \log 4.56 + \log 10^4 && \text{Log } uv = \log u + \log v \\ &= \log 4.56 + 4 && \text{Log } 10^4 = 4 \\ &= .6590 + 4 && \text{From Table 1} \\ &= 4.6590 \end{aligned}$$

**EXAMPLE 4** Find log 4.56.

$$\begin{aligned} \log 4.56 &= \log(4.56 \times 10^0) \\ &= \log 4.56 + \log 10^0 \\ &= .6590 + 0 \\ &= 0.6590 \end{aligned}$$

**EXAMPLE 5** Find log 0.00456.

$$\begin{aligned} \log 0.00456 &= \log(4.56 \times 10^{-3}) \\ &= \log 4.56 + \log 10^{-3} \\ &= .6590 + (-3) \\ &= .6590 - 3 && \textit{Not } -3.6590 \end{aligned}$$

In Example 5, notice that we left the characteristic separate from the mantissa. If we subtract, we have

$$.6590 - 3 = -2.3410.$$

Although $-2.3410$ is a correct form for the logarithm of .00456, it does not directly display the characteristic $(-3)$ and positive mantissa (.6590). When Table 1 is used, it is wise to leave the logarithm in the earlier form. However, if we use a calculator to find logarithms, the result is displayed in the latter form. That is, if you enter the number .00456 on a calculator with a log key and press the log button, the display will show $-2.3410352$ which rounds off to $-2.3410$.

You should be able to convert from one form of a logarithm to the other. Changing from

$$.6590 - 3 \quad \text{to} \quad -2.3410$$

is simply a matter of subtracting. Converting from

$$-2.3410 \quad \text{to} \quad .6590 - 3$$

is a bit more challenging. In this case we add and subtract 3 (any integer greater than the absolute value of −2.3410 will work, and 3 is the smallest such integer).

$$\begin{array}{r} 3.0000 - 3 \\ -2.3410 \phantom{{}-3} \\ \hline .6590 - 3 \end{array}$$

Suppose we add and subtract 10.

$$\begin{array}{r} 10.0000 - 10 \\ -2.3410 \phantom{{}-10} \\ \hline 7.6590 - 10 \end{array}$$

We clearly obtain the same result since

$$7.6590 - 10 = 7 + .6590 - 10 = .6590 + 7 - 10 = .6590 - 3.$$

EXAMPLE 6 Find log 0.0000287.

If you have a calculator with a log button, simply enter 0.0000287 and press the button. The display will show −4.5421181, which rounds to −4.5421 correct to four decimal places. To convert to the form which displays the positive mantissa and the characteristic, add and subtract 5.

$$\begin{array}{r} 5.0000 - 5 \\ -4.5421 \phantom{{}-5} \\ \hline .4579 - 5 \end{array}$$

You may also use Table 1.

$$\begin{aligned} \log 0.0000287 &= \log (2.87 \times 10^{-5}) \\ &= \log 2.87 + (-5) \\ &= .4579 - 5 \end{aligned}$$

Converting to the alternate form by subtraction, we obtain

$$-4.5421.$$

It is important to keep in mind that log 0.0000287 = −4.5421 means that $10^{-4.5421} = 0.0000287$.

EXAMPLE 7 Find log 0.00002869.

Using a calculator, we enter 0.00002869, press the log button, and obtain

$$-4.5422695.$$

Since Table 1 gives logs for three-digit numbers and our number has four significant digits (2, 8, 6, and 9), we have a special problem. We could simply round-off the given number to three places, obtaining

$$0.0000287,$$

and approximate log 0.00002869 by log 0.0000287, given in Example 6. This approximation is not far from the calculator value given above. Another approach to the problem is to make the approximation using the method of *linear interpolation.* This process enables us to approximate logarithms of four-place numbers using a table for three-digit numbers. However, with pocket calculators readily available, interpolation has lost some of its historical importance. As a result, we do not consider it in this text and we restrict numbers to three places so that calculator values and table values can be compared.

## EXERCISES 9.3

**1.** Common logarithms use as a base the number ________.

**2.** Instead of writing $\log_{10} 23.5$ we write ____________.

**3.** **(a)** $\log_{10} 1 =$ ________ **(b)** $\log_{10} 10 =$ ________ **(c)** $\log_{10} 10^2 =$ ________

**(d)** $\log_{10} 10^{-1} =$ ________ **(e)** $\log 1000 =$ ________ **(f)** $\log .01 =$ ________

**4.** **(a)** $\log_a a =$ ________ **(b)** $\log_a 1 =$ ________

**5.** The common logarithm of a number can always be expressed in two parts: an integer called the **(a)** ____________ of the logarithm and a decimal part between 0 and 1 called the **(b)** ____________ of the logarithm.

**6.** Given that $\log 73.2 = 1.8645$, the characteristic is **(a)** ________ and the mantissa is **(b)** ________.

**7.** Given that $\log .0732 = 0.8645 - 2$, the characteristic is **(a)** ________ and the mantissa is **(b)** ________.

**8.** Given that $\log 7.32 = 0.8645$, the characteristic is **(a)** ________ and the mantissa is **(b)** ________.

*Using Table 1 or your calculator, find the following logarithms.*

**9.** log 685

**10.** log 6.85

**11.** log .00759

**12.** log 7590

**13.** log .376

**14.** log 5
[*Hint:* 5 = 5.00]

**15.** log .5

**16.** log 500

*Convert each logarithm to the alternate form which displays the positive mantissa and the characteristic.*

**17.** −3.2751

**18.** −5.1258

**19.** −2.0013

**20.** −0.3176

*Express as a single logarithm.*

**21.** $\frac{1}{3}(\log_a x - 2 \cdot \log_a y)$

**22.** $3 \cdot \log_a x - \log_a y - 4 \cdot \log_a z$

*Express in terms of logarithms of x, y, and z.*

**23.** $\log_a \frac{x^3\sqrt{z}}{y^5}$

**24.** $\log_a \sqrt{xyz}$

**25.** **(a)** log 2 = ________ **(b)** log 3 = ________

**(c)** log 5 = ________ **(d)** log 6 = ________

**(e)** Does log (2 + 3) = log 2 + log 3?

**(f)** Does log (2 · 3) = (log 2)(log 3)?

**(g)** Does log (2 · 3) = log 2 + log 3?

**(h)** Does $\log \frac{6}{3} = \frac{\log 6}{\log 3}$?

**(i)** Does $\log \frac{6}{3} = \log 6 - \log 3$?

---

ANSWERS: **1.** 10 **2.** log 23.5 **3.** (a) 0 (b) 1 (c) 2 (d) −1 (e) 3 (f) −2 **4.** (a) 1 (b) 0 **5.** (a) characteristic (b) mantissa **6.** (a) 1 (b) .8645 **7.** (a) −2 (b) .8645 **8.** (a) 0 (b) .8645 **9.** 2.8357 **10.** 0.8357 **11.** 0.8802 − 3 or −2.1198 **12.** 3.8802 **13.** 0.5752 − 1 or −0.4248 **14.** 0.6990 **15.** 0.6990 − 1 or −0.3010 **16.** 2.6990 **17.** 0.7249 − 4 **18.** 0.8742 − 6 **19.** 0.9987 − 3 **20.** 0.6824 − 1 **21.** $\log_a \sqrt[3]{\frac{x}{y^2}}$ **22.** $\log_a \frac{x^3}{yz^4}$ **23.** $3 \cdot \log_a x + \frac{1}{2}\log_a z - 5 \cdot \log_a y$ **24.** $\frac{1}{2}(\log_a x + \log_a y + \log_a z)$ **25.** (a) 0.3010 (b) 0.4771 (c) 0.6990 (d) 0.7782 (e) no (f) no (g) yes (h) no (i) yes

## 9.4 Antilogarithms

In the preceding section, we concentrated on finding the logarithm of a given number. We are now interested in determining a number given its logarithm. A number that has a given logarithm is called an **antilogarithm** of the number. In the equation

$$\log n = x$$

$x$ is the logarithm of $n$, and $n$ is the antilogarithm of $x$. In exponential form,

$$n = 10^x.$$

Finding antilogarithms (antilogs) using a calculator is merely a matter of entering the known number $x$ and pressing the antilog key ($10^x$ key, $y^x$ key with $y = 10$, or inverse and then log keys). When Table 1 is used to find an antilog, the given logarithm must be in the standard form with positive mantissa and integer characteristic.

**EXAMPLE 1** Find the antilog of 2.7832. (That is, find $10^{2.7832}$.)

We must find $n$ in the equation

$$\begin{aligned}\log n &= 2.7832\\ &= .7832 + 2.\end{aligned}$$

The mantissa is .7832, and the characteristic is 2. In Table 1 we find .7832 in the row headed by 6.0 and in the column headed by 7.

| $n$ | 0 | 1 | 2 | 3 | 4 | 5 | 6 | 7 | 8 | 9 |
|---|---|---|---|---|---|---|---|---|---|---|
| 5.5 | .7404 | .7412 | .7419 | .7427 | .7435 | .7443 | .7451 | .7459 | .7466 | .7474 |
| 5.6 | .7482 | .7490 | .7497 | .7505 | .7513 | .7520 | .7528 | .7536 | .7543 | .7551 |
| 5.7 | .7559 | .7566 | .7574 | .7582 | .7589 | .7597 | .7604 | .7612 | .7619 | .7627 |
| 5.8 | .7634 | .7642 | .7649 | .7657 | .7664 | .7672 | .7679 | .7686 | .7694 | .7701 |
| 5.9 | .7709 | .7716 | .7723 | .7731 | .7738 | .7745 | .7752 | .7760 | .7767 | .7774 |
| 6.0 | .7782 | .7789 | .7796 | .7803 | .7810 | .7818 | .7825 | .7832 | .7839 | .7846 |
| 6.1 | .7853 | .7860 | .7868 | .7875 | .7882 | .7889 | .7896 | .7903 | .7910 | .7917 |
| 6.2 | .7924 | .7931 | .7938 | .7945 | .7952 | .7959 | .7966 | .7973 | .7980 | .7987 |
| 6.3 | .7993 | .8000 | .8007 | .8014 | .8021 | .8028 | .8035 | .8041 | .8048 | .8055 |
| 6.4 | .8062 | .8069 | .8075 | .8082 | .8089 | .8096 | .8102 | .8109 | .8116 | .8122 |

$$\begin{aligned}n &= 6.07 \times 10^2\\ &= 607\end{aligned}$$

Thus,

$$10^{2.7832} = 607.$$

When finding logarithms or antilogarithms, given or computed values may not be in Table 1. In such cases, a calculator is clearly desirable. However, Table 1 can still be used to make reasonable approximations by simply using the closest values.

**TO FIND THE ANTILOG OF A NUMBER IN A TABLE**

1. Separate the mantissa (always positive) from the characteristic.
2. Find the mantissa closest to the given mantissa in the table and record the corresponding number.
3. Multiply the number by 10 raised to the power equal to the characteristic.

**EXAMPLE 2** Find the antilog of 1.5765.

$$\begin{aligned} \log n &= 1.5765 \\ &= .5765 + 1 \\ n &= 3.77 \times 10^1 \qquad \text{.5765 is closest to .5763} \\ &= 37.7 \end{aligned}$$

Remember that finding the antilog of 1.5765 means finding $10^{1.5765} = 10^1 \cdot 10^{.5765}$. The table gives $10^{.5765} = 3.77$ which when multiplied by $10^1$ becomes 37.7.

**EXAMPLE 3** Find the antilog of $-2.4783$. Remember that $-2.4783 \neq 0.4783 - 2$.

$$\begin{aligned} \log n &= -2.4783 \\ &= 0.5217 - 3 \\ n &= 3.32 \times 10^{-3} \\ &= .00332 \end{aligned}$$

Add and subtract 3, as below:

$$\begin{array}{r} 3.0000 - 3 \\ -2.4783 \\ \hline .5217 - 3 \end{array}$$

Thus,

$$10^{-2.4783} = .00332.$$

In Example 3, we first expressed the logarithm in the form having the positive mantissa (.5217) and the characteristic (−3) readily identifiable. If a calculator is available, this step is unnecessary. We would simply enter −2.4783 and press the antilog key.

## EXERCISES 9.4

1. In the equation $\log m = y$, $y$ is called the **(a)** ______________ of $m$, and $m$ is called the **(b)** ______________ of $y$.
2. When using Table 1 to find the antilog of 3.9974, we find the number whose mantissa is **(a)** ________ and multiply it by 10 to the **(b)** ________ power.

*Using Table 1 or your calculator, find the antilogarithm of each number.*

3. 2.8887
4. .6571 − 1

**5.** 4.9754

**6.** .9091 − 3

**7.** 1.7545

**8.** −1.4321
[*Hint:* Convert first.]

**9.** −0.9511

**10.** 9.8400 − 10
[*Hint:* 9.8400 − 10 is the same as 0.8400 − 1.]

**11.** 7.5372 − 10

**12.** −7.5372

*Using Table 1 or your calculator, find the following logarithms.*

**13.** log .484

**14.** log 484

**15.** Given that log 27.5 = 1.4393,

(a) the mantissa is ________ (b) the characteristic is ________

(c) the antilog of 1.4393 is ________

(d) $10^{1.4393}$ = ________.

---

ANSWERS: 1. (a) logarithm (b) antilogarithm 2. (a) .9974 (b) 3rd 3. 774 4. .454 5. 94,500 6. .00811 (closest to .9090 in the table) 7. 56.8 8. .0370 9. .112 10. .692 11. .00345 12. .0000000290 13. .6848 − 1 or −0.3152 14. 2.6848 15. (a) .4393 (b) 1 (c) 27.5 (d) 27.5

## 9.5 Computation Using Logarithms

In the past, difficult and time-consuming calculations such as

$$\frac{(0.0325) \cdot (42.3)^2}{\sqrt[3]{1.07}}$$

were often done using logarithms. Products, quotients, and powers of decimals can be found by converting to logarithms and adding, subtracting, and multiplying the results. Since, for example, adding two decimals is easier than multiplying them, logarithmic calculation was a time-saving device. Today, however, the hand calculator has made logarithmic computation nearly obsolete. Nevertheless, in order to understand better the rules of logarithms and to learn more about logarithms in general, we shall briefly illustrate the techniques of logarithmic computation. The problems we have chosen cannot be readily solved on some calculators.

**EXAMPLE 1** Use logarithms to find $N$ if $N = \dfrac{(1.06)^{20}}{1.35}$.

$$N = \frac{(1.06)^{20}}{1.35}$$

$$\log N = \log \frac{(1.06)^{20}}{1.35} \qquad \text{If } x = y \text{ then } \log x = \log y$$

$$= \log (1.06)^{20} - \log 1.35 \qquad \text{Quotient rule}$$
$$= 20 \cdot \log 1.06 - \log 1.35 \qquad \text{Power rule}$$
$$= 20 \cdot (0.0253) - 0.1303$$
$$= 0.3757 \qquad \text{Remember that this is } \log N$$

$N$ is the antilog of 0.3757.

$$N = 2.38 \times 10^0 \qquad \text{Using the closest value}$$
$$= 2.38$$

**EXAMPLE 2** Use logarithms to find $N$ if $N = \sqrt[5]{21.5}\,(0.0349)^3$.

$$N = (21.5)^{1/5}\,(0.0349)^3 \qquad \text{Convert to fractional exponent}$$
$$\log N = \log [(21.5)^{1/5}\,(0.0349)^3] \qquad \text{Take log of both sides}$$
$$= \log (21.5)^{1/5} + \log (0.0349)^3 \qquad \text{Product rule}$$

$$= \frac{1}{5} \log 21.5 + 3 \cdot \log 0.0349 \qquad \text{Power rule}$$

$$= \frac{1}{5} (1.3324) + 3(0.5428 - 2) \qquad \text{Watch the characteristics}$$

$$= 0.2665 + 1.6284 - 6$$
$$= 1.8949 - 6$$
$$= 0.8949 - 5 \qquad \text{This is } \log N, \textit{ not } N$$
$$N = 7.85 \times 10^{-5} \qquad \text{Find the antilog: } 10^{.8949} = 7.85$$
$$= .0000785$$

The above examples show the usefulness of logarithms for power calculations when a calculator with a $y^x$ key is not available. When performing logarithmic calculations, it is important to write the problem in detail, as in these examples. If this is not done, we may find ourselves multiplying or dividing when we should be adding or subtracting, and we might not recognize that in the last step, an antilog must be found.

**EXAMPLE 3** Use logarithms and Table 1 to find $N$ if $N = \dfrac{\sqrt[3]{.0943}}{251}$.

$$N = \frac{(.0943)^{1/3}}{251}$$

$$\log N = \log \frac{(.0943)^{1/3}}{251}$$

$$= \log (.0943)^{1/3} - \log 251$$

$$= \frac{1}{3} \log (.0943) - \log 251$$

$$= \frac{1}{3} (0.9745 - 2) - 2.3997$$

If we divide $0.9745 - 2$ by 3, we no longer have an integer characteristic, since 2 is not evenly divisible by 3. Thus, we must convert $0.9745 - 2$ to $1.9745 - 3$ and then divide by 3 to obtain

$$\log N = (0.6582 - 1) - (2.3997).$$

Next, we change the form of the characteristic in $0.6582 - 1$ to $9.6582 - 10$ in order to subtract 2.3997 and keep the mantissa positive.

$$\begin{array}{r} 9.6582 - 10 \\ -\ 2.3997 \phantom{- 10} \\ \hline 7.2585 - 10 \end{array}$$

Then,

$$\begin{aligned} \log N &= 7.2585 - 10 \\ &= 0.2585 - 3 \\ N &= 1.81 \times 10^{-3} \qquad \text{Find the antilog: } 10^{.2585} = 1.81 \\ &= .00181. \end{aligned}$$

The two difficulties we met in the preceding example are not as bothersome when a calculator is used to find logarithms. In such cases, the positive mantissas necessary for using Table 1 need not be kept.

## EXERCISES 9.5

*Use logarithms to find N. (Follow the details of the examples. The first two problems are started.)*

**1.** $N = \dfrac{(1.07)^{30}}{5.13}$

$$\begin{aligned} \log N &= \log \frac{(1.07)^{30}}{5.13} \\ &= \log (1.07)^{30} - \log 5.13 \\ &= 30 \cdot \log 1.07 - \log 5.13 \\ &= 30(\qquad) - (\qquad) \end{aligned}$$

**2.** $N = \sqrt[7]{32.8}\,(0.491)^3$

$$\begin{aligned} \log N &= \log [\sqrt[7]{32.8}\,(0.491)^3] \\ &= \log (32.8)^{1/7} + \log (0.491)^3 \\ &= \frac{1}{7} \log (32.8) + 3 \cdot \log (0.491) \\ &= \end{aligned}$$

**3.** $N = (1.04)^{25}(3.78)$

**4.** $N = [(23.5)(1.08)]^{1/5}$

**5.** $N = \dfrac{(30700)^{1/3}}{(22.5)^3}$

**6.** $N = [\sqrt{1.03}(1.08)^3]^{2/3}$

**7.** $N = \dfrac{(0.00351)^{1/4}}{(4.12)^2}$

**8.** $N = \dfrac{0.005}{(0.0372)^{1/3}}$

**9.** Given that log 2.13 = 0.3284,

**(a)** $10^{0.3284} =$ ________ **(b)** the characteristic is ________

**(c)** the antilog of 0.3284 is ________ **(d)** the mantissa is ________.

---

ANSWERS: **1.** 1.48 **2.** .195 **3.** 10.1 **4.** 1.91 **5.** .00275 **6.** 1.18 **7.** .0143 **8.** .0150
**9.** (a) 2.13 (b) 0 (c) 2.13 (d) .3284

## 9.6 Logarithm Base Conversion

Logarithm bases other than 10 occur in science and mathematics. Next to 10, the base of common logarithms, the most widely used base is $e$. The number $e$ is an irrational number, approximately equal to 2.71828, and is important in calculus and higher mathematics. Since $e$ originates in a natural way in calculus, logarithms to the base $e$ are called **natural logarithms.** Some calculators and books use

$$\ln x \quad \text{for} \quad \log_e x.$$

Although 10 and $e$ are the most widely used bases, any positive number except 1 can be used as a base. To determine the relationship between logarithms to different bases, for example, $\log_a x$ and $\log_b x$, we let

$$u = \log_b x.$$

Then

$$b^u = x.$$

Take the logarithm to the base $a$ of both sides.

$$\log_a b^u = \log_a x \qquad \text{If } w = z \text{ then } \log_a w = \log_a z$$

$$u \log_a b = \log_a x \qquad \text{Power rule}$$

$$(\log_b x)(\log_a b) = \log_a x \qquad \text{Substitute } \log_b x \text{ for } u$$

$$\log_b x = \frac{\log_a x}{\log_a b} \qquad \textit{Not } \log_a \frac{x}{b} = \log_a x - \log_a b$$

A special case of this formula results by letting $x = a$.

$$\log_b x = \frac{\log_a x}{\log_a b}$$

$$\log_b a = \frac{\log_a a}{\log_a b} \qquad \text{Substitute } a \text{ for } x$$

$$\log_b a = \frac{1}{\log_a b} \qquad \text{Log}_a\ a = 1$$

If $a$ and $b$ are any two bases, and $x > 0$, then

$$\log_b x = \frac{\log_a x}{\log_a b} \quad \text{and} \quad \log_b a = \frac{1}{\log_a b}.$$

**EXAMPLE 1** Express $\log_{100} x$ in terms of $\log_{10} x$ and simplify.

$$\log_{100} x = \frac{\log_{10} x}{\log_{10} 100} = \frac{\log x}{\log 10^2} = \frac{\log x}{2} = \frac{1}{2} \log x$$

**EXAMPLE 2** Express $\log_{16} 4$ in terms of $\log_2 4$ and simplify.

$$\log_{16} 4 = \frac{\log_2 4}{\log_2 16} = \frac{\log_2 2^2}{\log_2 2^4} = \frac{2 \cdot \log_2 2}{4 \cdot \log_2 2} = \frac{2 \cdot 1}{4 \cdot 1} = \frac{2}{4} = \frac{1}{2}$$

**EXAMPLE 3** Express $\log_2 1000$ in terms of $\log_{10} 1000$ and simplify.

$$\log_2 1000 = \frac{\log_{10} 1000}{\log_{10} 2} = \frac{\log 10^3}{\log 2} = \frac{3}{\log 2}$$

**EXAMPLE 4** Express $\log_5 9$ in terms of base 9 logarithms.

$$\log_5 9 = \frac{1}{\log_9 5}$$

**EXAMPLE 5** Determine $\log_3 25$ by using common logarithms.

$$\log_3 25 = \frac{\log_{10} 25}{\log_{10} 3} = \frac{\log 25}{\log 3}$$

$$= \frac{1.3979}{0.4771} \qquad \textit{Divide} \text{ the logarithms; do not subtract}$$

$$\approx 2.93$$

The result in Example 5 does seem reasonable: $3^{2.93}$ would be close to 25 since $3^3 = 27$. This example points out that logarithms to any base can easily be determined by using a common logarithm table or a calculator with a log key.

## EXERCISES 9.6

**1.** Express $\log_{1000} x$ in terms of $\log_{10} x$ and simplify.

**2.** Express $\log_{32} 8$ in terms of $\log_2 8$ and simplify.

**3.** Express $\log_2 100$ in terms of $\log_{10} 100$ and simplify.

**4.** Express $\log_7 12$ in terms of base 12 logarithms.

**5.** Express $\log_7 12$ in terms of common logarithms and evaluate.

**6.** Determine $\log_5 30$ by using common logarithms.

**7.** Determine $\log_7 1.32$ by using common logarithms.

**8.** Express $\log_e x = \ln x$ in terms of common logarithms by using 2.72 as an approximate value for $e$.

*Use logarithms to find N.*

**9.** $N = (1.08)^{15}$

**10.** $N = \sqrt[6]{0.437}$

---

ANSWERS: **1.** $\frac{1}{3} \log x$ **2.** $\frac{3}{5}$ **3.** $\frac{2}{\log 2}$ **4.** $\frac{1}{\log_{12} 7}$ **5.** 1.28 **6.** 2.11 **7.** .143 **8.** $\frac{\log x}{.4346}$ **9.** 3.17 **10.** .871

## 9.7 Logarithmic and Exponential Equations

Equations that contain logarithms of expressions involving the variable are called **logarithmic equations.** Solving these equations requires a thorough knowledge of the properties of logarithms.

**EXAMPLE 1** Solve the logarithmic equation for $x$.

$$\log(x+10) - \log(x+1) = 1$$

$$\log \frac{x+10}{x+1} = 1 \qquad \text{Use quotient rule on left side}$$

$$\frac{x+10}{x+1} = 10^1 = 10 \qquad \text{Convert to exponential form}$$

$$x + 10 = 10(x+1) \qquad \text{Clear fraction}$$

$$x + 10 = 10x + 10$$

$$0 = 9x$$

$$0 = x$$

Check: $\log(0+10) - \log(0+1) \overset{?}{=} 1$

$$\log 10 - \log 1 \overset{?}{=} 1$$

$$1 - 0 \overset{?}{=} 1$$

$$1 = 1$$

The solution is 0.

Always check possible solutions in the original equation, and remember that any value which results in the logarithm of a negative number must be omitted.

**TO SOLVE A LOGARITHMIC EQUATION**

1. Obtain a single logarithmic expression using the same base on one side of the equation, or write each side as a logarithm using the same base.
2. Convert the result to an exponential equation or find the antilog of both sides and solve the result.
3. Check all possible solutions in the original equation. (Negative numbers do not have logarithms.)

**EXAMPLE 2** Solve the following logarithmic equation for $x$.

$$\log x + \log(x+1) = 0.3010$$

$$\log x(x+1) = 0.3010 \qquad \text{Product rule}$$

$$x(x+1) = 10^{0.3010} \qquad \text{Write in exponential form}$$

$$x(x+1) = 2 \qquad \text{From table 1, } \log 2 = 0.3010$$

$$x^2 + x - 2 = 0$$

$$(x+2)(x-1) = 0$$

$$x + 2 = 0 \quad \text{or} \quad x - 1 = 0$$

$$x = -2 \qquad\qquad x = 1$$

Check: $\log -2 + \log(-2+1) \stackrel{?}{=} 0.3010$

But log −2 is not defined so −2 is not a solution.

$$\log 1 + \log(1+1) \stackrel{?}{=} 0.3010$$
$$0 + \log 2 \stackrel{?}{=} 0.3010$$
$$0.3010 = 0.3010$$

The solution is 1.

Equations that contain expressions involving the variable as exponents are called **exponential equations.** Such equations can often be solved by taking the logarithm (using an appropriate base) of both sides and then applying the power rule.

**EXAMPLE 3** Solve the exponential equation for $x$.

$$10^{2x+1} = 100$$
$$\log 10^{2x+1} = \log 100 \qquad \text{Take common log of both sides}$$
$$(2x+1)\log 10 = \log 100 \qquad \text{Power rule}$$
$$(2x+1)(1) = 2 \qquad \text{Log } 10 = 1 \text{ and } \log 100 = 2$$
$$2x + 1 = 2$$
$$2x = 1$$
$$x = \frac{1}{2}$$

Check: $10^{2(1/2)+1} \stackrel{?}{=} 100$

$$10^{1+1} \stackrel{?}{=} 100$$
$$10^2 = 100$$

The solution is $\frac{1}{2}$.

When both sides of an exponential equation can be expressed as a power having the same base, it is much faster to **equate** the corresponding exponents (set them equal to each other) and solve. Suppose we work Example 3 using this method.

$$10^{2x+1} = 100$$
$$10^{2x+1} = 10^2 \qquad \text{Express as powers with same base}$$
$$2x + 1 = 2 \qquad \text{Equate exponents}$$
$$2x = 1$$
$$x = \frac{1}{2}$$

In such cases we can eliminate a few steps.

**TO SOLVE AN EXPONENTIAL EQUATION**

1. Try to express each side as a power using the same base, and equate the resulting exponents.
2. If Step 1 fails, take the logarithm of each side and use the power rule to eliminate the variable exponents.
3. Solve the resulting equation and check in the original.

**EXAMPLE 4** Solve $3^{x+1} = \frac{1}{27}$.

Since $\frac{1}{27} = \frac{1}{3^3} = 3^{-3}$, we must solve the equation below.

$$\begin{aligned} 3^{x+1} &= 3^{-3} && \text{Express as powers with same base} \\ x + 1 &= -3 && \text{Equate exponents} \\ x &= -4 \end{aligned}$$

Check: $3^{-4+1} \stackrel{?}{=} \frac{1}{27}$

$$3^{-3} \stackrel{?}{=} \frac{1}{27}$$

$$\frac{1}{27} = \frac{1}{27}$$

The solution is $-4$.

**EXAMPLE 5** Solve $5^{2x} = 8$.

Since 8 is not easily expressible as a power with base 5, we take the logarithm of both sides using 5 as a base.

$$\begin{aligned} \log_5 5^{2x} &= \log_5 8 \\ (2x) \log_5 5 &= \log_5 8 && \text{Power rule} \\ 2x &= \log_5 8 && \text{Log}_5\ 5 = 1 \\ x &= \frac{\log_5 8}{2} \end{aligned}$$

This is a perfectly good form for the solution to the equation, but if we prefer a decimal approximation, we must compute $\log_5 8$.

$$\log_5 8 = \frac{\log_{10} 8}{\log_{10} 5} = \frac{\log 8}{\log 5} = \frac{.9031}{.6990} \approx 1.29$$

Then

$$x \approx \frac{1.29}{2} = .645.$$

Since the logarithms we used were approximate, .645 is an approximation to the exact value of $x$ which is $\frac{\log_5 8}{2}$.

## EXERCISES 9.7

*Solve the following equations for x.*

**1.** $\log (9x + 1) - \log x = 1$

**2.** $\log x + \log (x - 1) = 0.3010$

**3.** $\log (x - 3) + \log x = 1$

**4.** $\log x - \log (7x + 6) = -1$

**5.** $10^{3x+5} = 100$

**6.** $2^{x+7} = 8$

**7.** $5^{x^2-1} = 125$

**8.** $3^{x^2} \cdot 9^x = \frac{1}{3}$

[*Hint:* Use $9^x = 3^{2x}$, then combine.]

**9.** $2^{3x} = 7$

**10.** $6^{x+1} = 27$

**11.** Express $\log_{27} 9$ in terms of $\log_3 9$ and simplify.

**12.** Express $\log_4 19$ in terms of base 19 logarithms.

**13.** Express $\log_4 19$ in terms of common logarithms and evaluate.

**14.** Determine $\log_3 19.2$ by using common logarithms.

ANSWERS: 1. 1 2. 2 3. 5 4. 2 5. −1 6. −4 7. ±2 8. −1 9. $\frac{\log_2 7}{3} \approx .936$ 10. $(\log_6 27) - 1 \approx .839$ 11. $\frac{2}{3}$ 12. $\frac{1}{\log_{19} 4}$ 13. 2.12 14. 2.69

## 9.8 Applied Problems

Numerous applied problems in science and business use formulas that involve logarithmic or exponential equations. The compound interest formula is one of the most common formulas in this category. If the number of compounding periods is great, or if the interest rate is unusual, then a good calculator or logarithms are almost necessary. The sum $S$ of an amount $A$ invested for $n$ compounding periods at a rate of $i$ interest per period is given by the formula

$$S = A(1 + i)^n.$$

Solving this formula for $A$ gives us the initial or present value necessary for accumulating with interest the future sum $S$.

$$A = S(1 + i)^{-n}$$

Without a $y^x$ button on a calculator, the next best method for solving such equations (especially if $n$ is large) is logarithmic computation.

**EXAMPLE 1** If \$1000 is invested at an annual rate of 8% compounded quarterly, what is the value of the account at the end of 10 years?

The formula we must use is

$$S = A(1 + i)^n,$$

where $A = \$1000$, the quarterly interest rate is $0.08/4 = 0.02$, and the number of compounding periods is $4(10) = 40$ (4 per year for 10 years).

$$\begin{aligned} S &= 1000(1 + 0.02)^{40} \\ &= 1000(1.02)^{40} \\ \log S &= \log [1000(1.02)^{40}] && \text{Take the log of both sides} \\ &= \log 1000 + \log (1.02)^{40} \\ &= 3 + 40 \cdot \log 1.02 && \text{Log } 1000 = \log 10^3 = 3 \\ &= 3 + 40(0.0086) \\ &= 3 + 0.3440 \\ &= 3.3440 \\ S &= 2210 && \text{The antilog of } 3.3440 \end{aligned}$$

Thus, in 10 years the investment will be worth about \$2210.

**EXAMPLE 2** What amount must be invested at 6% interest compounded semiannually in order to accumulate \$5440 at the end of 6 years?

The formula we must use is

$$A = S(1 + i)^{-n},$$

where $S = \$5440$, the semiannual interest rate is $0.06/2 = 0.03$, and the number of compounding periods is $2(6) = 12$.

$$\begin{aligned}
A &= 5440(1 + 0.03)^{-12} \\
&= 5440(1.03)^{-12} \\
\log A &= \log [5440(1.03)^{-12}] && \text{Take the log of both sides} \\
&= \log 5440 + \log (1.03)^{-12} \\
&= \log 5440 - 12 \cdot \log (1.03) \\
&= 3.7356 - 12(.0128) \\
&= 3.5820 \\
A &= 3820 && \text{The antilog of 3.5820}
\end{aligned}$$

Thus, an initial investment of \$3820 is required.

There are numerous physical applications that are described by an equation of the form

$$y = c \log_a \frac{x}{x_0}.$$

For sound and earthquakes, the base $a$ is 10, while for cell-generation problems, the base $a$ is 2.

**EXAMPLE 3** The equation that describes the number of cell fissions that occur in a given period of time is the number $y$ given by

$$y = \log_2 \frac{x}{x_0},$$

where $x_0$ is the initial cell count and $x$ is the final cell count. Find the number of cell fissions in a given time if the initial cell count is 25 and the final cell count is 630,000.

Since $x_0 = 25$ and $x = 630{,}000$, we evaluate the following expression.

$$\begin{aligned}
y &= \log_2 \frac{630{,}000}{25} \\
&= \log_2 25200 && \text{Simplify fraction} \\
&= \frac{\log 25200}{\log 2} \\
&= \frac{4.4014}{.3010} \approx 14.6
\end{aligned}$$

Thus, there are approximately 14.6 fissions during that period.

**EXAMPLE 4** The loudness of sound in decibels is given by the formula

$$y = 10 \cdot \log \frac{x}{x_0}$$

where $x_0$ is the weakest sound detected by the observer and $x$ is the intensity of a given sound. Find the value of $y$ if $x$ is 51,000 times $x_0$.

$$\begin{aligned} y &= 10 \cdot \log \frac{51000x_0}{x_0} \qquad x = 51000x_0 \\ &= 10 \cdot \log 51000 \\ &= 10(4.7076) \\ &\approx 47.1 \end{aligned}$$

Thus, the loudness of sound is about 47.1 decibels.

## EXERCISES 9.8

1. If \$1250 is invested at an annual rate of 12% compounded quarterly, what will be the value of the account at the end of 10 years?

2. What amount must be invested at 8% interest compounded semiannually in order to accumulate \$12,000 at the end of 15 years?

3. Use the equation $y = \log_2 \frac{x}{x_0}$ to determine the number of cell fissions that occur during the time of an experiment if there are 125 cells initially and 9525 cells finally.

4. Use the sound equation $y = 10 \cdot \log \frac{x}{x_0}$ to determine the loudness of a sound in decibels if the given sound has intensity 63,000 times the weakest sound detected.

**5.** If \$685 is invested at an annual rate of 8% compounded semiannually, what is the value of the account at the end of 25 years?

**6.** If \$685 is invested at an annual rate of 8% compounded quarterly, what is the value of the account at the end of 25 years?

**7.** What amount must be invested at 8% interest compounded quarterly in order to accumulate \$4000 at the end of 12 years?

**8.** Use the equation $y = \log_2 \frac{x}{x_0}$ to determine the number of cell fissions that occur in a given period of time if there are 30 cells initially and 437,000 cells finally.

**9.** Use the sound equation $y = 10 \cdot \log \frac{x}{x_0}$ to determine the loudness of a sound in decibels if the given sound has intensity 78,000 times the weakest sound detected.

**10.** The magnitude $y$ on the Richter scale of the intensity $x$ of an earthquake is given by

$$y = \log \frac{x}{x_0}$$

where $x_0$ is the minimum intensity used for comparison. If an earthquake has intensity 50,100 times the minimum intensity, what is its Richter scale magnitude?

**11.** Solve $\log x + \log (x - 2) = .4771$.

**12.** Solve $2^{3x+1} = 16$.

---

ANSWERS: **1.** \$4080 **2.** \$3700 **3.** 6.25 cell fissions **4.** 48.0 decibels **5.** \$4870 **6.** \$4960 **7.** \$1550 **8.** 13.8 cell fissions **9.** 48.9 decibels **10.** 4.7 **11.** 3 **12.** 1

## 9.9 Logarithmic and Exponential Functions

In Chapter 7 we considered constant, linear, and quadratic functions and their graphs. Two other important kinds of functions are exponential and logarithmic functions. A function defined by the equation

$$y = a^x, \qquad a > 0$$

is called an **exponential function,** and the number $a$ is called the **base.** It is not difficult to calculate the value of $y$ when $x$ is a rational number $p/q$ since $a^{p/q}$ is defined to be the $q$th root of $a$ raised to the $p$th power. Although we have not defined $a^x$ for irrational $x$ such as $x = \sqrt{2}$ or $x = \pi$, we shall assume that $a^{\sqrt{2}}$ can be approximated as closely as we wish by using rational approximations of $\sqrt{2}$. For example, since $\sqrt{2} \approx 1.414$,

$$a^{\sqrt{2}} \approx a^{1.414} = a^{1414/1000},$$

which does have meaning for us. The base $a$ to other irrational powers is approximated in a similar way.

When we plot the graph of a function such as $y = 2^x$, our assumption about irrational powers will become reasonable. Remember that in order to graph a function, we select values of $x$, calculate the corresponding values of $y$, place them in a table of values, and then plot the ordered pairs in a Cartesian coordinate system.

**EXAMPLE 1** Graph the function $y = 2^x$.

We select values of $x$ and calculate the corresponding values of $y$ to obtain the table of values in Figure 9.1. We can use 1.4 as an approximation for $\sqrt{2}$, and from the graph, we see that 2.7 is an approximation for $2^{\sqrt{2}}$. Similarly, using 3.1 as an approximation for $\pi$, we can estimate $2^{\pi}$ to be about 8.8.

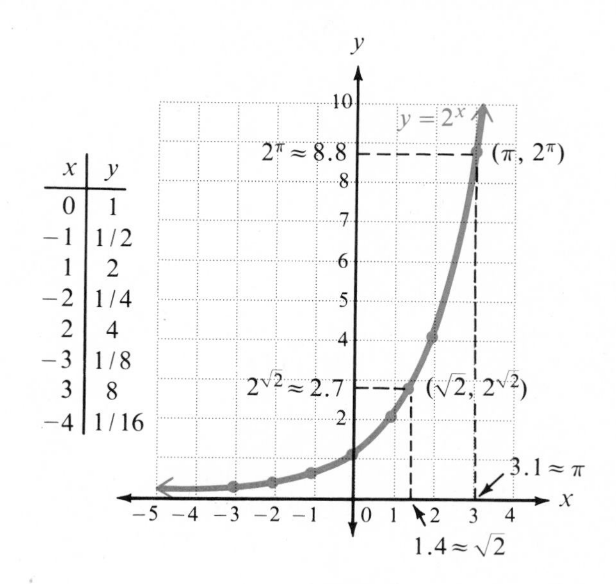


**Figure 9.1**

In order to compare exponential functions, in the next example we graph several such functions in the same coordinate system.

**EXAMPLE 2** Graph $y = 2^x$, $y = \left(\frac{1}{2}\right)^x$, $y = 3^x$, $y = \left(\frac{1}{3}\right)^x$, $y = 10^x$, and $y = \left(\frac{1}{10}\right)^x$.

First we construct a table of values.

| $x$ | 0 | −1 | 1 | −2 | 2 | −3 | 3 | −4 | 4 |
|---|---|---|---|---|---|---|---|---|---|
| $y = 2^x$ | 1 | $\frac{1}{2}$ | 2 | $\frac{1}{4}$ | 4 | $\frac{1}{8}$ | 8 | $\frac{1}{16}$ | 16 |
| $y = \left(\frac{1}{2}\right)^x$ | 1 | 2 | $\frac{1}{2}$ | 4 | $\frac{1}{4}$ | 8 | $\frac{1}{8}$ | 16 | $\frac{1}{16}$ |
| $y = 3^x$ | 1 | $\frac{1}{3}$ | 3 | $\frac{1}{9}$ | 9 | $\frac{1}{27}$ | 27 | $\frac{1}{81}$ | 81 |
| $y = \left(\frac{1}{3}\right)^x$ | 1 | 3 | $\frac{1}{3}$ | 9 | $\frac{1}{9}$ | 27 | $\frac{1}{27}$ | 81 | $\frac{1}{81}$ |
| $y = 10^x$ | 1 | $\frac{1}{10}$ | 10 | $\frac{1}{100}$ | 100 | $\frac{1}{1000}$ | 1000 | $\frac{1}{10000}$ | 10000 |
| $y = \left(\frac{1}{10}\right)^x$ | 1 | 10 | $\frac{1}{10}$ | 100 | $\frac{1}{100}$ | 1000 | $\frac{1}{1000}$ | 10000 | $\frac{1}{10000}$ |

The scale for our graphs in Figure 9.2 (on p. 490) does not allow us to plot all of the points listed in the table, but the values are included there for purposes of comparison.

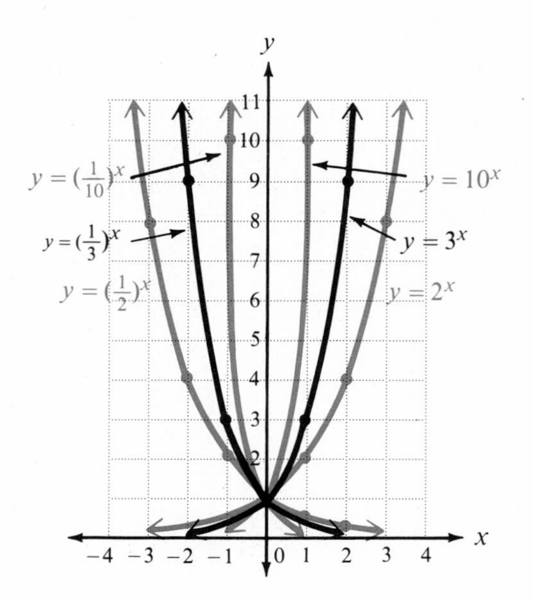


**Figure 9.2**

Several observations can be made from Example 2. In all cases the graph passes through the point (0, 1), since $a^0 = 1$ for any value of $a$. For every base $a$ greater than 1, the graph rises from left to right, and toward the left it gets closer and closer (but never crosses) the $x$-axis. For every base $a$ less than 1, the graph decreases from left to right, and toward the right it gets closer and closer (but never crosses) the $x$-axis.

A **logarithmic function** is a function defined by an equation of the form

$$y = \log_a x.$$

Since logarithmic equations and exponential equations are closely related, to graph $y = \log_a x$ we need only graph its equivalent exponential form

$$a^y = x.$$

When doing so, keep in mind that the roles of $x$ and $y$ have been interchanged from our previous work when we graphed $y = a^x$.

EXAMPLE 3 Graph $y = \log_2 x$ by plotting $x = 2^y$.

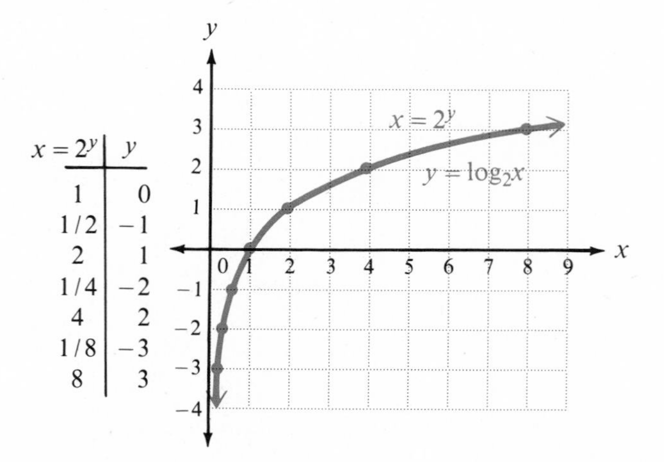

| $x = 2^y$ | $y$ |
|---|---|
| 1 | 0 |
| 1/2 | −1 |
| 2 | 1 |
| 1/4 | −2 |
| 4 | 2 |
| 1/8 | −3 |
| 8 | 3 |



**Figure 9.3**

We construct the table of values in Figure 9.3 by determining $x$ for a given value of $y$, plotting the points, and joining them with a smooth curve.

The logarithmic and exponential functions

$$y = \log_a x \qquad \text{and} \qquad y = a^x$$

are related in an important way. The graph of each is the reflection of the graph of the other across the line $y = x$. (See Figure 9.4.) Functions with this property are called **inverses** of each other, and if the graph of either is known, the graph of the other can be obtained.

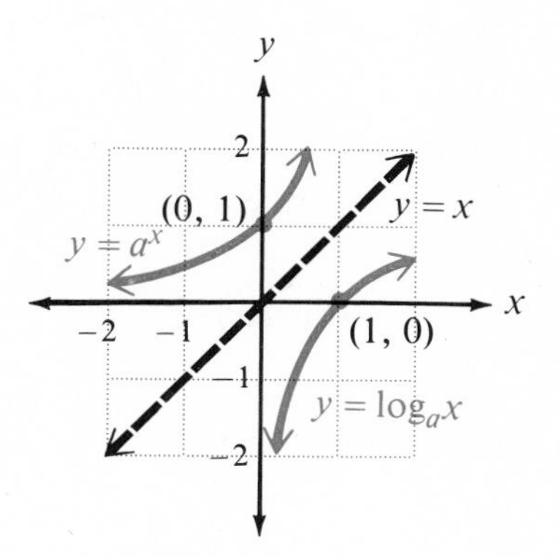


**Figure 9.4**

Generally, in mathematics, the most interesting exponential and logarithmic functions are those with base $a > 1$. The graph of every exponential function

$$y = a^x, \qquad a > 1$$

has the same basic shape, shown in Figure 9.5, and passes through the point (0, 1). The graph of every logarithmic function

$$y = \log_a x, \qquad a > 1$$

has the same basic shape, shown in Figure 9.6, and passes through the point (1, 0). With these two facts in mind, the graph of any exponential or logarithmic function can be quickly determined by plotting a few points.

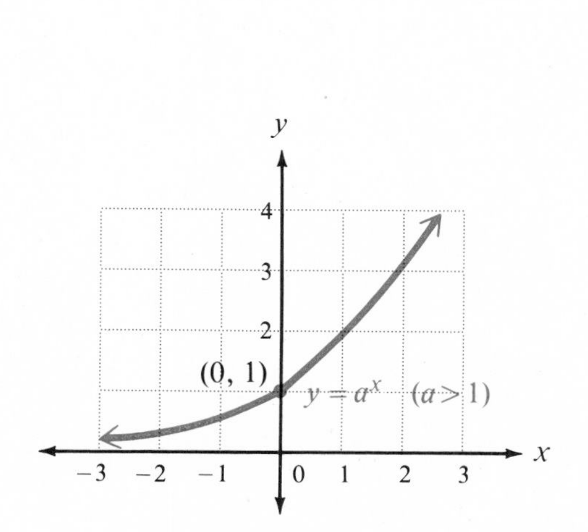


**Figure 9.5**

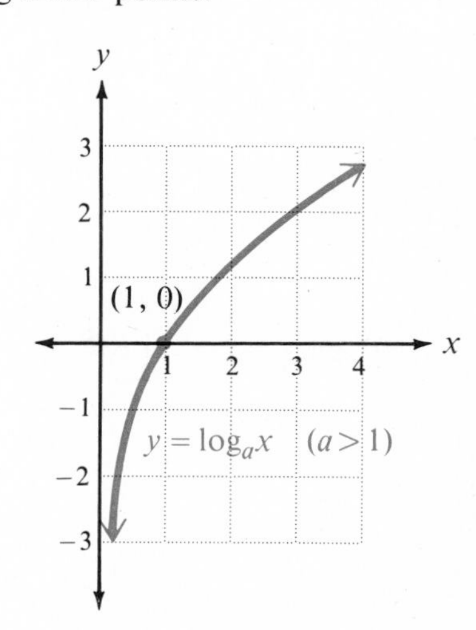


**Figure 9.6**

## EXERCISES 9.9

1. Given an exponential function $y = a^x$, $a$ is called the ______________.
2. The graph of the logarithmic function $y = \log_a x$ can be determined by plotting the points that satisfy the exponential equation ______________.
3. The graph of $y = \log_3 x$ can be obtained by reflecting the graph of $y = 3^x$ across the line with equation ______________.
4. The graph of $y = 5^x$ and $y =$ ______________ are reflections of each other across the line $y = x$.
5. Complete the following table and sketch the graphs of the functions $y = 2^x$, $y = \left(\frac{1}{3}\right)^x$, $y = 3^x$, $y = 4^x$ in the given coordinate system.

| $x$ | 0 | −1 | 1 | −2 | 2 | −3 | 3 |
|---|---|---|---|---|---|---|---|
| $y = 2^x$ | | | | | | | |
| $y = \left(\frac{1}{3}\right)^x$ | | | | | | | |
| $y = 3^x$ | | | | | | | |
| $y = 4^x$ | | | | | | | |

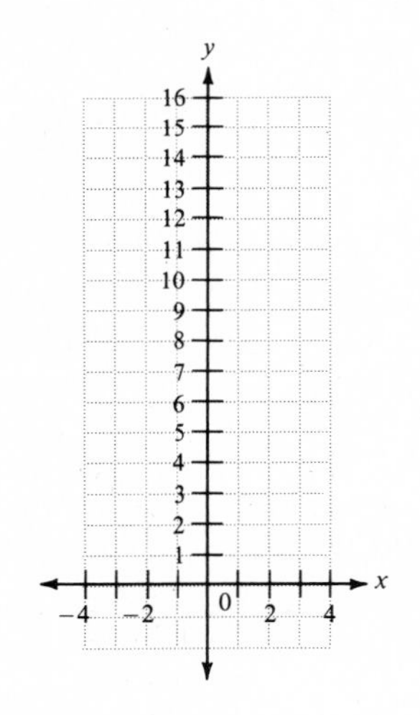


6. Sketch the graph of $y = \log_3 x$ by plotting $x = 3^y$.

| $x$ | | | | | |
|---|---|---|---|---|---|
| $y$ | 0 | 1 | −1 | 2 | −2 |

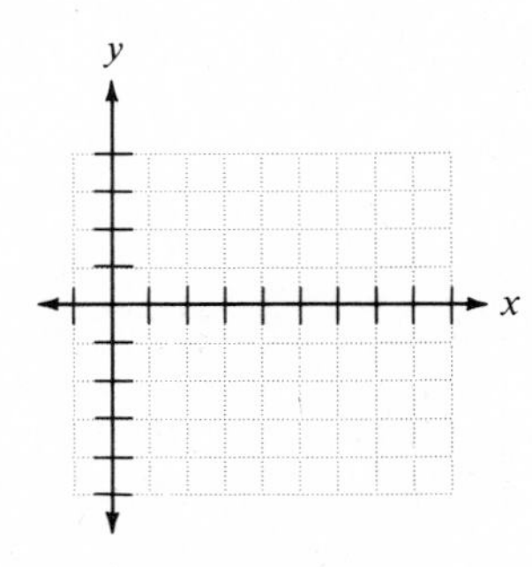

**7.** Having plotted $y = 3^x$ in Exercise 5, could you have used this information to sketch the graph required in Exercise 6? Explain.

**8.** If \$875 is invested at an annual rate of 4% compounded quarterly, what will the value of the account be at the end of 20 years?

**9.** What amount must be invested at 12% interest compounded semiannually in order to accumulate \$2750 at the end of 10 years?

**10.** Use the sound equation $y = 10 \cdot \log \frac{x}{x_0}$ to determine the loudness of a sound in decibels if the given sound has intensity 49,500 times the weakest sound detected.

---

ANSWERS: **1.** base **2.** $x = a^y$ **3.** $y = x$ **4.** $\log_5 x$ **5.**

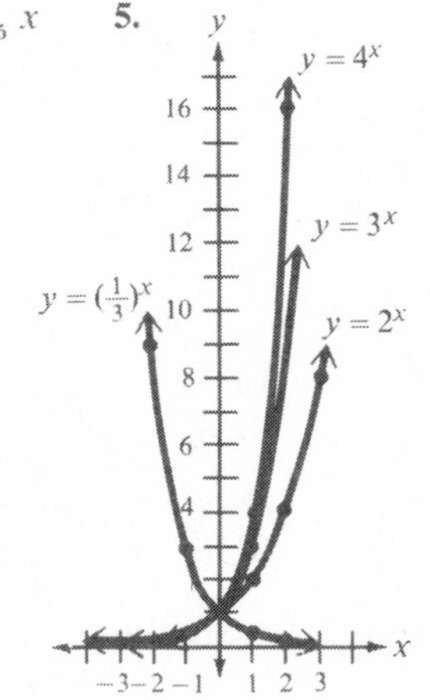


**6.**

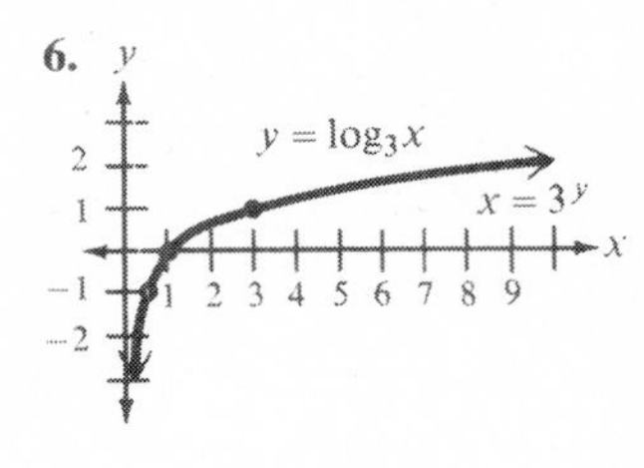


**7.** yes; by reflecting the graph across the line $y = x$
**8.** \$1940 **9.** \$857 **10.** 46.9 decibels

## Chapter 9 Summary

### Words and Phrases for Review

[9.1] base
exponential form
logarithmic form
[9.3] common logarithm
characteristic
mantissa
[9.4] antilogarithm
[9.6] natural logarithm
[9.7] logarithmic equation
exponential equation
[9.8] compound interest
cell fission
[9.9] exponential function
logarithmic function
inverse function

### Brief Reminders

[9.1] If $a$, $x$, and $y$ are numbers, $a > 0$, $a \neq 1$, then the exponential equation $y = a^x$ is equivalent to the logarithmic equation $x = \log_a y$.

[9.2] **1.** Product rule: $\log_a u \cdot v = \log_a u + \log_a v$ [which is *not* $\log_a (u + v)$].

**2.** Quotient rule: $\log_a \frac{u}{v} = \log_a u - \log_a v$ [which is *not* $\log_a (u - v)$].

**3.** Power rule: $\log_a u^c = c \cdot \log_a u$ [which is *not* $\log_a cu$].

[9.3] If $a$ is any base,

$$\log_a a = 1 \quad \text{and} \quad \log_a 1 = 0.$$

[9.3–9.4] **1.** $\log_{10} x$ is written $\log x$.

**2.** When using a log table, keep the mantissa positive. When using a calculator, we need not do this.

[9.5] Write out complete details when making logarithmic calculations. This will help you avoid errors such as multiplying when you should be adding, and dividing when you should be subtracting.

[9.6] If $a$ and $b$ are any two bases, and $x > 0$, then

$$\log_b x = \frac{\log_a x}{\log_a b} \quad \text{and} \quad \log_b a = \frac{1}{\log_a b}.$$

[9.7] **1.** To solve a logarithmic equation: obtain a single logarithmic expression using the same base on one side of the equation, or write each side as a logarithm using the same base; convert the result to an exponential equation or find the antilog of both sides.

**2.** To solve an exponential equation: try to express each side as a power using the same base and equate exponents; if this fails, take the log of both sides and use the power rule to eliminate the variable exponents.

**3.** When solving logarithmic and exponential equations, always check your answers in the original equation.

[9.9] The logarithmic function $y = \log_a x$ and the exponential function $y = a^x$ are inverses. The graph of each is the reflection of the graph of the other across the line $y = x$.

## CHAPTER 9 REVIEW EXERCISES

[9.1] **1.** Another term for the word *logarithm* is ______________.

**2.** The exponential form of $y = \log_a x$ is ______________.

**3.** The logarithmic form of $u = b^v$ is ______________.

**4.** Every positive number except ______________ can be used as a base for logarithms.

[9.2] **5.** If $a$ is any base and $u$ and $v$ are positive numbers, $\log_a u \cdot v =$ ______________.

**6.** If $a$ is any base and $u$ and $v$ are positive numbers, $\log_a \frac{u}{v} =$ ______________.

**7.** If $a$ is any base, $u$ is a positive number, and $c$ is any number, $\log_a u^c =$ ______________.

**8.** The rule given in Exercise 5 is called the ______________ rule.

**9.** The rule given in Exercise 6 is called the ______________ rule.

**10.** The rule given in Exercise 7 is called the ______________ rule.

[9.3] **11.** Base 10 logarithms are also called ______________ logarithms.

**12.** For any base $a$, $\log_a a =$ ______________.

**13.** For any base $a$, $\log_a 1 =$ ______________.

**14.** Given that $\log 0.523 = 9.7185 - 10$, the characteristic is **(a)** ______________ and the mantissa is **(b)** ______________.

**15.** It is best to express a number in ______________ before using Table 1 to find its logarithm.

[9.4] **16.** In the equation $\log n = x$, $n$ is the **(a)** ______________ of $x$, or 10 to the power **(b)** ______________, and $x$ is the **(c)** ______________ of $n$.

[9.6] **17.** Logarithms to the base $e$ are called ______________ logarithms.

**18.** If $a$ and $b$ are bases, and $x > 0$, then $\frac{\log_a x}{\log_a b} =$ **(a)** ______________ and $\frac{1}{\log_b a} =$ **(b)** ______________.

[9.9] **19.** A function defined by the equation $y = a^x$, where $a > 0$, is called a(n) ______________ function.

**20.** The graph of $y = \log_a x$ can be determined using the graph of $y = a^x$ by ______________.

[9.1] *Convert to logarithmic form.*

**21.** $2^x = 50$

**22.** $x^3 = 27$

**23.** $u^v = q$

*Convert to exponential form.*

**24.** $\log_3 a = b$ **25.** $\log_a a = 1$ **26.** $\log_c w = m$

*Determine the numerical value of x.*

**27.** $x^4 = 81$ **28.** $x^{-1} = \frac{1}{7}$ **29.** $x^{1/2} = 4$

**30.** $x = \log_2 8$ **31.** $\log_x 125 = 3$ **32.** $\log_3 x = -2$

[9.2] *Express in terms of logarithms of x, y, and z.*

**33.** $\log_a \frac{x^3 \sqrt{z}}{y^4}$ **34.** $\log_a \sqrt[5]{\frac{xy}{z^2}}$

*Express as a single logarithm.*

**35.** $\frac{1}{3} \log_a x + 5 \cdot \log_a y$ **36.** $\frac{1}{2} (\log_a x - \log_a z - 2 \cdot \log_a y)$

*Are the following true or false?*

**37.** $\log_a \frac{u}{v} = \frac{\log_a u}{\log_a v}$ **38.** $\log_a a = a$

**39.** $\log_a (u + v) = (\log_a u)(\log_a v)$ **40.** $\log_a u^c = u \log_a c$

[9.3] *Use Table 1 or your calculator to find the following logarithms.*

**41.** log .00378 **42.** log 13.2

*Convert each logarithm to the alternate form which displays the positive mantissa and the characteristic.*

**43.** $-2.1478$ **44.** $-0.5397$

[9.4] *Use Table 1 or your calculator to find the antilogarithm of each of the following.*

**45.** 1.4048

**46.** $0.6542 - 4$

[9.5] *Use logarithms to find N.*

**47.** $N = \sqrt[7]{25.2}$

**48.** $N = (1.09)^{80}(.265)$

[9.6] **49.** Determine $\log_5 2.59$ by using common logarithms.

**50.** Express $\log_9 53$ in terms of base 53 logarithms.

[9.7] *Solve the following equations.*

**51.** $\log (x + 3) + \log x = 1$

**52.** $\log x - \log (3x + 7) = -1$

**53.** $3^{5x-1} = 81$

**54.** $5^{2x} = 8$

[9.8] **55.** If \$8000 is invested at an annual rate of 14% compounded semi-annually, what is the value of the account at the end of 5 years?

**56.** Use the sound equation $y = 10 \cdot \log \frac{x}{x_0}$ to determine the loudness of a sound in decibels if the given sound has intensity 85,000 times the weakest sound detected.

[9.9] **57.** Sketch the graph of $y = 5^x$ and use it to sketch the graph of $y = \log_5 x$.

| $x$ | 0 | 1 | −1 | 2 | −2 |
|---|---|---|---|---|---|
| $y = 5^x$ | | | | | |

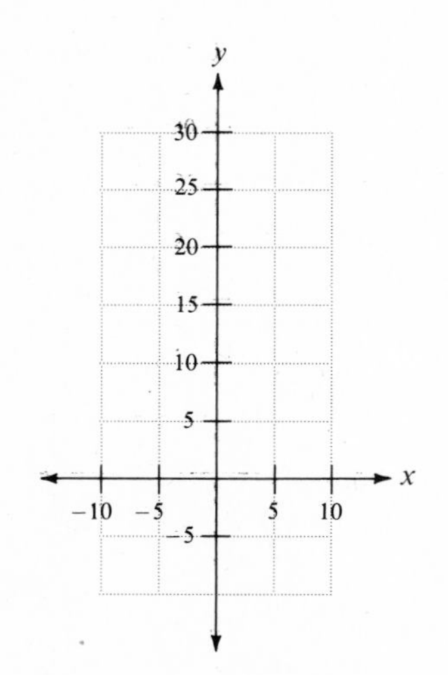

---

**ANSWERS:** **1.** exponent **2.** $x = a^y$ **3.** $v = \log_b u$ **4.** 1 **5.** $\log_a u + \log_a v$ **6.** $\log_a u - \log_a v$ **7.** $c \log_a u$ **8.** product **9.** quotient **10.** power **11.** common **12.** 1 **13.** 0 **14.** (a) $-1$ (b) .7185 **15.** scientific notation **16.** (a) antilogarithm (b) $x$ (c) logarithm **17.** natural **18.** (a) $\log_b x$ (b) $\log_a b$ **19.** exponential **20.** reflecting it across the line $y = x$ **21.** $x = \log_2 50$ **22.** $3 = \log_x 27$ **23.** $v = \log_u q$ **24.** $3^b = a$ **25.** $a^1 = a$ **26.** $c^m = w$ **27.** 3 **28.** 7 **29.** 16 **30.** 3 **31.** 5 **32.** $\frac{1}{9}$ **33.** $3 \cdot \log_a x + \frac{1}{2} \log_a z - 4 \cdot \log_a y$ **34.** $\frac{1}{5}(\log_a x + \log_a y - 2 \cdot \log_a z)$ **35.** $\log_a y^5 \sqrt[3]{x}$ **36.** $\log_a \sqrt{\frac{x}{zy^2}}$ **37.** false **38.** false **39.** false **40.** false **41.** $0.5775 - 3 = -2.4225$ **42.** 1.1206 **43.** $0.8522 - 3$ **44.** $0.4603 - 1$ **45.** 25.4 **46.** .000451 **47.** 1.59 **48.** 261 **49.** 0.5913 **50.** $\frac{1}{\log_{53} 9}$ **51.** 2 **52.** 1 **53.** 1 **54.** $\frac{\log_5 8}{2} \approx 0.646$ **55.** \$15,700 **56.** 49.3 decibels **57.** (Reflect the graph of $y = 5^x$ across the line $y = x$ to obtain the graph of $y = \log_5 x$.)

| $x$ | 0 | 1 | $-1$ | 2 | $-2$ |
|---|---|---|---|---|---|
| $y = 5^x$ | 1 | 5 | $\frac{1}{5}$ | 25 | $\frac{1}{25}$ |

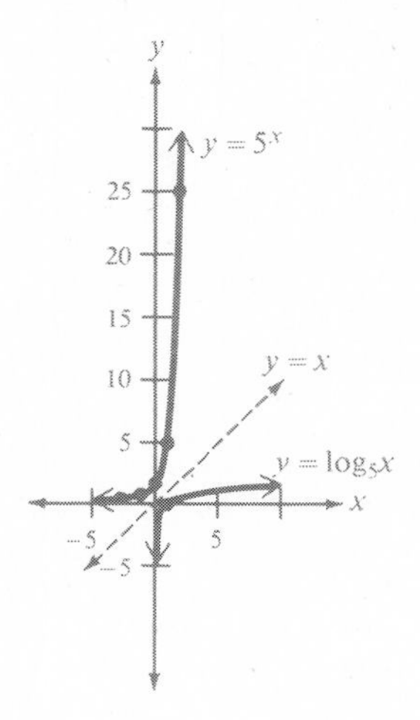

# 10

# Sequences and Series

## 10.1 Sequences and Series: Definitions

A collection of numbers, arranged in a particular order, is called a **sequence.** For example,

$$1, \quad 3, \quad 5, \quad 7, \quad 9 \cdots$$

and

$$-1, \quad -\frac{1}{2}, \quad -\frac{1}{4}, \quad -\frac{1}{8}, \quad -\frac{1}{16}, \cdots$$

are sequences. Three dots are used to indicate that the pattern continues. The numbers in a sequence are called **terms** of the sequence. If a sequence has a last term, it is called **finite,** and sequences which are not finite are called **infinite.** The two sequences above are infinite while

$$2, \quad 4, \quad 6, \quad 8, \quad 10$$

is finite with last term 10. For some sequences, it is possible to find a formula that describes a general term, called the ***n*th term.** The $n$th term in the preceding sequence is $2n$, where $n$ indicates the particular term.

| $n$ | 1 | 2 | 3 | 4 | 5 |
|---|---|---|---|---|---|
| $n$th term ($2n$) | $2 \cdot 1 = 2$ | $2 \cdot 2 = 4$ | $2 \cdot 3 = 6$ | $2 \cdot 4 = 8$ | $2 \cdot 5 = 10$ |

Sometimes we use a letter with a subscript to indicate a particular term of a sequence.

$$a_1 = \text{first term}, \qquad a_2 = \text{second term}, \qquad a_3 = \text{third term}$$

and in general

$$a_n = n\text{th term}.$$

Thus, in our example

$$a_1 = 2, \qquad a_2 = 4, \qquad a_3 = 6, \qquad a_4 = 8, \qquad a_5 = 10$$

and

$$a_n = 2n.$$

The $n$th term of a sequence is also called the **general term** of the sequence. The formula for the general term is used to generate a sequence by repeated substitution of counting numbers.

EXAMPLE 1 Write the first five terms of the infinite sequence with general term $a_n = 2n - 1$.

| 1 | 2 | 3 | 4 | 5 | $\cdots$ | $n$ |
|---|---|---|---|---|---|---|
| $a_1 = 2(1) - 1$ $= 1$ | $a_2 = 2(2) - 1$ $= 3$ | $a_3 = 2(3) - 1$ $= 5$ | $a_4 = 2(4) - 1$ $= 7$ | $a_5 = 2(5) - 1$ $= 9$ | $\cdots$ | $a_n = 2n - 1$ |

Thus, the first five terms are 1, 3, 5, 7, and 9, and the sequence is

$$1, \quad 3, \quad 5, \quad 7, \quad 9, \cdots, \quad 2n - 1, \cdots.$$

EXAMPLE 2 A finite sequence has four terms, and the formula for the $n$th term is $x_n = (-1)^n \frac{1}{2^{n-1}}$. What is the sequence?

When $n = 1$: $\qquad x_n = x_1 = (-1)^1 \frac{1}{2^{1-1}} = (-1)\frac{1}{2^0} = -1.$

When $n = 2$: $\qquad x_n = x_2 = (-1)^2 \frac{1}{2^{2-1}} = (-1)^2 \frac{1}{2^1} = \frac{1}{2}.$

When $n = 3$: $\qquad x_n = x_3 = (-1)^3 \frac{1}{2^{3-1}} = (-1)^3 \frac{1}{2^2} = -\frac{1}{4}.$

When $n = 4$: $\qquad x_n = x_4 = (-1)^4 \frac{1}{2^{4-1}} = (-1)^4 \frac{1}{2^3} = \frac{1}{8}.$

Thus, the sequence is

$$-1, \quad \frac{1}{2}, \quad -\frac{1}{4}, \quad \frac{1}{8}.$$

We are often interested in determining a sequence, given a formula for the general or $n$th term, as in the preceding examples. Also, the first few terms of a sequence may be given, and we must find a formula for the general or $n$th term. Usually the second type of problem is much more difficult to solve than the first. We shall only find the formula for a general term for two specific sequences, arithmetic and geometric progressions. These sequences will be studied in detail in following sections.

Associated with every sequence is a **series,** the indicated sum of the terms of the sequence. For example, the series associated with the sequence

$$2, 4, 6, 8, 10$$

is

$$2 + 4 + 6 + 8 + 10$$

and the series associated with

$$-1, \quad \frac{1}{2}, \quad -\frac{1}{4}, \quad \frac{1}{8}$$

is
$$(-1) + \left(\frac{1}{2}\right) + \left(-\frac{1}{4}\right) + \left(\frac{1}{8}\right).$$

These are examples of **finite series,** or finite sums of numbers.

The series associated with

$$1, 3, 5, 7, 9, \cdots$$

is the **infinite series**

$$1 + 3 + 5 + 7 + 9 + \cdots + (2n - 1) + \cdots.$$

An infinite series is not a number, but rather an expression which may or may not "sum up" to a number.

The Greek letter $\Sigma$ (sigma), called the **summation symbol,** is often used to abbreviate a series. For example, the series

$$2 + 4 + 6 + 8 + 10$$

which has general term $x_n = 2n$, can be written

$$\sum_{n=1}^{5} x_n \quad \text{or} \quad \sum_{n=1}^{5} 2n$$

and is read "the sum of the terms $x_n$ or $2n$ as $n$ varies over the counting numbers from 1 to 5." The letter $n$ is called the **index** on the summation while 1 and 5 are the **lower** and **upper limits of summation,** respectively. In general, if

$$x_1, x_2, x_3, x_4, \cdots, x_n$$

is a sequence, its associated series is

$$\sum_{k=1}^{n} x_k = x_1 + x_2 + x_3 + x_4 + \cdots + x_n.$$

**EXAMPLE 3** Write out the series $\sum_{k=1}^{5} (k^2 + 1)$ without using the sigma summation notation.

When $k = 1$: $x_1 = (1)^2 + 1 = 1 + 1 = 2.$

When $k = 2$: $x_2 = (2)^2 + 1 = 4 + 1 = 5.$

When $k = 3$: $x_3 = (3)^2 + 1 = 9 + 1 = 10.$

When $k = 4$: $x_4 = (4)^2 + 1 = 16 + 1 = 17.$

When $k = 5$: $x_5 = (5)^2 + 1 = 25 + 1 = 26.$

Thus,

$$\sum_{k=1}^{5} (k^2 + 1) = 2 + 5 + 10 + 17 + 26 = 60.$$

**EXAMPLE 4** Express $\sum_{k=2}^{4} (-1)^k \sqrt{k+1}$ without using the sigma summation notation.

When $k = 2$: $x_2 = (-1)^2 \sqrt{2+1} = (1)\sqrt{3} = \sqrt{3}.$

When $k = 3$: $x_3 = (-1)^3 \sqrt{3+1} = (-1)\sqrt{4} = -\sqrt{4} = -2.$

When $k = 4$: $x_4 = (-1)^4 \sqrt{4+1} = (1)\sqrt{5} = \sqrt{5}.$

Thus,

$$\sum_{k=2}^{4} (-1)^k \sqrt{k+1} = \sqrt{3} - 2 + \sqrt{5}.$$

When a sequence is infinite, such as

$$\frac{1}{2}, \frac{1}{4}, \frac{1}{8}, \frac{1}{16}, \frac{1}{32}, \ldots, \frac{1}{2^n}, \ldots$$

the infinite series

$$\frac{1}{2} + \frac{1}{4} + \frac{1}{8} + \frac{1}{16} + \frac{1}{32} + \cdots + \frac{1}{2^n} + \cdots$$

is written in sigma summation notation as

$$\sum_{n=1}^{\infty} \frac{1}{2^n}$$

and is read "the sum of $\frac{1}{2^n}$ as $n$ varies from 1 to infinity".

## EXERCISES 10.1

1. A collection of numbers arranged in a particular order is called a(n) ________________________________.
2. The numbers in a sequence are called the ______________________ of the sequence.
3. The indicated sum of the terms of a sequence is called the ______________ associated with the sequence.
4. When a sequence has a last term it is called a(n) ______________ sequence.
5. Sequences that do not have a last term but continue indefinitely are called ______________________ sequences.
6. The $n$th term of a sequence is also called the ______________ term of the sequence.
7. The Greek letter $\Sigma$ is called the ______________________ symbol.
8. In the sigma summation notation $\sum_{k=1}^{n} a_k$, $k$ is called the **(a)** ______________ of summation, 1 is called the **(b)** __________ limit, and $n$ the **(c)** __________ limit of summation.

**9.** To express $a_2 + a_3 + a_4 + a_5 + a_6$ using sigma summation notation we write

______________________.

**10.** To express $\sum_{k=3}^{7} b_k$ without using sigma summation notation, we write ________

______________________.

*In each of the following, the nth term of a sequence is given. Find the first four terms ($n = 1, 2, 3, 4$) and the seventh term ($n = 7$).*

**11.** $a_n = 3n$

**12.** $b_n = 3n - 1$

**13.** $x_n = \frac{(-1)^n}{n}$

**14.** $a_n = \frac{n+1}{n}$

**15.** $b_n = \left(-\frac{1}{2}\right)^n$

**16.** $x_n = n^2 - 2$

**17.** Which sequence is finite and which is infinite?

(a) $2, 1, \frac{1}{2}, \frac{1}{4}, \ldots$

(b) $4, 8, 12, 16, 20$

(c) $-1, 1, -1, 1, -1, 1, \cdots$

(d) $1, 100, 10{,}000, 1{,}000{,}000$

**18.** Find the associated series for each sequence in Exercise 17.

(a)

(b)

(c)

(d)

*Rewrite each series using the sigma summation notation.*

**19.** $2 + 5 + 8 + 11 + \cdots + (3n - 1)$

**20.** $1 + \sqrt{2} + \sqrt{3} + 2 + \cdots + \sqrt{n}$

**21.** $\frac{1}{2} + \frac{1}{3} + \frac{1}{4} + \cdots + \frac{1}{n+1} + \cdots$

**22.** $-1 + \frac{1}{4} - \frac{1}{9} + \frac{1}{16} + \cdots + (-1)^n \frac{1}{n^2} + \cdots$

*Rewrite each series without using the sigma summation notation.*

**23.** $\sum_{k=1}^{6} a_k$

**24.** $\sum_{k=1}^{3} x_k$

**25.** $\sum_{n=1}^{3} b_n$

**26.** $\sum_{i=2}^{8} x_i$

**27.** $\sum_{k=1}^{5} (2k+5)$

**28.** $\sum_{n=1}^{3} n\pi$

**29.** $\sum_{n=1}^{4} \frac{1}{n}$

**30.** $\sum_{k=4}^{5} (k^2+1)$

**31.** $\sum_{k=1}^{\infty} k$

**32.** $\sum_{n=1}^{\infty} (-1)^n 2^n$

---

ANSWERS: **1.** sequence **2.** terms **3.** series **4.** finite **5.** infinite **6.** general **7.** summation **8.** (a) index (b) lower (c) upper **9.** $\sum_{k=2}^{6} a_k$ **10.** $b_3 + b_4 + b_5 + b_6 + b_7$ **11.** $a_1 = 3, a_2 = 6, a_3 = 9, a_4 = 12, a_7 = 21$ **12.** $b_1 = 2, b_2 = 5, b_3 = 8, b_4 = 11, b_7 = 20$ **13.** $x_1 = -1, x_2 = \frac{1}{2}, x_3 = -\frac{1}{3}, x_4 = \frac{1}{4}, x_7 = -\frac{1}{7}$ **14.** $a_1 = 2, a_2 = \frac{3}{2}, a_3 = \frac{4}{3}, a_4 = \frac{5}{4}, a_7 = \frac{8}{7}$ **15.** $b_1 = -\frac{1}{2}, b_2 = \frac{1}{4}, b_3 = -\frac{1}{8}, b_4 = \frac{1}{16}, b_7 = -\frac{1}{128}$ **16.** $x_1 = -1, x_2 = 2, x_3 = 7, x_4 = 14, x_7 = 47$ **17.** (a) infinite (b) finite (c) infinite (d) finite **18.** (a) $2 + 1 + \frac{1}{2} + \frac{1}{4} + \cdots$ (b) $4 + 8 + 12 + 16 + 20$ (c) $-1 + 1 - 1 + 1 - 1 + \cdots$ (d) $1 + 100 + 10{,}000 + 1{,}000{,}000$ **19.** $\sum_{k=1}^{n} (3k-1)$ **20.** $\sum_{k=1}^{n} \sqrt{k}$ **21.** $\sum_{n=1}^{\infty} \frac{1}{n+1}$ **22.** $\sum_{n=1}^{\infty} (-1)^n \frac{1}{n^2}$ **23.** $a_1 + a_2 + a_3 + a_4 + a_5 + a_6$ **24.** $x_1 + x_2 + x_3$ **25.** $b_1 + b_2 + b_3$ **26.** $x_2 + x_3 + x_4 + x_5 + x_6 + x_7 + x_8$ **27.** $7 + 9 + 11 + 13 + 15$ **28.** $\pi + 2\pi + 3\pi$ **29.** $1 + \frac{1}{2} + \frac{1}{3} + \frac{1}{4}$ **30.** $17 + 26$ **31.** $1 + 2 + 3 + 4 + \cdots + k + \cdots$ **32.** $-2 + 4 - 8 + 16 - 32 + \cdots + (-1)^n 2^n + \cdots$

## 10.2 Arithmetic Progressions

In the sequence

$$2, 5, 8, 11, 14, \cdots$$

each term (after the first) can be obtained by adding 3 to the term immediately preceding it. That is,

| | |
|---|---|
| the second term = the first term plus 3 | $5 = 2 + 3$ |
| the third term = the second term plus 3 | $8 = 5 + 3$ |

and so forth. A sequence like this is given a special name. An **arithmetic progression** is a sequence in which every term after the first is the sum of the preceding term and a fixed number called the **common difference** of the progression. We use the following notation.

$a_1$ for the first term
$a_n$ for the $n$th term
$d$ for the common difference
$n$ for the number of terms from $a_1$ to $a_n$, inclusive
$S_n$ for the sum of the first $n$ terms

**EXAMPLE 1** For the arithmetic progression

$$1, 6, 11, 16, \cdots$$

$a_1 = 1$ and $d = 5$ (each term after the first is found by adding 5 to the preceding term). Thus,

$$a_2 = a_1 + d = 1 + 5 = 6,$$
$$a_3 = a_2 + d = 6 + 5 = 11,$$

and so forth. Notice that if we take any term and subtract the preceding term ($6 - 1 = 5$, $11 - 6 = 5$, $16 - 11 = 5$, etc.) the difference is always 5. This is why $d$ is called the *common difference* of the progression.

A general formula for calculating any particular term of an arithmetic progression is a useful tool. Suppose we calculate several terms of an arbitrary arithmetic progression.

$$\begin{aligned}
\text{1st term} &= a_1 = a_1 + 0d \\
\text{2nd term} &= a_2 = a_1 + d = a_1 + 1d \\
\text{3rd term} &= a_3 = a_2 + d = (a_1 + d) + d = a_1 + 2d \\
\text{4th term} &= a_4 = a_3 + d = (a_1 + 2d) + d = a_1 + 3d \\
\text{5th term} &= a_5 = a_4 + d = (a_1 + 3d) + d = a_1 + 4d
\end{aligned}$$

In each case, the $n$th term (5th for example) is the first term plus $(n - 1)$ $(5 - 1 = 4)$ times $d$. Thus, we have

$$n\text{th term} = a_n = a_1 + (n - 1)d$$

which is the formula for the general term of an arithmetic progression.

**EXAMPLE 2** Find the 5th and the 11th terms of the arithmetic progression with first term 3 and common difference 4.

We are given that $a_1 = 3$ and $d = 4$.

$$\begin{aligned}
a_5 &= a_1 + (5 - 1)d = 3 + (4)(4) = 19 \\
a_{11} &= a_1 + (11 - 1)d = 3 + (10)(4) = 43
\end{aligned}$$

Notice that the $n$ in $a_n$ is the same as the $n$ in $(n - 1)$.

The sum of the first $n$ terms in an arithmetic progression can also be calculated by a formula. Let $S_n$ denote the sum of the first $n$ terms of an arithmetic progression. Then

$$\begin{aligned}
S_n &= a_1 + a_2 + a_3 + a_4 + \cdots + a_n \\
&= a_1 + a_1 + d + a_1 + 2d + a_1 + 3d + \cdots + a_1 + (n - 1)d.
\end{aligned}$$

Reversing the order of addition, we obtain

$$\begin{aligned}
S_n &= a_n + a_{n-1} + a_{n-2} + \cdots + a_1 \\
&= a_1 + (n - 1)d + a_1 + (n - 2)d + a_1 + (n - 3)d + \cdots + a_1.
\end{aligned}$$

Add corresponding terms in both representations of $S_n$.

$$\begin{array}{rcccccccc}
S_n = & a_1 & + & a_1 + d & + & a_1 + 2d & + \cdots + & a_1 + (n-1)d \\
S_n = & a_1 + (n-1)d & + & a_1 + (n-2)d & + & a_1 + (n-3)d & + \cdots + & a_1 \\
\hline
2S_n = & 2a_1 + (n-1)d & + & 2a_1 + (n-1)d & + & 2a_1 + (n-1)d & + \cdots + & 2a_1 + (n-1)d
\end{array}$$

$$2S_n = n[2a_1 + (n-1)d] \qquad \text{There are } n \text{ terms of the form } 2a_1 + (n-1)d$$

$$S_n = \frac{n}{2}[2a_1 + (n-1)d]$$

**EXAMPLE 3** Find the 9th term and the sum of the first 9 terms of the arithmetic progression with $a_1 = -2$ and $d = 5$.

We must calculate $a_9$ and $S_9$.

$$a_9 = a_1 + (9-1)d = -2 + (8)5 = -2 + 40 = 38$$

$$S_9 = \frac{9}{2}[2a_1 + (9-1)d] = \frac{9}{2}[2(-2) + (8)(5)]$$

$$= \frac{9}{2}[-4 + 40] = \frac{9}{2}[36] = 162$$

An alternate form for $S_n$, which would have been useful in Example 3 when we calculated $a_9$, is easily derived from

$$S_n = \frac{n}{2}[2a_1 + (n-1)d].$$

By writing $2a_1$ as $a_1 + a_1$ and observing that $a_1 + (n-1)d = a_n$, we have

$$S_n = \frac{n}{2}[a_1 + \underbrace{a_1 + (n-1)d}_{a_n}]$$

$$S_n = \frac{n}{2}[a_1 + a_n].$$

In the preceding example we had already calculated $a_9 = 38$. When we know the $n$th term, it is easier to substitute in the second formula.

$$S_9 = \frac{9}{2}[a_1 + a_n] = \frac{9}{2}[-2 + 38] = \frac{9}{2}[36] = 162$$

**CAUTION:** Do not substitute $n$ (which is 9) for $a_n$ (which is 38). That is, $n \neq a_n$.

**EXAMPLE 4** Find the 20th term and the sum of the first 20 terms of the arithmetic progression

$$-7, -4, -1, 2, \cdots.$$

We have $a_1 = -7$. To find $d$, we take any term and subtract the one preceding it so that in this case, $d = -4 - (-7) = -4 + 7 = 3$.

$$a_{20} = a_1 + (20-1)d = -7 + (19)(3) = -7 + 57 = 50$$

$$S_{20} = \frac{20}{2}[a_1 + a_{20}] = \frac{20}{2}[-7 + 50] = 10[43] = 430$$

Let us summarize the important notions and formulas relative to arithmetic progressions.

A sequence in which each term after the first, $a_1$, is obtained by adding a fixed number $d$, called the common difference, to the preceding term is an **arithmetic progression.** The **general** or ***n*th term** of an arithmetic progression is given by

$$a_n = a_1 + (n-1)d.$$

The sum of the first $n$ terms of an arithmetic progression is given by

$$S_n = \frac{n}{2}[a_1 + a_n] \quad \text{or} \quad S_n = \frac{n}{2}[2a_1 + (n-1)d].$$

**EXAMPLE 5** Find the sum of the even integers from 2 through 100. Since there are fifty even integers from 2 through 100, we must calculate

$$S_{50} = 2 + 4 + 6 + 8 + \cdots + 100.$$

Since $a_1 = 2$, $d = 2$, $n = 50$, and $a_{50} = 100$,

$$S_{50} = \frac{50}{2}[a_1 + a_{50}] = 25[2 + 100] = 25(102) = 2550.$$

**EXAMPLE 6** A new car costs \$10,000. Assume that it depreciates 24% the first year, 20% the second year, 16% the third year, and continues in the same manner for 6 years. If all depreciations apply to the original cost, what is the value of the car in 6 years?

We calculate the sum of the depreciations with $a_1 = 24$, $a_2 = 20$, $a_3 = 16, \cdots$, and $d = -4$.

$$S_6 = \frac{6}{2}[2a_1 + (6-1)d] = \frac{6}{2}[(2)(24) + (5)(-4)]$$
$$= 3[48 - 20] = 3[28] = 84$$

The total percentage of depreciation is 84%, so the total depreciation is

$$84\% \text{ of } 10000 = (.84)(10000) = 8400.$$

Thus, the total value of the car in 6 years is

$$\$10000 - \$8400 = \$1600.$$

**EXAMPLE 7** A theater has 50 rows with 20 seats in the first row, 22 in the second, 24 in the third, and so forth. How many seats are in the theater?

We have $a_1 = 20$, $d = 2$, and $n = 50$, and we must calculate $S_{50}$.

$$S_{50} = \frac{50}{2}[2a_1 + (50-1)d] = 25[(2)(20) + (49)(2)]$$
$$= 25[40 + 98] = 25[138] = 3450$$

## EXERCISES 10.2

1. In an arithmetic progression, each term after the first is obtained by **(a)** ______ ______ a fixed number called the **(b)** ______ to the preceding term.

2. In an arithmetic progression, $a_1$ symbolizes the **(a)** ______, $a_n$ symbolizes the **(b)** ______, $d$ symbolizes the **(c)** ______ ______, and $S_n$ symbolizes the **(d)** ______.

3. The formula for calculating the $n$th term of an arithmetic progression is ______.

4. The formulas for calculating the sum of the first $n$ terms of an arithmetic progression are ______ and ______.

5. Are the following arithmetic progressions? If so, give the common difference.

   **(a)** $3, 6, 9, 12, \cdots$
   **(b)** $2, -2, 2, -2, \cdots$
   **(c)** $1, 1, 1, 1, \cdots$
   **(d)** $12, 5, -2, -9, -16, \cdots$
   **(e)** $-3, -1, 1, 3, \cdots$
   **(f)** $x, x+2, x+4, x+6, \cdots$

6. Write the first 5 terms of the arithmetic progression with $a_1 = -2$ and $d = 6$.

7. Write the first 5 terms of the arithmetic progression with $a_1 = 8$ and $d = -3$.

8. Write the first 6 terms of the arithmetic progression with $a_1 = 5$ and $a_2 = 9$.

9. Find the 20th term and the sum of the first 20 terms of the arithmetic progression with $a_1 = -7$ and $d = 2$.

10. Find the 15th term and the sum of the first 15 terms of the arithmetic progression with $a_1 = 18$ and $d = -3$.

11. Find the sum of the first 20 terms of the arithmetic progression with $a_1 = -12$ and $d = 3$.

**12.** Find the sum of the even integers from 2 through 200.
[*Hint:* $n = 100$]

**13.** A new car costs \$9000. Assume that it depreciates 21% the first year, 18% the second, 15% the third, and continues in the same manner for 5 years. If all depreciations apply to the original cost, what is the value of the car in 5 years?

**14.** An auditorium has 40 rows with 30 seats in the first row, 33 in the second row, 36 in the third row, and so forth. How many seats are in the auditorium?

**15.** A child puts 1¢ in her bank on the first day of June, 2¢ on the second day of June, 3¢ on the third day, and so forth. How much money will be in her bank at the end of the month?
[*Hint:* June has 30 days.]

**16.** The $n$th term of a general sequence (not necessarily an arithmetic progression) is $x_n = 3 + n^2$. Find $x_1$, $x_2$, $x_3$, and $x_5$.

*Rewrite each series using the sigma summation notation.*

**17.** $\frac{1}{2} + \frac{1}{5} + \frac{1}{10} + \cdots + \frac{1}{n^2 + 1}$

**18.** $\frac{1}{2} + \frac{\sqrt{2}}{3} + \frac{\sqrt{3}}{4} + \cdots + \frac{\sqrt{n}}{n + 1} + \cdots$

*Rewrite without using the sigma summation notation.*

**19.** $\sum_{n=1}^{4} \frac{2}{3n+1}$

**20.** $\sum_{k=1}^{\infty} (-1)^k a_k$

---

ANSWERS: **1.** (a) adding (b) common difference **2.** (a) first term (b) $n$th term (c) common difference (d) sum of the first $n$ terms **3.** $a_n = a_1 + (n-1)d$ **4.** $S_n = \frac{n}{2}[2a_1 + (n-1)d]$; $S_n = \frac{n}{2}[a_1 + a_n]$ **5.** (a) yes; $d = 3$ (b) no (c) yes; $d = 0$ (d) yes; $d = -7$ (e) yes; $d = 2$ (f) yes; $d = 2$ **6.** $-2, 4, 10, 16, 22$ **7.** $8, 5, 2, -1, -4$ **8.** $5, 9, 13, 17, 21, 25$ **9.** $a_{20} = 31$; $S_{20} = 240$ **10.** $a_{15} = -24$; $S_{15} = -45$ **11.** $S_{20} = 330$ **12.** 10,100 **13.** \$2250 **14.** 3540 seats **15.** \$4.65 **16.** $x_1 = 4$; $x_2 = 7$; $x_3 = 12$; $x_5 = 28$ **17.** $\sum_{k=1}^{n} \frac{1}{k^2+1}$ **18.** $\sum_{n=1}^{\infty} \frac{\sqrt{n}}{n+1}$ **19.** $\frac{2}{4} + \frac{2}{7} + \frac{2}{10} + \frac{2}{13}$ **20.** $-a_1 + a_2 - a_3 + a_4 - a_5 + \cdots$

## 10.3 Geometric Progressions

The second special type of sequence we study is called a geometric progression. In the sequence

$$1, 3, 9, 27, 81, \cdots$$

each term (after the first) can be found by multiplying the preceding term by 3. A **geometric progression** is a sequence for which every term after the first is the product of the preceding term and a fixed number called the **common ratio** of the progression. We use the following notation.

$a_1$ for the first term
$a_n$ for the $n$th term
$r$ for the common ratio
$n$ for the number of terms from $a_1$ to $a_n$, inclusive
$S_n$ for the sum of the first $n$ terms

EXAMPLE 1 For the geometric progression

$$1, \frac{1}{2}, \frac{1}{4}, \frac{1}{8}, \frac{1}{16}, \ldots$$

$a_1 = 1$ and $r = \frac{1}{2}$ (to obtain each succeeding term multiply the preceding one by $\frac{1}{2}$). Thus,

$$a_2 = a_1 \cdot r = 1 \cdot \frac{1}{2} = \frac{1}{2},$$

$$a_3 = a_2 \cdot r = \frac{1}{2} \cdot \frac{1}{2} = \frac{1}{4},$$

$$a_4 = a_3 \cdot r = \frac{1}{4} \cdot \frac{1}{2} = \frac{1}{8},$$

and so forth. Notice that if we divide any term by the preceding one $\left(\frac{1}{2} \div 1 = \frac{1}{2}, \frac{1}{4} \div \frac{1}{2} = \frac{1}{2}, \frac{1}{8} \div \frac{1}{4} = \frac{1}{2}, \text{ etc.}\right)$, the quotient or ratio is always $\frac{1}{2}$.
This is why we call $r$ the *common ratio* of the progression.

As with arithmetic progressions, there is a formula for calculating the $n$th (or general) term of a geometric progression. Let us calculate several terms of an arbitrary geometric progression.

$$\begin{aligned}
\text{1st term} &= a_1 = a_1 r^0 \\
\text{2nd term} &= a_2 = a_1 \cdot r = a_1 r^1 \\
\text{3rd term} &= a_3 = a_2 \cdot r = (a_1 r) \cdot r = a_1 r^2 \\
\text{4th term} &= a_4 = a_3 \cdot r = (a_1 r^2) \cdot r = a_1 r^3 \\
\text{5th term} &= a_5 = a_4 \cdot r = (a_1 r^3) \cdot r = a_1 r^4
\end{aligned}$$

In each case, the $n$th term (5th for example) is the first term times $r$ raised to the $(n - 1)$ power $(5 - 1 = 4)$. Thus, we have

$$n\text{th term} = a_n = a_1 r^{n-1}.$$

EXAMPLE 2 Find the 7th and 10th terms of the following geometric progression.

$$9, 3, 1, \frac{1}{3}, \frac{1}{9}, \ldots$$

We have $a_1 = 9$ and $r = \frac{1}{3}$.

$$a_7 = a_1 r^{7-1} = 9\left(\frac{1}{3}\right)^6 = 9\left(\frac{1}{9}\right)\left(\frac{1}{3^4}\right) = \frac{1}{3^4} = \frac{1}{81}$$

$$a_{10} = a_1 r^{10-1} = 9\left(\frac{1}{3}\right)^9 = 9\left(\frac{1}{9}\right)\left(\frac{1}{3^7}\right) = \frac{1}{3^7} = \frac{1}{2187}$$

Notice that the $n$ in $a_n$ is the same as the $n$ in the exponent $n - 1$.

We now derive a formula for calculating the sum of the first $n$ terms of a geometric progression. Let $S_n$ denote the sum of the first $n$ terms.

$$S_n = a_1 + a_1 r + a_1 r^2 + a_1 r^3 + \cdots + a_1 r^{n-2} + a_1 r^{n-1}$$

Multiply $S_n$ by $r$.

$$rS_n = a_1 r + a_1 r^2 + a_1 r^3 + \cdots + a_1 r^{n-2} + a_1 r^{n-1} + a_1 r^n$$

Subtract $rS_n$ from $S_n$.

$$\begin{aligned}
S_n &= a_1 + a_1 r + a_1 r^2 + a_1 r^3 + \cdots + a_1 r^{n-2} + a_1 r^{n-1} \\
-rS_n &= \quad\; - a_1 r - a_1 r^2 - a_1 r^3 - \cdots - a_1 r^{n-2} - a_1 r^{n-1} - a_1 r^n \\
\hline
S_n - rS_n &= a_1 \qquad\qquad\qquad\qquad\qquad\qquad\qquad\qquad - a_1 r^n \\
S_n(1 - r) &= a_1 - a_1 r^n \\
S_n &= \frac{a_1 - a_1 r^n}{1 - r}
\end{aligned}$$

**EXAMPLE 3** Find the 8th term and the sum of the first 8 terms of the geometric progression

$$-2, 1, -\frac{1}{2}, \frac{1}{4}, \cdots.$$

We have $a_1 = -2$ and $r = -\frac{1}{2}$. Since $r$ is negative, the signs of the terms alternate. We must calculate $a_8 (n = 8)$ and $S_8$.

$$a_8 = a_1 r^{8-1} = (-2)\left(-\frac{1}{2}\right)^7 = (-2)\left(-\frac{1}{2}\right)\left(-\frac{1}{2}\right)^6 = \frac{1}{2^6} = \frac{1}{64}$$

$$S_8 = \frac{a_1 - a_1 r^8}{1 - r} = \frac{(-2) - (-2)\left(-\frac{1}{2}\right)^8}{1 - \left(-\frac{1}{2}\right)} = \frac{(-2) - (-2)\left(-\frac{1}{2}\right)\left(-\frac{1}{2}\right)^7}{\frac{3}{2}}$$

$$= \frac{(-2) - \left(-\frac{1}{2^7}\right)}{\frac{3}{2}} = \frac{(-2) + \frac{1}{128}}{\frac{3}{2}}$$

$$= \frac{-256 + 1}{128} \cdot \frac{2}{3} = -\frac{255}{128} \cdot \frac{2}{3} = -\frac{85}{64}$$

An alternate form for $S_n$ can be derived from

$$S_n = \frac{a_1 - a_1 r^n}{1 - r}$$

by writing $a_1 r^n = r(a_1 r^{n-1}) = r a_n$.

$$S_n = \frac{a_1 - r a_n}{1 - r}$$

In Example 3 we calculated $a_8 = \frac{1}{64}$. It might have been easier to obtain $S_8$ by substituting $a_8$ in the formula above.

$$S_8 = \frac{a_1 - r a_8}{1 - r} = \frac{(-2) - \left(-\frac{1}{2}\right)\left(\frac{1}{64}\right)}{1 - \left(-\frac{1}{2}\right)} = \frac{-2 + \frac{1}{128}}{\frac{3}{2}} = -\frac{255}{128} \cdot \frac{2}{3} = -\frac{85}{64}$$

**EXAMPLE 4** Find the 10th term and the sum of the first 10 terms of the following geometric progression.

$$1, -2, 4, -8, \cdots$$

In this case, $a_1 = 1$ and $r = -2$.

$$a_{10} = a_1 r^{10-1} = (1)(-2)^9 = -2^9 = -512$$

$$S_{10} = \frac{a_1 - r a_{10}}{1 - r} = \frac{1 - (-2)(-512)}{1 - (-2)} = \frac{1 - 1024}{3} = -\frac{1023}{3} = -341$$

Let us summarize the important notions and formulas related to geometric progressions.

A sequence in which each term after the first, $a_1$, is obtained by multiplying the preceding term by a fixed number $r$, called the common ratio, is a **geometric progression.** The **general** or ***n*th term** of a geometric progression is given by

$$a_n = a_1 r^{n-1}.$$

The sum of the first $n$ terms of a geometric progression is given by

$$S_n = \frac{a_1 - ra_n}{1 - r} \quad \text{or} \quad S_n = \frac{a_1 - a_1 r^n}{1 - r}.$$

**EXAMPLE 5** Find $a_1$ and $r$ for a geometric progression that has $a_2 = 10$ and $a_5 = 80$.

$$80 = a_5 = a_1 r^{5-1} = a_1 r^4$$
$$10 = a_2 = a_1 r^{2-1} = a_1 r$$

Divide these terms.

$$\frac{80}{10} = \frac{a_5}{a_2} = \frac{a_1 r^4}{a_1 r} = r^3$$
$$8 = r^3$$
$$2 = r$$

Then

$$10 = a_2 = a_1 r = a_1 2$$
$$10 = a_1 2$$
$$5 = a_1.$$

Thus, $a_1 = 5$ and $r = 2$.

**EXAMPLE 6** A new car costing \$8000 depreciates 20% of its value each year. How much is the car worth at the end of 5 years?

When solving a word problem involving geometric progressions (arithmetic progressions also), it is wise to write out the first few terms. Once this is done, $a_1$ are $r$ are much easier to identify.

| At beginning of 1st year | At beginning of 2nd year | At beginning of 3rd year | At beginning of 4th year |
|---|---|---|---|
| $a_1 = 8000$ | $a_2 = (.80)(8000)$ | $a_3 = (.80)^2(8000)$ | $a_4 = (.80)^3(8000)$ |
| | $= 6400$ | $= (.80)(6400)$ | $= (.80)(5120)$ |
| | | $= 5120$ | $= 4096$ |

Notice that to obtain the next term in the sequence, the preceding term is multiplied by .80 (80%) giving the value of the car each succeeding year. Thus, $a_1 = 8000$ and $r = .8$, and we must find $a_6$ to obtain the value of the car at the end of 5 years (the value at the end of the fifth year is equal to the value at the beginning of the sixth year). Thus, $n = 6$ and

$$a_6 = a_1 r^{6-1} = (8000)(.80)^5 \approx 2621.$$

The value of the car after five years is approximately $2621.

CAUTION: Often, the preceding problem is solved incorrectly by assuming that $n = 5$ ("at end of five years"). However, when we write out a few terms, it is clear that $n$ must be 6. Similar remarks apply to interest (compounded annually) problems and population growth problems.

EXAMPLE 7 A woman borrows $1000.00 at 12% interest compounded annually. If she pays off the loan in full at the end of 3 years, how much does she pay?

| At beginning of 1st year | At beginning of 2nd year (or end of 1st year) | At beginning of 3rd year (or end of 2nd year) |
|---|---|---|
| $a_1 = 1000$ | $a_2 = 1000 + .12(1000)$ | $a_3 = (1.12)(1000) + (.12)(1.12)(1000)$ |
| | $= [1 + .12](1000)$ | $= [1 + .12](1.12)(1000)$ |
| | $= (1.12)(1000)$ | $= (1.12)^2(1000)$ |

To obtain the next term of the sequence, we multiply the preceding term by 1.12. To find the amount that would have to be repaid at the end of 3 years, we calculate $a_4$ (4 not 3). We have $a_1 = 1000$, $r = 1.12$, and

$$a_4 = a_1 r^{4-1} = (1000)(1.12)^3 \approx (1000)(1.40493) = 1404.93.$$

Thus, at the end of 3 years the woman would have to pay back $1404.93.

## EXERCISES 10.3

1. In a geometric progression, each term after the first is obtained by **(a)** ____________ a fixed number called the **(b)** ____________ by the preceding term.
2. In a geometric progression, $a_1$ symbolizes the **(a)** ____________, $a_n$ symbolizes the **(b)** ____________, $r$ symbolizes the **(c)** ____________, and $S_n$ symbolizes the **(d)** ____________.
3. The formula for calculating the $n$th term of a geometric progression is ____________.
4. The formula for calculating the $n$th term of an arithmetic progression is ____________.
5. The formulas for calculating the sum of the first $n$ terms of a geometric progression are ____________ and ____________.
6. The formulas for calculating the sum of the first $n$ terms of an arithmetic progression are ____________ and ____________.

**7.** Are the following geometric progressions? If so, give the common ratio.

(a) $5, 15, 45, 135, \cdots$

(b) $-1, 1, -1, 1, -1, 1, \cdots$

(c) $4, 2, 1, \frac{1}{2}, \cdots$

(d) $-3, 1, -\frac{1}{3}, \frac{1}{9}, \cdots$

(e) $1, 1, 1, 1, \cdots$

(f) $-5, -2, -\frac{4}{5}, -\frac{8}{25}, \cdots$

**8.** Write the first 5 terms of the geometric progression with $a_1 = 3$ and $r = 2$.

**9.** Write the first 5 terms of the geometric progression with $a_1 = \frac{1}{8}$ and $r = -2$.

**10.** Write the first 5 terms of the geometric progression with $a_1 = -15$ and $r = \frac{1}{3}$.

**11.** Write the first 5 terms of the arithmetic progression with $a_1 = 7$ and $d = -2$.

**12.** Write the first 5 terms of the geometric progression with $a_1 = 32$ and $r = \frac{1}{4}$.

**13.** Write the first 5 terms of the arithmetic progression with $a_1 = -\frac{1}{2}$ and $d = \frac{1}{2}$.

**14.** Find the 8th term and the sum of the first 8 terms of the geometric progression with $a_1 = 128$ and $r = -\frac{1}{2}$.

**15.** Find the sum of the first 6 terms of the geometric progression with $a_1 = -\frac{1}{9}$ and $a_6 = -27$.

**16.** Find $a_1$ and $r$ for the geometric progression which has $a_2 = \frac{1}{3}$ and $a_5 = -9$.

**17.** Find the sum of the first 8 terms of the arithmetic progression with $a_1 = -18$ and $a_8 = 38$.

**18.** A new car costing \$9000 depreciates 30% of its value each year. How much is the car worth at the end of 5 years?

**19.** A woman borrows \$1000 at 12% interest compounded annually. If she pays off the loan in full at the end of 4 years, how much does she pay?

**20.** Burford was offered a job for the month of June (30 days), and was told he would be paid 1¢ at the end of the first day, 2¢ at the end of the second day, 4¢ at the end of the third day, and so forth, doubling each previous day's salary. However, Burford refused the job thinking that the pay was inferior. Would you take the job? Why?

**21.** The population of a town is increasing 10% each year. If its present population is 1500, what will be its population in 5 years?
[*Hint:* This is a geometric progression problem.]

**22.** The population of a town is increasing by 100 each year. If its present population is 1500, what will be its population in 5 years?
[*Hint:* This is an arithmetic progression problem.]

**23.** What is the sum of the odd numbers from 1 through 99?
[*Hint:* $n = 50$]

**24.** A collection of dimes is arranged in a triangular array with 10 coins in the base row, 9 in the next, 8 in the next, and so forth. Find the value of the collection.

---

ANSWERS: **1.** (a) multiplying (b) common ratio **2.** (a) first term (b) $n$th term (c) common ratio (d) sum of the first $n$ terms **3.** $a_n = a_1 r^{n-1}$ **4.** $a_n = a_1 + (n-1)d$ **5.** $S_n = \frac{a_1 - ra_n}{1-r}$ and $S_n = \frac{a_1 - a_1 r^n}{1-r}$ **6.** $S_n = \frac{n}{2}[a_1 + a_n]$ and $S_n = \frac{n}{2}[2a_1 + (n-1)d]$ **7.** (a) yes; $r = 3$ (b) yes; $r = -1$ (c) yes; $r = \frac{1}{2}$ (d) yes; $r = -\frac{1}{3}$ (e) yes; $r = 1$ (f) yes; $r = \frac{2}{5}$ **8.** 3, 6, 12, 24, 48 **9.** $\frac{1}{8}, -\frac{1}{4}, \frac{1}{2}, -1, 2$ **10.** $-15, -5, -\frac{5}{3}, -\frac{5}{9}, -\frac{5}{27}$ **11.** 7, 5, 3, 1, $-1$ **12.** 32, 8, 2, $\frac{1}{2}, \frac{1}{8}$ **13.** $-\frac{1}{2}, 0, \frac{1}{2}, 1, \frac{3}{2}$ **14.** $a_8 = -1, S_8 = 85$ **15.** $-\frac{364}{9}$ **16.** $a_1 = -\frac{1}{9}, r = -3$ **17.** $S_8 = 80$ **18.** approximately $1500 **19.** $1573.52 **20.** Your answer should be *yes*, since you would be working for approximately $10,700,000 for the month. **21.** approximately 2416 people **22.** 2000 people **23.** 2500 **24.** $5.50

## 10.4 Infinite Geometric Series

We have formulas for the sum of the first $n$ terms of a geometric progression. In some instances, it is possible to find the sum of all the terms of a geometric progression even if there are infinitely many. To do this, we must generalize "finding a sum." Recall the story of the man who lives 1 mile from town and plans to walk from his home to town by walking one-half the remaining distance each day until he arrives. Perhaps this story can help us to see the significance of an infinite geometric series.

Distance walked first day: $a_1 = \frac{1}{2}$

Distance walked second day: $a_2 = \frac{1}{4}$

Distance walked third day: $a_3 = \frac{1}{8}$

$\vdots$

Distance walked $n$th day: $a_n = \left(\frac{1}{2}\right)^n$

The distances walked each day form the following geometric progression.

$$\frac{1}{2}, \frac{1}{4}, \frac{1}{8}, \ldots, \left(\frac{1}{2}\right)^n, \ldots \qquad \left(r = \frac{1}{2},\ a_1 = \frac{1}{2}\right)$$

Suppose that we calculate the sum of the first $n$ terms of this progression, for $n = 1, 2, 3, 4, 5, 6, 7$.

$$S_1 = \frac{1}{2}$$

$$S_2 = \frac{1}{2} + \frac{1}{4} = \frac{3}{4}$$

$$S_3 = \frac{1}{2} + \frac{1}{4} + \frac{1}{8} = \frac{7}{8}$$

$$S_4 = \frac{1}{2} + \frac{1}{4} + \frac{1}{8} + \frac{1}{16} = \frac{15}{16}$$

$$S_5 = \frac{1}{2} + \frac{1}{4} + \frac{1}{8} + \frac{1}{16} + \frac{1}{32} = \frac{31}{32}$$

$$S_6 = \frac{1}{2} + \frac{1}{4} + \frac{1}{8} + \frac{1}{16} + \frac{1}{32} + \frac{1}{64} = \frac{63}{64}$$

$$S_7 = \frac{1}{2} + \frac{1}{4} + \frac{1}{8} + \frac{1}{16} + \frac{1}{32} + \frac{1}{64} + \frac{1}{128} = \frac{127}{128}$$

When will $S_n$ equal 1? That is, when will the man reach town? Clearly, he will never actually reach town since on any given day, he walks only half the remaining distance. However, if we assume that the man continues to walk in the prescribed manner, the total distance that he has walked gets closer and closer to 1 (the sequence of sums $\frac{1}{2}, \frac{3}{4}, \frac{7}{8}, \frac{15}{16}, \frac{31}{32}, \frac{63}{64}, \frac{127}{128}, \cdots$ gets closer and closer to 1). In a sense, then, we could say that the total distance walked is 1 mile (or that the sum of the geometric progression $\frac{1}{2} + \frac{1}{4} + \frac{1}{8} + \frac{1}{16} + \cdots$ is 1). It is in this sense that we use the term *sum* when applying it to an infinite geometric progression. That is, if an infinite geometric progression

$$a_1, a_2, a_3, a_4, \cdots$$

has

$$S_1, S_2, S_3, S_4, \cdots$$

as its sequence of **partial sums** (sequence of sums of the first $n$ terms), and if the $S_n$'s approach some fixed number $S$ as $n$ gets larger and larger, we call $S$ the **sum of the geometric series**

$$a_1 + a_2 + a_3 + a_4 + \cdots.$$

Some infinite geometric series have sums and others do not. For example,

$$1 + 3 + 9 + 27 + 81 + \cdots \qquad (r = 3,\ a_1 = 1)$$

does not have a sum since the $S_n$'s

$$S_1 = 1, \quad S_2 = 4, \quad S_3 = 13, \quad S_4 = 40, \quad S_5 = 121, \cdots$$

get larger and do not approach some fixed number $S$. On the other hand,

$$1 + \frac{1}{3} + \frac{1}{9} + \frac{1}{27} + \frac{1}{81} + \cdots \qquad \left(r = \frac{1}{3},\ a_1 = 1\right)$$

does have a sum, since the $S_n$'s

$$S_1 = 1, \quad S_2 = 1\frac{1}{3}, \quad S_3 = 1\frac{4}{9}, \quad S_4 = 1\frac{13}{27}, \quad S_5 = 1\frac{40}{81}, \ldots$$

are approaching $1\frac{1}{2} = \frac{3}{2}$. In the first example, $r = 3$ while in the second, $r = \frac{1}{3}$. When the common ratio, $r$, in an infinite geometric series is such that $|r| < 1$, then the series has a sum in the sense described above. Considering the formula for the sum of the first $n$ terms in a geometric progression,

$$S_n = \frac{a_1 - a_1 r^n}{1 - r},$$

if $|r| < 1$, then $r^n$ gets smaller as $n$ gets larger. Thus, $a_1 r^n$ gets smaller and $S_n$ approaches

$$\frac{a_1}{1 - r}.$$

An infinite geometric progression with first term $a_1$ and common ratio $r$ satisfying $|r| < 1$ has a sum given by

$$S = \frac{a_1}{1 - r}.$$

**EXAMPLE 1** Find the sum of the infinite geometric progression $\frac{1}{2}, \frac{1}{4}, \frac{1}{8}, \frac{1}{16}, \frac{1}{32}, \ldots$.

Since $r = \frac{1}{2}$ and $|r| = \left|\frac{1}{2}\right| = \frac{1}{2} < 1$, the sum exists and is given by

$$S = \frac{a_1}{1 - r} = \frac{\frac{1}{2}}{1 - \frac{1}{2}} = \frac{\frac{1}{2}}{\frac{1}{2}} = 1.$$

(Compare with the discussion earlier.)

**EXAMPLE 2** Find the sum of the infinite geometric progression $1, \frac{1}{3}, \frac{1}{9}, \frac{1}{27}, \frac{1}{81}, \ldots$.

Since $r = \frac{1}{3}$ and $|r| = \left|\frac{1}{3}\right| = \frac{1}{3} < 1$, the sum exists and is given by

$$S = \frac{a_1}{1 - r} = \frac{1}{1 - \frac{1}{3}} = \frac{1}{\frac{2}{3}} = 1 \cdot \frac{3}{2} = \frac{3}{2}.$$

(Compare with the discussion earlier.)

**EXAMPLE 3** Find the sum of the infinite geometric progression $\frac{1}{64}, \frac{1}{32}, \frac{1}{16}, \frac{1}{8}, \ldots$.

Since $r = 2$ and $|r| = |2| = 2 > 1$, the sum does not exist. This becomes clear when the next few terms $\left(\frac{1}{4}, \frac{1}{2}, 1, 2, 4, 8, 16\right)$ are computed.

**EXAMPLE 4** Is there an infinite geometric progression with $a_1 = 2$ and $S = -4$?

Since $S = \frac{a_1}{1 - r}$, we have

$$-4 = \frac{2}{1 - r}$$

$$-4(1 - r) = 2 \qquad \text{Multiply by } 1 - r$$

$$-4 + 4r = 2$$

$$4r = 6$$

$$r = \frac{6}{4} = \frac{3}{2}.$$

But since $|r| = \left|\frac{3}{2}\right| = \frac{3}{2} > 1$, there cannot be an infinite geometric progression with first term 2 and sum $-4$.

**EXAMPLE 5** The tip of a pendulum moves back and forth in such a way that it sweeps out an arc 18 in long, and on each succeeding pass the length of the arc traveled is $\frac{8}{9}$ of the length of the preceding pass. What is the total distance traveled by the tip of the pendulum?

In reality, the pendulum will eventually stop because of friction. However, for simplicity we use an infinite geometric series to represent the total distance traveled.

$$18 + \frac{8}{9}(18) + \left(\frac{8}{9}\right)^2(18) + \left(\frac{8}{9}\right)^3(18) + \cdots$$

Then $a_1 = 18$, $r = \frac{8}{9}$ $\left(|r| = \left|\frac{8}{9}\right| = \frac{8}{9} < 1\right)$, so that

$$S = \frac{a_1}{1 - r} = \frac{18}{1 - \frac{8}{9}} = \frac{18}{\frac{1}{9}} = 18 \cdot 9 = 162.$$

Thus, the total distance traveled is approximately 162 in.

The formula for the sum of an infinite geometric progression is also used to convert repeating decimals to fractions.

**EXAMPLE 6** Convert $.\bar{3}$ to a fraction.

Since

$$\begin{aligned} .\bar{3} &= .33333 \cdots \\ &= .3 + .03 + .003 + .0003 + \cdots \\ &= .3 + (.1)(.3) + (.1)^2(.3) + (.1)^3(.3) + \cdots \end{aligned}$$

we see that $.\bar{3}$ is really an infinite geometric progression with $a_1 = .3$ and $r = .1$. Then

$$S = \frac{a_1}{1 - r} = \frac{.3}{1 - .1} = \frac{.3}{.9} = \frac{3}{9} = \frac{1}{3}.$$

Thus, $.\bar{3} = \frac{1}{3}$ (which can be checked by division).

EXAMPLE 7 Convert $.\overline{27}$ to a fraction.

Since

$$\begin{aligned} .\overline{27} &= .27 + .0027 + .000027 + .00000027 + \cdots \\ &= .27 + (.01)(.27) + (.01)^2(.27) + (.01)^3(.27) + \cdots \end{aligned}$$

we have $a_1 = .27$, $r = .01$, and

$$S = \frac{a_1}{1 - r} = \frac{.27}{1 - .01} = \frac{.27}{.99} = \frac{3}{11}.$$

Thus, $.\overline{27} = \frac{3}{11}$. (Check by division.)

## EXERCISES 10.4

*Find the sum of the given geometric progression.*

**1.** $1, \frac{3}{4}, \frac{9}{16}, \frac{27}{64}, \ldots$

**2.** $\frac{1}{50}, \frac{1}{10}, \frac{1}{2}, \frac{5}{2}, \ldots$

**3.** $9, 1, \frac{1}{9}, \frac{1}{81}, \ldots$

**4.** $16, -4, 1, -\frac{1}{4}, \ldots$

**5.** $25, 15, 9, \frac{27}{5}, \ldots$

**6.** $-36, 6, -1, \frac{1}{6}, -\frac{1}{36}, \ldots$

**7.** Find the first four terms of an infinite geometric progression with $a_1 = 15$ and $S = 30$.

**8.** Find the first four terms of an infinite geometric progression with $a_1 = 16$ and $S = 4$.

**9.** Evaluate $\sum_{k=1}^{\infty} \left(\frac{1}{4}\right)^k$.

**10.** Evaluate $\sum_{k=1}^{\infty} 5^{-k}$.

**11.** The tip of a pendulum moves back and forth in such a way that it sweeps out an arc 24 in long, and on each succeeding pass the length of the arc traveled is $\frac{7}{8}$ the length of the preceding pass. What is the total distance traveled by the tip of the pendulum?

**12.** A child on a swing traverses an arc of 21 ft. Each pass thereafter, the arc is $\frac{6}{7}$ the length of the previous arc. How far does he travel before coming to rest?

*Convert each decimal to a fraction and check your work by division.*

**13.** $.\overline{4}$

**14.** $.\overline{72}$

**15.** $.\overline{312}$

**16.** $1.\overline{3}$
[*Hint:* $1.\overline{3} = 1 + .\overline{3}$]

**17.** $2.\overline{15}$

**18.** $5.\overline{2}$

**19.** Find the 7th term and the sum of the first 7 terms of a geometric progression with $a_1 = \frac{1}{16}$ and $r = -4$.

**20.** A woman borrows \$2000 at 11% interest compounded annually. If she pays off the loan in full at the end of 5 years, how much does she pay?

---

ANSWERS: **1.** $4\left(r=\frac{3}{4}\right)$ **2.** no sum ($r=5$) **3.** $\frac{81}{8}\left(r=\frac{1}{9}\right)$ **4.** $\frac{64}{5}\left(r=-\frac{1}{4}\right)$ **5.** $\frac{125}{2}\left(r=\frac{3}{5}\right)$ **6.** $-\frac{216}{7}\left(r=-\frac{1}{6}\right)$ **7.** $15, \frac{15}{2}, \frac{15}{4}, \frac{15}{8}\left(r=\frac{1}{2}\right)$ **8.** No such infinite geometric series can exist since $r$ is $-3$. **9.** $\frac{1}{3}\left(a_1=\frac{1}{4}, r=\frac{1}{4}\right)$ **10.** $\frac{1}{4}\left(5^{-k}=\left(\frac{1}{5}\right)^k, a_1=\frac{1}{5}, r=\frac{1}{5}\right)$ **11.** 192 in **12.** 147 ft **13.** $\frac{4}{9}$ **14.** $\frac{8}{11}$ **15.** $\frac{104}{333}$ **16.** $\frac{4}{3}$ **17.** $\frac{71}{33}$ **18.** $\frac{47}{9}$ **19.** $a_7=256, S_7=\frac{3277}{16}$ **20.** \$3370.12

## 10.5 Binomial Expansion

When a binomial of the form $a + b$ is raised to a power, the resulting polynomial can be thought of as a series. Suppose we expand several such powers and search for a pattern.

$$
\begin{aligned}
(a+b)^0 &= 1\\
(a+b)^1 &= a+b\\
(a+b)^2 &= a^2+2ab+b^2\\
(a+b)^3 &= a^3+3a^2b+3ab^2+b^3\\
(a+b)^4 &= a^4+4a^3b+6a^2b^2+4ab^3+b^4\\
(a+b)^5 &= a^5+5a^4b+10a^3b^2+10a^2b^3+5ab^4+b^5\\
(a+b)^6 &= a^6+6a^5b+15a^4b^2+20a^3b^3+15a^2b^4+6ab^5+b^6
\end{aligned}
$$

In each case we observe the following:

There are always $n + 1$ terms in the expansion.

The exponents on $a$ start with $n$ and decrease to 0.

The exponents on $b$ start with 0 and increase to $n$.

The sum of the exponents in each term is always $n$.

If $a$ and $b$ are both positive, all terms are positive.

If $a$ is positive and $b$ is negative, the terms have alternating signs; those with odd powers of $b$ are negative.

If $a$ is negative and $b$ is positive, the terms have alternating signs; those with odd powers of $a$ are negative.

If $a$ and $b$ are both negative, all terms are positive if $n$ is even and negative if $n$ is odd.

To discover the pattern of the numerical coefficients of each term, we write the coefficients in the same arrangement as in the preceding expansions.

| | |
|---|---|
| Row 0 | 1 |
| Row 1 | 1 1 |
| Row 2 | 1 2 1 |
| Row 3 | 1 3 3 1 |
| Row 4 | 1 4 6 4 1 |
| Row 5 | 1 5 10 10 5 1 |
| Row 6 | 1 6 15 20 15 6 1 |

This triangular array forms what is known as **Pascal's triangle.** The row number corresponds to the exponent $n$ in the expansion of $(a + b)^n$. The numbers in any row, other than the first and the last which are always 1, can be determined by adding the two numbers immediately above and to the left and right of it. For example, as indicated, 15 is 5 + 10. Thus, Pascal's triangle gives us one way to determine the coefficients in the expansion of a given binomial.

**EXAMPLE 1** Expand $(2x + y)^4$.

In this example $n = 4$, $a = 2x$, and $b = y$.
Row 5 of Pascal's triangle has the following numbers for the coefficients.

$$1 \quad 4 \quad 6 \quad 4 \quad 1$$

We thus substitute $a = 2x$ and $b = y$ into

$$a^4 + 4a^3b + 6a^2b^2 + 4ab^3 + b^4$$

to obtain

$$\begin{aligned} &(2x)^4 + 4(2x)^3y + 6(2x)^2y^2 + 4(2x)y^3 + y^4 \\ &= 16x^4 + 4(8x^3)y + 6(4x^2)y^2 + 4(2x)y^3 + y^4 \\ &= 16x^4 + 32x^3y + 24x^2y^2 + 8xy^3 + y^4. \end{aligned}$$

**EXAMPLE 2** Expand $(z - 3)^5$.

In this example, $n = 5$, $a = z$, and $b = -3$ (since we are expanding $(z + (-3))^5$. The signs in the resulting series alternate since $b$ is negative. From Row 5 of Pascal's triangle, the coefficients are as follows.

$$1 \quad 5 \quad 10 \quad 10 \quad 5 \quad 1$$

If we substitute $a = z$ and $b = -3$ into

$$a^5 + 5a^4b + 10a^3b^2 + 10a^2b^3 + 5ab^4 + b^5$$

we obtain

$$\begin{aligned} &z^5 + 5z^4(-3) + 10z^3(-3)^2 + 10z^2(-3)^3 + 5z(-3)^4 + (-3)^5 \\ &= z^5 + 5z^4(-3) + 10z^3(9) + 10z^2(-27) + 5z(81) + (-243) \\ &= z^5 - 15z^4 + 90z^3 - 270z^2 + 405z - 243. \end{aligned}$$

**EXAMPLE 3** Expand $(a^2 - 2b)^6$.

Since we are expanding $(a^2 + (-2b))^6$, $n = 6$, $a = a^2$, and $b = -2b$. From Row 6 of Pascal's triangle, the coefficients are

$$1 \quad 6 \quad 15 \quad 20 \quad 15 \quad 6 \quad 1.$$

If we substitute $a^2$ for $a$ and $-2b$ for $b$ in

$$a^6 + 6a^5b + 15a^4b^2 + 20a^3b^3 + 15a^2b^4 + 6ab^5 + b^6$$

we obtain

$$(a^2)^6 + 6(a^2)^5(-2b) + 15(a^2)^4(-2b)^2 + 20(a^2)^3(-2b)^3 + 15(a^2)^2(-2b)^4 + 6(a^2)(-2b)^5 + (-2b)^6$$
$$= a^{12} - 12a^{10}b + 60a^8b^2 - 160a^6b^3 + 240a^4b^4 - 192a^2b^5 + 64b^6.$$

The student who continues in mathematics will encounter the *binomial theorem,* which provides a more sophisticated technique for expanding binomials when $n$ is very large and for determining only one particular term in an expansion.

## EXERCISES 10.5

**1.** Construct Pascal's triangle through Row 8.

*Expand each of the following binomials by using Pascal's triangle.*

**2.** $(x + 3)^5$

**3.** $(x - 3)^5$

**4.** $(x - y)^6$

**5.** $(x - 2y)^4$

**6.** $(x^2 + y)^6$

**7.** $(x^2 - 3y)^4$

**8.** $(x + y)^8$

**9.** $(a - b)^8$

**10.** $(x^{-1} + x)^5$

**11.** $(3 - 2)^5$

**12.** $(1 + 1)^7$

**13.** Find the sum of the infinite geometric progression with $a_1 = 7$ and $r = \frac{1}{8}$.

**14.** Convert each decimal to a fraction and check by division.

**(a)** $3.\overline{7}$ **(b)** $2.\overline{36}$

---

ANSWERS: **1.** Row 7: 1 7 21 35 35 21 7 1; Row 8: 1 8 28 56 70 56 28 8 1
**2.** $x^5 + 15x^4 + 90x^3 + 270x^2 + 405x + 243$ **3.** $x^5 - 15x^4 + 90x^3 - 270x^2 + 405x - 243$
**4.** $x^6 - 6x^5y + 15x^4y^2 - 20x^3y^3 + 15x^2y^4 - 6xy^5 + y^6$ **5.** $x^4 - 8x^3y + 24x^2y^2 - 32xy^3 + 16y^4$
**6.** $x^{12} + 6x^{10}y + 15x^8y^2 + 20x^6y^3 + 15x^4y^4 + 6x^2y^5 + y^6$ **7.** $x^8 - 12x^6y + 54x^4y^2 - 108x^2y^3 + 81y^4$
**8.** $x^8 + 8x^7y + 28x^6y^2 + 56x^5y^3 + 70x^4y^4 + 56x^3y^5 + 28x^2y^6 + 8xy^7 + y^8$ **9.** $a^8 - 8a^7b + 28a^6b^2 - 56a^5b^3 + 70a^4b^4 - 56a^3b^5 + 28a^2b^6 - 8ab^7 + b^8$ **10.** $\frac{1}{x^5} + \frac{5}{x^3} + \frac{10}{x} + 10x + 5x^3 + x^5$ **11.** 1 (note that $(3 - 2)^5 = 1^5 = 1$)
**12.** 128 (note that $(1 + 1)^7 = 2^7 = 128$) **13.** 8 **14.** (a) $\frac{34}{9}$ (b) $\frac{26}{11}$

## Chapter 10 Summary

### Words and Phrases for Review

[10.1] sequence
term
finite sequence
infinite sequence
$n$th (general) term
series
finite series
infinite series
summation symbol ($\Sigma$)
index

[10.2] arithmetic progression
common difference

[10.3] geometric progression
common ratio

[10.4] infinite geometric series

[10.5] binomial expansion
Pascal's triangle

### Brief Reminders

[10.1] The Greek letter $\Sigma$, called the summation symbol, is used to abbreviate a series. For example, $a_1 + a_2 + a_3 = \sum_{n=1}^{3} a_n$.

[10.2] Arithmetic progression: $a_n = a_1 + (n - 1)d$

$$S_n = \frac{n}{2}[2a_1 + (n - 1)d] = \frac{n}{2}[a_1 + a_n]$$

Do not confuse $n$, the *number* of terms, with $a_n$, the $n$th term.

[10.3] Geometric progression: $a_n = a_1r^{n-1}$

$$S_n = \frac{a_1 - ra_n}{1 - r} = \frac{a_1 - a_1r^n}{1 - r}$$

[10.4] The sum of an infinite geometric progression is given by

$$S = \frac{a_1}{1 - r} \quad \text{if} \quad |r| < 1.$$

[10.5] The coefficients in the expansion of a binomial of the form $a + b$ raised to a power $n$ can be found in the $n$th row of Pascal's triangle.

## CHAPTER 10 REVIEW EXERCISES

[10.1] **1.** A(n) **(a)** ______ is an ordered collection of numbers whose indicated sum is called its associated **(b)** ______.

**2.** Each member of a sequence is called a(n) ______ of the sequence.

**3.** Sequences that do not have a last term but continue on indefinitely are called ______ sequences.

**4.** The general term of a sequence is often called the ______ term of the sequence.

**5.** In the sigma summation notation $\sum_{n=1}^{m} a_n$, $n$ is called the **(a)** ______ of summation while 1 is the **(b)** ______ limit and $m$ is the **(c)** ______ limit of summation.

**6.** To express $b_4 + b_5 + b_6 + b_7$ using sigma summation notation we would write ______.

**7.** To express $x_1 + x_2 + x_3 + \cdots$ using sigma summation notation we would write ______.

[10.2] **8.** Each term after the first in an arithmetic progression is obtained by **(a)** ______ a fixed number called the **(b)** ______ to the preceding term.

**9.** The formula for calculating the $n$th term of an arithmetic progression is ______.

**10.** The formulas for calculating the sum of the first $n$ terms of an arithmetic progression are ______ and ______.

[10.3] **11.** Each term after the first in a geometric progression is obtained by **(a)** ______ a fixed number called the **(b)** ______ by the preceding term.

**12.** The formula for the $n$th term of a geometric progression is ______.

**13.** The formulas for calculating the sum of the first $n$ terms of a geometric progression are ______ and ______.

[10.4] **14.** The formula for calculating the sum of all terms of an infinite geometric progression is given by **(a)** ______ and only applies when the common ratio $r$ satisfies **(b)** ______.

[10.5] **15.** The coefficients in the expansion of a binomial can be determined by using ______.

[10.1] *In each exercise the nth term of a general sequence is given. Find the first 4 terms and the 7th term of each.*

**16.** $x_n = n^2 - n$

**17.** $a_n = \dfrac{(-1)^n}{3n}$

**18.** Rewrite without using the sigma summation notation.

$$\sum_{k=0}^{3} \sqrt{k+1}$$

**19.** Rewrite using the sigma summation notation.

$$\frac{1}{4} + \frac{4}{5} + \frac{9}{6} + \cdots + \frac{n^2}{n+3} + \cdots$$

[10.2] **20.** Write the first 5 terms of an arithmetic progression with $a_1 = -11$ and $d = 4$, and find $S_{10}$.

**21.** A new car costs $9500 and depreciates 25% the first year, 21% the second year, 17% the third, and so on for 5 years. If all depreciations apply to the original cost, what is the value of the car in 5 years?

**22.** A collection of nickels is arranged in a triangular array with 20 coins in the base row, 19 in the next, 18 in the next, and so forth. Find the value of the collection.

[10.3] **23.** Write the first 5 terms of a geometric progression with $a_1 = \frac{1}{5}$ and $r = -5$, and find $S_5$.

**24.** Find the sum of the first 8 terms of the geometric progression with $a_1 = 81$ and $a_8 = \frac{1}{27}$.

**25.** A man borrows $3000 at 8% interest compounded annually. If he pays off the loan in full at the end of 6 years, how much does he pay?

[10.4] **26.** Find the sum of the infinite geometric progression $18, -6, 2, -\frac{2}{3}, \ldots$.

**27.** Is there an infinite geometric progression with $a_1 = -3$ and $S = 10$?

**28.** A child on a swing traverses an arc of 20 ft. Each pass thereafter, he traverses an arc that is $\frac{16}{17}$ the length of the previous arc. How far does he travel before coming to rest?

*Convert each decimal to a fraction and check by division.*

**29.** $.\bar{7}$

**30.** $6.\overline{12}$

[10.5] *Expand each of the following binomials by using Pascal's triangle.*

**31.** $(y + 2)^6$

**32.** $(a - 2z)^5$

**33.** $(a - a^{-1})^4$

---

ANSWERS: **1.** (a) sequence (b) series **2.** term **3.** infinite **4.** $n$th **5.** (a) index (b) lower (c) upper **6.** $\sum_{k=4}^{7} b_k$ **7.** $\sum_{n=1}^{\infty} x_n$ **8.** (a) adding (b) common difference **9.** $a_n = a_1 + (n-1)d$ **10.** $S_n = \frac{n}{2}[a_1 + a_n]$; $S_n = \frac{n}{2}[2a_1 + (n-1)d]$ **11.** (a) multiplying (b) common ratio **12.** $a_n = a_1 r^{n-1}$ **13.** $S_n = \frac{a_1 - ra_n}{1-r}$, $S_n = \frac{a_1 - a_1 r^n}{1-r}$ **14.** (a) $S = \frac{a_1}{1-r}$ (b) $|r| < 1$ **15.** Pascal's triangle **16.** $x_1 = 0, x_2 = 2, x_3 = 6, x_4 = 12, x_7 = 42$ **17.** $a_1 = -\frac{1}{3}, a_2 = \frac{1}{6}, a_3 = -\frac{1}{9}, a_4 = \frac{1}{12}, a_7 = -\frac{1}{21}$ **18.** $1 + \sqrt{2} + \sqrt{3} + 2$ **19.** $\sum_{n=1}^{\infty} \frac{n^2}{n+3}$ **20.** $-11, -7, -3, 1, 5$; $S_{10} = 70$ **21.** \$1425 **22.** \$10.50 **23.** $\frac{1}{5}, -1, 5, -25, 125$; $S_5 = \frac{521}{5}$ **24.** $\frac{3280}{27}$ **25.** \$4760.62 **26.** $\frac{27}{2}$ **27.** No such infinite series exists since $r$ is $\frac{13}{10}$. **28.** 340 ft **29.** $\frac{7}{9}$ **30.** $\frac{202}{33}$ **31.** $y^6 + 12y^5 + 60y^4 + 160y^3 + 240y^2 + 192y + 64$ **32.** $a^5 - 10a^4z + 40a^3z^2 - 80a^2z^3 + 80az^4 - 32z^5$ **33.** $a^4 - 4a^2 + 6 - \frac{4}{a^2} + \frac{1}{a^4}$

## FINAL REVIEW EXERCISES

### Chapter 1

**1.** $\left(-\frac{3}{8}\right)+\frac{1}{4}=$

**2.** $\left(-\frac{4}{5}\right)-\left(-\frac{3}{10}\right)=$

**3.** $\left(-\frac{4}{3}\right)\cdot\left(\frac{3}{16}\right)=$

**4.** $\left(-\frac{5}{8}\right)\div\left(-\frac{15}{2}\right)=$

**5.** $|(-4)(5)-0\cdot 3|=$

**6.** Evaluate $-4a^2+(-2a)^3$ for $a=-2$.

**7.** Remove parentheses and simplify.

(a) $4[5y-(3y-1)]$

(b) $a+[2-2(3a-5)]$

**8.** Convert .4% to a decimal.

**9.** Convert $\frac{7}{8}$ to a percent.

**10.** Factor $3x+15y$.

**11.** (a) 3.52 dl = ________ kl

(b) 3 $\text{ft}^2$ = ________ $\text{in}^2$

**12.** Simplify and write without negative exponents.

(a) $\frac{5^0a^3b^{-7}}{a^{-2}b^5}$

(b) $3^0x^{-4}x^{-2}x^7$

(c) $(2y^{-2}z^4)^{-3}$

**13.** Find the area of a triangle with base 8 ft and altitude 13 ft.

**14.** Find the perimeter of a rectangle with length 2.1 cm and width 1.5 cm.

## Chapter 2

*Solve the following.*

**15.** $3 - 3(x - 1) = 4x - 1$

**16.** $(2y + 1) - (y + 5) = 0$

**17.** $\dfrac{x}{\frac{1}{3}} = 12$

**18.** $2y - 7 \leq 5y + 2$

**19.** $|4 - 3x| = 7$

**20.** $P = q + 50r$ for $r$

**21.** $|2y + 1| > 5$

**22.** $|2x - 3| \leq 7$

**23.** After a 6% increase, a man's new salary is \$9752. What was his old salary?

**24.** Two cars leave the same city at 9:00 A.M., one heading north and the other south. If one is traveling 50 mph and the other 35 mph, at what time will they be 340 miles apart?

**25.** Find the measures of each angle of a triangle if the second is four times the first and the third is 5° more than twice the first.

**26.** The sum of two consecutive odd integers is 96. Find the two integers.

## Chapter 3

**27.** The polynomial $4x^5 + 2x^3 - x^2 + x - 15$ has **(a)** ________ terms, and the coefficient of the term $2x^3$ is **(b)** ________.

**28.** In any factoring problem, the first step is to factor out any ________ factors.

**29.** When factoring $ax^2 + bxy + cy^2$ with $a > 0$, $b < 0$, and $c > 0$, the signs of the coefficients of the $y$ terms in the binomial factors are ________.

**30.** The FOIL method is helpful when multiplying two ________.

**31.** The term $-2x^4y$ and the term ________ in the polynomial $x^5y^3 - 2x^4y^3 + 8x^4y - 3x^3 + 7$ are like terms.

**32.** $(xy + y^2 - 2x^2) + (2xy - y^2 + 4x^2) =$

**33.** $(5a^2 - 4ab - 3b^2) - (9a^2 - 4ab + 2b^2) =$

**34.** $(2x + y)(x - 4y) =$

**35.** $(a - 3b)^2 =$

*Factor.*

**36.** $3x^2 - 12$

**37.** $a^2 + 10a + 25$

**38.** $4y^3 - 32$

**39.** $3x^2 - 8x + 4$

**40.** $3x^3 + 3$

**41.** $-6u^2 + 10uv + 4v^2$

## Chapter 4

**42.** Simplify.

(a) $\sqrt{3xy}\sqrt{12x^3y^4}$

(b) $\dfrac{\sqrt[3]{40xy^3}}{\sqrt[3]{5x}}$

(c) $\sqrt[3]{27a^5b^9} - 3b\sqrt[3]{125a^5b^6}$

**43.** Rationalize the denominator of $\dfrac{\sqrt{3}-1}{\sqrt{3}+1}$.

**44.** Simplify $3\sqrt{27} - 8\sqrt{75} - 6\sqrt{12}$.

**45.** Simplify $\left(\dfrac{9x^2y^{-1}}{x^{-2}y^5}\right)^{3/2}$.

**46.** Solve $\sqrt{y+11} - 1 = \sqrt{y+4}$.

**47.** Solve $\sqrt{2x+1} - \sqrt{x+6} = 0$.

**48.** Solve $x^{1/3}y^{1/3} = a$ for $y$.

## Chapter 5

*Solve.*

**49.** $y^2 - 6y - 7 = 0$

**50.** $4x^2 = 8x$

**51.** $x^2 - 2x - 1 = 0$

**52.** $x^2 - 1 = 3(x + 1)$

**53.** $x^{2/3} - x^{1/3} - 6 = 0$

**54.** $y^4 - 3y^2 + 2 = 0$

**55.** $\sqrt{x^2 + 2} - \sqrt{x + 1} = 0$

**56.** $\sqrt{3y + 1} - \sqrt{y - 4} = 3$

**57.** Twice the product of two consecutive odd integers is 126. Find the integers.

**58.** The length of a rectangle is 7 cm more than the width and the area is 44 $cm^2$. Find its dimensions.

**59.** Perform the indicated operations.

**(a)** $(2 + 3i) + (8 - i) =$

**(b)** $(3 + i)(2 - 4i) =$

**(c)** $\dfrac{2 - i}{3 + 2i} =$

**60.** Use the discriminant to determine the nature of the solutions.

**(a)** $3x^2 + x - 2 = 0$ **(b)** $2x^2 - 3x + 7 = 0$ **(c)** $x^2 - 6x + 9 = 0$

**61.** Determine a quadratic equation that has the following solutions.

**(a)** $3, -4$ **(b)** $\frac{1}{2}, -2$

*Solve.*

**62.** $x^2 + 2x - 3 \geq 0$

**63.** $2x^2 - 3x - 2 < 0$

## Chapter 6

*Perform the indicated operations.*

**64.** $\dfrac{4}{x+y} + \dfrac{2x+y}{x^2-y^2} =$

**65.** $\dfrac{2}{a+b} - \dfrac{2a}{a^2+2ab+b^2} =$

**66.** $\dfrac{x-y}{4x+4y} \cdot \dfrac{x^2-y^2}{x^2-2xy+y^2} =$

**67.** $\dfrac{a^2-9}{a^2-8a+16} \div \dfrac{3a-9}{a-4} =$

**68.** Simplify $\dfrac{\dfrac{x}{y}-\dfrac{y}{x}}{\dfrac{1}{y}+\dfrac{1}{x}}$.

**69.** Divide and indicate the remainder.

$y-3\overline{)y^3+3y^2-2y+6}$

**70.** The speed of a river is 5 mph. If a boat travels 75 mi downstream in the same time that it takes to travel 45 mi upstream, find the speed of the boat in still water.

**71.** Pipe $A$ can fill a reservoir in 5 days and pipe $B$ can fill it in 3 days. How long would it take to fill it if they are turned on together?

**72.** Solve $\dfrac{1}{y}=\dfrac{y-2}{24}$.

**73.** Solve $\dfrac{a}{a+3}-\dfrac{18}{a^2-9}=1$.

## Chapter 7

**74.** Give the $x$-intercept, $y$-intercept, slope, and graph of the line with equation $2x-5y+10=0$.

$x$-intercept = ______________

$y$-intercept = ______________

slope = ______________

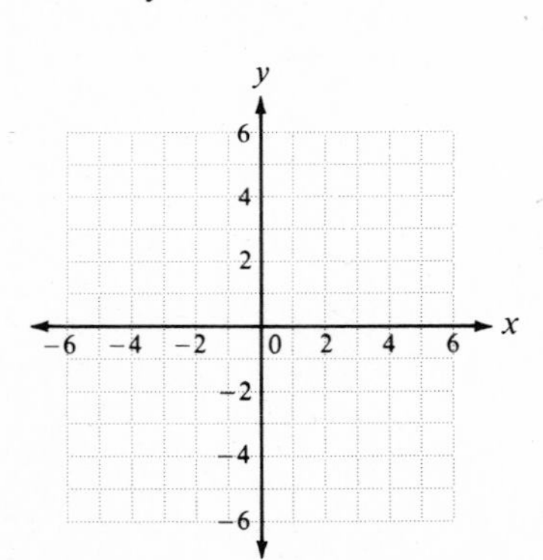

**75.** Are the lines in the system parallel, perpendicular, or neither?

$$2x + y = 3$$
$$x + 2y = -3$$

**76.** Find the equation, *in general form,* of the following lines.

**(a)** through $(3, -1)$ with slope $m = 4$

**(b)** through $(4, -2)$ and $(-1, 3)$

**77.** Write the equation, *in general form,* of the line containing the point $(2, 5)$ and perpendicular to $4x + 3y - 8 = 0$.

**78.** Find the distance between the points $(3, 1)$ and $(1, -2)$.

**79.** Find the midpoint of the line segment joining the points $(4, 7)$ and $(-2, -3)$.

**80.** Given the function $y = x^2 + 3$, find the following.

**(a)** $f(-2)$

**(b)** $f(1)$

**(c)** $f(b - 1)$

*Graph the following equations.*

**81.** $(x+1)^2 + (y-3)^2 = 4$

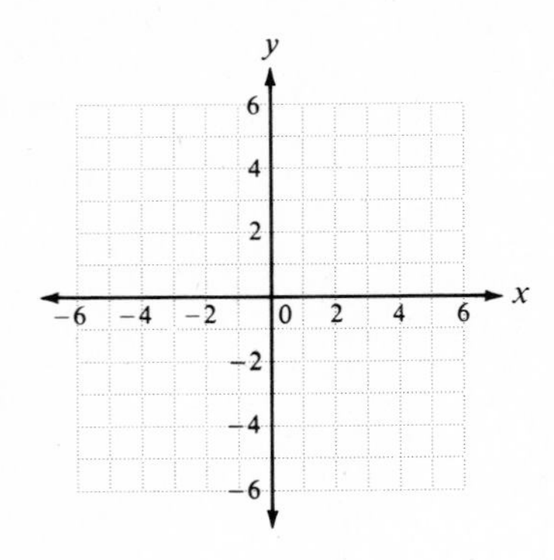


**82.** $\dfrac{x^2}{4} - \dfrac{y^2}{9} = 1$

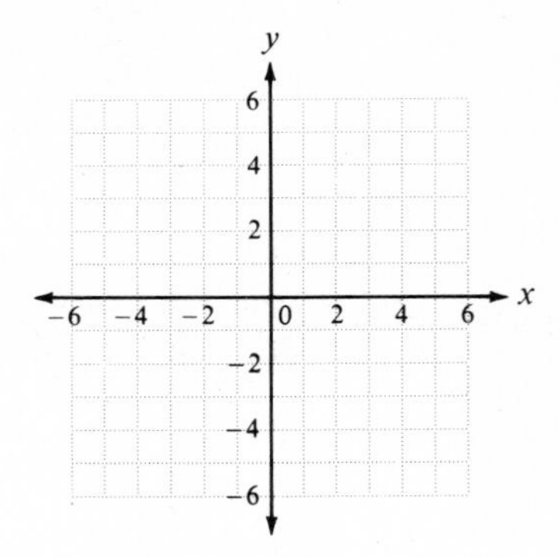


**83.** Is the following a function?

1 → 5
2 → 6
3 → 7

**84.** Is the following the graph of a function?

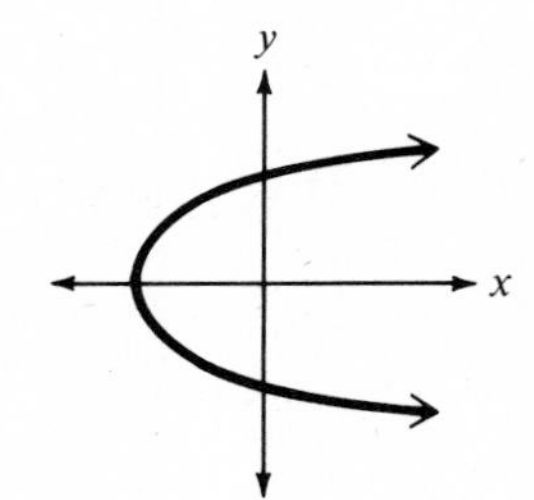


**85.** Find the $x$-intercepts, the vertex, and graph of $f(x) = -2x^2 - 10x - 8$.

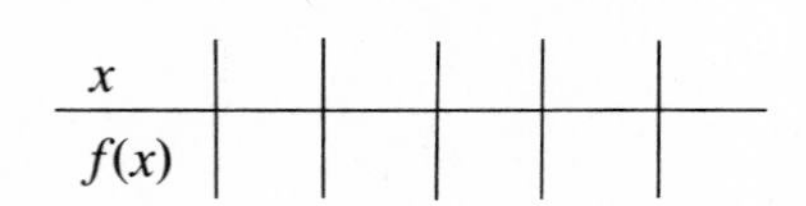

| $x$ | | | | | |
|---|---|---|---|---|---|
| $f(x)$ | | | | | |

$x$-intercepts:
vertex:

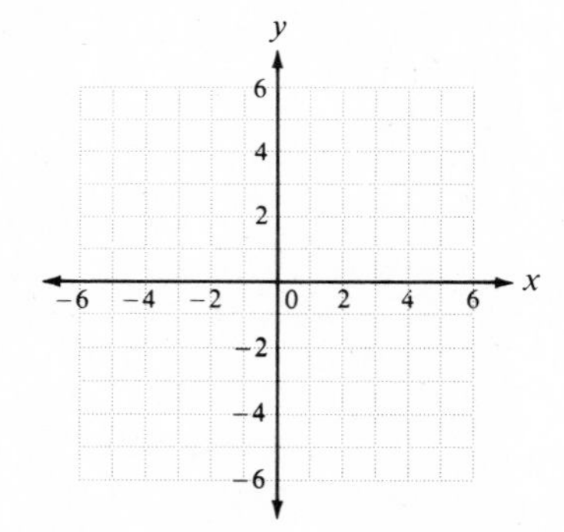


## Chapter 8

*Solve the following systems.*

**86.** $\begin{aligned} 3x + 5y &= -2 \\ x - 3y &= 4 \end{aligned}$

**87.** $\begin{aligned} x + y - z &= 4 \\ 2x - y + 3z &= -3 \\ x + 2y - z &= 6 \end{aligned}$

**88.** A woman has a candy mix that sells for \$1.30 per pound. If the mix is composed of one kind of candy worth 90¢ per pound and a second kind worth \$1.50 per pound, how many pounds of each are there in 60 pounds of the mix?

**89.** A collection of dimes and quarters is worth \$3.75. If there are 27 coins in the collection, how many of each are there?

**90.** The average of a student's three scores is 79. The sum of the first and second scores is 57 more than the third, and the first is 7 less than the second. Find the three scores.

**91.** Graph $4x + 2y > 8$.

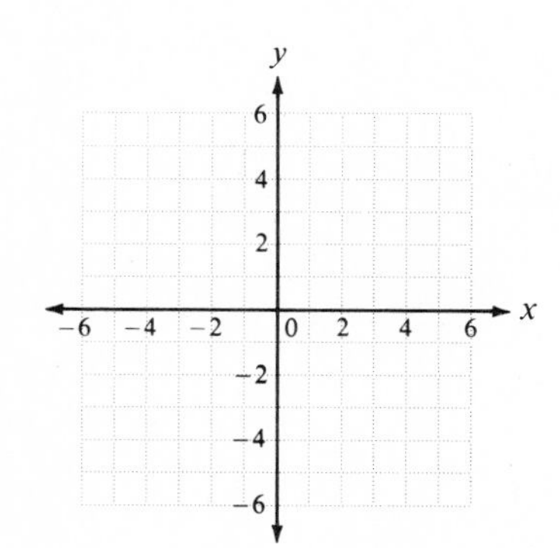


**92.** Graph the system.

$$\begin{aligned} 5x - 2y &\leq 10 \\ x + 2 &> 0 \end{aligned}$$

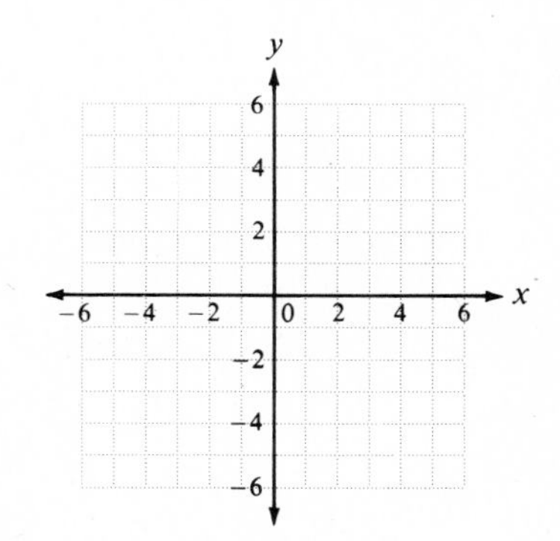


## Chapter 9

**93.** Convert $\log_a 25 = 6$ to exponential form.

**94.** Convert $7^x = 3$ to logarithmic form.

**95.** Express $\log_a \frac{x^2}{y\sqrt{z}}$ as a combination of logarithms of $x$, $y$, and $z$.

**96.** Express $-3 \cdot \log_a x + \frac{1}{2} \log_a y$ as a single logarithm.

**97.** Given that $\log_a 3 = 0.4771$ and $\log_a 5 = 0.6990$, find the following logarithms.

**(a)** $\log_a 9$ **(b)** $\log_a \frac{3}{5}$ **(c)** $\log_a 15$

**98.** Express $\log_7 24$ in terms of common logarithms.

**99.** Solve $\log x - \log (x + 9) = -1$.

**100.**. Solve $3^{2x+6} = 27$.

**101.** Use the sound equation $y = 10 \cdot \log \frac{x}{x_0}$ to determine the loudness of a sound in decibels if the given sound has intensity 75,000 times the weakest sound detected.

**102.** If \$1000 is invested at an annual rate of 12% compounded quarterly, what will be the value of the account at the end of 8 years?
[*Hint:* $S = A(1 + i)^n$.]

## Chapter 10

**103.** Express $b_1 + b_2 + b_3 + b_4$ using sigma summation notation.

**104.** Express $\sum_{k=1}^{5} x_k$ without using sigma summation notation.

**105.** The $n$th term of a general sequence is $x_n = 3^n + 1$; find the first three terms $x_1$, $x_2$, and $x_3$.

**106.** An arithmetic progression has $a_1 = 11$ and $d = -4$; find $a_6$ and $S_6$.

**107.** A geometric progression has $a_1 = 32$ and $r = -\frac{1}{2}$; find $a_5$ and $S_5$.

**108.** An auditorium has 30 rows with 20 seats in the first row, 23 in the second row, 26 in the third row, and so forth. How many seats are in the auditorium?

**109.** A new car costing \$10,000 depreciates 25% of its value each year. How much is the car worth at the end of 6 years?

**110.** Convert $3.\overline{45}$ to a fraction.

**111.** The tip of a pendulum moves back and forth in such a way that it sweeps out an arc 24 in long, and on each succeeding pass, the length of the arc traveled is $\frac{9}{10}$ the length of the preceding pass. What is the total distance traveled by the tip of the pendulum?

**112.** Expand $(x + 3)^5$ using Pascal's triangle.

---

ANSWERS: **1.** $-\frac{1}{8}$ **2.** $-\frac{1}{2}$ **3.** $-\frac{1}{4}$ **4.** $\frac{1}{12}$ **5.** 20 **6.** 48 **7.** (a) $8y + 4$ (b) $-5a + 12$ **8.** .004 **9.** 87.5% **10.** $3(x + 5y)$ **11.** (a) .000352 (b) 432 **12.** (a) $\frac{a^5}{b^{12}}$ (b) $x$ (c) $\frac{y^6}{8z^{12}}$ **13.** 52 ft² **14.** 7.2 cm **15.** 1 **16.** 4 **17.** 4 **18.** $y \geq -3$ **19.** $-1, \frac{11}{3}$ **20.** $r = \frac{P - q}{50}$ **21.** $y > 2$ or $y < -3$ **22.** $-2 \leq x \leq 5$ **23.** \$9200 **24.** 1:00 P.M. **25.** 25°, 100°, 55° **26.** 47, 49 **27.** (a) 5 (b) 2 **28.** common **29.** both negative **30.** binomials **31.** $8x^4y$ **32.** $3xy + 2x^2$ **33.** $-4a^2 - 5b^2$ **34.** $2x^2 - 7xy - 4y^2$ **35.** $a^2 - 6ab + 9b^2$ **36.** $3(x - 2)(x + 2)$ **37.** $(a + 5)(a + 5)$ **38.** $4(y - 2)(y^2 + 2y + 4)$ **39.** $(3x - 2)(x - 2)$ **40.** $3(x + 1)(x^2 - x + 1)$ **41.** $-2(u - 2v)(3u + v)$ **42.** (a) $6x^2y^2\sqrt{y}$ (b) $2y$ (c) $-12ab^3\sqrt[3]{a^2}$ **43.** $2 - \sqrt{3}$ **44.** $-43\sqrt{3}$ **45.** $\frac{27x^6}{y^9}$ **46.** 5 **47.** 5 **48.** $\frac{a^3}{x}$ **49.** $-1$; 7 **50.** 0; 2 **51.** $1 \pm \sqrt{2}$ **52.** $-1$; 4 **53.** $-8$; 27 **54.** $\pm\sqrt{2}$; $\pm 1$ **55.** $\frac{1 \pm i\sqrt{3}}{2}$ **56.** 5; 8 **57.** $-9$ and $-7$; 7 and 9 **58.** 4 cm by 11 cm **59.** (a) $10 + 2i$ (b) $10 - 10i$ (c) $\frac{4 - 7i}{13}$ **60.** (a) two real (b) two complex (c) one real **61.** (a) $x^2 + x - 12 = 0$ (b) $2x^2 + 3x - 2 = 0$ **62.** $x \leq -3$ or $x \geq 1$ **63.** $-\frac{1}{2} < x < 2$ **64.** $\frac{3(2x - y)}{(x - y)(x + y)}$ **65.** $\frac{2b}{(a + b)(a + b)}$ **66.** $\frac{1}{4}$ **67.** $\frac{a + 3}{3(a - 4)}$ **68.** $x - y$ **69.** $y^2 + 6y + 16$ remainder 54 **70.** 20 mph **71.** $\frac{15}{8}$ days **72.** 6; $-4$ **73.** no solution **74.** $(-5, 0)$; $(0, 2)$; $\frac{2}{5}$

$2x - 5y + 10 = 0$

**75.** neither **76.** (a) $4x - y - 13 = 0$ (b) $x + y - 2 = 0$ **77.** $3x - 4y + 14 = 0$ **78.** $\sqrt{13}$ **79.** (1, 2) **80.** (a) 7 (b) 4 (c) $b^2 - 2b + 4$

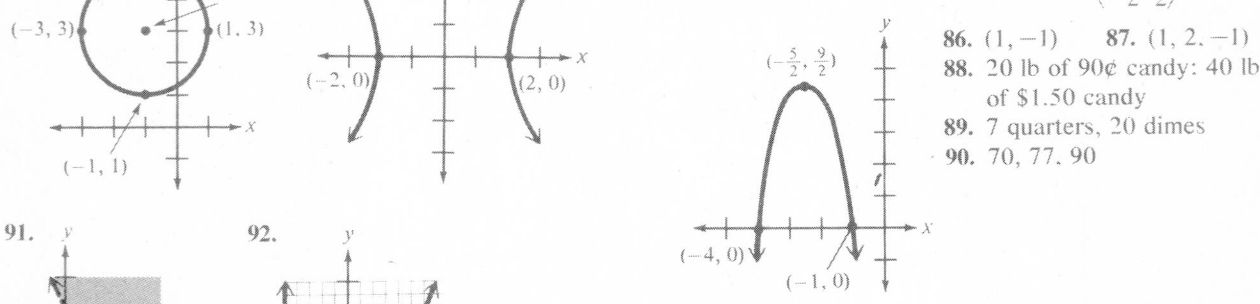


**83.** not a function **84.** not a function

**85.** $x$-intercepts: $(-1, 0)$ and $(-4, 0)$; vertex $\left(-\frac{5}{2}, \frac{9}{2}\right)$

**86.** $(1, -1)$ **87.** $(1, 2, -1)$ **88.** 20 lb of 90¢ candy: 40 lb of \$1.50 candy **89.** 7 quarters, 20 dimes **90.** 70, 77, 90

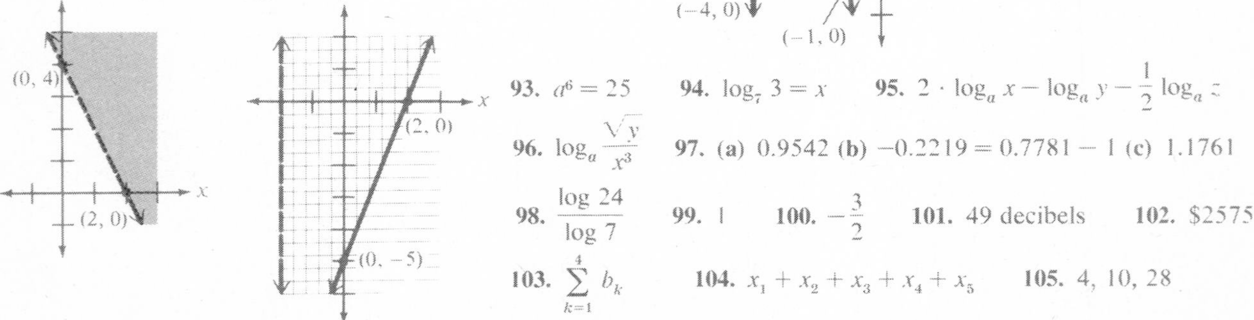


**93.** $a^6 = 25$ **94.** $\log_7 3 = x$ **95.** $2 \cdot \log_a x - \log_a y - \frac{1}{2}\log_a z$ **96.** $\log_a \frac{\sqrt{y}}{x^3}$ **97.** (a) 0.9542 (b) $-0.2219 = 0.7781 - 1$ (c) 1.1761 **98.** $\frac{\log 24}{\log 7}$ **99.** 1 **100.** $-\frac{3}{2}$ **101.** 49 decibels **102.** \$2575 **103.** $\sum_{k=1}^{4} b_k$ **104.** $x_1 + x_2 + x_3 + x_4 + x_5$ **105.** 4, 10, 28 **106.** $a_6 = -9$; $S_6 = 6$ **107.** $a_5 = 2$; $S_5 = 22$ **108.** 1905 **109.** \$1780 **110.** $\frac{38}{11}$ **111.** 240 in **112.** $x^5 + 15x^4 + 90x^3 + 270x^2 + 405x + 243$

## Table 1 Common Logarithms

| n | 0 | 1 | 2 | 3 | 4 | 5 | 6 | 7 | 8 | 9 |
|---|---|---|---|---|---|---|---|---|---|---|
| 1.0 | .0000 | .0043 | .0086 | .0128 | .0170 | .0212 | .0253 | .0294 | .0334 | .0374 |
| 1.1 | .0414 | .0453 | .0492 | .0531 | .0569 | .0607 | .0645 | .0682 | .0719 | .0755 |
| 1.2 | .0792 | .0828 | .0864 | .0899 | .0934 | .0969 | .1004 | .1038 | .1072 | .1106 |
| 1.3 | .1139 | .1173 | .1206 | .1239 | .1271 | .1303 | .1335 | .1367 | .1399 | .1430 |
| 1.4 | .1461 | .1492 | .1523 | .1553 | .1584 | .1614 | .1644 | .1673 | .1703 | .1732 |
| 1.5 | .1761 | .1790 | .1818 | .1847 | .1875 | .1903 | .1931 | .1959 | .1987 | .2014 |
| 1.6 | .2041 | .2068 | .2095 | .2122 | .2148 | .2175 | .2201 | .2227 | .2253 | .2279 |
| 1.7 | .2304 | .2330 | .2355 | .2380 | .2405 | .2430 | .2455 | .2480 | .2504 | .2529 |
| 1.8 | .2553 | .2577 | .2601 | .2625 | .2648 | .2672 | .2695 | .2718 | .2742 | .2765 |
| 1.9 | .2788 | .2810 | .2833 | .2856 | .2878 | .2900 | .2923 | .2945 | .2967 | .2989 |
| 2.0 | .3010 | .3032 | .3054 | .3075 | .3096 | .3118 | .3139 | .3160 | .3181 | .3201 |
| 2.1 | .3222 | ,3243 | .3263 | .3284 | .3304 | .3324 | .3345 | .3365 | .3385 | .3404 |
| 2.2 | .3424 | .3444 | .3464 | .3483 | .3502 | .3522 | .3541 | .3560 | .3579 | .3598 |
| 2.3 | .3617 | .3636 | .3655 | .3674 | .3692 | .3711 | .3729 | .3747 | .3766 | .3784 |
| 2.4 | .3802 | .3820 | .3838 | .3856 | .3874 | .3892 | .3909 | .3927 | .3945 | .3962 |
| 2.5 | .3979 | .3997 | .4014 | .4031 | .4048 | .4065 | .4082 | .4099 | .4116 | .4133 |
| 2.6 | .4150 | .4166 | .4183 | .4200 | .4216 | .4232 | .4249 | .4265 | .4281 | .4298 |
| 2.7 | .4314 | .4330 | .4346 | .4362 | .4378 | .4393 | .4409 | .4425 | .4440 | .4456 |
| 2.8 | .4472 | .4487 | .4502 | .4518 | .4533 | .4548 | .4564 | .4579 | .4594 | .4609 |
| 2.9 | .4624 | .4639 | .4654 | .4669 | .4683 | .4698 | .4713 | .4728 | .4742 | .4757 |
| 3.0 | .4771 | .4786 | .4800 | .4814 | .4829 | .4843 | .4857 | .4871 | .4886 | .4900 |
| 3.1 | .4914 | .4928 | .4942 | .4955 | .4969 | .4983 | .4997 | .5011 | .5024 | .5038 |
| 3.2 | .5051 | .5065 | .5079 | .5092 | .5105 | .5119 | .5132 | .5145 | .5159 | .5172 |
| 3.3 | .5185 | .5198 | .5211 | .5224 | .5237 | .5250 | .5263 | .5276 | .5289 | .5302 |
| 3.4 | .5315 | .5328 | .5340 | .5353 | .5366 | .5378 | .5391 | .5403 | .5416 | .5428 |
| 3.5 | .5441 | .5453 | .5465 | .5478 | .5490 | .5502 | .5514 | .5527 | .5539 | .5551 |
| 3.6 | .5563 | .5575 | .5587 | .5599 | .5611 | .5623 | .5635 | .5647 | .5658 | .5670 |
| 3.7 | .5682 | .5694 | .5705 | .5717 | .5729 | .5740 | .5752 | .5763 | .5775 | .5786 |
| 3.8 | .5798 | .5809 | .5821 | .5832 | .5843 | .5855 | .5866 | .5877 | .5888 | .5899 |
| 3.9 | .5911 | .5922 | .5933 | .5944 | .5955 | .5966 | .5977 | .5988 | .5999 | .6010 |
| 4.0 | .6021 | .6031 | .6042 | .6053 | .6064 | .6075 | .6085 | .6096 | .6107 | .6117 |
| 4.1 | .6128 | .6138 | .6149 | .6160 | .6170 | .6180 | .6191 | .6201 | .6212 | .6222 |
| 4.2 | .6232 | .6243 | .6253 | .6263 | .6274 | .6284 | .6294 | .6304 | .6314 | .6325 |
| 4.3 | .6335 | .6345 | .6355 | .6365 | .6375 | .6385 | .6395 | .6405 | .6415 | .6425 |
| 4.4 | .6435 | .6444 | .6454 | .6464 | .6474 | .6484 | .6493 | .6503 | .6513 | .6522 |
| 4.5 | .6532 | .6542 | .6551 | .6561 | .6571 | .6580 | .6590 | .6599 | .6609 | .6618 |
| 4.6 | .6628 | .6637 | .6646 | .6656 | .6665 | .6675 | .6684 | .6693 | .6702 | .6712 |
| 4.7 | .6721 | .6730 | .6739 | .6749 | .6758 | .6767 | .6776 | .6785 | .6794 | .6803 |
| 4.8 | .6812 | .6821 | .6830 | .6839 | .6848 | .6857 | .6866 | .6875 | .6884 | .6893 |
| 4.9 | .6902 | .6911 | .6920 | .6928 | .6937 | .6946 | .6955 | .6964 | .6972 | .6981 |
| 5.0 | .6990 | .6998 | .7007 | .7016 | .7024 | .7033 | .7042 | .7050 | .7059 | .7067 |
| 5.1 | .7076 | .7084 | .7093 | .7101 | .7110 | .7118 | .7126 | .7135 | .7143 | .7152 |
| 5.2 | .7160 | .7168 | .7177 | .7185 | .7193 | .7202 | .7210 | .7218 | .7226 | .7235 |
| 5.3 | .7243 | .7251 | .7259 | .7267 | .7275 | .7284 | .7292 | .7300 | .7308 | .7316 |
| 5.4 | .7324 | .7332 | .7340 | .7348 | .7356 | .7364 | .7372 | .7380 | .7388 | .7396 |
| n | 0 | 1 | 2 | 3 | 4 | 5 | 6 | 7 | 8 | 9 |

## Table 1 Common Logarithms (continued)

| *n* | 0 | 1 | 2 | 3 | 4 | 5 | 6 | 7 | 8 | 9 |
|---|---|---|---|---|---|---|---|---|---|---|
| 5.5 | .7404 | .7412 | .7419 | .7427 | .7435 | .7443 | .7451 | .7459 | .7466 | .7474 |
| 5.6 | .7482 | .7490 | .7497 | .7505 | .7513 | .7520 | .7528 | .7536 | .7543 | .7551 |
| 5.7 | .7559 | .7566 | .7574 | .7582 | .7589 | .7597 | .7604 | .7612 | .7619 | .7627 |
| 5.8 | .7634 | .7642 | .7649 | .7657 | .7664 | .7672 | .7679 | .7686 | .7694 | .7701 |
| 5.9 | .7709 | .7716 | .7723 | .7731 | .7738 | .7745 | .7752 | .7760 | .7767 | .7774 |
| 6.0 | .7782 | .7789 | .7796 | .7803 | .7810 | .7818 | .7825 | .7832 | .7839 | .7846 |
| 6.1 | .7853 | .7860 | .7868 | .7875 | .7882 | .7889 | .7896 | .7903 | .7910 | .7917 |
| 6.2 | .7924 | .7931 | .7938 | .7945 | .7952 | .7959 | .7966 | .7973 | .7980 | .7987 |
| 6.3 | .7993 | .8000 | .8007 | .8014 | .8021 | .8028 | .8035 | .8041 | .8048 | .8055 |
| 6.4 | .8062 | .8069 | .8075 | .8082 | .8089 | .8096 | .8102 | .8109 | .8116 | .8122 |
| 6.5 | .8129 | .8136 | .8142 | .8149 | .8156 | .8162 | .8169 | .8176 | .8182 | .8189 |
| 6.6 | .8195 | .8202 | .8209 | .8215 | .8222 | .8228 | .8235 | .8241 | .8248 | .8254 |
| 6.7 | .8261 | .8267 | .8274 | .8280 | .8287 | .8293 | .8299 | .8306 | .8312 | .8319 |
| 6.8 | .8325 | .8331 | .8338 | .8344 | .8351 | .8357 | .8363 | .8370 | .8376 | .8382 |
| 6.9 | .8388 | .8395 | .8401 | .8407 | .8414 | .8420 | .8426 | .8432 | .8439 | .8445 |
| 7.0 | .8451 | .8457 | .8463 | .8470 | .8476 | .8482 | .8488 | .8494 | .8500 | .8506 |
| 7.1 | .8513 | .8519 | .8525 | .8531 | .8537 | .8543 | .8549 | .8555 | .8561 | .8567 |
| 7.2 | .8573 | .8579 | .8585 | .8591 | .8597 | .8603 | .8609 | .8615 | .8621 | .8627 |
| 7.3 | .8633 | .8639 | .8645 | .8651 | .8657 | .8663 | .8669 | .8675 | .8681 | .8686 |
| 7.4 | .8692 | .8698 | .8704 | .8710 | .8716 | .8722 | .8727 | .8733 | .8739 | .8745 |
| 7.5 | .8751 | .8756 | .8762 | .8768 | .8774 | .8779 | .8785 | .8791 | .8797 | .8802 |
| 7.6 | .8808 | .8814 | .8820 | .8825 | .8831 | .8837 | .8842 | .8848 | .8854 | .8859 |
| 7.7 | .8865 | .8871 | .8876 | .8882 | .8887 | .8893 | .8899 | .8904 | .8910 | .8915 |
| 7.8 | .8921 | .8927 | .8932 | .8938 | .8943 | .8949 | .8954 | .8960 | .8965 | .8971 |
| 7.9 | .8976 | .8982 | .8987 | .8993 | .8998 | .9004 | .9009 | .9015 | .9020 | .9025 |
| 8.0 | .9031 | .9036 | .9042 | .9047 | .9053 | .9058 | .9063 | .9069 | .9074 | .9079 |
| 8.1 | .9085 | .9090 | .9096 | .9101 | .9106 | .9112 | .9117 | .9122 | .9128 | .9133 |
| 8.2 | .9138 | .9143 | .9149 | .9154 | .9159 | .9165 | .9170 | .9175 | .9180 | .9186 |
| 8.3 | .9191 | .9196 | .9201 | .9206 | .9212 | .9217 | .9222 | .9227 | .9232 | .9238 |
| 8.4 | .9243 | .9248 | .9253 | .9258 | .9263 | .9269 | .9274 | .9279 | .9284 | .9289 |
| 8.5 | .9294 | .9299 | .9304 | .9309 | .9315 | .9320 | .9325 | .9330 | .9335 | .9340 |
| 8.6 | .9345 | .9350 | .9355 | .9360 | .9365 | .9370 | .9375 | .9380 | .9385 | .9390 |
| 8.7 | .9395 | .9400 | .9405 | .9410 | .9415 | .9420 | .9425 | .9430 | .9435 | .9440 |
| 8.8 | .9445 | .9450 | .9455 | .9460 | .9465 | .9469 | .9474 | .9479 | .9484 | .9489 |
| 8.9 | .9494 | .9499 | .9504 | .9509 | .9513 | .9518 | .9523 | .9528 | .9533 | .9538 |
| 9.0 | .9542 | .9547 | .9552 | .9557 | .9562 | .9566 | .9571 | .9576 | .9581 | .9586 |
| 9.1 | .9590 | .9595 | .9600 | .9605 | .9609 | .9614 | .9619 | .9624 | .9628 | .9633 |
| 9.2 | .9638 | .9643 | .9647 | .9652 | .9657 | .9661 | .9666 | .9671 | .9675 | .9680 |
| 9.3 | .9685 | .9689 | .9694 | .9699 | .9703 | .9708 | .9713 | .9717 | .9722 | .9727 |
| 9.4 | .9731 | .9736 | .9741 | .9745 | .9750 | .9754 | .9759 | .9763 | .9768 | .9773 |
| 9.5 | .9777 | .9782 | .9786 | .9791 | .9795 | .9800 | .9805 | .9809 | .9814 | .9818 |
| 9.6 | .9823 | .9827 | .9832 | .9836 | .9841 | .9845 | .9850 | .9854 | .9859 | .9863 |
| 9.7 | .9868 | .9872 | .9877 | .9881 | .9886 | .9890 | .9894 | .9899 | .9903 | .9908 |
| 9.8 | .9912 | .9917 | .9921 | .9926 | .9930 | .9934 | .9939 | .9943 | .9948 | .9952 |
| 9.9 | .9956 | .9961 | .9965 | .9969 | .9974 | .9978 | .9983 | .9987 | .9991 | .9996 |
| *n* | 0 | 1 | 2 | 3 | 4 | 5 | 6 | 7 | 8 | 9 |

# Index